Adrenergic Dysfunction and Psychobiology

Adrenergic Dysfunction and Psychobiology

Edited by
Oliver G. Cameron, M.D., Ph.D.
Department of Psychiatry
University of Michigan
Ann Arbor, Michigan

American Psychiatric Press, Inc.
Washington, DC
London, England

Note: The authors have worked to ensure that all information in this book concerning drug dosages, schedules, and routes of administration is accurate as of the time of publication and consistent with standards set by the U.S. Food and Drug Administration and the general medical community. As medical research and practice advance, however, therapeutic standards may change. For this reason and because human and mechanical errors sometimes occur, we recommend that readers follow the advice of a physician who is directly involved in their care or the care of a member of their family.

Manufactured in the United States of America on acid-free paper
97 96 95 94 4 3 2 1
First Edition

American Psychiatric Press, Inc.
1400 K Street, N.W., Washington, DC 20005

Library of Congress Cataloging-in-Publication Data
Adrenergic dysfunction and psychobiology / [edited by] Oliver G. Cameron.
p. cm.
Includes bibliographical references and index.
ISBN 0-88048-409-8
1. Adrenergic mechanisms. 2. Neuropsychiatry. 3. Medicine, Psychosomatic. I. Cameron, Oliver G.
[DNLM: 1. Receptors, Adrenergic—physiology. 2. Autonomic Nervous System—drug effects. 3. Autonomic Nervous System Diseases—physiopathology. 4. Mental Disorders—physiopathology. WL 600 A242 1994]
RC343.A44 1994
616.8—dc20
DNLM/DLC
for Library of Congress 93-46870
CIP

British Library Cataloguing in Publication Data
A CIP record is available from the British Library.

Contents

Section III: Psychiatric Disorders

Contributors

James L. Abelson, M.D., Ph.D.
Assistant Professor, Department of Psychiatry, University of Michigan Medical Center, Ann Arbor, Michigan

Stephen M. Aronson, M.D.
Acting Director, Geriatric Psychiatry Program, and
Assistant Professor, Department of Psychiatry, Wayne State University School of Medicine, Detroit, Michigan

Gary Aston-Jones, Ph.D.
Professor, Department of Mental Health Sciences, Director, Division of Behavioral Neurobiology, Hahnemann University, Philadelphia, Pennsylvania

Oliver G. Cameron, M.D., Ph.D.
Professor and Associate Chair, Department of Psychiatry, University of Michigan Medical Center, Ann Arbor, Michigan

Qiang Chang, B.S.
Graduate Student, Graduate School of Biomedical Sciences, University of Medicine and Dentistry, Newark, New Jersey; Neurobehavioral Unit, VA Medical Center and Department of Neurosciences, New Jersey Medical School, East Orange, New Jersey

Robert S. Conway, Ph.D.
Assistant Professor of Surgery, Department of Surgery, Robert Wood Johnson Medical School of Medicine, and Neurobehavioral Unit VA Medical Center, New Brunswick, New Jersey

Joel E. Dimsdale, M.D.
Professor, Department of Psychiatry, University of California, San Diego Medical Center, San Diego, California

Guy Drolet, Ph.D.
Associate Professor, Université Laval, Départements de Pharmacologie and Médecine, Québec, Québec, Canada

Robert R. Freedman, Ph.D.
Professor and Director of Behavioral Medicine Laboratory, Department of Psychiatry, Wayne State University School of Medicine, Detroit, Michigan

Roy A. Glover, Ph.D.
Associate Professor, Department of Anatomy and Cell Biology, University of Michigan Medical Center, Ann Arbor, Michigan

Paul Glue, M.D., Ph.D.
Senior Associate Director, Clinical Pharmacology, Schering-Plough Research Institute, Kenilworth, New Jersey

David S. Goldstein, M.D., Ph.D.
Chief, Clinical Neurochemistry Section, Clinical Neuroscience Branch, National Institute of Neurological Diseases and Stroke, Bethesda, Maryland

John A. Gurklis Jr., M.D.
Assistant Professor, Highland Drive VA Medical Center and Western Psychiatric Institute and Clinic, University of Pittsburgh School of Medicine, Pittsburgh, Pennsylvania

Sally K. Guthrie, Pharm.D.
Associate Professor, College of Pharmacy and Department of Psychiatry, University of Michigan Medical Center, Ann Arbor, Michigan

Peggie J. Hollingsworth, Ph.D.
Assistant Research Scientist, Department of Pharmacology, University of Michigan Medical Center, Ann Arbor, Michigan

Walter H. Kaye, M.D.
Professor of Psychiatry, Director, Eating Disorder Module, Department of Psychiatry, Western Psychiatric Institute and Clinic, University of Pittsburgh School of Medicine, Pittsburgh, Pennsylvania

Markku Linnoila, M.D., Ph.D.
Scientific Director, National Institute of Alcohol Abuse and Alcoholism, Bethesda, Maryland

Husseini K. Manji, M.D., F.R.C.P.C.
Visiting Scientist, Section on Clinical Pharmacology, Experimental Therapeutics Branch, National Institute of Mental Health, Bethesda, Maryland

Alan M. Mellow, M.D., Ph.D.
Chief, Psychiatry Service, Ann Arbor VA Medical Center, and Assistant Professor, Department of Psychiatry, University of Michigan Medical Center, Ann Arbor, Michigan

Benjamin H. Natelson, M.D.
Professor of Neuroscience, Chief, Neurobehavioral Unit, VA Medical Center and Department of Neurosciences, New Jersey Medical School, East Orange, New Jersey

Richard A. Nelesen, Ph.D.
Postdoctoral Fellow, Department of Psychiatry, University of California, San Diego Medical Center, San Diego, California

David J. Nutt, M.D., Ph.D.
Director of Psychopharmacology Unit, School of Medical Sciences, Bristol, United Kingdom

John E. Ottenweller, Ph.D.
Associate Professor of Neurosciences, Neurobehavioral Unit, VA Medical Center and Department of Neurosciences, New Jersey Medical School, East Orange, New Jersey

William Z. Potter, M.D., Ph.D.
Chief, Section on Clinical Pharmacology, Experimental Therapeutics Branch, National Institute of Mental Health, Bethesda, Maryland

Matthew V. Rudorfer, M.D.
Assistant Chief, Clinical Treatment Research Branch, Division of Clinical and Treatment Research, National Institute of Mental Health, Bethesda, Maryland

Robert H. Schneider, M.D.
Associate Professor, Director of Center for Health and Aging Studies, Department of Physiological and Biological Sciences, Maharishi International University, Fairfield, Iowa

Charles B. Smith, M.D., Ph.D.
Professor, Department of Pharmacology, University of Michigan Medical Center, Ann Arbor, Michigan

Walter N. Tapp, Ph.D.
Professor of Neurosciences, Neurobehavioral Unit, VA Medical Center and Department of Neurosciences, New Jersey Medical School, East Orange, New Jersey

Rita J. Valentino, Ph.D.
Associate Professor, Department of Mental Health Sciences, Division of Behavioral Neurobiology, Hahnemann University, Philadelphia, Pennsylvania

Daniel P. van Kammen, M.D., Ph.D.
Professor and Chief of Staff, Highland Drive VA Medical Center and Western Psychiatric Institute and Clinic, University of Pittsburgh School of Medicine, Pittsburgh, Pennsylvania

Theodore E. Weltzin, M.D.
Assistant Professor, Department of Psychiatry, Medical Director, Eating Disorder Module, Western Psychiatric Institute and Clinic, University of Pittsburgh School of Medicine, Pittsburgh, Pennsylvania

Michael G. Ziegler, M.D.
Professor of Medicine, Department of Medicine, Division of Nephrology, University of California, San Diego Medical Center, San Diego, California

Preface

Another book on catecholamines, stress, locus coeruleus activity, and so on—is this really necessary? Obviously, I believe the answer is yes. Although the number of volumes on these subjects in aggregate is quite large, most concern themselves mainly with animal research and/or with physiology, anatomy, and pharmacology, but not with the interface of these subjects with clinical issues. Even those that do address clinical problems often do not have psychiatric and psychosomatic issues as their primary focus of concern. Furthermore, of those few that have focused on these psychobiological problems, none are up to date as of 1993. The very fact that there is so much research and interest in adrenergic psychobiology makes it difficult for any but the most diligent to keep up with this literature without the periodic availability of comprehensive reviews. As readers of this book will discover, adrenergic abnormalities are present not just in depression and anxiety (the "usual suspects") but in other syndromes as well. It is not yet known of what etiologic significance these abnormalities are in any of these disorders; thus it would be premature to judge the importance of any dysfunction based either on its apparent magnitude of effect or on the number of times that it has been demonstrated. What is clear is that the ubiquitousness of adrenergic neurons, synapses, and receptors throughout the human nervous system makes adrenergic systems prime candidates for loci of dysfunction in psychobiological disorders.

As editor of this volume, I have recruited recognized experts to prepare each of the chapters. In every case, the authors are active researchers who have published extensively in the areas covered by the respective chapters.

The material contained herein will be of interest to anyone working at the boundaries of mental health problems, behavioral research, and neuroscience. Thus psychiatrists and clinical psychologists with a biological perspective, behavioral pharmacologists and other behavioral biologists interested in clinical problems, and neuroscientists of many stripes (pharmacologists, physiologists, neuroanatomists, biochemists, molecular biologists—have I missed anyone?) will find this book of interest. So will medical students, residents in psychiatry and neurology, and graduate students in each of the above areas.

I wish to convey my appreciation to the many individuals who have been involved in preparing this book. First and foremost, of course, I thank the authors themselves; an edited book is truly a collaborative effort. Second, I found the assistance of the editorial staff at the American Psychiatric Press invaluable. Third, I found the comments of two anonymous reviewers of the book extremely helpful—really, the critiques were welcome! I hope that the reader will learn as much from reading this book it as I have in preparing it. Bon appetit!

Oliver G. Cameron, M.D., Ph.D.

1

Introduction

OLIVER G. CAMERON, M.D., PH.D.

In this book, the authors review research that has evaluated adrenergic functioning in disorders that are usually considered to be at the interface between biology and psychology (i.e., psychobiological). As noted in the preface, this is a large area of study, and periodic review serves to allow clinicians and investigators to obtain a broad view of the field from time to time, a perspective otherwise difficult for one individual to maintain due to the extent of the research being done. Review specifically of adrenergic function and dysfunction is of particular interest for several reasons. First, as already stated, there is a considerable amount of research in this area, perhaps more than in any other well-defined anatomical-physiological-pharmacological system. Second, adrenergic mechanisms are broadly represented in the nervous system, both centrally and peripherally, making it likely that adrenergic activity will be involved in a wide variety of nervous system functions. Third, there is evidence that adrenergic dysfunction is evident in many psychiatric disorders and in several often considered to have an important psychosomatic component—

witness the extraordinary body of research related to catecholamine changes and stress. In this introductory chapter, I provide a brief historical background of the development of interest in adrenergic psychobiology and then briefly describe the actual content of the rest of this volume.

The physiology and pharmacology of the autonomic nervous system (ANS) have been studied for at least 100 years. Around the turn of the century, it was observed that injections of adrenal gland extracts produced effects similar to sympathetic nerve stimulation. In the 1920s, foreshadowing future interest in adrenergic psychobiology, Marañon (1924) injected what we now know was epinephrine (adrenaline) into humans and studied the symptoms produced. Also in the 1920s, the research of Cannon (1929) laid the foundation for a tremendous number of subsequent studies into the relationship between peripheral sympathetic activation and psychophysiological responses such as the "fight-flight" reaction. In the same decade, Loewi (1924) identified *"Vagusstoff"* (acetylcholine), the primary neurotransmitter of the parasympathetic part of the ANS, and was the first to provide a convincing demonstration of neurohumoral involvement in the transmission of nerve impulses.

More recently, researchers have developed techniques to study the central nervous system (CNS) mechanisms related to ANS functioning. The same adrenergic and cholinergic neurotransmitter substances that were observed in the ANS (as well as many others) have been identified in the CNS, and the CNS mechanisms controlling ANS activity are beginning to be understood. Additionally, recent research has demonstrated the importance of afferent as well as efferent ANS-CNS connections.

There have been biochemical as well as anatomical and physiological advances in the understanding of ANS functioning. Not only are investigators now able to quantify levels of a wide variety of presumptive neurotransmitter and neurohormone substances in body fluids and tissues, but ligand assay techniques permit specific identification of the receptors for these substances. Molecular biology techniques have been used to identify receptors as well. Finally, techniques including scanning procedures have been developed to evaluate nervous system function in vivo.

Based on the results of studies using these techniques, there is evidence of changes in adrenergic-noradrenergic dysfunction in a variety of medical disorders; among these are several that are characterized as psychiatric or psychosomatic. Although there are many summaries of information about adrenergic function and dysfunction in humans and experimental animals (e.g., Heal and Marsden 1990; Insel 1987; Mason 1984; Usdin et al. 1980; Ziegler and Lake 1984), general reviews of this information from a psychiatric or psychosomatic perspective (e.g., Lake and Ziegler 1985) have rarely been assembled previously in one place.

In Chapters 2 through 7, the authors provide basic information relevant to the subsequent sections. Glover, in Chapter 2, outlines anatomy of the adrenergic-noradrenergic nervous system. The two central noradrenergic systems (locus coeruleus and lateral tegmental) and the central adrenergic system in the brain are described. The peripheral sympathoadrenal system is also described, and several functional aspects of these systems are considered. In Chapter 3, Valentino et al. review CNS interactions between noradrenergic and peptidergic systems. They focus on anatomical considerations, locus coeruleus physiology, and corticotropin-releasing-factor and opioid neurotransmitter systems.

Pharmacological issues are emphasized in Chapters 4 and 5. In Chapter 4, Smith and Hollingsworth classify adrenergic receptors. (The reader interested in background material on "receptorology" is referred elsewhere for a basic introduction [Cooper et al. 1991], more detailed information [Insel 1987; Nemeroff 1992; Yamamura et al. 1990], and intensely mathematical presentations [Kenakin 1987; Levitzki 1984].) In Chapter 5, Guthrie reviews the effects of various psychotropic drugs on noradrenergic function. Medications considered include antidepressants, lithium, stimulants, and neuroleptics.

In Chapter 6, Natelson et al. examine animal models. This is an area of research too large to review in one chapter; indeed, whole volumes and even multivolume sets have addressed animal studies of psychopathology (e.g., Bond 1984; Hanin and Usdin 1977), adrenergic function (e.g., Belmaker et al. 1988; Heal and Marsden 1990; Mason 1984; Ziegler and Lake 1984), and/or stress (e.g., Usdin et al. 1980; Van Loon et al. 1989). Therefore, the authors have provided data from their own laboratory as an example of one such model, including the effect of tranquilizers on the outcome of the procedure being studied. The potential involvement of adrenergic mechanisms in the model is described, including the possibility that the benzodiazepine used (alprazolam) might have specific effects on stress mediated through its modification of systemic catecholamine levels (see also Stratton and Halter 1985). This research is especially germane to the material in the second and third sections of the book, which focus on human studies, because it provides an example of the links one hopes to be able to draw between an animal model and a specific clinical condition present in humans (in this case, the so-called "syndrome X").

In Chapter 7, Nelesen and Ziegler discuss methods of adrenergic evaluation. An entire chapter was devoted to this subject to highlight the importance of methodology. There is no single measure of adrenergic function, and different measures often do not agree; what is measured will determine what conclusions can be drawn. These methods include evaluation of end-organ response, measurement of catecholamine levels in various fluids, receptor assay techniques, and challenge studies with agonist and antagonist substances.

The authors of the next five chapters focus on topics of relevance to psychosomatic medicine. As an introduction to this section, Goldstein, in Chapter 8, presents a broad historical and conceptual review of the concept of stress. The nature of the involvement of adrenergic-noradrenergic systems in the stress response has been a focus of intense research and theorizing for several decades.

In Chapter 9, I provide a summary of issues and studies relevant to adrenergic symptomatology per se. A brief historical introduction is followed by discussions of cardiorespiratory symptoms and symptoms produced by hypoglycemia; studies of symptoms produced by exogenous administration of adrenergic substances are then described. Finally, the potential relationship between adrenergic symptoms and stress is noted, and issues related to CNS mechanisms are discussed.

Hypertension has been considered by many investigators to have an important psychosomatic component. In Chapter 10, Nelesen and Dimsdale review the relationship between hypertension and adrenergic-noradrenergic functioning. Catecholamines under resting and nonresting conditions, sympathetic nervous system fiber activity, and other issues also are addressed.

Two other chapters of relevance to psychosomatic considerations are provided. In Chapter 11, Schneider evaluates the potential involvement of adrenergic-noradrenergic mechanisms in the type A behavior pattern, including catecholamines, newer and novel markers of adrenergic function, and the efficacy of antiadrenergic medications in the treatment of type A behavior. In Chapter 12, Freedman describes studies that evaluate adrenergic factors in other areas of behavioral medicine such as the relaxation response, menopausal "hot flashes," and Raynaud's phenomenon.

The authors of Chapters 13 through 18 focus on potential adrenergic-noradrenergic changes in psychiatric disorders. In Chapter 13, Mellow and Aronson describe adrenergic changes associated with aging and discuss changes seen in Alzheimer's disease. In Chapter 14, Gurklis and van Kammen describe norepinephrine and abnormal adrenergic-noradrenergic functioning in schizophrenia, including autopsy studies, catecholamine and catecholamine metabolite levels, platelet adrenergic receptor changes, responses to noradrenergically active drugs, and animal studies of relevance.

Affective and anxiety disorders are two of the most common groups of psychiatric disorders. In Chapter 15, Manji et al. outline findings in affective disorders, including catecholamines and metabolites, adrenergic receptors, and adrenergic challenge studies. In Chapter 16, Abelson and I describe studies of these same aspects of adrenergic functioning in anxiety disorders.

In Chapter 17, Kaye and Weltzin address possible adrenergic changes seen in eating disorders. In the final chapter, Chapter 18, Glue et al. address adrenergic changes seen in association with substance abuse. Drugs of abuse reviewed include ethanol, opioids, stimulants, and other miscellaneous drugs.

No single volume can fully encompass all issues potentially relevant to the overall topic being considered. For example, although sleep abnormalities have been described in association with some of the disorders addressed here (e.g., depression), and there is evidence of adrenergic mechanisms involved in the sleep-wake cycle and levels of alertness (Gaillard 1985; Hilakivi 1987; Jones 1991; Monti 1982; Siegel and Rogawski 1988), a separate chapter on adrenergic function and sleep is not included. Nevertheless, the interested reader will find most of the material of direct relevance to the overall topic of this volume in the chapters provided.

References

Belmaker RH, Dahlstrom A, Sandler M: Progress in Catecholamine Research, Vols A (Basic Aspects and Peripheral Mechanisms), B (Central Aspects), and C (Clinical Aspects). New York, Alan R Liss, 1988

Bond NW: Animal Models in Psychopathology. Sydney, Academic Press, 1984

Cannon WB: Bodily Changes in Pain, Hunger, Fear, and Rage. Appleton, NY, 1929

Cooper JR, Bloom FE, Roth RH: The Biochemical Basis of Neuropharmacology. New York, Oxford University Press, 1991

Gaillard JM: Neurochemical regulation of the states of alertness. Annals of Clinical Research 17:175–184, 1985

Hanin I, Usdin E: Animal Models in Psychiatry and Neurology. Oxford, Pergamon, 1977

Heal DJ, Marsden CA: The Pharmacology of Noradrenaline in the Central Nervous System. New York, Oxford University Press, 1990

Hilakivi I: Biogenic amines in the regulation of wakefulness and sleep. Medical Biology 65:97–104, 1987

Insel P: Adrenergic Receptors in Man. New York, Marcel Dekker, 1987

Jones BE: Paradoxical sleep and its chemical/structural substrates in the brain. Neuroscience 40:637–656, 1991

Kenakin TP: Pharmacologic Analysis of Drug-Receptor Interaction. New York, Raven, 1987

Lake CR, Ziegler MG: The Catecholamines in Psychiatric and Neurologic Disorders. Boston, MA, Butterworth Publishers, 1985

Levitzki A: Receptors: A Quantitative Approach. Menlo Park, CA, Benjamin/Cummings Publishing, 1984

Loewi O: Uber humorale ubertragbarkeit der herznervenwi rkung. Pflugers Achives ges Physiologie 189:239–242, 1921

Marañon G: Contribution a l'étude de l'action émotive de l'adrenline. Revue Française d'Endcrinologie 2:301–325, 1924

Mason ST: Catecholamines and Behavior. Cambridge, Cambridge University Press, 1984

Monti JM: Catecholamines and the sleep-wake cycle; I: EEG and behavioral arousal. Life Sci 30:1145–1157, 1982

Nemeroff CB: Neuroendocrinology. Boca Raton, FL, CRC Press, 1992

Siegel JM, Rogawski R: A function for REM sleep: regulation of noradrenergic receptor sensitivity. Brain Res 472:213–233, 1988

Stratton JR, Halter JB: Effect of a benzodiazepine (alprazolam) on plasma epinephrine and norepinephrine levels during exercise stress. Am J Cardiol 56:136–139, 1985

Usdin E, Kvetnansky R, Kopin IJ: Catecholamines and Stress: Recent Advances. New York, Elsevier, 1980

Van Loon GR, Kvetnansky R, McCarty R, et al: Stress: Neurochemical and Humoral Mechanisms, Vols 1 & 2. New York, Gordon & Breach Science Publishers, 1989

Yamamura HI, Enna SJ, Kuhar MJ: Methods in Neurotransmitter Receptor Analysis. New York, Raven, 1990

Ziegler MG, Lake RC: Norepinephrine. Baltimore, MD, Williams & Wilkins, 1984

Section I

Basic Science

2

The Functional Anatomy of Adrenergic Neuron Systems

ROY A. GLOVER, PH.D.

Traditional neuroanatomy textbooks have for decades emphasized descriptive morphology. This can easily be seen mirrored in their extensive coverage of topics such as neuronal cytoarchitecture, nuclear organization, and hodology (the study of neural pathways). Function, in this context, is less highly emphasized and is presented primarily through the correlation of clinical findings with damage to specific nuclear groups and fiber tracts. Contemporary topics such as chemical neuroanatomy are either nonexistent or, if present, are usually greatly abbreviated.

Neuroscientists have since begun to realize that descriptive neuroanatomy, although inherently useful, must be understood within the broader context of neurotransmitter interactions and the neuromodulator-neurotransmitter relationships that mediate interneuronal communication. Much of the rapid expansion of our knowledge in this area has been brought about primarily through the development of powerful new immunocytochemical methods that have revolutionized our ability both to identify and to localize chemical neurotransmitter, neuromodulator, and neuroregulator molecules within the nervous system. Evidence of such changes in our thinking and understanding can be seen both in contemporary neuroscience books that now usually contain at least one chapter devoted to chemical neuroanatomy and in the proliferation of journals and monographs dealing with the topic of chemical neurotransmission.

Along with this wealth of new chemical knowledge has come important evidence implicating neurotransmitter abnormalities in the pathogenesis of many diseases of the nervous system, including Parkinson's disease, Huntington's disease, Alzheimer's disease, schizophrenia, depression, and several other central affective disorders. Indeed, many of these disorders are now being effectively treated using drugs and other measures designed to correct chemical imbalances within the nervous system resulting from biochemical abnormalities in neurotransmitter synthesis, storage, release, metabolism, and reuptake.

Currently, neurotransmitters alone include several different classes of molecules and more than 20 individual substances. Most chemical neurotransmitters fall into one of four different groups of compounds: acetylcholine, monoamines, amino acids, and short-chain neuropeptides. In this chapter, I describe the location, course, and synaptic connections of a chemical system made up of neurons that synthesize and release the monoamine neurotransmitters norepinephrine (noradrenaline) and epinephrine (adrenaline). This information is intended to provide a structural framework within which to understand later chapters, whose primary focus is on the clinical implications of alterations in adrenergic function.

Central Noradrenergic Systems

A description of the regional distribution of noradrenergic (NA) neurons in the brain dates back to the observations of Vogt (1954). However, it was not until 1964 that central NA neurons were extensively described and mapped out through the pioneering work of Dahlstrom and Fuxe.

Using fluorescence histochemistry, they described a longitudinal column composed of several NA cell groups, which they designated A1–A7, extending through the lateral tegmentum of the rodent hindbrain. Subsequently, most of these same cell groups have been recognized within the brain stem of primates (Felten and Sladek 1983; Felten et al. 1974; Garver and Sladek 1975; Hubbard and DiCarlo 1973, 1974; Jacobwitz and MacLean 1978; Schofield and Dixson 1982), including humans (Bogerts 1981; Nobin and Bjorklund 1973; Olson et al. 1973).

Today, information about central catecholamine systems has accumulated to a point where our knowledge of their organization undoubtedly exceeds that of most central neuronal systems (Fillenz 1990). Most of this information, specifically related to NA neurons and their projections, has come from the study of the brains of rabbits, cats, monkeys, and, most importantly, the albino rat. However, comparative analysis of data obtained from these different animals appears to indicate that NA systems are relatively stable phylogenetically. Thus although studies of human NA systems are somewhat sparse, the observations and descriptions that have come from the extensive study of nonhuman species can apparently be generalized to humans with relative safety.

Although the cell bodies of NA neurons within the mammalian brain appear to form a single longitudinal cell column within the medulla and pons, a more careful study of these neurons and their fiber projections has led investigators to divide and organize them into two distinct component systems: 1) a locus coeruleus (LC) system and 2) a noncoerulear lateral tegmental (LT) system. These systems differ in that fibers of the LC system arise exclusively within the pons and ramify throughout the entire neuraxis, whereas the fibers that make up the LT system arise both within the medulla and pons and are much more restricted in their distribution. On this basis, separate functional roles have been postulated for these two systems. In the following section, these central NA systems are described and compared.

Locus Coeruleus System

Cell bodies. The LC, which contains about half of the total number of norepinephrine-synthesizing neurons, is the largest accumulation of such cells found within the mammalian brain. It is located within the periventricular gray of the pontine isthmus at the rostral end of the fourth ventricle. At this level it can be seen slightly rostral to the main sensory nucleus and ventromedial to the mesencephalic root and nucleus of the trigeminal nerve. The name *locus coeruleus,* Latin for "blue spot," reflects the presence of cytoplasmic neuromelanin pigment, which is synthesized by most of its neurons. This pigment allows the LC to be seen macroscopically in unstained brain stem sections and is the reason why it is sometimes referred to as the *nucleus pigmentus pontis.*

The neuron cell bodies that make up the LC correspond to the cell group designated A6 by Dahlstrom and Fuxe (1964). Although organized into a relatively small compact nucleus, LC neurons display extensive axonal collaterals that ramify and synapse within a variety of nuclei throughout the neuraxis (Nygren and Olson 1977; Olson and Fuxe 1971; Tohyama et al. 1974; Ungerstedt 1971). In this respect they illustrate several of the features of brain stem reticular neurons, namely a diffuseness and nonspecificity of synaptic connections. The location of the synaptic endings of select LC neurons appears to indicate that it has the ability to exert its functional influence through both vascular and cerebrospinal fluid routes of action (Figure 2–1).

Fiber pathways. The axons and collaterals of LC neurons collectively form both ascending and descending projection bundles that distribute in a complex fashion throughout the neuraxis. These efferent fibers contribute to two important ascending fiber systems: 1) the dorsal NA bundle and 2) the rostral limb of the dorsal periventricular pathway.

The dorsal NA bundle or dorsal tegmental bundle (Lindvall and Bjorklund 1974) is a compact fiber bundle that forms a component of the central tegmental fasciculus (Brodal 1957; Nauta and Kuypers 1958) and the major longitudinal catecholamine bundle (Jones and Friedman 1983). Its fibers traverse the mesencephalic reticular formation lateroventral to the periaqueductal gray, where they are responsible for innervating a number of nuclei within the midbrain. On reaching the level of the fasciculus retroflexus (habenulointerpeduncular tract), medial to the red nucleus of the rostral midbrain, it gives rise to two diverging groups of fibers that innervate distinct areas of the forebrain.

The larger branch of the dorsal NA bundle turns ventrolaterally, traverses the prerubral field of Forel medial to the zona incerta, and joins the medial forebrain bundle at the level of the caudal hypothalamus. It then accompanies the medial forebrain bundle rostrally through the lateral hypothalamus before giving off fibers that enter the ansa peduncularis-ventral amygdalofugal complex. After traversing the temporal stem, these fibers eventually reach the basal forebrain and lateral neocortex. When the medial forebrain bundle reaches the level of the anterior commissure and septum, it undergoes a terminal division into many smaller fiber bundles that provide NA innervation to areas of the diencephalon, basal forebrain, and frontal, dorsal, and lateral neocortex (Shimizu et al. 1974; Tohyama et al. 1974). Distribution to these areas is primarily via the stria medullaris thalami, the stria terminalis, the fornix, the cingulum bundle, the diagonal band of Broca, the external capsule, and the rostral continuation of the medial forebrain bundle into the frontal pole (Morrison et al. 1981).

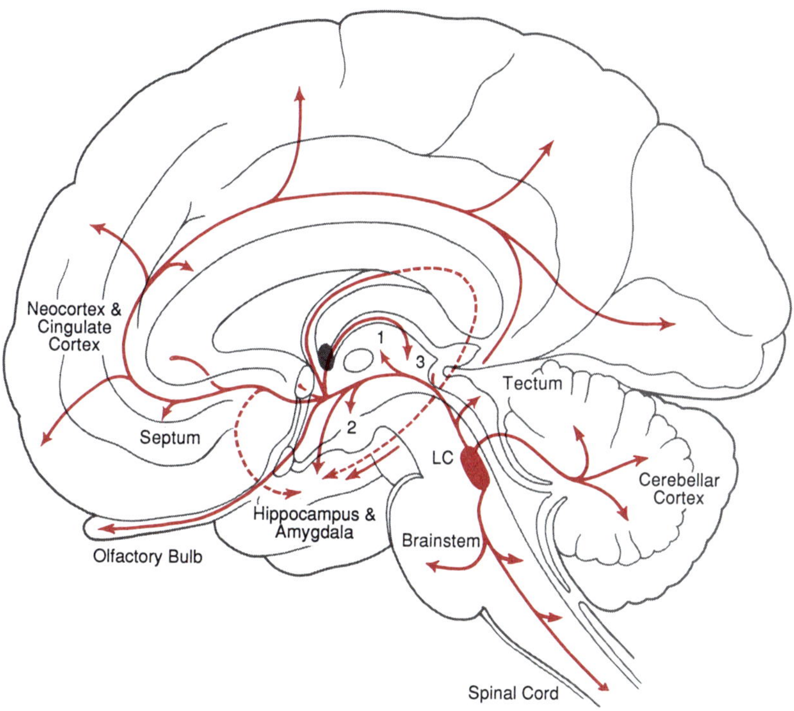

Figure 2–1. Schematic diagram localizing the norepinephrine cell bodies of the locus coeruleus (LC) system and their major efferent pathways and projection areas. 1 = dorsal thalamus; 2 = hypothalamus; 3 = epithalamus.

A smaller branch of the dorsal NA bundle enters the forebrain by turning dorsally along the fasciculus retroflexus toward the habenular complex. Eventually these fibers ascend within both the internal and external medullary laminae to provide a substantial input into the dorsal thalamus (Lindvall et al. 1974).

The second major group of ascending efferent projections from the LC form a major component of the rostral limb of the dorsal periventricular pathway. Fibers in this pathway run through the ventromedial part of the mesencephalic periaqueductal gray matter where they form part of the dorsal longitudinal fasciculus complex (Felten and Sladek 1983; Schofield and Everitt 1981). Rostrally these fibers continue forward as a part of the diencephalic periventricular fiber plexus of the third ventricle and eventually distribute both to the dorsal thalamus and to selected hypothalamic centers (Jones and Moore 1977).

In addition to its ascending projections, efferent fibers originating in the LC also project laterally into the cerebellum and caudally to nuclei of the lower brain stem and spinal cord. Those fibers entering the cerebellum accompany the superior cerebellar peduncle. There is indirect evidence that many of these fibers are collateral branches of axons that also project widely to the cerebral cortex (Hoffer et al. 1973). Within the cerebellum they distribute to the deep cerebellar nuclei and form a dense terminal plexus within the molecular layer of the cortex in close proximity to Purkinje cell dendrites (Bloom and Battenberg 1976).

LC projections to the caudal brain stem and spinal cord follow two different routes: 1) the caudal limb of the dorsal periventricular pathway and 2) the caudal limb of the dorsal NA bundle. The caudal limb of the dorsal periventricular pathway, which like its rostral counterpart forms a part of the dorsal longitudinal fasciculus, distributes primarily to primary sensory and association nuclei of the pons and medulla (Westlund and Coulter 1980). Fibers within the caudal limb of the dorsal NA bundle follow along the path of the central tegmental tract. They not only reinforce the innervation of rhombencephalic nuclei (Levitt and Moore 1979; Pearson et al. 1983) but also distribute to areas of gray matter as far caudal as the sacral spinal cord (Westlund and Coulter 1980).

Although the LC projects primarily to the ipsilateral side of the brain and cord, there are a number of commissures that allow fibers to reach structures on the contralateral side of the neuraxis. Such fibers cross at points along the entire length of the cord (Commissiong 1981) and brain stem as well as within the medullary core of the cerebellum. Commissural fibers that interconnect areas of the mesencephalon and forebrain cross at a number of points to distribute primarily in the opposite dorsal NA bundle. Points of crossing include the isthmic tegmentum of the mesencephalon, the posterior commissure, the supraoptic commissure of the hypothalamus, the anterior commissure, and the genu of the corpus callosum (Moore and Bloom 1979) (Figure 2–1).

Terminal projections. A list of all the structures innervated by the LC would be lengthy and therefore impractical. Several review articles provide a comprehensive list of such projections (Hokfelt et al. 1984; Jacobwitz 1978; Moore and Bloom 1979; Moore and Card 1984). Some of the more important projection areas are highlighted in the following paragraphs and summarized in Table 2–1.

Telencephalon. The LC provides the exclusive NA innervation for the entire neocortex and hippocampal formation (dentate gyrus, Ammon's horn, and subiculum). Terminals have been described within all cortical laminae but are most dense in layers I, IV, V, and VI. Areas of the cingulate, frontal, visual, auditory, and sensorimotor cortex show a particularly high density of fiber terminals. Many areas of the basal forebrain are either moderately or sparsely innervated, with fibers having been described within the pyriform lobe, the septum, the amygdaloid complex, and olfactory structures including the olfactory tubercle, bulb, and anterior olfactory nucleus (Moore and Bloom 1979).

Table 2–1. Locus coeruleus system

Brain area	Summary of projections
Spinal cord	Nuclei at all levels, especially the dorsal and ventral gray
Brain stem	Nuclei at all levels, especially primary sensory and association nuclei
Cerebellum	All areas of the cortex; also the deep nuclei
Diencephalon	All areas, especially the dorsal thalamus
Basal forebrain	The septum, amygdala, and olfactory bulbs; exclusively to the hippocampus
Neocortex	Exclusively to all areas

Diencephalon. The LC is the major source of NA input to the dorsal thalamus. Significant numbers of terminals have been demonstrated within the anterior, lateral, ventral, and intralaminar nuclei and the medial and lateral geniculate bodies; a sparser number of terminals have been described within medial and midline nuclear groups. The LC appears to provide a somewhat sparser innervation to both the epithalamus and hypothalamus. Here the greatest number of LC terminals have been described within the habenular nuclear complex and the dorsomedial, paraventricular, supraoptic, and periventricular hypothalamic nuclei.

Cerebellum. The LC provides the largest NA projection into the cerebellum. Fibers enter the cerebellum along the superior cerebellar peduncle and pass into the molecular layer of the cortex where a large number of synaptic terminals have been identified in the vicinity of Purkinje cell dendrites. A lesser number of terminals have been identified within the deep cerebellar nuclei.

Brain stem. The highest density of projections from the LC has been identified within primary sensory and association nuclei. Included in this group are the superior and inferior colliculi (tectum), the interpeduncular nucleus, the pontine nuclei, the sensory nuclei of the trigeminal nerve, the nucleus of the solitary tract, and the cochlear nuclei. Terminals have also been identified within the trigeminal motor nucleus, the dorsal vagal nucleus, the mesencephalic central gray, the dorsal raphe nucleus, and the entire rhombencephalic reticular formation.

Spinal cord. Descending fibers from the LC enter the spinal cord and distribute along its entire length primarily within the lateral and ventral funiculi. Many of these fibers arise as collaterals of ascending axonal projections. Distribution is primarily to the ventral laminae of the dorsal and ventral gray, with a differential distribution to the lateral aspect of the ventral gray in the cervical and lumbar enlargements. Fewer fibers appear to terminate within the intermediate gray; those that do appear to end in relation to parasympathetic preganglionic neurons in the sacral cord.

Lateral Tegmental System

Cell bodies. The LT system comprises axonal projections of those NA neurons that lie outside of the LC within the lateral tegmentum of the brain stem. Although these neurons are somewhat scattered, they are usually subdivided into two major nuclear groups: 1) a medullary nuclear group and 2) a pontine nuclear group.

The medullary nuclear group, which is in the caudal medulla, has itself been subdivided by Dahlstrom and Fuxe (1964) into two nuclear groups, which they designated A1 and A2. Group A1, referred to as the tegmental group, lies adjacent to the lateral reticular nucleus, whereas group A2, referred to as the dorsal medullary group, lies more dorsally and medially intermingled with cells of the solitary nucleus. Both A1 and A2, however, have a similar longitudinal extent that runs from the caudal border of the medulla to the caudal portion of the facial nucleus.

The pontine nuclear group runs the entire length of the pons and consists of a column of scattered neurons within the lateral reticular formation in line with medullary group A1. This group is also subdivided into two separate groups that are designated A5 and A7 in Dahlstrom and Fuxe's (1964) nomenclature. Group A5 lies along the medial border of the facial root and includes cells located ventral to the LC (subcoeruleus group), whereas group A7 lies further forward medial to the lateral lemniscus and ventrolateral to the brachium conjunctivum of the caudal mesencephalon. This point marks the most rostral extent of NA neurons within the brain stem (Figure 2–2).

Fiber pathways. The (significantly) larger ascending fiber projection of the LT system originates within the ventrolateral group of LT nuclei (A1, A5, A7). These cells, according to the classical mapping study of Ungerstedt (1971), project their axons in a rostral direction within a component of the central tegmental tract (Lindvall and Bjorklund 1974) known as the ventral NA bundle.

The LT fibers within the ventral NA bundle are chiefly responsible for the innervation for the cerebellum, mesencephalon (areas of the central gray and reticular formation), and eventually areas of the forebrain via the medial forebrain bundle. Fibers that enter the cerebellum exit the ventral NA bundle laterally in the rostral pons and project primarily to the cortex of the anterior and posterior lobes and to the deep cerebellar nuclei by following the superior cerebellar peduncle (Hokfelt and Fuxe 1969; Olson and Fuxe 1971; Pasquier et al. 1980). Those fibers that ultimately enter the medial forebrain bundle eventually join with fibers of the dorsal NA bundle and, in most instances, distribute along with them to areas of the hypothalamus and basal forebrain. Within these regions the two systems overlap somewhat, with the most substantial innervation of both the hypothalamus and basal thalamus appearing to come from the LT system. The exceptions, as previously mentioned, include the hippocampal formation and the neocortex, whose innervation comes exclusively from the LC system.

In addition to the ventral NA bundle, a second, somewhat smaller group of long ascending LT projections, which originate primarily from lower medullary cell groups A1 and A2, enter the dorsal periventricular pathway and distribute rostrally through the periventricular fiber plexus of the diencephalon to areas of the hypothalamus, dorsal thalamus, and epithalamus (Lindvall and Bjorklund 1983; Lindvall et al. 1974; McKellar and Loewy 1982).

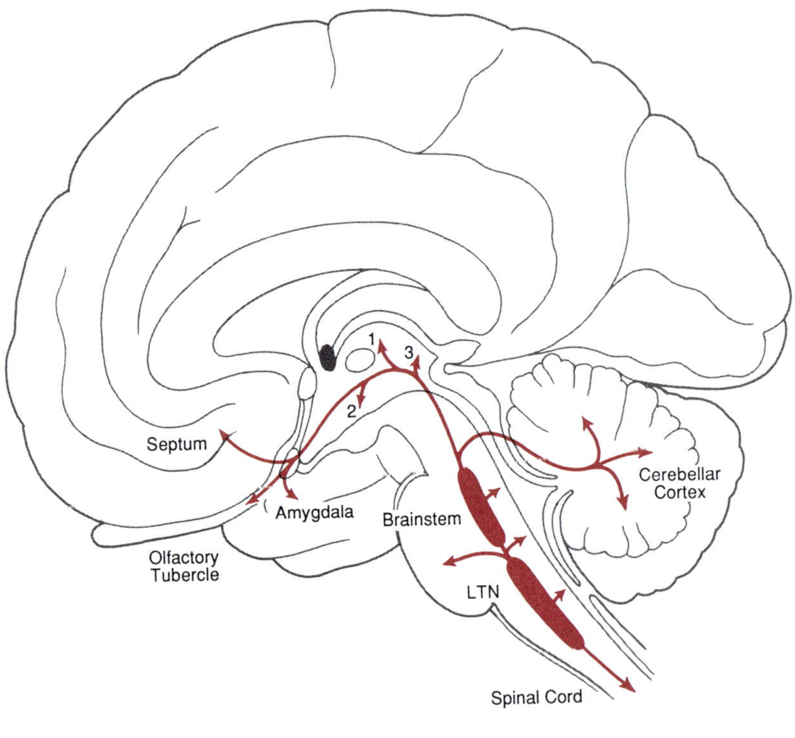

Figure 2–2. Schematic diagram localizing the norepinephrine cell bodies of the lateral tegmental (LT) system and their major efferent pathways and projection areas. 1 = dorsal thalamus; 2 = hypothalamus; 3 = epithalamus.

As with the LC system, most ascending fibers of the LT system project bilaterally with an ipsilateral predominance; this appears to be most true of fibers that originate from groups A1 and A2 within the caudal medulla (Blessing et al. 1982; Sawchenko and Swanson 1981).

Descending LT projections appear to originate largely if not exclusively from pontine nuclear groups A5 and A7. They pass through the pontine and medullary regions within the caudal limb of the dorsal NA bundle and innervate brain stem nuclear groups, particularly primary motor and visceral nuclei.

A significant number of these fibers continue caudally to enter the deeper part of the lateral funiculus of the spinal cord through which they project to all its major segments.

However, their main distribution appears to be to the ventral horn and the superficial layers of the dorsal horn (i.e., the substantia gelatinosa), with a preferential distribution within the lateral sympathetic gray (Blessing et al. 1981; Satoh et al. 1977; Westlund et al. 1983, 1984).

Descriptive studies comparing the LC and LT systems have shown that the dorsal and ventral NA bundles cannot be sharply separated but anatomically overlap at points to form one complex that has been termed the *central tegmental tract* (Moore and Bloom 1979; Swanson et al. 1981) or the *major longitudinal catecholamine bundle* (Jones and Friedman 1983). Thus although the ascending and descending fiber projections of both systems are usually discussed separately, in reality many of their major projections, including those projections within the dorsal periventricular pathway, become intermingled in one complex and to some extent distribute together both rostrally and caudally. This fact not only makes the accurate mapping of each system's efferent projections more difficult, but it also prevents the exclusive linking of any fiber bundle with any single NA system (see Figure 2–2).

Terminal projections. A detailed and comprehensive list of projections is provided elsewhere (Moore and Bloom 1979; Moore and Card 1984). The major projection areas of the LT system are highlighted in the following paragraphs and summarized in Table 2–2.

Telencephalon. The LT system does not innervate the neocortex but instead appears to provide the major innervation to structures of the basal forebrain, with the exception of the hippocampal formation. The greatest concentration of terminals have been described within the nucleus of the diagonal band, the amygdaloid complex, the septum, and the olfactory tubercle.

Diencephalon. NA fibers of the LT system innervate the entire hypothalamus, with the greatest density of nerve terminals being described within the medial preoptic, dorsomedial, periventricular, infundibular, supraoptic, and paraventricular nuclei; anterior and posterior hypothalamic areas; and the internal layer of the median eminence. A sparser innervation is provided to areas of the dorsal thalamus and epithalamus.

Cerebellum. The LT system innervates the cerebellum via the superior cerebellar peduncle. Its fibers distribute preferentially to cortical areas of the anterior and posterior lobes and to the deep nuclei.

Brain stem. The LT system provides the innervation to both primary motor and visceral nuclei of the brain stem. Particularly heavy projections have been described within the dorsal vagal nucleus, the nucleus of the solitary tract, the parabranchial nuclei, and the motor nuclei of the facial and trigeminal nerves. A significant but less dense innervation has been described for the hypoglossal nucleus, the pontine nuclei, the LC, and the raphe nuclei of the pontine and medullary reticular formation.

Spinal cord. A projection originating primarily from pontine LT nuclei descends in the ventral part of the lateral funiculus and provides a rather homogeneous innervation to the spinal cord gray at all segmental levels. Areas of the ventral gray appear to be more densely innervated than areas of the dorsal gray but, within the dorsal gray, a particularly dense projection has been identified within its more superficial laminae. The greatest num-

Table 2–2. Lateral tegmental system

Brain area	Summary of projections
Spinal cord	Nuclei at all levels, especially the intermediate autonomic gray
Brain stem	Nuclei at all levels, especially primary motor and visceral nuclei
Cerebellum	The cortex, particularly of the anterior and posterior lobes; also the deep nuclei
Diencephalon	The dorsal thalamus and epithalamus, especially the hypothalamus
Basal forebrain	The septum, amygdala, and olfactory tubercle
Neocortex	None

ber of terminals, however, appear to be concentrated within the lateral sympathetic gray of the thoracic cord.

Central Epinephrine System

The existence of central catecholamine neurons that synthesize epinephrine has been examined and debated by many investigators since first suggested by Vogt (1954). However, since the purification of the epinephrine-forming enzyme phenylethanolamine *N*-methyltransferase (Axelrod 1962) and the production of antibodies to the enzyme (Goldstein et al. 1973), many studies, particularly those on the rat brain by Hokfelt et al. (1973, 1974), have provided strong support for the existence of central epinephrine-producing neurons within the brain.

Cell bodies. The number of central epinephrine-producing neurons appears to be sparse when compared with their norepinephrine-producing counterparts. Those that have been identified are located primarily within three separate nuclear groups within the medullary portion of the brain stem. These nuclear groups, by analogy with the nomenclature of Dahlstrom and Fuxe (1964), have been designated C1, C2, and C3 (Hokfelt et al. 1974; Howe et al. 1980). The largest and most ventral of these nuclear groups, designated C1, is located in the ventrolateral medulla medial to the inferior olivary nucleus. The other two groups are more dorsally positioned, with C2 being located adjacent to the nucleus of the solitary tract and C3 adjacent to the hypoglossal root and medial longitudinal fasciculus (see Figure 2–3).

Fiber pathways. One well-defined bundle, the central epinephrine bundle, appears to contain both ascending and descending axons. This bundle, formed by projections of all three cell groups (C1–C3), can be traced rostrally through the reticular formation, the ventral tegmental area of the midbrain, and into the lateral hypothalamus (Hokfelt et al. 1974). Along its course, the central epinephrine bundle is closely related to the ventral NA bundle, the central tegmental tract, and, rostrally, at hypothalamic levels, eventually contributes fibers to the medial forebrain bundle (Lindvall et al. 1974). Central epinephrine fibers that arise from this rostral projection distribute primarily to the brain stem, cerebellum, diencephalon, and regions of the basal forebrain (Hokfelt et al. 1974). A caudal group of epinephrine-containing fibers, arising primarily from C1, descends within the central epinephrine bundle to enter the dorsal part of the lateral funiculus and distributes to nuclei at all segmental levels of the spinal cord (Figure 2–3).

Terminal projections. A particularly comprehensive list of projections is provided by Hokfelt et al. (1984). The major projection areas of the central epinephrine system are highlighted in the following paragraphs and summarized in Table 2–3.

Telencephalon. Telencephalic structures are, in general, poorly innervated by fibers of the central epinephrine system. No terminal fields have been identified within any area of the neocortex, and only a sparse number of projections have been identified within the basal forebrain. Those that

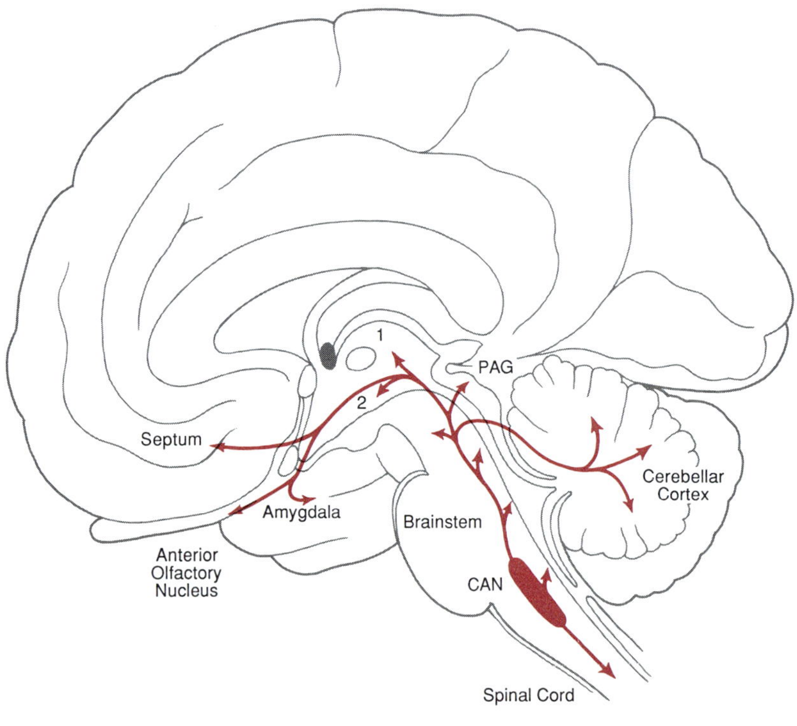

Figure 2–3. Schematic diagram localizing the cell bodies of central epinephrine neurons (CEN) and their major efferent pathways and projection areas. PAG = periaqueductal gray; 1 = dorsal thalamus; 2 = hypothalamus.

have been identified seem to be concentrated within the anterior olfactory nucleus, the nucleus accumbens, the septal region, and the amygdaloid nuclear complex. By far the greatest concentration of fibers has been described within the bed nucleus of the stria terminalis.

Diencephalon. Central epinephrine fibers to areas of the hypothalamus far exceed those to any other part of the diencephalon. The greatest density of terminals has been described within the arcuate nucleus-median eminence; the infundibular, supraoptic, paraventricular, dorsomedial, and preoptic nuclei; as well as the lateral hypothalamus and mammillary perifornical, periventricular, and retrochiasmatic areas. By comparison, areas of the dorsal thalamus receive a much sparser innervation. Here most terminals appear to be concentrated within the periventricular and rhomboid nuclei, the midline nuclei, the posterior thalamic nucleus, and the lateral geniculate nucleus.

Cerebellum. A projection from the central epinephrine system to the cerebellum still is not completely accepted. However, those terminals that have been described appear to be localized within the Purkinje cell layer of the cortex, particularly the cortex of phylogenetically older parts of the cerebellum.

Brain stem. Nerve terminals of the central epinephrine system have been identified within the periventricular gray; the ventral tegmental area; and several nuclei related to the reticular formation, including parts of the raphe complex and, most significantly, the LC. Other nuclear groups containing a relatively large number of terminals are the dorsal vagal complex,

Table 2–3. Central epinephrine system

Brain area	Summary of projections
Spinal cord	Nuclei at all levels, especially the sympathetic intermediate gray
Brain stem	Nuclei at all levels, especially visceral nuclei, the locus coeruleus, and the periaqueductal gray
Cerebellum	The cortex, especially phylogenetically older areas
Diencephalon	The dorsal thalamus, especially the hypothalamus
Basal forebrain	The septum, amygdala, and anterior olfactory nucleus, especially the bed nucleus of the stria terminalis
Neocortex	None

the nucleus of the solitary tract, and the inferior olive.

Spinal cord. Central epinephrine fibers project to the dorsal and ventral gray at all segmental levels of the cord. However, epinephrine positive terminals appear to be preferentially distributed at their highest concentration within the intermediate thoracic sympathetic gray (Ross et al. 1981).

Peripheral Sympathoadrenal System

Aside from the adrenergic neurons found within the central nervous system, all of the remaining adrenergic neurons appear to form peripheral components of the sympathetic division of the autonomic nervous system. Therefore a consideration of peripheral adrenergic nerves is equivalent to discussing the structure and function of this major subdivision of the autonomic nervous system.

With the exception of a small number of sympathetic postganglionic neurons that are cholinergic (Burnstock 1969), all peripheral sympathetic postganglionic neurons are adrenergic. The majority of these are NA neurons whose cell bodies are located within both sympathetic paravertebral (chain) and prevertebral (collateral) ganglia (Figure 2–4). The axons of these sympathetic postganglionic neurons innervate involuntary visceral effectors (i.e., smooth muscle, cardiac muscle, glands, and adipose tissue). Like all autonomic postganglionic fibers, these NA fibers can either excite or inhibit these effector structures through the interaction of the neurotransmitter they release with different postsynaptic receptors (Reis 1974).

Postganglionic NA neurons are responsible for the sympathetic innervation of visceral structures throughout the body. As a general rule, those neurons arising from the paravertebral chain ganglia of the sympathetic trunk provide the innervation to the body wall and thoracic viscera. Those arising from prevertebral or collateral ganglia (i.e., celiac, superior mesenteric, inferior mesenteric) provide the innervation to either the abdominal or the pelvic viscera.

The remaining peripheral adrenergic neurons form a collection of sympathetic postganglionic cells that reside within the adrenal medulla. These neurons synthesize both norepinephrine and epinephrine and are unique in that they possess no axonal processes. Instead these glandlike nerve cells release their secretions directly into the extracellular fluid from which they pass into the systemic circulation. Adrenal adrenergic cells are able to regulate the activity of widely dispersed visceral effectors through this specialized neuroendocrine pathway (see Figure 2–4).

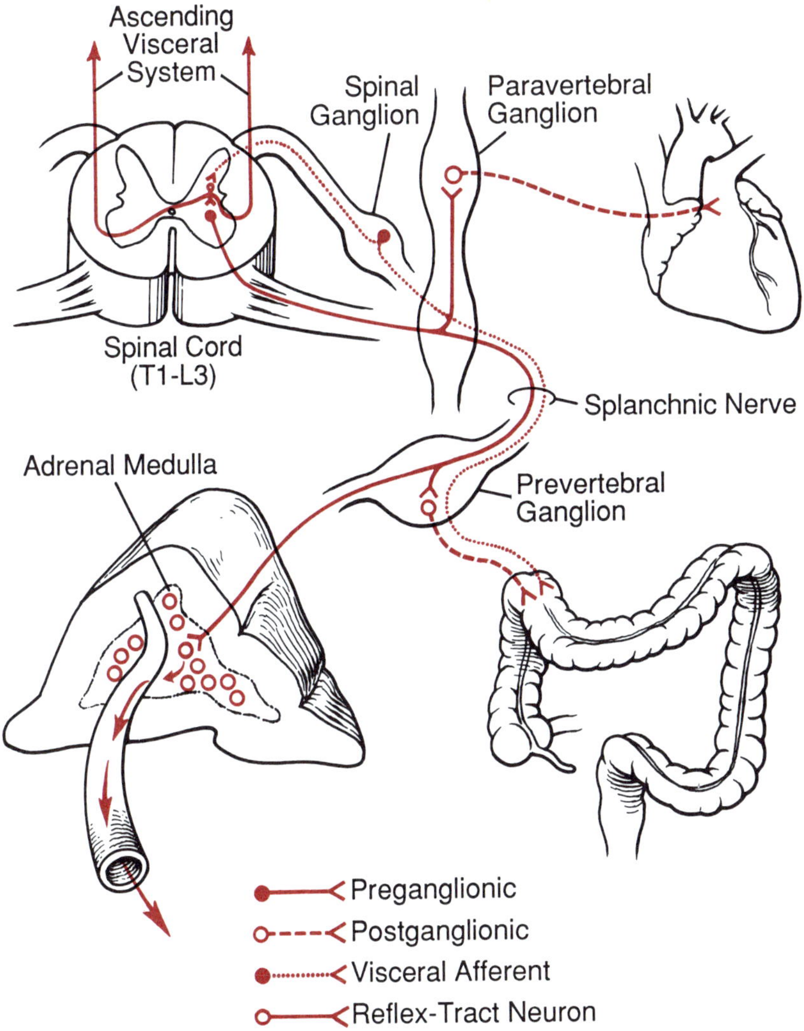

Figure 2–4. Schematic diagram showing the important noradrenergic and adrenergic components of the peripheral sympathoadrenal system. Included are the visceral afferent fibers that form major components of both reflex and central ascending visceral circuitry.

Two other important points should be remembered. First, all peripheral sympathetic adrenergic postganglionic neurons are innervated by central preganglionic neurons (cholinergic) that originate in the intermediate gray horn of the thoracic and upper lumbar (T1–L3) spinal cord (Figure 2–4). Therefore, descending central adrenergic fiber bundles, as well as other supraspinal pathways, are capable of regulating the functional activity of the peripheral sympathoadrenal system through their innervation of this central preganglionic neuron population.

Second, autonomic visceral afferent fibers that originate within the body organs provide important sensory information necessary for both the initiation and regulation of viscerovisceral and viscerosomatic reflexes mediated through the spinal cord and brain stem reticular formation (Figure 2–4). Therefore, they too are able to regulate the functional activity of the peripheral sympathoadrenal system through the essential input that they provide to central adrenergic centers (i.e., central adrenergic nuclei).

Functional Considerations

Even though a detailed consideration of the function of central and peripheral adrenergic systems is outside of the scope of this chapter, still one cannot escape making a few general observations.

It should be noted that, in contrast with other central catecholamine systems, such as the dopamine system, that show a discrete topography and restricted terminal distribution, central adrenergic systems have widespread projections throughout the neuraxis and show little topographic organization. It is this apparent lack of organization that has led neuroscientists to implicate the adrenergic neuron system in functions that are state dependent and not discretely stimulus-response in nature. The following is an abbreviated list of a number of the functions that have been attributed to the adrenergic nervous system.

Because of the extensive connections that the LC makes with the neocortex and hippocampus, many investigators have implicated the adrenergic nervous system in mechanisms underlying cortical activation (Jones et al. 1973); paradoxical sleep (Chu and Bloom 1973; Lindbrink 1974); and the control of learning, memory, and attention (Van Dongen 1981). In addition, since so many of the connections of the system are made with vegetative areas of the basal forebrain, hypothalamus, brain stem, and cord, the involvement of adrenergic pathways in mechanisms concerned with homeostatic and behavioral adaptation of the organism to the environment has had considerable support.

Reinforcing this idea are the numerous studies implicating the adrenergic system in neuroendocrine regulation of hypothalamic-pituitary function (Kalra and McCann 1973), in the regulation of food and water intake (Leibowitz 1970), in temperature control (Feldberg and Lotti 1967), in stress adaptation (Saaverda 1980), and in the regulation of cardiovascular function (Struyker-Boudier and Bekers 1975). In regard to hypothalamic-pituitary function, central adrenergic systems have been shown to be involved in the regulation of oxytocin and vasopressin secretion (Ricardo and Koh 1978), as well as gonadotropin, growth hormone, and adrenocorticotropic hormone production and secretion, most likely through their innervation of the medial hypothalamus and median eminence (Moore and Bloom 1979; Weiner and Ganong 1978).

Central adrenergic projections also appear to be involved in the regulation and integration of peripheral elements of the sympathetic spinal system (Loewy et al. 1979a, 1979b). It is many of these same descending adrenergic fibers—particularly those that distribute primarily to sensory nuclear groups, such as the substantia gelatinosa of the dorsal horn and the spinal trigeminal nuclear complex—that appear to underlie the proposed functional involvement of central adrenergics in the facilitation and inhibition of afferent sensory signals (Sasa and Takaori 1973).

Finally, one cannot overlook the growing number of studies that hypothesize a putative relationship between the loss of adrenergic neurons and a number of important diseases of the nervous system, including Alzheimer's disease, Parkinson's disease, and senile dementia of the Alzheimer's type (Mann et al. 1984; McGeer 1984; Van Dongen 1981). It is intriguing findings like these that make a comprehensive understanding of the functional anatomy of adrenergic neuron systems equally important and relevant for both the basic scientist and the practicing clinician.

References

Axelrod J: Purification and properties of phenylethanolamine *N*-methyltransferase. J Biol Chem 237:1657–1660, 1962

Blessing WW, Goodchild AK, Dampney RAL, et al: Cell groups in the lower brain stem of the rabbit projecting to the spinal cord, with special reference to catecholamine-containing neurons. Brain Res 221:35–55, 1981

Blessing WW, Jaeger CB, Ruggiero DA, et al: Hypothalamic projections of medullary catecholamine neurons in the rabbit: a combined catecholamine fluorescence and HRP transport study. Brain Res Bull 9:279–286, 1982

Bloom FE, Battenberg ELF: A rapid, simple and sensitive method for the demonstration of central catecholamine-containing neurons and axons by glyoxylic acid-induced fluorescence, II: a detailed description of methodology. J Histochem Cytochem 24:561–571, 1976

Bogerts B: A brainstem atlas of catecholaminergic neurons in man, using melanin as a natural marker. J Comp Neurol 197:63–80, 1981

Brodal A: The Reticular Formation of the Brain Stem: Anatomical Aspects and Functional Correlations. Springfield, IL, Charles C Thomas, 1957

Burnstock G: Evolution of the autonomic innervation of visceral and cardiovascular systems in vertebrates. Pharmacol Rev 21:247–324, 1969

Chu NS, Bloom FE: Norepinephrine-containing neurons: changes in spontaneous discharge patterns during sleep and waking. Science 179:908–910, 1973

Commissiong JW: Evidence that the noradrenergic coerulospinal projection decussates at the spinal level. Brain Res 212:145–151, 1981

Dahlstrom A, Fuxe K: Evidence for the existence of monoamine-containing neurons in the central nervous system; I: demonstration of monoamines in the cell bodies of brain stem neurons. Acta Physiol Scand Suppl 62:1–55, 1964

Feldberg W, Lotti VJ: Temperature responses to monoamines and an inhibitor of MAO injected into the cerebral ventricles of rats. Br J Pharmacol 31:152–161, 1967

Felten DL, Sladek JR Jr: Monoamine distribution in primate brain; V: monoaminergic nuclei: anatomy, pathways and local organization. Brain Res Bull 10:171–284, 1983

Felten DL, Laties AM, Carpenter MB: Monoamine-containing cell bodies in the squirrel monkey brain. Am J Anat 153–166, 1974

Fillenz M: Noradrenergic Neurons. New York, Cambridge University Press, 1990

Garver DL, Sladek JR Jr: Monoamine distribution in primate brain; I: catecholamine-containing perikarya in the brain stem of Macaca speciosa. J Comp Neurol 159:289–304, 1975

Goldstein M, Anagnoste B, Freedman LS, et al: Characterization, localization and regulation of catecholamine synthesizing enzymes, in Frontiers in Catecholamine Research. Edited by Usdin E, Snyder S. New York, Pergamon, 1973, pp 69–78

Hoffer BJ, Siggins GR, Oliver AP, et al: Activation of the pathway from locus ceruleus to rat cerebellar Purkinje neurons: pharmacological evidence of noradrenergic central inhibition. J Pharmacol Exp Ther 184:553–569, 1973

Hokfelt T, Fuxe K: Cerebellar monoamine nerve terminals, a new type of afferent fibers to the cortex cerebelli. Exp Brain Res 9:63–72, 1969

Hokfelt T, Fuxe K, Goldstein M: Evidence for adrenaline neurons in the rat brain. Acta Physiol Scand 89:286–288, 1973

Hokfelt T, Fuxe K, Goldstein M, et al: Immunohistochemical evidence for the existence of adrenaline neurons in the rat brain. Brain Res 66:235–251, 1974

Hokfelt T, Johansson O, Goldstein M: Central catecholamine neurons with special reference to adrenaline neurons, in Handbook of Chemical Neuroanatomy, Vol 2. Edited by Bjorklund A, Hokfelt T. New York, Elsevier, 1984, pp 157–276

Howe PRC, Costa M, Furness JB, et al: Simultaneous demonstration of phenylethanolamine *N*-methyltransferase immunofluorescent and catecholamine fluorescent nerve cell bodies in the rat medulla oblongata. Neuroscience 5:2229–2238, 1980

Hubbard JE, DiCarlo V: Fluorescence histochemistry of monoamine-containing cell bodies in the brain stem of the squirrel monkey (Saimiri sciureus): the locus coeruleus. J Comp Neurol 147:553–566, 1973

Hubbard JE, DiCarlo V: Fluorescence histochemistry of monoamine-containing cell bodies in the brain stem of squirrel monkey (Saimiri sciureus); II: catecholamine-containing groups. J Comp Neurol 153:369–384, 1974

Jacobowitz DM: Monoaminergic pathways in the central nervous system, in Psychopharmacology: A Generation of Progress. Edited by Lipton MA, DiMascio A, Killam KF. New York, Raven, 1978, pp 119–129

Jacobowitz DM, MacLean PD: A brainstem atlas of catecholaminergic neurons and serotonergic perikarya in a pygmy primate (Cebulla pygmaea). J Comp Neurol 177:397–416, 1978

Jones BE, Friedman L: Atlas of catecholamine perikarya, varicosities and pathways in the brainstem of the cat. J Comp Neurol 215:382–396, 1983

Jones BE, Moore RY: Ascending projections of the locus coeruleus in the rat; II: autoradiographic study. Brain Res 127:23–53, 1977

Jones BE, Bobillier P, Pin C, et al: The effects of lesions of catecholamine-containing neurons upon monoamine content of the brain and EEG and behavioral waking in the cat. Brain Res 58:157–177, 1973

Kalra SP, McCann SM: Effects of drugs modifying catecholamine synthesis of plasma LH and ovulation in the rat. Neuroendocrinology 15:79–91, 1973

Leibowitz SF: Hypothalamic beta-adrenergic "satiety" system antagonizes an alpha-adrenergic "hunger" system in the rat. Nature 226:963–964, 1970

Levitt P, Moore RY: Origin and organization of brainstem catecholamine innervation in the rat. J Comp Neurol 186:505–528, 1979

Lindbrink P: The effects of lesions of ascending noradrenaline pathways on sleep and waking in the rat. Brain Res 74:19–40, 1974

Lindvall O, Bjorklund A: The organization of the ascending catecholamine neuron systems in the rat brain as revealed by the glyoxylic acid fluorescence method. Acta Physiol Scand Suppl 412:1–48, 1974

Lindvall O, Bjorklund A: Dopamine- and norepinephrine-containing neuron systems: their anatomy in the rat brain, in Clinical Neuroanatomy. Edited by Emson PC. New York, Raven, 1983, pp 229–255

Lindvall O, Bjorklund A, Nobin A, et al: The adrenergic innervation of the rat thalamus as revealed by the glyoxylic acid and fluorescent method. J Comp Neurol 154:317–348, 1974

Loewy AD, McKellar S, Saper CB: Direct projections from the A5 catecholamine cell group to the intermediolateral cell column. Brain Res 174:309–314, 1979a

Loewy AD, Gregorie EM, McKellar S, et al: Electrophysiological evidence that the A5 catecholamine cell group is a vasomotor center. Brain Res 178:196–200, 1979b

Mann DMA, Yates PO, Marcyniuk B: A comparison of changes in the nucleus basalis and locus coeruleus in Alzheimer's disease. J Neurol Neurosurg Psychiatry 47:201–203, 1984

McGeer PL: The 12th JAF Stevenson memorial lecture: aging, Alzheimer's disease and the cholinergic system. Can J Physiol Pharmacol 62:741–754, 1984

McKellar S, Loewy AD: Efferent projections of the A1 catecholamine cell group in the rat: an autoradiographic study. Brain Res 241:11–29, 1982

Moore RY, Bloom FE: Central catecholamine neuron systems: anatomy and physiology of the norepinephrine and epinephrine systems. Annu Rev Neurosci 2:113–168, 1979

Moore RY, Card PJ: Noradrenaline-containing neuron systems, in Handbook of Chemical Neuroanatomy, Vol 2. Edited by Bjorklund A, Hokfelt T. New York, Elsevier, 1984, pp 123–156

Morrison JH, Molliver ME, Grzanna R, et al: The intra-cortical trajectory of the coerulo-cortical projection in the rat: a tangentially organized cortical afferent. Neuroscience 6:139–158, 1981

Nauta WJH, Kuypers HGJM: Some ascending pathways in the brainstem reticular formation, in Reticular Formation of the Brain (Henry Ford Hospital International Symposium). Edited by Jasper HH, Proctor LD, Knighton RS, et al. Boston, MA, Little, Brown, 1958, pp 3–30

Nobin A, Bjorklund A: Topography of the monoamine neuron systems in the human brain as revealed in fetuses. Acta Physiol Scand 388:1–40, 1973

Nygren LG, Olson L: A new major projection from locus coeruleus: the main source of noradrenergic nerve terminals in the ventral and dorsal columns of the spinal cord. Brain Res 132:85–93, 1977

Olson L, Fuxe K: On the projections from the locus coeruleus noradrenaline neurons: the cerebellar innervation. Brain Res 28:165–171, 1971

Olson L, Borens LO, Seiger A: Histochemical demonstration and mapping of 5-hydroxytryptamine and catecholamine-containing neuron systems in the human fetal brain. Zeitschrift Fyer Anatomie und Entwicklungsgeschichte 139:259–282, 1973

Pasquier DA, Gold MA, Jacobowitz DM: Noradrenergic perikarya (A5-A7, subcoeruleus) projections to the rat cerebellum. Brain Res 196:270–275, 1980

Pearson J, Goldstein M, Markey K, et al: Human brainstem catecholamine neuronal anatomy as indicated by immunocytochemistry with antibodies to tyrosine hydroxylase. Neuroscience 8:3–32, 1983

Reis DJ: Central noradrenergic neurons. Adv Neurol 5:45–56, 1974

Ricardo JA, Koh ET: Anatomical evidence of direct projections from the nucleus of the solitary tract to the hypothalamus, amygdala and other forebrain structures in the rat. Brain Res 153:1–26, 1978

Ross CA, Armstrong DM, Ruggiero DA, et al: Adrenaline neurons in the rostral ventrolateral medulla innervate thoracic spinal cord: a combined immunocytochemical and retrograde transport demonstration. Neurosci Lett 25:257–262, 1981

Saaverda JM: Brain stem adrenergic neurons participate in the regulation of the stress response and in genetic experimental hypertension, in Central Adrenaline Neurons: Basic Aspects and Their Role in Cardiovascular Functions. Edited by Fuxe K, Goldstein M, Hokfelt B, et al. Elmsford, NY, Pergamon, 1980, pp 235–244

Sasa M, Takaori S: Influence of the locus coeruleus on transmission in the spinal trigeminal nucleus neurons. Brain Res 55:203–208, 1973

Satoh K, Tohyama M, Yamamoto K, et al: Noradrenaline innervation of the spinal cord studied by the horseradish peroxidase method combined with monoamine oxidase staining. Exp Brain Res 30:175–186, 1977

Sawchenko PE, Swanson LW: Central noradrenergic pathways for the integration of hypothalamic neuroendocrine and autonomic responses. Science 214:685–687, 1981

Schofield SPM, Dixson AF: Comparative anatomy of brain monoaminergic neurons in new world and old world monkeys. American Journal of Primatology 2:3–19, 1982

Schofield SPM, Everitt BJ: The organization of catecholamine-containing neurons in the brain of the rhesus monkey (Macaca mulatta). J Anat 132:391–418, 1981

Shimizu N, Ohnishi S, Toyahama M, et al: Demonstration by degeneration silver method of the ascending projection from the locus ceruleus. Exp Brain Res 20:181–192, 1974

Struyker-Boudier HAJ, Bekers AD: Adrenaline-induced cardiovascular changes after intrahypothalamic administration to rats. Eur J Pharmacol 31:153–155, 1975

Swanson LW, Sawchenko PE, Berod A, et al: An immunohistochemical study of the organization of catecholaminergic cells and terminal fields in the paraventricular and supraoptic nuclei of the hypothalamus. J Comp Neurol 196:271–285, 1981

Tohyama M, Maeda T, Shimizu N: Detailed noradrenaline pathways of locus coeruleus neurons to the cerebral cortex with use of 6-hydroxydopa. Brain Res 79:139–144, 1974

Ungerstedt U: Stereotaxic mapping of the monoamine pathways in the rat. Acta Physiol Scand Suppl 367:1–49, 1971

Van Dongen PAM: The human locus coeruleus in neurology and psychiatry. Prog Neurobiol 17:97–139, 1981

Vogt M: The concentration of sympathin in different parts of the central nervous system under normal conditions and after the administration of drugs. J Physiol (Lond) 123:451–481, 1954

Weiner RI, Ganong WF: Role of brain monoamines and histamine in regulation of anterior pituitary secretion. Physiol Rev 58:905–976, 1978

Westlund KN, Coulter JD: Descending projections of the locus coeruleus and subcoeruleus/medial parabrachial nuclei in the monkey: axonal transport studies and dopamine-beta-hydroxylase immunocytochemistry. Brain Res Rev 2:235–264, 1980

Westlund KN, Bowker RM, Ziegler MG, et al: Noradrenergic projections to the spinal cord of the rat. Brain Res 263:15–31, 1983

Westlund KN, Bowker RM, Ziegler MG, et al: Origins and terminations of descending noradrenergic projections to the spinal cord of monkey. Brain Res 292:1–16, 1984

3

Central Nervous System Noradrenergic–Peptide Interactions

RITA J. VALENTINO, PH.D.
GUY DROLET, PH.D.
GARY ASTON-JONES, PH.D.

In this chapter, we focus on the specific peptides corticotropin-releasing factor (CRF) and enkephalin, and their interaction with noradrenergic neurons of the nucleus locus coeruleus (LC). The pontine noradrenergic nucleus LC is considered to be the major central nervous system (CNS) noradrenergic nucleus primarily because of its extensive projections throughout the neuraxis (Dahlstrom and Fuxe 1964; Swanson and Hartman 1976). This characteristic feature has led to numerous investigations into the role of the LC in normal and pathological behavior, many of which are reviewed in this volume. The identification of specific afferents (Aston-Jones et al. 1986) to the LC has provided a firmer base on which to test hypotheses of LC function in normal and pathological states. Thus, in addition to what has been learned about the LC-norepinephrine system by characterizing the effects of drugs, peptides, neurotransmitters, or behavioral manipulations on LC activity or norepinephrine turnover, new knowledge is being generated from studies of the effects of stimulating, lesioning, or manipulating known afferents to the LC. Elucidation of the circuitry and neurotransmitters by which stimuli act to alter activity of the LC-norepinephrine system will help to refine our hypotheses of the role of the LC in various states. By the integration of what has been learned about afferent control of LC with what is known about the effects of LC activity on its target cells, we will make an attempt to go beyond a description of peptide-adrenergic interactions to providing a model of the circuitry underlying these interactions.

Numerous peptides have been visualized in fibers within the LC, and many have effects on LC discharge. For example, immunohistochemical studies would suggest that the LC receives projections from neurons containing substance P (Ljungdahl et al. 1978; Pickel et al. 1979), neurotensin (Uhl et al. 1979a), β-endorphin (Bloom et al. 1978), enkephalin (Haber and Elde 1982), angiotensin II (Fuxe et al. 1976), vasoactive intestinal peptide (Eiden et al. 1982; Sutin and Jacobowitz 1988), and CRF (Sakanaka et al. 1987). Likewise, receptors for many peptides have been reported to be localized in the LC (Atweh and Kuhar 1977; DeSouza 1987; DeSouza et al. 1985; Mendelsohn et al. 1984), and the effects of peptides, including substance P (Cheeseman et al. 1983; Guyenet and Aghajanian 1977), neurotensin (Young et al. 1978), met-enkephalin (Guyenet and Aghajanian 1979), angiotensin II (Xiong and Marshall 1990), vasopressin (Olpe and Baltzer 1981), vasoactive intestinal peptide (Wang and Aghajanian 1989), and CRF (Valentino et al. 1983), on LC spontaneous discharge have been described. Rather than provide a survey of these peptides, their effects, and their localization, we describe the effects and function of two endogenous peptides, CRF and enkephalin, in the LC. The rationale for concentrating on these specific peptides is that these are probably the most well-studied peptides in the LC, and reasonable evidence exists to support a neurotransmitter role for both in the LC. Moreover, the putative neurotransmitter role of both peptides may be important in specific physiological or psychiatric states. Thus, we focus on 1) CRF in the LC and the role of this interaction in stress and depression; and 2) the localization, source, effects, and possible functions of enkephalin in the LC. We first provide a brief overview of the anatomical and physiological features of the LC that are important for understanding its potential functions. We then review anatomical and physiological studies that implicate the peptides as neurotransmitters in the LC. Finally, we integrate the results of these and pharmacological studies into workable hypotheses of the role of peptide-noradrenergic interactions in behavior or psychiatric disorders.

Anatomical Considerations

The details of the anatomy of LC and other adrenergic nuclei are the subject of Chapter 2 of this volume. However, three important anatomical features of the LC are addressed here: 1) its location in the brain, 2) its efferent projections, and 3) afferent projections to the LC. These specific features are addressed because each contributes to the importance of the LC as an amplification mechanism in the CNS. Moreover, it is because of these anatomical features of the LC that the interaction of certain peptides with LC neurons becomes important for their function in the CNS.

Location

In the primate and rat, the LC is a homogeneous cluster of cells (Grzanna and Molliver 1980; Swanson 1976) lying adjacent to the fourth ventricle. In contrast to the rat and primate LC, the cat (Léger and Hernandez-Nicaise 1980), rabbit (Blessing et al. 1981), and guinea pig (G. Aston-Jones, unpublished observations, September 1977) LC are heterogeneous with respect to cell type and morphology. The localization of the LC near the ventricle as well as its close approximation to blood vessels (K. H. Finley and Cobb 1940) implies that LC neurons are exposed to a multitude of circulating factors such as hormones, metabolites, or drugs, in addition to the neurotransmitters making synaptic contacts with LC dendrites. Indeed, dopamine β-hydroxylase immunoreactive cells have been described as lying on the roof of the fourth ventricle and having dendrites that extend between ependymal cells to the ventricular surface (Grzanna and Molliver 1980).

Locus Coeruleus Efferent Projections

As previously mentioned, its efferent projection system is a characteristic feature that has sparked much interest in the LC. From this relatively small cluster of cells (numbering approximately 1,600 per hemisphere in the rat; Swanson 1976) emerges a divergent efferent system that innervates all regions of the neuraxis. Anterograde and retrograde transport techniques indicate that the primary noradrenergic innervation of all layers of neocortex originates from the LC (Freedman et al. 1975). Likewise, the sole source of norepinephrine in the hippocampus derives from the LC (Loy et al. 1980). In contrast, the LC provides a minor source of norepinephrine to the hypothalamus, primarily innervating the medial periventricular zone of the paraventricular nucleus (Sawchenko and Swanson 1982). Because these cells project to the median eminence, LC innervation here may be important in control of neuroendocrine function and may be involved in a feedforward mechanism for regulating CRF release in stress (see later in this chapter). Finally, the LC also innervates other brain-stem nuclei and regions of the spinal cord.

These descriptive studies indicate that this small cluster of norepinephrine cells innervates regionally, morphologically, neurochemically, and functionally diverse targets. Moreover, injection of retrograde tracers into different LC targets has revealed that individual LC neurons often diverge to different targets (Ader et al. 1980; Nagai et al. 1981; Steindler 1981). This feature of the LC gives it the potential to serve as an amplifier in the CNS. For example, any agent or manipulation that alters LC activity would have the potential to alter simultaneously activity in targets of diverse function (i.e., cortex, hippocampus, spinal cord). This type of control may come into play during situations that require a coordinated, multicomponent response to a stimulus. This hypothesis is detailed below, specifically in regard to the stress response.

Afferents to Locus Coeruleus

Investigations into afferent control of the LC have revealed restricted sources of afferent inputs to LC. These studies combined anterograde and retrograde tract tracing techniques (Aston-Jones et al. 1986, 1990; Pieribone and Aston-Jones 1991; Pieribone et al. 1988), as well as electrophysiological identification of LC afferents (Aston-Jones and Ennis

1988; Ennis and Aston-Jones 1986, 1987, 1988, 1989). The primary sources of LC innervation as determined from these studies are two medullary nuclei: the medial nucleus prepositus hypoglossi (PrH) and the nucleus paragigantocellularis (PGi). Retrogradely labeled neurons are also found in the dorsal cap of the paraventricular nucleus of the hypothalamus and in the intermediate zone of the cervical spinal cord after injection of retrograde tracers into the LC, although the sparse labeling in these latter regions suggests that they are relatively minor sources. However, LC innervation from the dorsal cap of the paraventricular nucleus may be important in the role of the LC in stress responses (see later). Another minor afferent to LC as determined by both anatomical and electrophysiological studies is the periaqueductal gray (Ennis et al. 1991). That the PrH and PGi are the primary sources of afferents to the LC has been confirmed by demonstrations of antidromic activation of these cells by LC stimulation and by the inability to drive cells antidromically in other regions by LC stimulation that were previously thought to provide input to LC (i.e., dorsal raphe and nucleus tractus solitarii) (Ennis and Aston-Jones 1987, 1989).

The neurochemical identity of LC afferents is now being investigated. As discussed later, some PGi and PrH neurons are immunoreactive for CRF and enkephalin. Substantial evidence exists supporting an excitatory amino acid projection from PGi to the LC (Ennis and Aston-Jones 1988). For example, electrical stimulation of PGi produces activation as well as inhibition in a majority of LC neurons. The excitatory effect is prevented by prior administration of the amino acid antagonist kynurenic acid (Ennis and Aston-Jones 1988), and the underlying inhibition is prevented in a majority of neurons by the α_2-antagonist idazoxan (Astier and Aston-Jones 1989), implicating both an excitatory amino acid input and an inhibitory adrenergic input to the LC from PGi. In contrast to PGi stimulation, stimulation of PrH generally inhibits LC discharge, and this effect appears to be mediated by gamma-aminobutyric acid–A (GABA-A) receptors because it is sensitive to antagonism by bicuculline and picrotoxin (Ennis and Aston-Jones 1989).

It is of interest that a nucleus with such widespread influence receives such a restricted input, particularly because the LC is responsive to many modes of sensory and autonomic stimuli (see later). However, information about these various stimuli could reach the LC via PrH or PGi because cardiovascular, respiratory, and nociceptive functions have been implicated for the nucleus PGi, and a major function of the PrH is oculomotor control, which comes into play when animals are orienting to external sensory stimuli. These concepts are reviewed in an article by Aston-Jones and colleagues (1990).

Locus Coeruleus Physiology

Two characteristics of LC physiology have provided a basis for hypotheses of LC function in arousal and vigilance. The first involves the discharge characteristics of LC neurons, and the second involves the effects of LC activation on activity of LC target cells. Studies of LC discharge characteristics in unanesthetized rats (Aston-Jones and Bloom 1981a, 1981b), cats (Rasmussen et al. 1986), and monkeys (Foote et al. 1980; Grant et al. 1988) indicate that the spontaneous discharge of LC neurons is labile, corresponding to, and often anticipating, changes in behavioral state. For example, LC activity is relatively slow during slow wave sleep, absent in rapid eye movement sleep, and relatively high when animals are aroused. Novel, phasic stimuli of many sensory modalities (visual, olfactory, auditory, tactile) elicit LC discharge in unanesthetized rats, and this is often associated with an orienting response (Aston-Jones and Bloom 1981a, 1981b). Additionally, a variety of stressors increase LC discharge in unanesthetized animals (see discussion later). In anesthetized rats, LC discharge is relatively stable, and only potentially noxious stimuli elicit LC discharge compared with the nonnoxious modalities, which elicit LC discharge in unanesthetized rats. Interestingly, even in halothane-anesthetized rats, electroencephalographic activity is correlated to LC activity such that manipulations that increase LC discharge (direct application of muscarinic agonists, foot shock, hypotensive stress) are associated with the electroencephalogram (EEG) activation, characterized by desynchronization and decreased low amplitude activity recorded from cortical leads and theta rhythm recorded from the hippocampus (Berridge and Foote 1991; Valentino et al. 1991). Conversely, direct application of α_2-agonists, such as clonidine, onto LC neurons is associated with decreased LC discharge and EEG synchronization with high-amplitude, low-frequency activity (Berridge et al. 1993). These observations are consistent with an important role of the LC in arousal. Whether LC activation is necessary or sufficient for EEG arousal is under investigation.

The observation that LC activity is positively correlated with arousal and that phasic sensory stimuli activate the LC has led to the suggestion that the LC is important in initiating state changes that are associated with orienting toward novel stimuli (for reviews, see Aston-Jones 1985; Foote et al. 1983). The influence of the LC on its target cells may be critical in producing these behavioral state changes. These effects have been studied both in vivo and in vitro in several targets including hippocampus (Segal and Bloom 1974a, 1974b, 1976), cortex (Foote et al. 1975; Waterhouse et al. 1980, 1982), cerebellum (Freedman et al. 1977; Moises and Woodward 1980; Moises et al. 1979, 1981), and spinal cord (White and Neuman 1980) by recording activity of the appropriate target cell during electrical stimulation of the LC or iontophoretic application of norepinephrine. During electrical stimulation of the LC or iontophoresis of norepinephrine, spontaneous discharge rates of LC target cells are decreased or unaffected while activity evoked by stimulation of inputs to these same targets is enhanced. Likewise, the effects of iontophoresing either excitatory (i.e., acetylcholine) or inhibitory agents (i.e., GABA) are enhanced (Waterhouse et al. 1980).

Thus, when the LC is active and presumably releasing norepinephrine in target areas, activity of target cells is altered such that evoked activity is enhanced relative to basal discharge; that is, signal-to-noise is increased. This would be predicted to bias brain activity to favor processing information conveyed by strongly driven interactions. Considering that an individual LC cell may innervate functionally different targets, the consistent effect of LC activation on target cells would be predicted to result in widespread simultaneous alterations in cell sensitivity throughout the CNS. The biasing of target activity toward receipt of synaptic information may be adaptive in situations such as the stress response, where multiple systems (endocrine, autonomic, motor, behavior) are simultaneously engaged in the response to a stimulus. Thus, the anatomical feature of a divergent efferent system, its effects on target cells, and its responsiveness to stimuli give the LC the potential to serve as an amplification mechanism whose function may be to change the state of the animal toward responding to a stimulus.

Corticotropin-Releasing Factor as a Neurotransmitter in the Locus Coeruleus

Since the identification and characterization of CRF by Vale et al. (1981) as the primary neurohormone responsible for adrenocorticotropic hormone (ACTH) release during stress, much evidence has accumulated suggesting that CRF has a more global role in the stress response. This is supported by the extrahypothalamic distribution of CRF cells and fibers (Merchenthaler et al. 1982; Olschowka et al. 1982; Sakanaka et al. 1987; Swanson et al. 1983) and CRF binding sites (DeSouza 1987; DeSouza et al. 1985), and the finding that central administration of CRF mimics many of the autonomic (Brown et al. 1982; Fisher et al. 1982; Tache et al. 1983) and behavioral (D. R. Britton et al. 1982; K. Britton et al. 1985; Eaves et al. 1985; Kalin 1985; Sutton et al. 1982) aspects of stress even when administered to hypophysectomized rats. These data suggest that CRF functions outside of the hypothalamic-pituitary axis as a neurotransmitter. Consistent with this, many autonomic (Brown et al. 1985; Lenz et al. 1988) and behavioral (Kalin et al. 1988; Tazi et al. 1987) aspects of stress can be attenuated by pretreatment with a CRF antagonist α-helical $CRF_{9\text{-}41}$. Taken together, these data are consistent with a role for CRF as a neurohormone in hypophyseal circuits to mediate the endocrine branch of the stress response, and as a neurotransmitter in extrahypophyseal circuits to mediate autonomic and behavioral components of stress. If each of these CRF circuits were equisensitive to stressors, a single stressor would be predicted to initiate coordinated, simultaneous release of CRF throughout the brain, resulting in a coordinated, multicomponent response.

Anatomical Considerations

The LC is one region where substantial evidence implicates CRF as a neurotransmitter. Several anatomical studies of CRF distribution have demonstrated the existence of CRF-immunoreactive (CRF-ir) fibers in rat LC (Cummings et al. 1983; Sakanaka et al. 1987), although some studies do not support this. Part of the discrepancy lies in the use of different sources of CRF antibody. Our own studies have demonstrated CRF-ir fibers in LC, although the density is only moderate in LC proper

(Valentino et al. 1992). A high density of CRF-ir fibers is visualized just lateral to the anterior pole of LC and medial to the mesencephalic nucleus of the fifth nerve (Valentino et al. 1992) (Figure 3–1). This area is of interest because the dendrites of LC neurons overlap this region (Shipley et al., in press). Without ultrastructural analysis it is impossible to determine whether CRF neurons synapse with LC cells or dendrites in this region. Although the electrophysiological studies described later suggest that CRF serves as a neurotransmitter in the LC, it is possible that CRF from nearby (pericoerulear) regions may act on LC neurons in a neurohormone capacity. Alternatively, CRF may act as a neurotransmitter on other pericoerulear neurons that project to LC. In addition to having been found in rat LC, CRF-ir fibers have been described in monkey (Foote and Cha 1988) and human (Pammer et al. 1990) LC.

The source of CRF in the LC has been a recent subject of investigation in our laboratory. Numerous CRF-ir neurons are visualized in PrH and PGi, the medullary nuclei that are primary sources of LC afferents (Sakanaka et al. 1987; Valentino et al. 1992). Likewise, CRF-ir neurons are abundant in the dorsal cap of the paraventricular nucleus of the hypothalamus, which Aston-Jones et al. (1986) found to be a minor source of LC afferents. Double labeling studies utilizing injections of retrograde tract tracers in LC and CRF immunohistochemistry have suggested that the CRF input to LC derives in part from PGi, from the dorsal cap of the paraventricular nucleus, and perhaps from pericoerulear sources (Valentino et al. 1992). In contrast, the population of CRF neurons in PrH is distinct from PrH neurons that project to LC (Valentino et al. 1992). CRF afferents to LC from PGi are of interest because hypotensive stress elicits CRF release within the LC (see later), and this also activates neurons in PGi (Brown and Guyenet 1986). Thus, CRF afferents from PGi to LC may be an important circuit for the relay of information about autonomic stressors to the LC. A CRF input to LC originating from the paraventricular nucleus would suggest that the same stimuli that elicit ACTH release also simultaneously release CRF within the LC. Two groups of CRF neurons in close proximity to LC are potential sources of CRF afferents in LC. These include Barrington's nucleus, which lies just rostral and ventral to the LC, and is made up of CRF-ir neurons. Barrington's nucleus projects to the spinal cord and is involved in micturition (Barrington 1925; Satoh et al. 1978; Vincent and Satoh 1984). In addition to Barrington's nucleus, CRF-ir cells lie along the lateral border

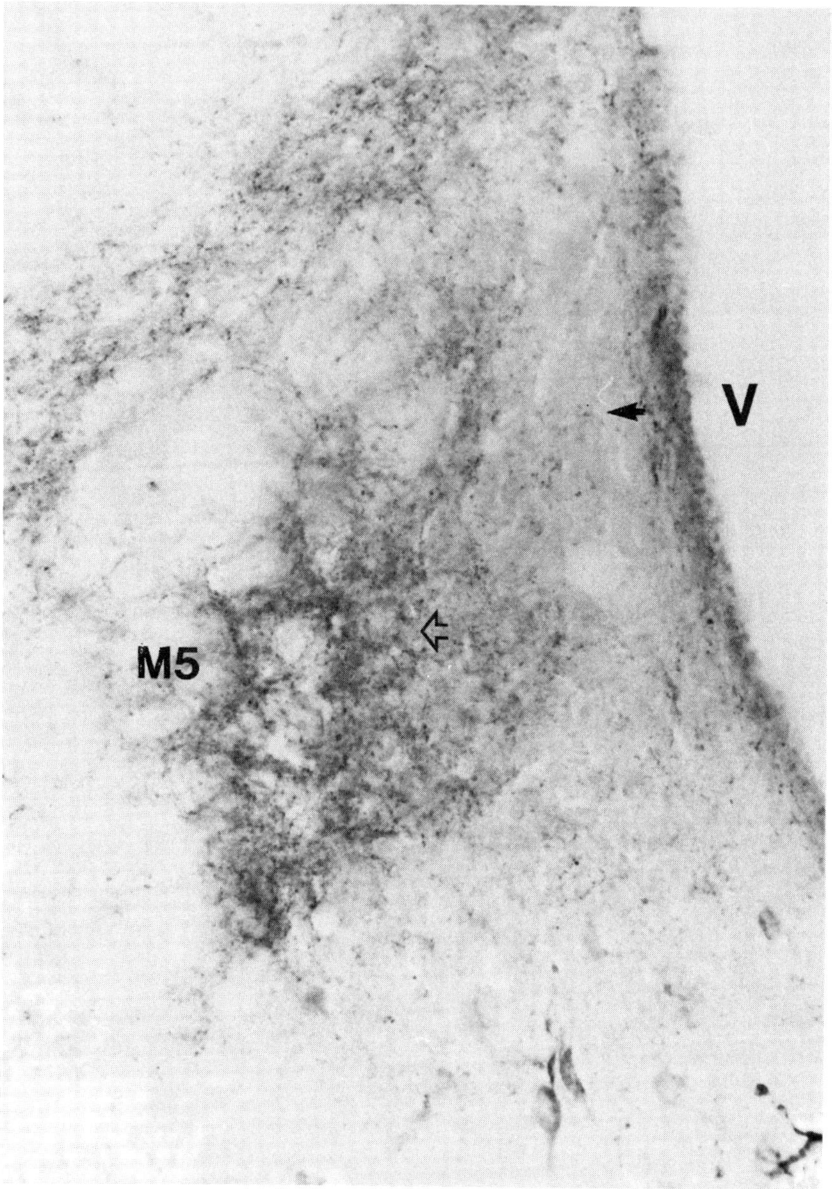

Figure 3–1. Corticotropin-releasing factor (CRF) innervation of the locus coeruleus (LC) region. Brightfield photomicrograph of a 40-μm thick coronal section through the rostral pole of the rat LC. CRF-immunoreactive (CRF-ir) fibers can be seen in the nucleus LC (solid arrow). However, a dense plexus of CRF-ir fibers (open arrow) is visualized lateral to LC and medial to the mesencephalic nucleus of the trigeminus (M5). Top is dorsal. V = ventricle.

of the LC, medial to the mesencephalic nucleus of the trigeminus. However, the general spread of retrograde tracers into these pericoerulear CRF-containing regions after injection into the LC makes it difficult to determine whether these local sources send projections to the LC.

Corticotropin-Releasing Factor Effects on Locus Coeruleus Discharge

Prior to the isolation and characterization of CRF, the LC had been implicated in stress. This was partly based on studies that showed that stressors increase norepinephrine turnover in brain regions whose sole source of norepinephrine was from the LC (Cassens et al. 1980; Korf et al. 1973). These findings, taken together with more recent studies demonstrating that LC spontaneous discharge rate is increased by certain stressors (Abercrombie and Jacobs 1987a, 1987b; Elam et al. 1981, 1984; Morilak et al. 1987a, 1987b; Svensson 1987; Svensson and Thoren 1979), indicate that stressors activate the LC-norepinephrine system. However, the circuitry and neurotransmitters involved in LC activation remain to be elucidated. CRF is a potential mediator of LC activation because of its role in stress and its localization in fibers in LC.

The effects of CRF on LC activity have now been characterized in anesthetized and unanesthetized rats. Central CRF administration (icv) increases LC spontaneous discharge rate in a dose-dependent manner (Valentino and Foote 1987, 1988; Valentino et al. 1983). CRF is more potent and efficacious in unanesthetized versus halothane-anesthetized rats (Valentino and Foote 1988). An interesting phenomenon concerning the time course of CRF effects has been reported. In unanesthetized, but not anesthetized, rats the dose-response curve for CRF activation of LC determined 40 minutes after injection is shifted to the left of that determined 10 minutes after injection (Valentino and Foote 1988). This increased effect with time has been reported for other actions of CRF, although the mechanism for this is not clear. Direct application of CRF to LC neurons by pressure application also increases LC spontaneous discharge rate, at least in anesthetized rats (Valentino et al. 1983). The structural and regional specificity of the effects of CRF on LC cells has been verified. For example, structural analogues of CRF that do not release ACTH from anterior pituitary cells, CRF-OH and Ala_{14}CRF, do not increase LC spontaneous discharge rate (Valentino and Foote 1987; Valentino et al. 1983). Other studies have shown that CRF does not increase spontaneous discharge rates of all cells in the brain stem (Valentino et al. 1983).

Responsiveness to phasic sensory stimuli is an important characteristic of LC discharge that has been part of the evidence implicating the LC in arousal (see previous paragraph). Central CRF administration alters LC responses to phasic sensory stimuli in both anesthetized and unanesthetized rats. In anesthetized rats where repeated sciatic nerve stimulation is used to elicit LC discharge, CRF increases nonstimulated discharge and decreases stimulus-evoked discharge (Valentino and Foote 1987) (Figure 3–2A). The net effect is that LC neurons discharge at a greater frequency throughout a trial of stimulus presentations, and discharge is not temporally correlated to the stimulus. When administered to unanesthetized rats, presented with repeated auditory stimuli (20-msec tones), nonstimulated LC discharge increases to a much greater degree than evoked discharge (Valentino and Foote 1988). The net effect is similar to that observed in anesthetized rats; that is, LC discharge is greater throughout a trial of sensory stimulation and shows less temporal correlation to the sensory stimulus. The effect of CRF on LC sensory responses is difficult to interpret because the pattern of LC discharge to phasic sensory stimuli has yet to be translated to norepinephrine release in targets, to effects on target cell discharge, or to any function. Presumably, the burst of LC discharge that is elicited by a sensory stimulus would cause a pulse of norepinephrine release in targets, the end result of which may be to enhance processing of information about the sensory stimulus. The production by CRF of a more general increase in LC discharge rate that is not specifically correlated to a phasic sensory stimulus would be predicted to result in a more persistent release of norepinephrine in targets and a general tuning of all targets toward receipt of information that is not limited to a specific sensory stimulus. These effects may be adaptive in stress because it may be more important to respond to the stressor (which is usually a tonic stimulus) and to be less responsive to more phasic stimuli, which may be less important.

Stress Effects on Locus Coeruleus Discharge

The effects of various physiological and environmental stressors on LC discharge have been well studied in unanesthetized cats (Abercrombie and Jacobs 1987a, 1987b; Morilak et al. 1987a, 1987b). Most of the challenges presented increased LC spontaneous discharge rate as well as

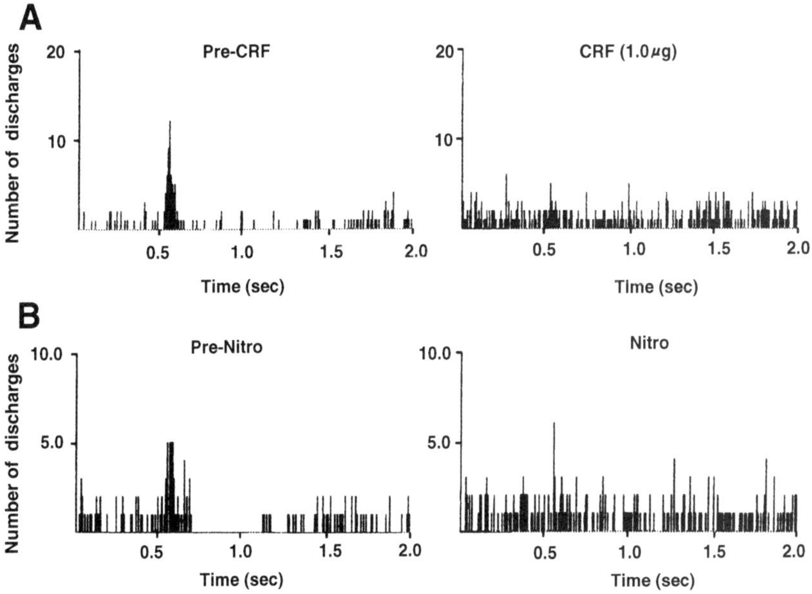

Figure 3–2. Comparison of corticotropin-releasing factor (CRF) and stress effects on locus coeruleus (LC) sensory-evoked discharge. A and B are pairs of poststimulus time histograms (PSTHs) from two different halothane-anesthetized rats (A and B). PSTHs show the cumulative discharge of a single LC cell during repeated presentation of sciatic nerve stimulation (1.3 mA, 0.5 msec duration, 0.1 Hz; 50 trials). The stimulus is presented at 0.5 second. Activity is recorded for 0.5 second prior to the stimulus and for the following 1.5 seconds. The control histograms on the left show the typical response of an LC neuron to this type of stimulation; that is, LC discharge is greatly increased for 80–100 msec following the stimulus (onset 10–20 msec), and this is followed by a relatively quiescent period. The A PSTH shows the effect of icv administered CRF on this response. CRF causes the cell to discharge at a higher frequency that is not temporally correlated to the stimulus. The B PSTH was generated during infusion of nitroprusside. Nitroprusside has a similar effect on LC sensory responses as does CRF.
Source. Reprinted with permission from Valentino RJ, Wehby RG: "Corticotropin-Releasing Factor: Evidence for a Neurotransmitter Role in the Locus Coeruleus During Hemodynamic Stress." *Neuroendocrinology* 48:674–677, 1988. Copyright 1988, S. Karger AG, Basel, Switzerland.

autonomic activity, indicating that stress-induced LC activation is accompanied by autonomic activation. This has not been as thoroughly studied in the rat. However, hypercarbia, hypovolemia, hypotension, and bladder distention all increase LC discharge rates of anesthetized rats (Elam et al. 1981, 1984; Svensson 1987; Svensson and Thorén 1979). This is consistent with the findings of increased norepinephrine turnover in LC target areas during stress.

CRF is one likely mediator of stress-induced LC activation because of its localization and its effects on LC discharge. In support of this, CRF levels have been demonstrated to increase in the LC region in acutely and chronically stressed rats, although it is not known whether the CRF levels measured reflect increases in processing or changes in CRF release (Chappel et al. 1986). Moreover, certain stressors induce tyrosine hydroxylase expression in the LC, and these effects are prevented by the CRF antagonist α-helical $CRF_{9\text{-}41}$, suggesting that stressors activate LC by a CRF-dependent mechanism (Melia et al. 1990).

Electrophysiological evidence strongly supports the hypothesis that CRF is a mediator of LC activation by at least one stressor: hemodynamic stress. Nitroprusside infusion, which has been used as an autonomic stress because it elicits hypophyseal CRF release (Plotsky and Vale 1984; Plotsky et al. 1986), mimics many of the neuronal effects of CRF on LC discharge. For example, nitroprusside increases LC spontaneous discharge by approximately 30%, and the time course of this effect is correlated with the time course of hypotension (Valentino 1989; Valentino and Wehby 1988a). Interestingly, immediately following nitroprusside infusion, blood pressure overshoots for a few minutes, and LC discharge falls below prenitroprusside discharge rates at the same time (Valentino 1989). The LC sensory response to sciatic nerve stimulation during nitroprusside infusion is disrupted in a manner similar to that observed after CRF injection; that is, unstimulated discharge is enhanced, and evoked discharge is decreased (Valentino and Wehby 1988a) (Figure 3–2B). These changes in LC discharge elicited by nitroprusside are now thought to be mediated by CRF release within the LC because LC activation is completely prevented by prior icv administration of α-helical $CRF_{9\text{-}41}$ (Valentino and Wehby 1988a) or by direct application of α-helical $CRF_{9\text{-}41}$ (100–150 ng) into the LC (Valentino et al. 1991) (Figure 3–3). The antagonism by α-helical $CRF_{9\text{-}41}$ is specific in that the CRF antagonist has no effect on LC responses to sciatic nerve stimulation or LC spontaneous discharge rates (Valentino et al. 1991). In contrast to administration of α-helical CRF, prior administration of dexamethasone, which prevents hypophyseal CRF release by hemodynamic stress, does not prevent LC activation by nitroprusside (Valentino and Wehby 1988a). Likewise, icv administration of kynurenic

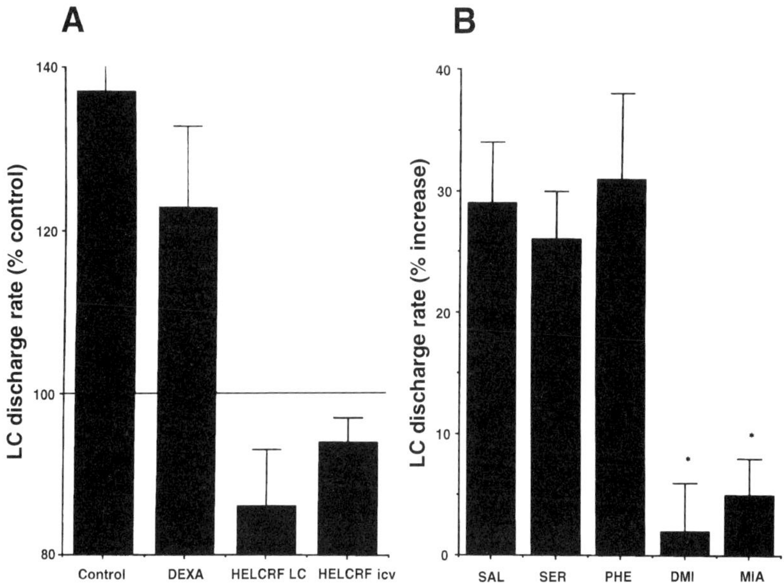

Figure 3–3. Locus coeruleus (LC) activation by nitroprusside requires endogenous corticotropin-releasing factor (CRF) and is attenuated by chronic administration of some antidepressants. A: Bars indicate the effect of nitroprusside infusion on LC spontaneous discharge rate expressed as a percentage of the mean rate determined during a 9-minute period prior to nitroprusside. Control = nitroprusside alone ($n = 15$). DEXA = dexamethasone (0.1 mg/kg, im, administered 6 hours prior to recording; $n = 6$). HELCRF icv = α-helical $CRF_{9\text{-}41}$ was administered icv (50 µg in 5 µl saline) 10–20 minutes prior to nitroprusside ($n = 7$). HELCRF LC = α-helical $CRF_{9\text{-}41}$ was directly infused into the LC (150 ng in 150 nl saline in a 6-minute period), 10–30 minutes prior to nitroprusside ($n = 7$). LC activation by nitroprusside appears to require CRF release into the LC. B: Bars indicate the effect of nitroprusside on LC spontaneous discharge rate in rats chronically administered saline (SAL; 1.0 ml/kg/day, ip, 21 days), sertraline (SER; 10 mg/kg/day, ip, 21 days), phenelzine (PHE; 10 mg/kg/day, ip, 21 days), desmethylimipramine (DMI; 10 mg/kg/day, ip, 21 days), and mianserin (MIA; 10 mg/kg/day, ip, 21 days). Chronic administration of DMI and MIA attenuated LC activation by hemodynamic stress.

acid in a dose sufficient to block LC activation by sciatic nerve stimulation or PGi stimulation has no effect on LC activation by nitroprusside (Valentino et al. 1991). These results strongly suggest that LC activation by this hemodynamic stress is mediated by CRF release within the LC. Interestingly, nitroprusside is associated with a decrease in LC spontaneous discharge rate when infused after direct injection of the CRF antagonist into the LC (Valentino et al. 1991). This could be explained if the stressor activated both excitatory and inhibitory inputs to LC. In this case, antagonism of excitatory effects would unmask inhibitory components of the response.

To date, these findings are the most convincing evidence supporting a neurotransmitter role for CRF in the LC. It is critical to elucidate the circuitry underlying CRF release in the LC during stress. Hypotensive stress elicited by nitroprusside has been reported to activate neurons in PGi (Brown and Guyenet 1986), and this is a major input to LC. However, it is not known whether the PGi cells activated by hemodynamic stress project to LC, or whether they are CRF-containing neurons.

Another important question to address is the function of LC activation during hemodynamic stress, particularly because this activation is not dramatic (i.e., approximately 30% increase above baseline discharge rates). It is not known if an increase of this magnitude for even a sustained length of time would result in increased norepinephrine release in LC targets. However, this is likely, because increases in LC discharge rates of halothane-anesthetized rats administered nitroprusside are temporally correlated to changes in cortical and hippocampal EEG (Valentino et al. 1991). Thus, cortical EEG becomes desynchronized and is characterized by low-amplitude, high-frequency activity characteristic of arousal. Consistent with this, a strong theta rhythm becomes apparent in EEG activity recorded from the hippocampus. The onset of the EEG changes is temporally correlated to the onset of LC activation. Local bilateral LC injection of CRF antagonists prevents both LC activation by nitroprusside and the associated EEG changes (Page et al. 1993). These studies are of interest because they suggest that even relatively small increases in LC discharge rate are sufficient to produce EEG activation and that LC activation by CRF is necessary for forebrain EEG activation during stress. This is consistent with studies demonstrating that EEG activation can be produced by locally infusing carbachol into the LC (Berridge and Foote 1991). These findings also suggest that LC activation associated with nitroprusside infusion may be important in increasing or maintaining arousal in situations where blood pressure is decreased, and that CRF may be involved in this effect.

Microdialysis studies that parallel electrophysiological studies would be needed to confirm that increases in LC discharge rate such as those that are produced by hemodynamic stress are translated to norepinephrine release in targets. Similar studies have demonstrated increased norepinephrine release in the hippocampus during restraint and tailshock-induced stress, indicating that LC activation by these stressors is associated with norepinephrine release in targets (Abercrombie et al. 1988).

Another function of LC activation during hypotensive stress may be to counteract against hypotension. However, antagonism of LC activation is not accompanied by a change in the magnitude of hypotension elicited by nitroprusside (Valentino 1989). These findings argue against an autonomic function for LC activation during hypotension.

Some behavioral studies suggest that the LC-norepinephrine system is important for certain behavioral effects of CRF. For example, the β-adrenergic receptor antagonist, propranolol, inhibits the potentiation of the conditioned emotional response by CRF (Cole and Koob 1988). This effect of CRF is thought to be representative of its anxiogenic potential. Consistent with this, microinjection of CRF into the LC produces anxiogenic behaviors at doses much lower than equieffective icv doses, indicating that the LC is a possible site of action of anxiogenic effects of CRF (Butler et al. 1990).

Corticotropin-Releasing Factor–Locus Coeruleus Link in Stress

The finding of CRF-ir fibers in LC, the effects of CRF on LC neurons, and evidence linking hypotensive-induced LC activation with CRF release in the LC implicate CRF as the neurotransmitter that mediates LC activation by stressors. The anatomical and physiological features of the LC give it the potential to serve as an amplifier for CRF or a way for CRF simultaneously to alter activity of targets whose function varies from autonomic, motor, or behavioral. This may be adaptive when the response to a stimulus requires the coordinated activity of endocrine, autonomic, motor, and behavioral systems. Based on the effects of LC activation on target cells, LC activation by CRF would be predicted to bias the CNS toward focusing on the stressor.

At this point, there are several important questions. What is the origin of CRF fibers in LC? What kinds of stressors activate the LC, and is CRF release in the LC responsible for activation in each case? Do similar neuronal circuits underlie LC activation by different stressors? What happens to release of norepinephrine and target cell activity when the LC is activated by stressors or CRF? Answers to these questions are crucial to understanding the role of CRF-adrenergic interactions in stress.

Corticotropin-Releasing Factor–Locus Coeruleus Link in Depression

Abnormalities in hypothalamic-pituitary function are characteristic of depression (for a review, see Stokes and Sikes 1987), and several studies have implicated CRF as the major defect responsible for hypercortisolism observed in depression. The findings that pituitary corticotrophes are responsive to inhibition by glucocorticoid administration (Gold et al. 1984), that CRF levels in the CSF of depressed patients are increased (Nemeroff et al. 1984), and that chronic infusion of CRF can mimic the same neuroendocrine profile as that observed in depression (Schulte et al. 1985) suggest that CRF may be hypersecreted in depression. It is not clear whether CRF hypersecretion in depression would be limited to the hypophyseal portal system or whether it would be a more general feature of CRF neurons in depression. CRF hypersecretion in the LC would be predicted to result in profound changes in LC physiology and consequent changes throughout the brain via LC targets. For example, persistent exposure of LC cells to CRF should result in higher baseline spontaneous discharge rates and decreased responses to phasic sensory stimuli, assuming that LC cells do not adapt to chronic CRF exposure. This in turn could result in increased arousal and altered responses to discrete stimuli. Interestingly, these are characteristic features of depression.

The hypothesis that CRF hypersecretion in the LC leads to behavioral or physiological changes that characterize depression has not been directly tested. However, a corollary of this hypothesis, that antidepressants act by interfering with CRF in the LC, is being tested. In characterizing and quantifying the acute and chronic effects of antidepressants on LC discharge, it was found that some antidepressants attenuate LC activation by stress, whereas others alter LC discharge in a manner opposite to CRF (Curtis and Valentino 1991; Valentino et al. 1990). Thus far, the effects of four pharmacologically distinct antidepressants on LC spontaneous discharge, LC sensory-evoked discharge, LC activation by stress, and LC activation by CRF have been quantified. Chronic administration of the norepinephrine reuptake inhibitor desmethylimipramine, and the atypical antidepressant mianserin, results in an attenuation of LC activation by nitroprusside infusion, but not by icv CRF, suggesting that CRF release in the LC might be affected by chronic antidepressant administration (Curtis and Valentino 1991; Valentino et al. 1990) (Figure 3–3). In contrast, the serotonin reuptake inhibitor sertraline and the monoamine oxidase inhibitor phenelzine did not share these effects. However, in rats chronically administered sertraline and phenelzine, LC responses in phasic sensory stimuli were enhanced relative to chronic vehicle rats (Valentino and Curtis 1991; Valentino et al. 1990). This effect is opposite to that produced by CRF and may be sufficient to antagonize functionally CRF that is hypersecreted in the LC. Thus, four pharmacologically distinct antidepressants have in common the potential to interfere with the CRF-LC link, by either attenuating CRF release or by functionally antagonizing CRF. It is not known whether this is a common feature of all antidepressants. Further tests of the hypothesis will reveal whether this is an important mechanism of antidepressant action and whether CRF hypersecretion in the LC is the link between neuroendocrine and catecholamine abnormalities in depression. In support of these electrophysiological studies, others have shown that chronic administration of certain antidepressants attenuates tyrosine hydroxylase expression in the LC (Melia et al. 1990; Nestler et al. 1990). Thus, the CRF-LC link may be important in the normal physiological response to a stressor and also in the pathology underlying depression.

Opioid–Locus Coeruleus Interactions

A neurotransmitter role for endogenous opioids in the LC has been proposed based on the substantial evidence that has linked opiates to the LC. For example, the LC contains a high density of opiate receptors (Atweh and Kuhar 1977; Pert et al. 1975; Tempel and Zukin 1987), and the opioid peptides have potent effects on LC discharge (see later). The anatomical data presented later describe a dense enkephalinergic innervation of the LC. Taken together with the demonstration, at the ultrastructural level, of enkephalin-immunoreactive axons and terminals on tyrosine hydroxylase immunoreactive dendrites in the LC (Pickel et al. 1979), these findings convincingly support a neurotransmitter role for endogenous opioids in the LC. The primary argument against enkephalinergic neurotransmission in the LC is based on the lack of synaptically evoked opioid responses in LC as measured by sensitivity to opiate antagonists (Ennis and Aston-Jones 1989; Williams et al. 1987). This apparent lack of effect of endogenous opioid input is in sharp contrast to the potent effects induced by exogenously administered opiates on LC neurons. This discrepancy could be explained if enkephalin release only occurred under conditions or stimulus frequencies that have not been mimicked in the laboratory, or if the opiate antagonists used to unmask an opiate effect were not of the appropriate specificity. Interestingly, opiate antagonists have potent effects on LC discharge of opiate-dependent animals. Although these effects may have little to do with any neurotransmitter role for enkephalins in the LC, they have sparked great interest in opiate peptide-norepinephrine interactions on both a basic science and clinical level, which is reviewed in the following subsections.

Opiate Dependence and the Locus Coeruleus

The acute effect of opiates and opioid peptides on LC neurons is to decrease spontaneous discharge rate. This effect is mediated by an interaction with μ-opiate receptors that are coupled to K^+ channels (see later). Opiate-induced K^+ channel opening results in hyperpolarization and consequently decreased spontaneous discharge rate. The rate-suppressant effects of opiates can be demonstrated in vivo (Bird and Kuhar 1977; Korf et al. 1974; Valentino and Wehby 1988b) and in in vitro slice preparations containing the LC (North and Williams 1985; North et al. 1987; Williams and North 1984; Williams et al. 1982). With repeated administration of opiates, tolerance develops to the inhibition of LC discharge (Andrade et al. 1983; Christie et al. 1987; Valentino and Wehby 1989). In addition, a state of dependence develops, such that systemic or central (icv) administration of opiate antagonists dramatically increases LC discharge (Valentino and Wehby 1989). This effect is thought to be a cellular correlate of opiate withdrawal. Consistent with this, an induction of c-fos protein in the LC during antagonist-precipitated opiate withdrawal has also been demonstrated (Akaoka et al. 1992). The physiologically measured effects of opiate antagonists on LC neurons in dependent animals is attenuated by α_2-agonists such as clonidine (Aghajanian 1978; Taylor et al. 1988). The mechanism for clonidine's attenuation is based on its interaction with α_2-receptors on LC neurons that are linked to the same K^+ channels as μ-opiate receptors (Aghajanian and Wang 1987; North and Williams 1985). The efficacy of clonidine in attenuating mild clinical opiate withdrawal (Charney et al. 1981; Gold et al. 1978, 1980) has been attributed to these cellular effects and has supported the hypothesis that LC hyperactivity during opiate withdrawal is responsible for much of the corresponding clinical symptomatology.

However, the site of action of naloxone or clonidine in altering LC discharge and behavioral symptoms of opiate-dependent subjects is not clear. The efficacy of local LC infusions of naloxone (Esposito et al. 1987) and clonidine (Taylor et al. 1988) to precipitate and attenuate opiate withdrawal, respectively, suggests that these agents act directly on LC neurons. However, both of these studies used relatively large doses of the drugs, and not all signs of opiate withdrawal were attenuated by LC injection of clonidine. Moreover, the finding that tolerance, but not dependence or withdrawal, can be demonstrated in vitro (Andrade et al. 1983; Christie et al. 1987) suggests that distal afferents to the LC are integral to the expression of opiate withdrawal in the LC. In this regard, it has been demonstrated that lesion of one of the primary afferents to the LC, the PGi, attenuates LC activation by naloxone (Rasmussen and Aghajanian 1989) as does the excitatory amino acid antagonist kynurenic acid (Akaoka et al. 1991; Rasmussen and Aghajanian 1989). This latter finding is of interest because of the evidence supporting the existence of an excitatory amino acid pathway from PGi to the LC. This is also consistent with our findings that microinfusion of naloxone into LC of morphine-dependent rats in vivo does not alter their activity, whereas similar infusion of kynurenate and, to a lesser extent, the selective *N*-methyl-D-aspartate (NMDA)–receptor antagonist AP5, attenuates activation of LC induced by systemic naloxone (Akaoka et al. 1991). There are also reports that such antagonists attenuate some behavioral signs of opiate withdrawal (Rasmussen and Aghajanian 1989; Tung et al. 1990), although other studies (Trujillo and Akil 1991), including our own (Aston-Jones et al. 1991), disagree. Thus, although data implicate a link between LC activation and opiate withdrawal, several points have yet to be established. First, it appears that LC activation is not a direct consequence of opiate antagonist effects on LC neurons. Second, if LC activation is not due to direct effects of opiate antagonists on LC neurons, the circuitry for activating LC should be determined. Third, the involvement of specific LC projections in specific withdrawal signs should be determined. Nonetheless, this is one example in which proposed opiate-adrenergic interactions have been used to explain clinical findings and have led to potential treatments.

Opioid Peptides as Neurotransmitters in the Locus Coeruleus

Anatomical considerations. The LC is innervated by all three major endogenous opioids (Bloom et al. 1978; Fallon and Leslie 1986; Pickel et al. 1979). Additionally, the LC is enriched with opiate receptors, primarily of the mu and kappa subtypes (McFadzean et al. 1987; Pert et al. 1975; Tempel and Zukin 1987), providing anatomical support for an interaction between the LC and endogenous opioids. As described earlier, ultrastructural studies demonstrating enkephalin-immunoreactive axons and terminals on tyrosine hydroxylase–positive dendrites in the LC constitute strong evidence for a neurotransmitter role of endogenous opioids in LC (Pickel et al. 1979). It is noteworthy that the substrates of such an interaction vary between species. For example, in the neurochemically heterogenous cat LC, enkephalin neurons represent a substantial percentage of the total population of LC neurons (Charnay et al. 1982; Reddy et al. 1990). In contrast, rat LC (which consists entirely of noradrenergic neurons) does not contain enkephalin-immunoreactive neurons (Drolet et al. 1992; Murakami et al. 1989). This difference is important in interpreting comparative anatomical and physiological results of opiate studies performed in the cat and rat.

Early studies of opioid anatomy described the presence of moderately dense, β-endorphin-like immunoreactive fibers (Bloom et al. 1978; J. C. W. Finley et al. 1981b), enkephalin fibers and terminals (Fallon and Leslie 1986; J. C. W. Finley et al. 1981a; Khachaturian et al. 1983; Murakami et al. 1987; Uhl et al. 1979b), as well as dynorphin-immunoreactive fibers in LC (Fallon and Leslie 1986). The pattern of the enkephalinergic fiber innervation in the rat LC area was investigated in detail using an antibody directed against the extended enkephalin peptide (Met-enkephalin-Arg^6-Gly^7-Leu^8) (Drolet et al. 1992). This analysis revealed that the LC is strongly innervated by fine, varicose, enkephalin-immunoreactive fibers that appear to ramify in the nucleus in a rostrocaudal orientation. Moreover, enkephalin-immunoreactive fibers were also densely located in the rostromedial and caudal juxtaependymal peri-coerulear regions (Figure 3–4), where the majority of LC neurons have extranuclear dendrites extending from the nucleus proper (Shipley et al.

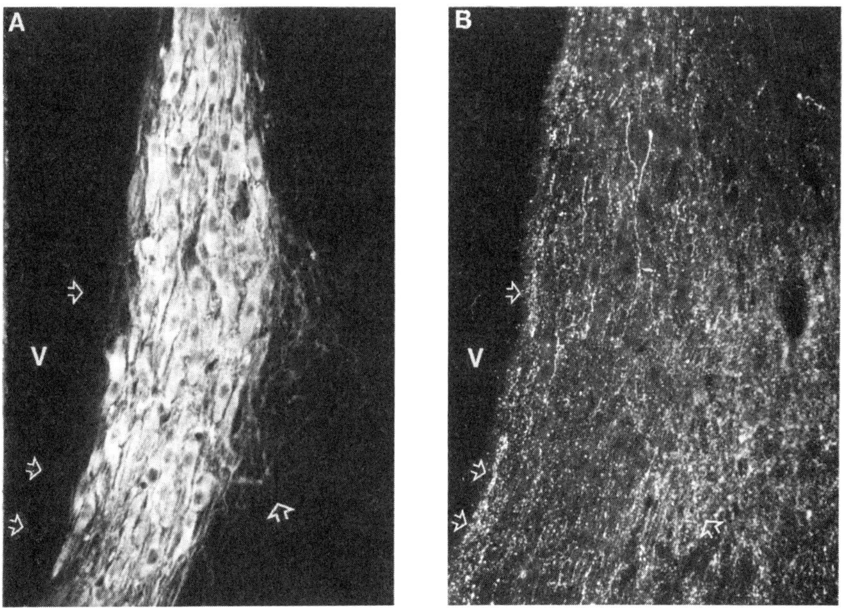

Figure 3–4. Fluorescent photomicrographs of a horizontal section (16-μm thick), showing the locus coeruleus (LC) and peri-LC regions. The photomicrograph on the left (A) shows, with fluoroscein (FITC) illumination, tyrosine hydroxylase (TH) immunoreactivity, which revealed the noradrenergic LC cell bodies and processes. The right panel (B) is a double exposure of the same section, with rhodamine (RITC) illumination, revealing enkephalin immunoreactivity. Note moderately dense plexus of enkephalin fibers and terminals in the LC proper. Note also the dense enkephalin immunoreactivity medial and lateral to LC (open arrows), in the area of LC dendrites. Top is caudal; right is lateral. V = 4th ventricle.

1990, in press). This suggests that, in addition to influencing LC perikarya, enkephalin fibers may also innervate LC dendrites outside of the LC cell body region. This is consistent with previous studies demonstrating numerous axodendritic contacts in the LC by enkephalin terminals (Pickel et al. 1979).

Until recently, the origin of opioid innervation to LC was unknown. The β-endorphin fibers in LC were thought to originate from the arcuate nucleus because electrical stimulation of the arcuate nucleus produced a naloxone-reversible inhibition of LC spontaneous discharge (Strahlendorf et al. 1980). However, anatomical studies using retrograde tract tracers indicated that the arcuate nucleus was not a source of projections to the nucleus LC in rat (Aston-Jones et al. 1986), although projections to the rostral peri-LC area containing LC dendrites are possible (Bloom et al. 1978). We addressed the issue of the origin of enkephalin input to LC by combining retrograde transport from LC with immunohistochemical techniques (Drolet et al. 1992). Injection of the retrograde tracer, a lectin-protein-gold conjugate (WGA-apoHRP-Au), into the nucleus LC confirmed previous reports that major afferents to the LC derive from the nucleus PGi in the rostral ventrolateral medulla and the PrH in the dorsomedial medulla (Aston-Jones et al. 1986). Interestingly, the distribution of enkephalin cells in the PrH was confined to the mediodorsal perifasicular region, which also contains LC afferents. In the ventrolateral medulla, the enkephalin cells were aggregated into groups, including one in the retrofacial portion of PGi, which contains LC-projecting neurons (Figure 3–5). Examination of enkephalin-immunoreactivity and retrograde labeling in the same tissue sections revealed an unexpected high incidence of doubly labeled cells in PGi. Overall, 57% of the retrogradely labeled, LC-projecting neurons in both PrH and PGi were immunoreactive for enkephalin. Systematic studies of the effects of stimulating enkephalinergic pathways to LC from the rostral medulla on LC discharge and sensitivity to opiate antagonists must be done to elucidate the function of these enkephalinergic pathways. They may provide the anatomical substrate for some of the proposed opioid functions in LC described later.

Physiological Effects of Opioids on Locus Coeruleus Activity

Because of the technical restraints involved in the use of peptides, most in vivo studies of opioid effects on LC neurons have utilized morphine or other synthetic opiate analogues. Systemic administration of opiates generally decreases LC spontaneous discharge rates in a naloxone-sensitive manner (Korf et al. 1974). Iontophoretic studies have verified the local inhibitory effects of opiates on LC neurons, their stereospecificity, and sensitivity to naloxone (Bird and Kuhar 1977). In vivo studies comparing morphine effects in anesthetized versus unanesthetized rats indicate that morphine (administered icv) is more potent in anesthetized animals (Valentino and Wehby 1988b). Moreover, in both anesthetized and unanesthetized rats, LC spontaneous discharge is more sensitive to inhibition by morphine than is LC sensory-evoked discharge. These findings suggest that the degree of arousal and the availability of sensory stimuli are important determinants of morphine effects on LC discharge and can, in fact, antagonize opiate effects. Interestingly, LC neurons of cats are excited by systemically administered morphine (Rasmussen and Jacobs 1985). However, the anatomical differences between cat and rat LC, both

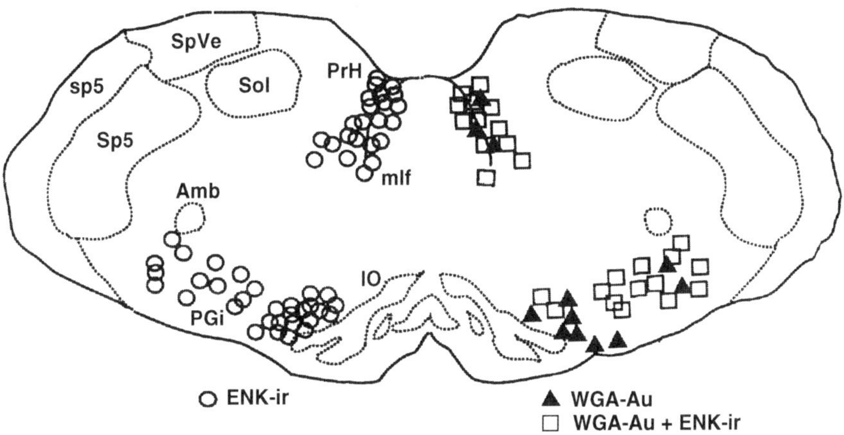

Figure 3–5. Camera lucida drawing of a coronal section through the rostral medulla illustrating the distribution of neurons immunoreactive for enkephalin (ENK-ir, ❍) on the left side of the section, and of retrogradely labeled (WGA-Au, ▲) and doubly labeled (WGA-Au + ENK-ir, ❑) neurons on the right side of the section. This section, which corresponds to the midlevel of paragigantocellularis (PGi) and the caudal part of prepositus hypoglossi (PrH), was plotted from a typical rat with a WGA-Au injection in the LC. Amb = nucleus ambiguus; IO = inferior olive; mlf = medial longitudinal fasciculus; Sol = nucleus of solitary tract; sp5 = nucleus of the spinal tract of trigeminal nerve; Sp5 = spinal trigeminalnucleus; SpVe = spinal vestibularnucleus.

in the difficulty in identifying noradrenergic neurons in cat recordings and in enkephalin localization (see earlier), make it difficult to generalize these findings to other species without further systematic studies.

The pharmacology and mechanism of opioid peptide effects on LC neurons have been characterized in vitro using the rat LC slice preparation. In this preparation, opioid peptides and synthetic opiates hyperpolarize LC neurons, and the order of potency of the different agonists and sensitivity to specific opiate antagonists indicate that this effect is mediated specifically by μ-opiate receptors (North and Williams 1985; North et al. 1987; Williams and North 1984; Williams et al. 1982). Opiate-induced hyperpolarization is the result of an increase in K^+ current (North and Williams 1985; Williams et al. 1982). High concentrations of opiates that produce a large hyperpolarization in LC neurons also decrease Ca^{++} action potentials in these cells, and this effect is a result of the opening of K^+ channels (North and Williams 1983). The μ-opiate receptor involved in this ionotropic effect is thought to be coupled to a G protein, and the coupling does not appear to involve cyclic adenosine monophosphate-dependent protein kinase or protein kinase C (North et al. 1987). Interestingly, α_2-agonists such as clonidine share effects, but not receptors, with opiates on LC neurons. For example, α_2-agonists produce a similar hyperpolarization of LC neurons that is due to opening of the same class of potassium channels as opiates. This has been suggested by the lack of additivity of the effect with opiates, and sensitivity to similar agents (Aghajanian and Wang 1987; North and Williams 1985). The ability of α_2-agonists to activate the same cellular effector in LC neurons as opiates, although through a different receptor, is thought to be an important feature in the efficacy of α_2-agonists in the treatment of opiate withdrawal.

Although most opiates have similar direct effects on LC neurons and act exclusively through an interaction with μ-receptors, a unique effect of kappa opiate receptor activation has been reported. Selective kappa agonists decrease excitatory postsynaptic potentials induced by focal stimulation in the LC slice preparation, and these effects are antagonized by kappa selective antagonists (McFadzean et al. 1987). Although most opiate receptor binding in LC is to μ-receptors, a proportion of binding has been described as kappa, and it has been suggested that the function of these receptors may be to inhibit excitatory input to LC (McFadzean et al. 1987). As of this writing, there have been no reports of naloxone effects on LC neurons in vitro, suggesting that opioid input to LC is not tonically active in the slice preparation. In addition, opioid-mediated synaptic potentials have not yet been detected in LC cells.

Possible Functions of Opioid Peptide–Locus Coeruleus Interactions

The findings discussed suggest that the LC-noradrenergic system is important in vigilance and adaptive responding to salient environmental stimuli, such that increased LC activity in response to a sensory event is predicted to lead to a state of heightened vigilance or surveillance of, and response to, salient stimuli in the environment. This LC function may be under tonic control of endogenous opioids, as indicated by the results described for opioid innervation of, and influence on, LC neurons. Additional studies demonstrate an interaction between opioid and LC-noradrenergic systems in the regulation of environmental stimulus-directed behavior (Arnsten et al. 1981). In these studies, the opioid antagonist naloxone was demonstrated to increase the interaction of rats with novel stimuli in their environment, and this was opposite to the effect produced by opiate agonists. Moreover, the effect of naloxone required the presence of an intact noradrenergic LC system, as chemical destruction of LC efferent projections by local injection of 6-hydroxydopamine abolished the naloxone effect and mimicked a morphine-like reduction in stimulus contact (Arnsten et al. 1981). Interestingly, exposure to stressors also produces an opioid-like alteration of investigatory behavior in rats, and this effect is sensitive to antagonism by naloxone (Arnsten et al. 1985). These results are consistent with an inhibitory action of endogenous opioids on LC-noradrenergic transmission. It is noteworthy that, as opioid receptors are located on LC terminals in target areas (Llorens et al. 1978), as well as within the nucleus LC, the effect of endogenous opioids could take place presynaptically on LC noradrenergic terminals, at the level of LC somata, or in both sites.

Endogenous opioids and LC activity may also be linked in stress. Evidence presented earlier supports the concept that the LC-noradrenergic system is activated during stress. Similarly, endogenous opioid systems are activated during stress as indicated by naloxone-sensitive, stress-induced analgesia (Olson et al. 1989a, 1989b, 1990; Terman et al. 1984). One function of opiates may be to alter sensitivity to noxious stimuli during stress. In addition, opioids appear to influence norepinephrine turnover in the LC during stress. For example, immobilization stress-induced increases in norepinephrine turnover in the LC, and the associated emotional responses, are potentiated by naloxone (Tanaka et al. 1988, 1990). In contrast, morphine (iv) or met-enkephalin administered (icv) immediately before stress attenuated both stress-elicited increases in norepinephrine turnover and emotional changes in a naloxone-reversible manner (Tanaka et al. 1983, 1989). These findings suggest that endogenous opioids might act at the level of the LC to attenuate stress-evoked hyperemotionality. In this regard, it is of interest that naloxone, which is usually without effect on LC discharge characteristics in opiate-naive animals (Valentino and Wehby 1989), selectively increases LC discharge in unanesthetized cats exposed to stressors (Abercrombie and Jacobs 1988). In more general terms, endogenous opioid innervation of LC during stress might serve to moderate potentially deleterious effects of other stress-related neuromodulators such as CRF and norepinephrine by maintaining the stress response within limits compatible with homeostasis.

Summary

The anatomical and physiological characteristics of the pontine noradrenergic nucleus LC implicate this as a potential amplification mechanism or effector by which stimuli can influence global brain function. Although numerous studies suggest that many different peptides utilize the LC-norepinephrine system to this end, we have focused on the link between CRF and endogenous opioids and the LC. Findings supporting a neurotransmitter role for these peptides in the LC have been reviewed, and possible functions of their neurotransmitter role have been discussed. Future studies designed to characterize further the neurochemical identity of specific inputs to the LC and the effects of manipulating inputs on LC activity may lead to the elucidation of other peptides as neurotransmitters in the LC. Additional studies that relate LC output to target cell activity and, ultimately, to function will be necessary to determine the role of peptide-LC interactions in normal behavior and psychiatric disorders.

References

Abercrombie ED, Jacobs BL: Single unit response of noradrenergic neurons in locus coeruleus of freely moving cats, I: acutely presented stressful and non-stressful stimuli. J Neurosci 7:2837–2843, 1987a

Abercrombie ED, Jacobs BL: Single unit response of noradrenergic neurons in locus coeruleus of freely moving cats, II: adaptation to chronically presented stressful stimuli. J Neurosci 7:2844–2848, 1987b

Abercrombie ED, Jacobs BL: Systemic naloxone administration potentiates locus coeruleus noradrenergic neuronal activity under stressful but not non-stressful conditions. Brain Res 441:362–366, 1988

Abercrombie ED, Keller RW, Zigmond MJ: Characterization of hippocampal norepinephrine release as measured by microdialysis perfusion: pharmacological and behavioral studies. Neuroscience 27:897–904, 1988

Ader JP, Room P, Postrema F, et al: Bilaterally diverging axon collaterals and contralateral projections from rat locus coeruleus neurons, demonstrated by fluorescent retrograde double labeling and norepinephrine metabolism. J Neural Transm 49:207–218, 1980

Aghajanian GK: Tolerance of locus coeruleus neurons to morphine and suppression of withdrawal response by clonidine. Nature 276:186–188, 1978

Aghajanian GK, Wang Y-Y: Common alpha$_2$ and opiate effector mechanisms in the locus coeruleus: intracellular studies in brain slices. Neuropharmacology 26:793–799, 1987

Akaoka H, Aston-Jones G: Opiate withdrawal-induced hyperactivity of locus coeruleus neurons is substantially mediated by augmental excitatory amino acid input. J Neurosci 11:3830–3839, 1991

Akaoka H, Zhu Y, Shipley MT, et al: Expressionn of fos protein in central catecholamine neurons during opiate withdrawal. Society for Neuroscience Abstracts 18, 1992

Andrade R, Vandermaelen CP, Aghajanian GK: Morphine tolerance and dependence in the locus coeruleus: single cell studies in brain slices. Eur J Pharmacol 91:161–169, 1983

Arnsten AFT, Segal DS, Loughlin SE, et al: Evidence for an interaction of opioid and noradrenergic locus coeruleus systems in the regulation of environmental stimulus-directed behavior. Brain Res 222:351–363, 1981

Arnsten AFT, Berridge C, Segal DS: Stress produces opioid-like effects on investigatory behavior. Pharmacol Biochem Behav 22:803–809, 1985

Astier B, Aston-Jones G: Electrophysiological evidence for medullary adrenergic inhibition of rat locus coeruleus. Society for Neuroscience Abstracts 15:1012, 1989

Aston-Jones G: Behavioral functions of locus coeruleus derived from cellular attributes. Physiol Psych 13:118–126, 1985

Aston-Jones G, Bloom FE: Activity of norepinephrine-containing locus coeruleus neurons in behaving rats anticipates fluctuations in the sleep-waking cycle. J Neurosci 1:876–886, 1981a

Aston-Jones G, Bloom FE: Norepinephrine-containing locus coeruleus neurons in behaving rats exhibit pronounced responses to non-noxious environmental stimuli. J Neurosci 1:887–900, 1981b

Aston-Jones G, Ennis M: Sensory-evoked activation of locus coeruleus may be mediated by a glutamate pathway from the rostral ventrolateral medulla, in Frontiers in Excitatory Amino Acid Research. Edited by Cavalheiro A, Lehmann J, Turski L. New York, Alan R Liss, 1988, pp 471–478

Aston-Jones G, Ennis M, Pieribone VA, et al: The brain nucleus locus coeruleus: restricted afferent control of a broad efferent network. Science 234:734–737, 1986

Aston-Jones G, Shipley MT, Ennis M, et al: Restricted afferent control of locus coeruleus neurons revealed by anatomical, physiological and pharmacological studies, in The Pharmacology of Noradrenaline in the Central Nervous System. Edited by Marsden CA, Heal DJ. Oxford, Oxford University Press, 1990, pp 187–247

Aston-Jones G, Shipley MT, Chouvet G, et al: Afferent regulation of locus coeruleus neurons: anatomy, physiology and pharmacology. Prog Brain Res 88:47–75, 1991

Atweh SF, Kuhar MJ: Autoradiographic localization of opiate receptors in rat brain, II: the brainstem. Brain Res 129:1–12, 1977

Barrington FJT: The effect of lesion of the hind- and mid-brain on micturition in the cat. Q J Exp Physiol 15:81–102, 1925

Berridge CW, Foote SL: Effects of locus coeruleus activation on electroencephalographic activity in the neocortex and hippocampus. J Neurosci 11:3135–3145, 1991

Berridge CW, Page ME, Valentino RJ, et al: Effects of locus coeruleus inactivation on electroencephalographic activity in neocortex and hippocampus. Neuroscience 55:381–393, 1993

Bird SJ, Kuhar MJ: Iontophoretic application of opiates to the locus coeruleus. Brain Res 122:523–533, 1977

Blessing WW, Goodchild AK, Dampney RAL, et al: Cell groups in the lower brain stem of the rabbit projecting to the spinal cord, with special reference to catecholamine-containing neurons. Brain Res 221:35–55, 1981

Bloom FE, Battenberg ELF, Rossier J, et al: Neurons containing β-endorphin in rat brain exist separately from those containing enkephalin: immunocytochemical studies. Proc Natl Acad Sci U S A 75:1591–1595, 1978

Britton DR, Koob GF, Rivier J, et al: Intraventricular corticotropin-releasing factor enhances behavioral effects of novelty. Life Sci 31:363–367, 1982

Britton K, Morgan J, Rivier J, et al: Chlordiazepoxide attenuates CRF-induced response suppression in the conflict test. Psychopharmacology (Berl) 86:170–174, 1985

Brown DL, Guyenet PG: Cardiovascular neurons of brainstem with projections to spinal cord. Am J Physiol 247:R1009–R1026, 1986

Brown MR, Fisher LA, Spiess J, et al: Corticotropin-releasing factor (CRF): actions on the sympathetic nervous system and metabolism. Endocrinology 111:928–931, 1982

Brown MR, Fisher LA, Webb V, et al: Corticotropin-releasing factor: a physiologic regulator of adrenal epinephrine secretion. Brain Res 328:355–357, 1985

Butler PD, Weiss JM, Stout JC, et al: Corticotropin-releasing factor produces fear-enhancing and behavioral activating effects following infusion into the Locus Coeruleus. J Neurosci 10:176–183, 1990

Cassens G, Roffman G, Kuruc A, et al: Alterations in brain norepinephrine metabolism induced by environmental stimuli previously paired with inescapable shock. Science 209:1138–1139, 1980

Chappell PB, Smith MA, Kilts CD, et al: Alterations in corticotropin-releasing factor-like immunoreactivity in discrete rat brain regions after acute and chronic stress. J Neurosci 6:2908–2914, 1986

Charnay YL, Léger F, Dray A, et al: Evidence for the presence of enkephalin in catecholaminergic neurons of the cat locus coeruleus. Neurosci Lett 30:147–151, 1982

Charney DS, Sternberg DE, Kleber HD, et al: The clinical use of clonidine in abrupt withdrawal from methadone. Arch Gen Psychiatry 38:1273–1277, 1981

Cheeseman SJ, Pinnock RD, Henderson G: Substance P excitation of rat locus coeruleus neurons. Eur J Pharmacol 94:93–99, 1983

Christie MJ, Williams JT, North RA: Cellular mechanisms of opioid tolerance: studies in single brain neurons. J Pharmacol Exp Ther 32:633–638, 1987

Cole BJ, Koob GF: Propranolol antagonizes the enhanced conditioned fear produced by corticotropin-releasing factor. J Pharmacol Exp Ther 247:902–910, 1988

Cummings S, Elde R, Ells J, et al: Corticotropin-releasing factor immunoreactivity is widely distributed within the central nervous system of the rat: an immunohistochemical study. J Neurosci 3:1355–1368, 1983

Curtis AL, Valentino RJ: Acute and chronic effects of the atypical antidepressant, mianserin on brain noradrenergic neurons. Psychopharmacology (Berl) 103:330–338, 1991

Dahlstrom A, Fuxe K: Evidence for the existence of monoamine-containing neurons in the central nervous system, I: demonstration of monoamines in the cell bodies of brainstem neurons. Acta Physiol Scand Suppl 232:1–55, 1964

DeSouza EB: Corticotropin-releasing factor receptors in the rat central nervous system: characterization and regional distribution. J Neurosci 7:88–100, 1987

DeSouza EB, Insel TR, Perrin MH, et al: Corticotropin-releasing factor receptors are widely distributed within the rat central nervous system: an autoradiographic study. J Neurosci 5:3189–3203, 1985

Drolet G, Van Bockstaele EB, Aston-Jones G: Robust enkephalin innervation of the locus coeruleus from the rostral medulla. J Neurosci 12:3162–3174, 1992

Eaves M, Thatcher-Britton K, Rivier J, et al: Effects of corticotropin-releasing factor on locomotor activity in hypophysectomized rats. Peptides 6:923–926, 1985

Eiden LE, Nilaver G, Palkovits M: Distribution of vasoactive intestinal polypeptide (VIP) in the rat brain stem nuclei. Brain Res 231:472–477, 1982

Elam M, Yao T, Thoren P, et al: Hypercapnia and hypoxia: chemoreceptor-mediated control of locus coeruleus neurons and splanchnic, sympathetic nerves. Brain Res 222:373–381, 1981

Elam M, Yao T, Svensson TH, et al: Regulation of locus coeruleus neurons and splanchnic, sympathetic nerves. Brain Res 290:281–287, 1984

Ennis M, Aston-Jones G: A potent excitatory input to the nucleus locus coeruleus from the ventrolateral medulla. Neurosci Lett 71:299–305, 1986

Ennis M, Aston-Jones G: Two physiologically distinct populations of neurons in the ventrolateral medulla innervate the locus coeruleus. Brain Res 425:275–282, 1987

Ennis M, Aston-Jones G: Activation of locus coeruleus from nucleus paragigantocellularis: a new excitatory amino acid pathway in brain. J Neurosci 8:3644–3657, 1988

Ennis M, Aston-Jones G: GABA-mediated inhibition of locus coeruleus from the dorsomedial rostral medulla. J Neurosci 9:2973–2981, 1989

Ennis M, Shipley M, Van Bockstaele EJ, et al: Projections from the periaqueductal gray to the rostromedial pericoeralear region and nucleus locus coeruleus: anatomic and physiologic studies. J Comp Neurol 306:480–494, 1991

Esposito E, Kruszewska A, Ossowska G, et al: Noradrenergic and behavioral effects of naloxone injected in the locus coeruleus of morphine-dependent rats and their control by clonidine. Psychopharmacology (Berl) 93:393–396, 1987

Fallon JH, Leslie FM: Distribution of dynorphin and enkephalin peptides in the rat brain. J Comp Neurol 249:293–336, 1986

Finley JCW, Lindström P, Petrusz P: Immunohistochemical localization of β-endorphin-containing neurons in the rat brain. Neuroendocrinology 33:28–42, 1981a

Finley JCW, Maderdrut JL, Petrusz P: The immunocytochemical localization of enkephalin in the central nervous system of the rat. J Comp Neurol 198:541–565, 1981b

Finley KH, Cobb S: The capillary bed of the locus coeruleus. J Comp Neurol 73:49–58, 1940

Fisher LA, Rivier J, Rivier C, et al: Corticotropin-releasing factor (CRF): central effects on mean arterial pressure and heart rate in rats. Endocrinology 110:2222–2224, 1982

Foote SL, Cha CI: Distribution of corticotropin-releasing factor-like immunoreactivity in brainstem of two monkey species (Saimiri sciureus and Macaca Facsicularis): an immunohistochemical study. J Comp Neurol 276:239–264, 1988

Foote SL, Freedman R, Oliver AP: Effects of putative neurotransmitter on neuronal activity in monkey auditory cortex. Brain Res 86:229–242, 1975

Foote SL, Aston-Jones G, Bloom FE: Impulse activity of locus coeruleus neurons in awake rats and monkeys is a function of sensory stimulation and arousal. Proc Natl Acad Sci U S A 77:3033–3037, 1980

Foote SL, Bloom FE, Aston-Jones G: Nucleus locus coeruleus: new evidence of anatomical and physiological specificity. Physiol Rev 63:844–914, 1983

Freedman R, Foote SL, Bloom FE: Histochemical characterization of a neocortical projection of the nucleus locus coeruleus in the squirrel monkey. J Comp Neurol 164:209–232, 1975

Freedman R, Hoffer BJ, Woodward DJ, et al: Interaction of norepinephrine with cerebellar activity evoked by mossy and climbing fibers. Exp Neurol 55:269–288, 1977

Fuxe K, Ganten D, Hokfelt T, et al: Immunohistochemical evidence for the existence of angiotensin II-containing nerve terminals in the brain and spinal cord in the rat. Neurosci Lett 2:229–234, 1976

Gold MS, Redmond DE, Kleber HD: Clonidine in opiate withdrawal. Lancet 2:599–602, 1978

Gold MS, Pottash AC, Sweeney DR, et al: Opiate withdrawal using clonidine. JAMA 243:343–346, 1980

Gold PW, Chrousos GP, Kellner CH: Overview: psychiatric implications of basic and clinical studies with corticotropin-releasing factor. Am J Psychiatry 141:619–627, 1984

Grant SJ, Huang YH, Redmond DE Jr: Behavior of monkeys during opiate withdrawal and locus coeruleus stimulation. Pharmacol Biochem Behav 30:13–19, 1988

Grzanna R, Molliver ME: The locus coeruleus in the rate: an immunohistochemical delineation. Neuroscience 5:21–40, 1980

Guyenet PG, Aghajanian GK: Excitation of neurons in the nucleus locus coeruleus by substance P and related peptides. Brain Res 136:178–184, 1977

Guyenet PG, Aghajanian GK: ACh, substance P and met-enkephalin in the locus coeruleus: pharmacological evidence for independent sites of action. Eur J Pharmacol 53:319–328, 1979

Haber S, Elde R: The distribution of enkephalin immunoreactive fibers and terminals in the monkey central nervous system: an immunohistochemical study. Neuroscience 7:1049–1095, 1982

Kalin NH: Behavioral effects of ovine corticotropin-releasing factor administered to rhesus monkeys. Federation Proceedings 44:249–254, 1985

Kalin NH, Sherman JE, Takahashi LK: Antagonism of endogenous corticotropin-releasing hormone systems attenuates stress-induced freezing behaviors in rats. Brain Res 457:130–135, 1988

Khachaturian H, Lewis ME, Watson SJ: Enkephalin systems in the diencephalon and brainstem of the rat. J Comp Neurol 220:310–320, 1983

Korf J, Aghajanian GK, Roth R: Increased turnover of norepinephrine in the rat cerebral cortex during stress: role of the locus coeruleus. Neuropharmacology 12:933–938, 1973

Korf J, Bunney BS, Aghajanian GK: Noradrenergic neurons: morphine inhibition of spontaneous activity. Eur J Pharmacol 25:165–169, 1974

Léger L, Hernandez-Nicaise M-L: The cat locus coeruleus: light and electron microscopic study of the neuronal somata. Anat Embryol (Berl) 159:181–198, 1980

Lenz HJ, Raedler A, Geten H, et al: Stress-induced gastrointestinal secretory and motor responses in rats are mediated by endogenous corticotropin-releasing factor. Gastroenterology 95:1510–1517, 1988

Ljungdahl A, Hokfelt T, Nilsson G: Distribution of substance P-like immunoreactivity in the central nervous system of the rat; I: cell bodies and nerve terminals. Neuroscience 3:861–943, 1978

Llorens C, Martres MP, Baudry M, et al: Hypersensitivity to noradrenaline in cortex after chronic morphine: relevance to tolerance and dependence. Nature 274:603–605, 1978

Loy R, Koziell DA, Lindsey JD, et al: Noradrenergic innervation of the adult rat hippocampal formation. J Comp Neurol 189:699–710, 1980

McFadzean I, Lacey MG, Hill RG, et al: Kappa opioid receptor activation depresses excitatory synaptic input to rat locus coeruleus neurons in vitro. Neuroscience 20:231–239, 1987

Melia KR, Nestler EJ, Haycock J, et al: Regulation of tyrosine hydroxylase (TH) in the locus coeruleus (LC) by corticotropin-releasing factor (CRF): relation to stress and depression. Society for Neuroscience Abstracts 16:444, 1990

Mendelsohn FAO, Quirion R, Saavedra JM, et al: Autoradiographic localization of angiotension II receptors in rat brain. Proc Natl Acad Sci U S A 81:1575–1579, 1984

Merchenthaler I, Hynes MA, Vigh S, et al: Immunocytochemical localization of corticotropin-releasing factor (CRF) in rat brain. Am J Anat 165:383–396, 1982

Moises HC, Woodward DJ: Potentiation of GABA inhibitory action in cerebellum by locus coeruleus stimulation. Brain Res 182:327–344, 1980

Moises HC, Woodward DJ, Hoffer BJ, et al: Interactions of norepinephrine with Purkinje cell responses to putative amino acid neurotransmitters applied microiontophoresis. Exp Neurol 64:493–515, 1979

Moises HC, Waterhouse BD, Woodward DJ: Locus coeruleus stimulation potentiates Purkinje cell responses to afferent input: the climbing fiber system. Brain Res 222:43–64, 1981

Morilak DA, Fornal C, Jacobs BL: Effects of physiological manipulations on locus coeruleus neuronal activity in freely moving cats, I: thermoregulatory challenge. Brain Res 422:17–23, 1987a

Morilak DA, Fornal C, Jacobs BL: Effects of physiological manipulations on locus coeruleus neuronal activity in freely moving cats, II: glucoregulatory challenge. Brain Res 422:24–31, 1987b

Murakami S, Okamura H, Yanaihara C, et al: Immunocytochemical distribution of met-enkephalin-Arg^6-Gly^7-Leu^8 in the rat lower brainstem. J Comp Neurol 261:193–208, 1987

Murakami SH, Okamura H, Pelletier G, et al: Differential colocalization of neuropeptide Y- and methionine-enkephalin-arg^6-gly^7-leu^8-like immunoreactivity in catecholaminergic neurons in the rat brain stem. J Comp Neurol 281:532–544, 1989

Nagai T, Satoh K, Imamoto K, et al: Divergent projections of catecholamine neurons of the locus coeruleus as revealed by fluorescent retrograde double labeling technique. Neurosci Lett 23:117–123, 1981

Nemeroff C, Widerlov E, Bissette GT, et al: Elevated concentrations of CSF corticotropin-releasing factor-like immunoreactivity in depressed patients. Science 226:1342–1344, 1984

Nestler EJ, McMahon A, Sabban EL, et al: Chronic antidepressant administration decreases the expression of tyrosine hydroxylase in the rat locus coeruleus. Proc Natl Acad Sci U S A 87:7522–7526, 1990

North RA, Williams JT: Opiate activation of potassium conductance inhibits calcium action potentials in rat locus coeruleus neurons. Br J Pharmacol 80:225–228, 1983

North RA, Williams JT: On the potassium conductance increased by opioids in rat locus coeruleus neurons. J Physiol (Lond) 364:265–280, 1985

North RA, Williams JT, Surprenant A, et al: Mu and delta receptors belong to a family of receptors that are coupled to potassium channels. Proc Natl Acad Sci U S A 84:5487–5491, 1987

Olpe H-R, Baltzer V: Vasopressin activates noradrenergic neurons in the rat locus coeruleus: a microiontophoretic investigation. Eur J Pharmacol 73:377–378, 1981

Olschowka JA, O'Donahue TL, Mueller GP, et al: The distribution of corticotropin-releasing factor-like immunoreactive neurons in rat brain. Peptides 3:995–1015, 1982

Olson GA, Olson RD, Kastin AJ: Endogenous opioid: 1987. Peptides 10:205–236, 1989a

Olson GA, Olson RD, Kastin AJ: Endogenous opioid: 1988. Peptides 10:1252–1280, 1989b

Olson GA, Olson RD, Kastin AJ: Endogenous opioid: 1989. Peptides 11:1277–1304, 1990

Page ME, Berridge CW, Foote SL, et al: Corticotropin-releasing factor in the locus coeruleus mediates EEG activation associated with hypotensive stress. Neuroscience Letters 164:81–84, 1993

Pammer C, Gorcs T, Palkovits M: Peptidergic innervation of the locus coeruleus cells in the human brain. Brain Res 515:247–255, 1990

Pert CB, Kuhar MJ, Snyder SH: Autoradiographic localization of the opiate receptor in rat brain. Life Sci 16:1849–1854, 1975

Pickel VM, Joh TH, Reis DJ, et al: Electron microscopic localization of substance P and enkephalin in axon terminals related to dendrites of catecholaminergic neurons. Brain Res 160:387–400, 1979

Pieribone VA, Aston-Jones G: Adrenergic innervation of the rat nucleus locus coeruleus arises predominantly from the C1 and C3 cell groups in the rostral medulla: an anatomic study combining retrograde transport and immunofluorescence. Neuroscience 41:525–542, 1991

Pieribone VA, Aston-Jones G, Bohn MC: Adrenergic and non-adrenergic neurons of the C1 and C3 areas project to locus coeruleus: a fluorescent double labeling study. Neurosci Lett 85:297–303, 1988

Plotsky PM, Vale W: Hemorrhage-induced secretion of corticotropin-releasing factor-like immunoreactivity into the rat hypophyseal portal circulation and its inhibition by glucocorticoids. Endocrinology 114:164–169, 1984

Plotsky PM, Otto S, Sapolsky RM: Inhibition of immunoreactive corticotropin-releasing factor secretion into the hypophyseal-portal circulation by delayed glucocorticoid feedback. Endocrinology 119:1126–1130, 1986

Rasmussen K, Aghajanian GK: Withdrawal-induced activation of locus coeruleus neurons in opiate-dependent rats: attenuation by lesions of the nucleus paragigantocellularis. Brain Res 505:346–350, 1989

Rasmussen K, Jacobs BL: Locus coeruleus activity in freely moving cats is increased following systemic morphine administration. Brain Res 344:240–248, 1985

Rasmussen K, Morilak DA, Jacobs BL: Single unit activity of locus coeruleus neurons in the freely moving cat; I: during naturalistic behaviors and in response to simple and complex stimuli. Brain Res 371:324–328, 1986

Reddy VK, Fung SJ, Zhuo H, et al: Localization of enkephalin neurons in the dorsolateral pontine tegmentum projecting to the spinal cord of the cat. J Comp Neurol 291:195–202, 1990

Sakanaka M, Shibasaki T, Lederes K: Corticotropin-releasing factor-like immunoreactivity in the rat brain as revealed by a modified cobalt-glucose oxide-diaminobenzidene method. J Comp Neurol 260:256–298, 1987

Satoh K, Shimizu N, Tohyama M, et al: Localization of the micturition reflex center at dorsolateral pontine tegmentum of the rat. Neurosci Lett 8:27–33, 1978

Sawchenko PE, Swanson LW: The organization of noradrenergic pathways from the brainstem to the paraventricular and supraoptic nuclei in the rat. Brain Research Review 4:275–325, 1982

Schulte HM, Chrousos GP, Gold PW, et al: Continuous administration of synthetic ovine corticotropin-releasing factor in man: physiological and pathophysiological implications. J Clin Invest 75:1781–1785, 1985

Segal M, Bloom FE: The action of norepinephrine in the rat hippocampus; I: iontophoretic studies. Brain Res 72:79–97, 1974a

Segal M, Bloom FE: The action of norepinephrine in the rat hippocampus; II: activation of the input pathway. Brain Res 72:99–114, 1974b

Segal M, Bloom FE: The action of norepinephrine in the rat hippocampus; III: hippocampal cellular responses to locus coeruleus stimulation in the awake rat. Brain Res 107:499–511, 1976

Shipley MT, Harris G, Williams J, et al: Asymmetric orientation of locus coeruleus (LC) dendrites in the pericoerulear region: biocytin-filled LC neurons in vitro. Society for Neuroscience Abstracts 16:1177, 1990

Shipley MT, Fu L, Ennis M, et al: Distribution of locus coeruleus extranuclear dendrites; I: immunocytochemical LM and EM studies. J Comp Neurol (in press)

Steindler DA: Locus coeruleus neurons have axons that branch to the forebrain and cerebellum. Brain Res 223:367–373, 1981

Stokes PE, Sikes CR: Hypothalamic-pituitary-adrenal axis in affective disorders, in Psychopharmacology: The Third Generation of Progress. Edited by Meltzer HY. New York, Raven, 1987, pp 589–607

Strahlendorf HK, Strahlendorf JC, Barnes CD: Endorphin-mediated inhibition of locus coeruleus neurons. Brain Res 191:284–288, 1980

Sutin EL, Jacobowitz DM: Immunocytochemical localization of peptides and other neurochemicals in the rat laterodorsal tegmental nucleus and adjacent area. J Comp Neurol 270:243–270, 1988

Sutton RE, Koob GF, LeMoal M, et al: Corticotropin-releasing factor produces behavioral activation in rats. Nature 297:331–333, 1982

Svensson TH: Peripheral, autonomic regulation of locus coeruleus noradrenergic neurons in brain: putative implications for psychiatry and psychopharmacology. Psychopharmacology (Berl) 92:1–7, 1987

Svensson TH, Thorén P: Brain noradrenergic neurons in the locus coeruleus: inhibition by blood volume load through vagal afferents. Brain Res 172:174–178, 1979

Swanson LW: The locus coeruleus: a cytoarchitectonic, Golgi and immunohistochemical study in the albino rat. Brain Res 110:39–56, 1976

Swanson LW, Hartman BK: The central adrenergic system: an immunofluorescence study of the location of cell bodies and their efferent connections in the rat using dopamine-β-hydroxylase as a marker. J Comp Neurol 163:467–506, 1976

Swanson LW, Sawchenko PE, Rivier J, et al: Organization of ovine corticotropin-releasing factor immunoreactive cells and fibers in rat brain: an immunohistochemical study. Neuroendocrinology 36:165–186, 1983

Tache Y, Goto Y, Gunion M, et al: Inhibition of gastric acid secretion in rats by intracerebral injection of corticotropin-releasing factor (CRF). Science 222:935–937, 1983

Tanaka M, Kohno Y, Tsuda A, et al: Differential effects of morphine on noradrenaline release in brain regions of stressed and non-stressed rats. Brain Res 275:105–115, 1983

Tanaka M, Ida Y, Tsuda A: Naloxone, given before but not after stress exposure, enhances stress-induced increases in regional brain noradrenaline release. Pharmacol Biochem Behav 29:613–616, 1988

Tanaka M, Ida Y, Tsuda A, et al: Met-enkephalin, injected during the early phase of stress, attenuates stress-induced increases in noradrenaline release in rat brain regions. Pharmacol Biochem Behav 32:791–795, 1989

Tanaka M, Tsuda A, Yokoo H, et al: Involvement of the brain noradrenaline system in emotional changes caused by stress in rats. Ann N Y Acad Sci 597:159–174, 1990

Taylor JR, Elsworth JD, Garcia EJ, et al: Clonidine infusions into the locus coeruleus attenuate behavioral and neurochemical changes associated with naloxone-precipitated withdrawal. Psychopharmacology (Berl) 96:121–134, 1988

Tazi A, Dantzer R, LeMoal M, et al: Corticotropin-releasing factor antagonists blocks stress-induced fighting in rats. Regul Pept 18:37–42, 1987

Tempel A, Zukin RS: Neuroanatomical patterns of the μ, d and k opioid receptors of rat brain as determined by quantitative in vitro autoradiography. Proc Natl Acad Sci U S A 84:4308–4312, 1987

Terman GW, Shavit Y, Lewis JW, et al: Intrinsic mechanisms of pain inhibition: activation by stress. Science 226:1270–1277, 1984

Trujillo KA, Akil H: Inhibition of morphine tolerance and dependence by the NMDA receptor antagonist MK-801. Science 251:85–87, 1991

Tung C-S, Grenhoff J, Svensson TH: Morphine withdrawal responses of rat locus coeruleus neurons are blocked by an excitatory amino-acid antagonist. Acta Physiol Scand 138:581–582, 1990

Uhl GR, Goodman RR, Kuhar MJ, et al: Immunohistochemical mapping of enkephalin containing cell bodies, fibers and nerve terminals in the brainstem of the rat. Brain Res 166:75–94, 1979a

Uhl GR, Goodman RR, Snyder SH: Neurotensin-containing cell bodies, fibers and nerve terminals in the brain stem of the rat: immunohistochemical mapping. Brain Res 167:77–91, 1979b

Vale W, Spiess J, Rivier C: Characterization of a 41-residue ovine hypothalamic peptide that stimulates secretion of corticotropin and β-endorphin. Science 213:1394–1397, 1981

Valentino RJ: Corticotropin-releasing factor: putative neurotransmitter in the noradrenergic nucleus locus coeruleus. Psychopharmacol Bull 25:306–311, 1989

Valentino RJ, Curtis AL: Antidepressant interactions with corticotropin-releasing factor in the noradrenergic nucleus locus coeruleus. Psychopharm Bull 27:263–269, 1991

Valentino RJ, Foote SL: Corticotropin-releasing factor disrupts sensory responses of brain noradrenergic neurons. Neuroendocrinology 45:28–36, 1987

Valentino RJ, Foote SL: Corticotropin-releasing hormone increases tonic but not sensory-evoked activity of noradrenergic locus coeruleus neurons in unanesthetized rats. J Neurosci 8:1016–1025, 1988

Valentino RJ, Wehby RG: Corticotropin-releasing factor: evidence for a neurotransmitter role in the locus coeruleus during hemodynamic stress. Neuroendocrinology 48:674–677, 1988a

Valentino RJ, Wehby RG: Morphine effects on locus coeruleus neurons are dependent on the state of arousal and availability of external stimuli: studies in anesthetized and unanesthetized rats. J Pharmacol Exp Ther 244:1178–1186, 1988b

Valentino RJ, Wehby RG: Locus coeruleus discharge characteristics of morphine-dependent rats: effects of naltrexone. Brain Res 488:126–134, 1989

Valentino RJ, Foote SL, Aston-Jones G: Corticotropin-releasing factor activates noradrenergic neurons of the locus coeruleus. Brain Res 270:363–367, 1983

Valentino RJ, Curtis AL, Parris D, et al: Antidepressant actions on brain noradrenergic neurons. J Pharmacol Exp Ther 253:833–840, 1990

Valentino RJ, Page ME, Curtis AL: Activation of noradrenergic locus coeruleus neurons by hemodynamic stress is due to local release of corticotropin-releasing factor. Brain Res 555:25–34, 1991

Valentino RJ, Page M, Van Bockstaele EJ, et al: Corticotropin-releasing factor innervation of the locus coeruleus region: distribution of fibers and sources of input. Neuroscience 48:689–705, 1992

Vincent SR, Satoh K: Corticotropin-releasing factor (CRF) immunoreactivity in the dorsolateral pontine tegmentum: further studies on the micturition reflex system. Brain Res 308:387–391, 1984

Wang Y-Y, Aghajanian GK: Excitation of locus coeruleus neurons by vasoactive intestinal peptide: evidence for a G-protein-mediated inward current. Brain Res 500:107–118, 1989

Waterhouse BD, Moises HC, Woodward DJ: Noradrenergic modulation of somatosensory cortical neuronal responses to iontophoretically applied putative neurotransmitters. Exp Neurol 69:30–49, 1980

Waterhouse BD, Moises HC, Yeh HH, et al: Norepinephrine enhancement of inhibitory synaptic mechanisms in cerebellum and cerebral cortex: mediation by β-adrenergic receptors. J Pharmacol Exp Ther 221:495–506, 1982

White SR, Neuman RS: Facilitation of spinal motoneurone excitability by 5-hydroxytryptamine and noradrenaline. Brain Res 188:119–127, 1980

Williams JT, North RA: Opiate receptor interactions on single locus coeruleus neurons. Mol Pharmacol 26:489–497, 1984

Williams JT, Egan TM, North RA: Enkephalin opens potassium channels on mammalian central neurons. Nature 299:74–77, 1982

Williams JT, Christie MJ, North RA, et al: Potentiation of enkephalin action by peptidase inhibitors in rat locus coeruleus in vitro. J Pharmacol Exp Ther 243:397–401, 1987

Xiong H, Marshall KC: Angiotension II modulation of glutamate excitation of locus coeruleus neurons. Neurosci Lett 118:261–264, 1990

Young WS III, Uhl GR, Kuhar MJ: Iontophoresis of neurotensin in the area of the locus coeruleus. Brain Res 150:431–435, 1978

4

Classification of Adrenergic Receptors

CHARLES B. SMITH, M.D., PH.D.
PEGGIE J. HOLLINGSWORTH, PH.D.

In the mid-1960s the biogenic amine theory of depression (melancholia), independently proposed by Bunney and Davis (1965) and by Schildkraut (1965), focused attention on monoaminergic neuronal synapses in the central nervous system as possible sites of dysfunction in certain behavioral disorders and as sites of action of drugs and other treatments used to manage such disorders. In general terms, this theory stated that abnormalities of synaptic transmission at monoaminergic synapses (noradrenergic and serotonergic synapses, primarily) resulted in behavioral dysfunction and that successful therapies reversed or normalized these processes. This "monoamine theory of depression" was based on three general observations. First, reserpine, a drug that depletes endogenous stores of many neurotransmitters, including norepinephrine and serotonin, causes a behavioral disorder quite similar to melancholia.

Second, inhibitors of monoamine oxidase, such as phenelzine and tranylcypromine, are in some instances effective in the treatment of certain forms of depression. These monoamine oxidase inhibitors also prevent the oxidative deamination of norepinephrine and serotonin. Third, tricyclic antidepressant drugs, such as imipramine and desipramine, inhibit the transport processes by which monoamines are inactivated through neuronal and extraneuronal uptake from within the synapse. This inhibition of uptake could lead to increased intrasynaptic neurotransmitter content. Over the decade that followed the proposal of the biogenic amine theory of depression, intense study failed to provide conclusive support for the theory, and as a result interest waned.

Despite the fact that major classes of both noradrenergic and serotonergic receptors had been identified and classified nearly 20 years earlier, this biogenic amine theory of depression and most of the research it stimulated did not address to any significant degree the possibility that neurotransmitter receptor function might be abnormal in individuals with behavioral dysfunctions or the possibility that receptors for neurotransmitters might be the sites of action of the various treatments for such dysfunctions. In 1948, Ahlquist postulated the existence of two classes of noradrenergic receptors that he designated α- and β-adrenergic receptors; and in 1957, Gaddum and Picarelli described two classes of serotonergic receptors that they designated as "M" and "D" receptors. In the late 1970s, the observations that long-term administration of tricyclic antidepressant drugs altered the ability of β-adrenergic receptor agonists to stimulate adenylate cyclase in rat brain (Vetulani and Sulser 1975) and decreased the sensitivity to certain α-adrenergic receptor agonists of the newly described presynaptic autoregulatory α-adrenergic receptors in rat heart (Crews and Smith 1978) stimulated the interest of psychiatrists, psychologists, and behavioral scientists in abnormal neurotransmitter receptors as the basis for behavioral dysfunctions in humans and as sites of action of psychotherapeutic agents and treatments. In subsequent years, knowledge of neurotransmitter receptors has literally exploded to the extent that recently the identification of putative receptors, first through the development of radioligand binding techniques and then through the development of molecular biological techniques, has dramatically outpaced the development of knowledge of receptor function and possible receptor dysfunction in individuals with behavioral disorders. In this chapter, we review current concepts related to 1) the identification and classification of adrenergic receptors and 2) the evaluation of certain aspects of central nervous system adrenergic receptor function in humans.

Approaches to Classification of Adrenergic Receptors

The division by Ahlquist (1948) of adrenergic receptors into two classes—α-adrenergic receptors and β-adrenergic receptors—was based on differences in the relative potencies of six adrenergic agonists in stimulating adrenergic receptors in a variety of tissues and organs. The discovery of antagonists that selectively blocked the actions of these and other agonists completed this functional approach to receptor classification. The classical α-adrenergic receptor is a receptor at which epinephrine and norepinephrine are most potent, and isoproterenol is least potent, as agonists. The classical β-adrenergic receptor is a receptor at which isoproterenol and epinephrine are most potent, and norepinephrine is less potent, as agonists. The α-adrenergic receptor is blocked by phenoxybenzamine, phentolamine, and tolazoline, whereas the β-adrenergic receptor is blocked by propranolol. Thus, this functional approach to receptor identification and classification is based on the relative potencies (as estimates of affinity of the compounds for their receptors) of agonists and antagonists, usually measured in a variety of different tissues and organs. This approach, however, is only one of three currently in use.

A second approach to adrenergic receptor classification involves the use of radioactively labeled drugs (ligands) that bind reversibly to adrenergic receptors (Bylund 1992). Such ligands can either be agonists or antagonists and can bind either selectively or nonselectively to receptors located in tissues or on isolated cells or cell membranes. For highly selective radioligands, the classification of the receptor depends on the relative affinities of the ligands for the binding sites, whereas for nonselective radioligands, the classification of the receptor depends on the relative potencies of highly selective agonists and antagonists in displacing the nonselective radioligand from the receptor site(s). Although this approach does not enable one to determine receptor function directly, it does allow one to quantitate directly the number of receptors present and to characterize biochemically the recognition site of the receptor.

The most recent approach involves the molecular cloning of adrenergic receptor cDNAs or genes. A variety of biochemical techniques have been used to isolate adrenergic receptors and to characterize their structure. These include membrane solubilization, affinity and high performance liquid chromatography, and photoaffinity labeling (Insel 1989a, 1989b). Once a receptor subtype has been isolated and purified, cDNA probes can be synthesized that can be used to isolate receptor cDNA or receptor genes from tissues.

This powerful approach will eventually lead to an understanding of regulation of receptor biosynthesis and degradation as well as receptor sensitization and desensitization (Emorine et al. 1991; Harrison et al. 1991). The current classification of adrenergic receptors is based primarily on radioligand binding studies, and to a lesser extent on functional studies and studies involving the molecular cloning of such receptors. However, none of the three approaches alone provides total understanding of the nature and function of adrenergic receptors, and some, such as the functional approach first used by Ahlquist (1948), might yet lend themselves better than others to the evaluation of adrenergic receptor function and dysfunction in individuals with behavioral disorders.

Types of Adrenergic Receptors

Currently there appear to be three major types of adrenergic receptors: the α_1-, the α_2-, and the β-adrenergic receptors. By use of the techniques described earlier, these major types have been classified further into a number of subtypes (Figure 4–1). There have now been described at least three subtypes of α_1-adrenergic receptors (α_{1A}, α_{1B}, and α_{1C}), four subtypes of α_2-adrenergic receptors (α_{2A}, α_{2B}, α_{2C}, and α_{2D}), and three subtypes of β-adrenergic receptors (β_1, β_2, and β_3). The agonists and antagonists used to define these receptor subtypes are listed in Table 4–1 and Table 4–2. There are a number of fairly extensive reviews of current knowledge of adrenergic neurotransmitter receptors (Bylund 1988, 1992; Harrison et al. 1991; Insel 1989a, 1989b; Lomasney et al. 1991; Roth et al. 1991; Ruffolo et al. 1991).

α_1-Adrenergic receptors. Agonists that stimulate selectively the α_1-family of adrenergic receptors include phenylephrine and methoxamine. At this time there is no consistent evidence that these agonists possess relative selectivity for any of the receptor subtypes. Prazosin and terazosin, α_1-adrenergic receptor selective antagonists, have essentially the same affinity for all three subtypes. However, phentolamine and WB-4101 have high affinity for the α_{1A}-subtype and even higher affinity for the α_{1C}-subtype, but low affinity for the α_{1B}-subtype. 5-Methyl-urapidil and (+)-niguldipine also have high affinity for the α_{1A}-subtype and low affinity for the α_{1B}-subtype. α_{1A}-Adrenergic receptors are found in the rat renal artery and are believed to be linked to a signal transduction mechanism that regulates Ca^{++} channels (Table 4–3). α_{1B}-Adrenergic receptors have been described in rat liver, rat spleen, and a variety of cells isolated from rats and hamsters and grown in tissue culture. Stimulation of α_{1B}-adrenergic receptors leads to a rapid increase in the formation of inositol 1,4,5-triphosphate and facilitates the release of Ca^{++} at intracellular sites. Both α_{1A}- and α_{1B}-adrenergic receptors have been described in rat and human hippocampus. The α_{1C}-adrenergic receptor has only been cloned from bovine brain and has not yet been identified in tissues by either radioligand binding studies or by functional studies with agonists and antagonists (Schwinn et al. 1990). The

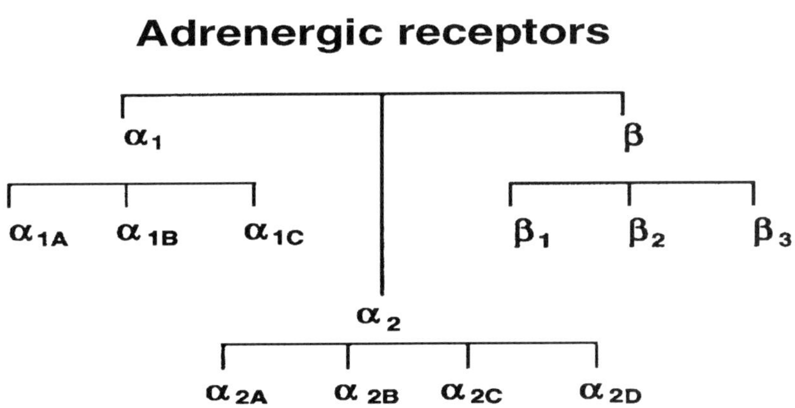

Figure 4–1. Schematic illustration of currently recognized classes and subclasses of adrenergic receptors.

α_1-family of adrenergic receptors are localized at postsynaptic sites and mediate responses of various types of smooth muscle to adrenergic agonists.

α_2-Adrenergic receptors. Agonists that stimulate selectively the α_2-family of adrenergic receptors include clonidine, α-methylnorepinephrine, and brimonidine. There is some evidence that these agonists might differ in their selectivity for α_2-adrenergic receptor subtypes (Yoshioka et al. 1991). Yohimbine, rauwolfscine, and idazoxan are antagonists relatively selective for α_2-adrenergic receptors. The differentiation of the four receptor subtypes at this time, however, rests primarily on the binding affinities of antagonists. Oxymetazoline is an antagonist that has significantly higher affinity for α_{2A}-adrenergic receptors than for α_{2B}- or α_{2C}-adrenergic receptors. BAM-1303 is an antagonist that has the highest affinity for α_{2C}-

Table 4–1. Classification of α-adrenergic receptors

Receptor type	Receptor subtype	Agonists	Antagonists
α		Epinephrine, norepinephrine	Phenoxybenzamine, phentolamine, tolazoline
α_1		Phenylephrine, methoxamine	Prazosin, terazosin
	α_{1A}		Phentolamine, WB-4101, 5-methyl-urapidil, (+)-niguldipine
	α_{1B}		Selectively inactivated by chloroethylclonidine; lower affinity for antagonists that bind to other subtypes
	α_{1C}		Phentolamine, WB-4101
α_2		Clonidine, α-methylnorepinephrine, bromonidine	Rauwolfscine, yohimbine, idazoxan
	α_{2A}		Oxymetazoline, BAM-1303
	α_{2B}		Prazosin, ARC-239, spiroxatrine
	α_{2C}		Prazosin, ARC-239, spiroxatrine, WB-4101, BAM-1303
	α_{2D}		Lower affinity for rauwolfscine than other subtypes; prazosin, BAM-1303, mianserin

adrenergic receptors, less affinity for α_{2A}-adrenergic receptors, and least affinity for α_{2B}-adrenergic receptors. Prazosin and spiroxatrine are antagonists that have the highest affinity for α_{2B}-adrenergic receptors, lesser affinity for α_{2C}-adrenergic receptors, and virtually no affinity for α_{2A}-adrenergic receptors. WB-4101 is an antagonist that among the various α_2-adrenergic receptors is relatively selective for the α_{2C}-subtype. The most recently identified α_2-receptor subtype is the α_{2D}-adrenergic receptor. This receptor has a lower affinity for rauwolfscine than do the other three subtypes and has affinity for prazosin, mianserin, and BAM-1303, which differentiates it from the other three (Bylund 1992).

The α_2-adrenergic receptor was first identified as an autoregulatory receptor located on noradrenergic nerve terminals. Stimulation of this receptor results in an inhibition of norepinephrine release. In the central nervous system, α_2-adrenergic receptor stimulation is reflected by decreased levels of 3-methoxy-4-hydroxyphenylglycol (MHPG), a major metabolite of norepinephrine in the brain. Subsequently, this class of receptors has been shown to regulate neurotransmitter release from a variety of neurons in addition to noradrenergic neurons, including serotonergic, dopa-

Table 4–2. Classification of β-adrenergic receptors

Receptor type	Receptor subtype	Agonists	Antagonists
β		Isoproterenol, epinephrine, norepinephrine	Propranolol, nadolol, pindolol
	β_1	Epinephrine greater than or equal in potency to norepinephrine; salbutamol is a partial agonist	Metoprolol, atenolol, bisoprolol, betaxolol, esmolol
	β_2	Epinephrine always greater in potency than norepinephrine; salbutamol is a full agonist	IC-118,551
	β_3	Norepinephrine is equipotent to epinephrine; salbutamol is equipotent to isoproterenol	β_1 and β_2 selective antagonists have much lower affinity for this receptor

minergic, and cholinergic neurons. This subject has been reviewed extensively (Langer 1980; Starke 1981; Westfall 1977). α_2-Adrenergic receptors are located at both pre- and postsynaptic sites and on human platelets. Some pharmacological effects of α_2-adrenergic receptor stimulation include contraction of vascular smooth muscle, stimulation of platelet aggregation, inhibition of lipolysis, inhibition of insulin release, inhibition of renin release, and inhibition of growth hormone release. Little is known about either the distribution of the α_2-adrenergic receptor subtypes or their function in humans. α_{2A}-Adrenergic receptors appear to be the sole α-adrenergic recep-

Table 4–3. Some responses mediated by α- and β-adrenergic receptors

α_1	α_2	β_1	β_2
Smooth muscle contraction (vascular, genitourinary, and intestinal)	Smooth muscle contraction (vascular)	Heart: increase in rate and force of contraction	Heart: increase in rate and force of contraction
			Relaxation of vascular, bronchial, and uterine smooth muscle
	Inhibition of lipolysis	Stimulation of lipolysis	
	Inhibition of renin release	Stimulation of renin release	
	Inhibition of insulin release		Stimulation of insulin release
		Stimulation of amylase secretion	Increased glycogenolysis and gluconeogenesis
	Increased platelet aggregation		Decreased platelet aggregation
	Inhibition of neurotransmitter release		Facilitation of neuronal norepinephrine release
Activation of PI turnover	Inhibition of adenylate cyclase	Activation of adenylate cyclase	Activation of adenylate cyclase
Mobilization of intracellular Ca^{++}	Regulation of K^+ channels		

tor located on human platelets. The α_{2B}-subtype has been identified in neonatal rat lung and the α_{2C}-subtype in the opossum kidney. The α_{2D}-subtype was first described in the bovine pineal gland and is probably the same subtype as that found in the rat submaxillary gland. The α_2-adrenergic receptors are known to act through inhibition of adenylate cyclase. In addition, some appear to regulate K^+ channels, an action more likely related to the regulation of neurotransmitter release. Because of their important role in the regulation of neurotransmitter release and their sensitivity to antidepressant drugs, electroconvulsive therapy, and other psychotherapeutic modalities, the α_2-adrenergic receptor family has stimulated the greatest interest among neurobiologists. Nevertheless, our knowledge of the function of these receptors is just in its infancy.

β-Adrenergic receptors. The β-adrenergic receptor family was the first to be divided into subtypes, by Lands et al. (1967). This group of receptors was divided into β_1- and β_2-subtypes (Brodde et al. 1990; Emorine et al. 1991; Frielle et al. 1989). Although isoproterenol is an agonist considerably more potent than either epinephrine or norepinephrine on both receptor subtypes, in some instances (e.g., in the heart) epinephrine is equipotent to norepinephrine in its actions on the β_1-adrenergic receptor subtype, whereas epinephrine is always more potent than norepinephrine in its actions on the β_2-adrenergic receptor subtype. Antagonists relatively selective for the β_1-adrenergic receptor subtype include metoprolol, atenolol, bisoprolol, and betaxolol. ICI 118,551 is an antagonist relatively selective for the β_2-subtype. Recently, a third type of β-adrenergic receptor has been identified that does not fully meet criteria for either β_1- or β_2-adrenergic receptors. This β_3-adrenergic receptor subtype is found on human adipocytes (Emorine et al. 1991, 1992; Zaagsma and Nahorski 1990).

Stimulation of both β_1- and β_2-adrenergic receptors increases the rate and force of contraction of the heart. β_1-Adrenergic receptors are located at postsynaptic sites. Responses mediated by β_1-adrenergic receptors include stimulation of lipolysis, stimulation of renin release, and stimulation of amylase secretion. β_2-Adrenergic receptors are located at both pre- and postsynaptic sites and on human lymphocytes and granulocytes. Some responses to β_2-adrenergic receptor stimulation include relaxation of vascular, bronchial, and uterine smooth muscle; stimulation of insulin release; increased glycogenolysis and gluconeogenesis; decreased platelet aggregation; and hypokalemia. The significance of β_2-adrenergic receptors in the facilitation of norepinephrine release from noradrenergic nerve terminals is highly debated (Nedergaard and Abrahamsen 1990). The β_3-adrenergic receptor is now thought to mediate increases in lipolysis in adipocytes. All subtypes of β-adrenergic receptors apparently increase adenylate cyclase activity.

Evaluation of Adrenergic Receptor Function in Humans

Although the actions of adrenergic drugs on a variety of physiological systems have been studied extensively for nearly a century, little is known about the adrenergic receptors that mediate the actions of these drugs within the central nervous system. Despite the rapidly expanding knowledge of adrenergic receptor subtypes, most recent information with regard to these receptors is based on radioligand binding studies and on molecular biological studies and lends little insight into the functions of these receptors, if any, within the nervous system. Many of the peripheral actions of various receptor-specific adrenergic drugs have been evaluated in individuals with psychiatric disorders both before and after treatment. The results of such investigations are highly variable from one study to the next, and their interpretation is often problematical. There are a number of rather widely used approaches, however, to the evaluation of the function of the major classes of adrenergic receptors found in the human brain.

Studies of α_1-Adrenergic Receptor Function

Although α_1-adrenergic receptors are found in the human brain, most studies in humans of α_1-adrenergic receptor function have involved drug-induced changes in vascular resistance, since this receptor is thought to be the primary adrenergic receptor that regulates the tone of vascular smooth muscle. Such studies have not led to an understanding of α_1-adrenergic receptor function in patients with behavioral disorders. It has been proposed that the corticotropin response to adrenergic stimulation in humans is an indicator of hypothalamic α_1-adrenergic receptor function (Al-Damluji 1991). α_1-Adrenergic receptors are found in high density in the human hypothalamus along with α_2-adrenergic receptors. Methoxamine, a highly selective α_1-adrenergic receptor agonist, stimulates corticotropin secretion, and this effect is abolished by the concomitant administration of prazosin. In contrast, activation of α_2-adrenergic receptors inhibits corticotropin secretion produced by hypoglycemia in humans. Further studies are needed to validate the use of corticotropin secretion as an index of α_1-adrenergic receptor function in humans.

Studies of α_2-Adrenergic Receptor Function

Clonidine, a prototypical α_2-adrenergic receptor agonist, causes an increase in plasma growth hormone levels. This response is blocked by α_2-adrenergic receptor antagonists such as yohimbine and idazoxan. In depressed patients, the clonidine-induced increase in growth hormone levels is blunted, an observation that has been attributed to a decrease in postsynaptic α_2-adrenergic receptor sensitivity (Checkley et al. 1981; Matussek et al. 1980; Siever and Uhde 1982). Yohimbine causes increases in plasma cortisol levels, an observation that suggests that stimulation of postsynaptic α_2-adrenergic receptors decreases plasma cortisol levels. Basal plasma cortisol levels have been reported to be higher in depressed patients than in nondepressed subjects, and the increase in plasma cortisol produced by yohimbine is greater in these patients than in nondepressed subjects (Price et al. 1986), an observation that suggests a subsensitivity of postsynaptic α_2-adrenergic receptors in the brains of depressed subjects.

Presynaptic α_2-adrenergic receptor function in the human brain has been assessed by measuring changes in the plasma levels of MHPG in response to the administration of either clonidine or yohimbine. Plasma MHPG is thought to be a good index of norepinephrine turnover in the central nervous system. Clonidine administration decreases and yohimbine administration increases norepinephrine turnover. Changes in MHPG levels in response to either clonidine or yohimbine have been found to be unaltered in melancholic patients, observations that have been interpreted to indicate that presynaptic α_2-receptors are unaltered in depression (Heninger et al. 1988). Other studies conducted on changes in heart rate and blood pressure induced by clonidine and yohimbine have proven to be equivocal (Siever and Uhde 1984).

Human blood platelets have been used rather extensively to assess α_2-adrenergic receptor status in a variety of psychiatric conditions and after various psychotherapeutic treatments. This receptor is thought to be an α_{2A}-adrenergic receptor. This is the same α_2-adrenergic receptor subtype that regulates that release of norepinephrine from neurons in the rat brain. The status of this platelet receptor has been evaluated in radioligand binding studies that used either agonists or antagonists. A fairly consistent finding is that on platelet membranes from melancholic patients the number of binding sites for ^{3}H-clonidine are increased when compared with platelets from control subjects, whereas the number of binding sites for ^{3}H-yohimbine are unchanged. Treatment with tricyclic antidepressants and electroconvulsive therapy tends to decrease to control values of the number of ^{3}H-clonidine binding sites on the platelet membranes (Pandey et al. 1990; Smith and Hollingsworth 1988).

Studies of β-Adrenergic Receptor Function

Many studies of β-adrenergic receptor function have been performed in humans, especially to evaluate cardiovascular status. Such studies have involved the responsiveness of the heart (primarily β_1-adrenergic receptors) and the blood vessels (primarily β_2-adrenergic receptors) in control subjects and in patients with cardiovascular disorders. Most studies of β-adrenergic receptor function in humans have involved the use of lymphocytes or leukocytes. The leukocyte β-adrenergic receptor belongs to the β_2-adrenergic receptor subtype. Studies that involve the binding of radioligands such as ^{125}iodocyanopindolod or ^{3}H-labeled dihydroalprenolol indicate that the density of such receptors are decreased on leukocyte membranes from melancholic patients (Pandey et al. 1990; Werstiuk et al. 1990). Furthermore, the stimulation of ^{3}H-labeled cyclic adenosine monophosphate accumulation by adrenergic agonists is less in melancholic patients than in control subjects, an observation consistent with decreased numbers of β_2-adrenergic receptors on leukocytes from melancholic patients. Another possible model for the study of β-adrenergic receptor function in humans is the adipocyte. This receptor is atypical when compared with β_1- and β_2-adrenergic receptors and has been called a β_3-adrenergic receptor. Studies to evaluate the function of this receptor on adipocytes need to be carried out in patients with behavioral disorders.

There appears to be a β-adrenergic receptor located at noradrenergic nerve endings that regulate the release of neurotransmitter. Stimulation of this receptor with β-adrenergic receptor agonists such as isoproterenol enhance norepinephrine release. This receptor is thought to be of the β_2 subtype (Nedergaard and Abrahamsen 1990). Nevertheless, the significance of this receptor in the regulation of neurotransmitter release is highly controversial, and studies in humans to evaluate the function of this receptor, similar to those performed to evaluate the function of the presynaptic α_2-adrenergic receptor, need to be conducted.

The functional characterization of the many subtypes of adrenergic receptors, identified through radioligand binding studies or by means of molecular biological techniques, both in humans and in experimental animals remains in large part to be determined. There are numerous problems that must first be resolved. Most studies of adrenergic receptor function in humans have been carried out with agonists and antagonists that lack specificity for the various receptor subtypes. Considerable time will most likely elapse before receptor-selective drugs are developed that are both safe and potentially therapeutically efficacious when administered to humans. There is, however, sufficient current knowledge to allow for the development of such agents. Also, the interpretation of observations in intact humans will always remain extremely complex. Numerous factors, possibly including multiple types of adrenergic receptors, probably influence the outcome of the various tests used currently and anticipated in the future to assess receptor function. Even the development of such promising techniques as positron-emission tomography will not provide the necessary insights into adrenergic receptor function in the human brain. Nevertheless, with the rapid expansion of knowledge that is occurring with regard to the classification of adrenergic receptors and their function, one might anticipate new insights into the role that such receptors play in the human nervous system.

References

Ahlquist RP: A study of the adrenotropic receptors. Am J Physiol 153:586–600, 1948

Al-Damluji S: Measuring the activity of brain adrenergic receptors in man. J Endocrinol Invest 14:245–254, 1991

Brodde OE, Daul A, Michel MC: Subtype-selective modulation of human β_1- and β_2-adrenoceptor function by β-adrenoceptor agonists and antagonists. Clin Physiol Biochem 2 (suppl 8):11–17, 1990

Bunney WE, Davis JM: Norepinephrine in depressive reactions. Arch Gen Psychiatry 13:483–494, 1965

Bylund DB: Subtypes of α_2-adrenoceptors: pharmacological and molecular biological evidence coverage. Trends Pharmacol Sci 9:356–361, 1988

Bylund DB: Subtypes of α_2-adrenergic receptors. FASEB J 6:832–839, 1992

Checkley SA, Slade AP, Shur E: Growth hormone and other responses to clonidine in patients with endogenous depression. Br J Psychiatry 138:51–55, 1981

Crews FT, Smith CB: Presynaptic *alpha*-receptor subsensitivity after long-term antidepressant treatment. Science 202:322–324, 1978

Emorine LJ, Feve B, Pairault J, et al: Structural basis for functional diversity of β_1-, β_2- and β_3-adrenergic receptors. Biochem Pharmacol 41:853–859, 1991

Emorine LJ, Feve B, Pairault J, et al: The human β_3-receptor: relationship with atypical receptors. Am J Clin Nutr 55 (suppl 1):215S–218S, 1992

Frielle T, Caron MG, Lefkowitz RJ: Properties of the β_1- and β_2-adrenergic receptor subtypes revealed by molecular cloning. Clin Chem 35:721–725, 1989

Gaddum JH, Picarelli ZP: Two kinds of tryptamine receptor. Br J Pharmacol 12:323–328, 1957

Harrison JK, Pearson WR, Lynch KR: Molecular characterization of α_1- and α_2-adrenoceptors. Trends Pharmacol Sci 12:62–67, 1991

Heninger GR, Charney DS, Price LH: α_2-Adrenergic receptor sensitivity in depression. Arch Gen Psychiatry 45:718–726, 1988

Insel PA: Adrenergic receptors: evolving concepts on structure and function. Am J Hypertens 2:112S–118S, 1989a

Insel PA: Structure and function of α-adrenergic receptors. Am J Med 87:12S–18S, 1989b

Langer SZ: Presynaptic regulation of the release of catecholamines. Pharmacol Rev 32:337–362, 1980

Lands AM, Arnold A, McAuliff JP, et al: Differentiation of receptor systems activated by sympathomimetic amines. Nature 214:597–598, 1967

Lomasney JW, Cotecchia S, Lefkowitz RJ, et al: Molecular biology of α-adrenergic receptors: implications for receptor classification and for structure-function relationships. Biochim Biophys Acta 1095:127–139, 1991

Matussek N, Ackenhell M, Hippius H, et al: Effects of clonidine on growth hormone release in psychiatric patients and controls. Psychiatry Res 2:25–36, 1980

Nedergaard OA, Abrahamsen J: Modulation of noradrenaline release by activation of presynaptic β-adrenoceptors in the cardiovascular system. Ann N Y Acad Sci 604:528–544, 1990

Pandey GN, Pandey SC, Davis JM: Peripheral adrenergic receptors in affective illness and schizophrenia. Pharmacol Toxicol 3 (suppl 66):13–36, 1990

Price LH, Charney DS, Rubin AL, et al: Alpha$_2$-adrenergic receptor function in depression: the cortisol response to yohimbine. Arch Gen Psychiatry 43:849–858, 1986

Roth NS, Lefkowitz RJ, Caron MG: Structure and function of the adrenergic receptor family. Adv Exp Med Biol 308:223–238, 1991

Ruffolo RR Jr, Nichols AJ, Stadel JM, et al: Structure and function of α-adrenoceptors. Pharmacol Rev 43:475–505, 1991

Schildkraut JJ: The catecholamine hypothesis of affective disorders: a review of supporting evidence. Am J Psychiatry 122:509–522, 1965

Schwinn DA, Cotecchia S, Lorenz W, et al: The α_{1C}-adrenergic receptor: a new member in the α_1-adrenergic receptor family. Trans Assoc Am Physicians 103:112–118, 1990

Siever LJ, Uhde TW: The growth hormone response to clonidine as a probe of noradrenergic receptor responsiveness in affective disorder patients and controls. Psychiatry Res 6:171–183, 1982

Siever LJ, Uhde TW: New studies and perspectives on the noradrenergic receptor system in depression: effects of the alpha2-adrenergic agonist clonidine. Biol Psychiatry 19:131–156, 1984

Smith CB, Hollingsworth PJ: Platelet alpha$_2$ receptor binding studies in depressive and anxiety patients, in Progress in Catecholamine Research, Part C: Clinical Aspects, Neurology and Neurobiology, Vol 42C. Edited by Belmaker RH, Sandler S, Dahlstrom A. New York, Alan R Liss, 1988, pp 349–354

Starke K: Presynaptic receptors. Ann Rev Pharmacol Toxicol 21:7–30, 1981

Vetulani J, Sulser F: Action of various antidepressant treatments reduces reactivity of noradrenergic cyclic AMP-generating system in limbic forebrain. Nature 257:495–496, 1975

Werstiuk ES, Steiner M, Burns T: Studies on leukocyte β-adrenergic receptors in depression: a critical appraisal. Life Sci 47:85–105, 1990

Westfall TC: Local regulation of adrenergic neurotransmission. Physiol Rev 57:659–728, 1977

Yoshioka M, Jacob AC, Hollingsworth PJ, et al: Junctional evidence that α_{2B}-adrenergic receptors regulate ^{3}H-serotonin release in rat brain slices. FASEB J 5 (A4950):175, 1991

Zaagsma J, Nahorski SR: Is the adipocyte beta-adrenoceptor a prototype for the recently cloned atypical 'β_3-adrenoceptor'? Trends Pharmacol Sci 11:3–7, 1990

5

Adrenergic Effects of Psychotropic Drugs

SALLY K. GUTHRIE, PHARM.D.

Studies to determine the effect of psychotropic drugs on the norepinephrine (NE) system have been conducted in both humans and animals (Table 5–1). Most studies have determined either endogenous concentrations of NE and its metabolites, or the amounts and proportions of these compounds that are formed after administration of labeled NE or a labeled precursor, such as tyrosine. There are two crucial questions: Which substance should be studied, and in which fluid should it be determined—cerebrospinal fluid (CSF), plasma, or urine? Since the majority of NE that is released from neurons is quickly taken up by the presynaptic neuron or metabolized, NE may not yield the most information concerning the overall activity of the noradrenergic system. In both rodents and humans, 3-methoxy-4-hydroxyphenylglycol (MHPG) is the most prevalent NE metabolite in the central nervous system (CNS). Consequently, MHPG is the NE metabolite that has been most commonly

Table 5–1. Effects of psychotropic drugs on norepinephrine and its metabolites

Drug	Reference	Treatment duration	Tissue	Norepinephrine effects
Antidepressants				
Desipramine	Roffman et al. 1977	Acute/chronic (2 weeks)	Rat brain	Acute: decrease in MHPG-SO_4 Chronic: increase in MHPG-SO_4
	Tang et al. 1978, 1979	Acute/chronic (21 days)	Rat brain	Acute: decrease in MHPG Chronic: increase in MHPG
	Bareggi et al. 1978b	Acute	Rat brain	Decrease in MHPG-SO_4
	Racagni et al. 1982	Acute/chronic (18 days)	Rat brain	Acute: increase in NM in hypothalamus and cortex Chronic: decrease in NM
	Rosloff and Davis 1978	Acute/chronic (23 days)	Rat brain	Acute: decrease in tritiated NE following tritiated tyrosine
	Schubert et al. 1970	Acute	Mouse brain	Reduction in accumulation of labeled NE following labeled tyrosine infusion
	Moret and Briley 1992	Acute	Rat brain	Reduction (45%) in accumulation of dopa following aromatic acid decarboxylase inhibitor
	Nielsen and Braestrup 1977	Acute/chronic (20 days)	Rat brain	Acute: decrease in conversion to NE following tritiated tyrosine Chronic: labeled DHPG-SO_4, MHPG, and NE decreased following tritiated tyrosine
	Karoum et al. 1984	Chronic (2 weeks)	Rat brain	Decrease in MHPG in hypothalamus
	Potter et al. 1981, 1984, 1985	Chronic (4 weeks)	Human CSF and urine	Decrease in MHPG in CSF; decrease in urinary excretion rates of NE, MHPG, VMA, and sum[a]

	Dahl et al. 1982	Chronic (6 weeks)	Human CSF	No change in MHPG in CSF
	Donnelly et al. 1986	Chronic (2 weeks)	Human plasma and urine	Decrease in urinary excretion rates of NE, MHPG, and VMA; decrease in supine plasma MHPG; increase in standing plasma NE concentrations
Imipramine	Roffman et al. 1977	Acute/chronic (2 weeks)	Rat brain	Acute: decrease in $MHPG\text{-}SO_4$ Chronic: increase in $MHPG\text{-}SO_4$
	Schildkraut et al. 1970	Acute/chronic (3 weeks)	Rat brain	Acute: increase in labeled NE, 270 minutes following tritiated NE injection Chronic: lower endogenous NE, also increased labeled NE but decreased deaminated O-methylated metabolites
	Moret and Briley 1992	Acute	Rat brain	Reduction (30%) in accumulation of dopa following aromatic acid decarboxylase inhibitor
	Schildkraut et al. 1964	Chronic (3 weeks)	Human urine	Decrease in urinary VMA excretion rate
	Roccatagliata et al. 1981	Chronic (30 days)	Human CSF and urine	Significant decreases in MHPG in CSF, in patients showing clinical response
	Beckmann and Goodwin 1975	Chronic (2–3 weeks)	Human urine	Decrease in urinary MHPG excretion rate
	Bowden et al. 1985	Chronic (20 days)	Human CSF	Decrease in MHPG in CSF
Amitriptyline	Roffman et al. 1977	Acute/chronic (2 weeks)	Rat brain	No effects on $MHPG\text{-}SO_4$
	Tang et al. 1978	Acute	Rat brain	No effects on $MHPG\text{-}SO_4$
	Schubert et al. 1970	Acute	Mouse brain	No effect on accumulation of labeled NE following labeled tyrosine infusion
	Beckmann and Goodwin 1975	Chronic (2–3 weeks)	Human urine	Decrease in urinary MHPG excretion rate
	Bowden et al. 1985	Chronic (20 days)	Human CSF	Decrease in MHPG in CSF

continued

Table 5–1. Effects of psychotropic drugs on norepinephrine and its metabolites *(continued)*

Drug	Reference	Treatment duration	Tissue	Norepinephrine effects
	Sacchetti et al. 1976	Chronic (20 days)	Human urine	Some patients showed decrease and some showed increase in urinary MHPG excretion rate
	Modai et al. 1979	Chronic (5 weeks)	Human urine	Increase in urinary MHPG excretion rate
	Robinson et al. 1983	Chronic (6 weeks)	Human plasma	Significant reduction in plasma DHPG
Nortriptyline	Roffman et al. 1977	Acute/chronic (2 weeks)	Rat brain	Acute: no change in MHPG-SO_4 Chronic: no change in MHPG-SO_4
	Schubert et al. 1970	Acute	Mouse brain	Reduction in labeled NE accumulation following labeled tyrosine infusion
	Bertilsson et al. 1974	Chronic (3 weeks)	Human CSF	Decrease in MHPG in CSF
	Thorén et al. 1980	Chronic (3 weeks)	Human CSF	Decrease in MHPG in CSF
Clomipramine	Bertilsson et al. 1974	Chronic (3 weeks)	Human CSF	Decrease in MHPG in CSF
	Åsberg et al. 1977	Chronic (3 weeks)	Human CSF	Decrease in MHPG in CSF
	Träskman et al. 1979	Chronic (3 weeks)	Human CSF	Decrease in MHPG in CSF
	Thorén et al. 1980	Chronic (3 weeks)	Human CSF	Decrease in MHPG in CSF
	Mårtensson et al. 1991	Chronic (3 and 6 weeks)	Human CSF	Decrease in MHPG in CSF
Zimelidine	Siwers et al. 1977	Chronic (20 days)	Human CSF	Decrease in MHPG in CSF
	Bertilsson et al. 1980	Chronic (3 weeks)	Human CSF	Decrease in MHPG in CSF
	Wålinder et al. 1981	Chronic (3 weeks)	Human CSF	Decrease in MHPG in CSF
	Linnoila et al. 1982a; Potter et al. 1981, 1984, 1985	Chronic (4 weeks)	Human CSF and urine	Decrease in MHPG in CSF; decrease in urinary excretion rates of MHPG, NE, VMA, and sum[a]
	Åberg-Wistedt et al. 1985	Chronic (3 weeks)	Human CSF	Decrease in MHPG in CSF
Alaproclate	Åberg-Wistedt et al. 1985	Chronic (3 weeks)	Human CSF	No change in MHPG in CSF
Femoxetine	Dahl et al. 1982	Chronic (6 weeks)	Human CSF	No change in MHPG in CSF

Citalopram	Moret and Briley 1992	Acute/chronic (21 days)	Rat brain	Acute/chronic: no change in dopa accumulation following an aromatic acid decarboxylase inhibitor
	Bjerkenstedt et al. 1985	Chronic (4 weeks)	Human CSF	Decrease in MHPG in CSF
Fluoxetine	Mårtensson et al. 1989	Chronic (3–6 weeks)	Human CSF	Decrease in MHPG in CSF
Pheniprazine	Sjoerdsma et al. 1959	Chronic (5–10 days)	Human urine	No change in urinary NE excretion rate
Iproniazid	Studnitz 1959	Chronic (3–4 days)	Human urine	Decrease in urinary VMA excretion rate
Phenelzine	Schildkraut et al. 1964	Chronic (3 weeks)	Human urine	Decrease in urinary VMA excretion rate
	Beckmann and Murphy 1977	Chronic	Human urine (4 weeks)	Decrease in urinary MHPG excretion rate
	Robinson et al. 1983	Chronic (6 weeks)	Human plasma	Decrease in DHPG and NE in plasma
	Sharma et al. 1990	Chronic (4 weeks)	Human urine	Decrease in urinary MHPG excretion rate
Tranylcypromine	Schildkraut 1970	Acute	Rat brain	Decrease in MHPG, DHPG, VMA; increase in NM; decrease in tritiated NE uptake
	Zametkin et al. 1985b	Chronic (2 weeks)	Human CSF and plasma	(Results combined clorgyline with tranylcypromine) decrease in urinary excretion rates of MHPG and VMA; increase in NM excretion rate; decrease in MHPG and NE in plasma
Pargyline	Schildkraut 1970	Acute	Rat brain	Decrease in MHPG, DHPG, and VMA; increase in NM
	Murphy et al. 1979	Chronic (4 weeks)	Human CSF and plasma	Decrease in MHPG in CSF but no change in VMA or NE; no change in plasma NE
	Pickar et al. 1981	Chronic (4 weeks)	Human plasma	Decrease in MHPG in plasma
Deprenyl	Yang and Neff 1974	Acute	Rat brain	No effect on NE concentration
	Braestrup et al. 1975	Acute	Rat brain	Reduction in MHPG, but only at high dose
	Pickar et al. 1981	Chronic (4 weeks)	Human plasma	Slight (not significant) decrease in MHPG in plasma
	Sunderland et al. 1985	Chronic (3 weeks)	Human plasma	Dose-dependent decrease in MHPG in plasma

continued

Table 5–1. Effects of psychotropic drugs on norepinephrine and its metabolites *(continued)*

Drug	Reference	Treatment duration	Tissue	Norepinephrine effects
Clorgyline	Yang and Neff 1974	Acute	Rat brain	Increase in NE concentration
	Braestrup et al. 1975	Acute	Rat brain	Decrease in MHPG
	Moret and Briley 1992	Acute	Rat brain	Decrease (80%) in dopa accumulation following an aromatic acid decarboxylase inhibitor
	Murphy et al. 1979	Chronic (4 weeks)	Human plasma and CSF	Decrease in MHPG in CSF; decrease in supine and standing NE in plasma
	Pickar et al. 1981	Chronic (4 weeks)	Human plasma	Decrease in MHPG in plasma
	Linnoila et al. 1982b	Chronic (3–4 weeks)	Human urine	Decrease in urinary excretion rates of MHPG and VMA; increase in urinary NM excretion rate
	Potter et al. 1984, 1985	Chronic (3–8 weeks)	Human CSF and urine	Decrease in urinary excretion rate of sum[a]; decrease in MHPG in CSF
	Zametkin et al. 1985b	Chronic (2 weeks)	Human CSF and plasma	(Results combined clorgyline with tranylcypromine) decrease in urinary excretion rates of MHPG and VMA; increase in NM excretion rate; decrease in MHPG and NE in plasma
	Ross et al. 1985	Chronic (6–10 weeks)	Human plasma	Decrease in supine plasma NE concentration
Moclobemide	Dajas et al. 1984	Chronic (14 days)	Human urine	Decrease in urinary NE excretion rate
	Berlin et al. 1990	Chronic (1 week)	Human urine and plasma	Decrease in plasma DHPG AUC; increase in urine NM

Toloxatone	Berlin et al. 1990	Chronic (1 week)	Human urine and plasma	Decrease in plasma DHPG AUC; no significant change in urine NM
Buproprion	Golden et al. 1988	Chronic (4–6 weeks)	Human CSF, urine, and plasma	No change in plasma NE concentration; urinary excretion rate of sum[a] decreased; nonsignificant decrease of MHPG in CSF
Lithium				
Lithium	Schildkraut et al. 1966	Acute	Rat brain	No change in NE; increase in deaminated catechols; decrease in NM
	Schildkraut et al. 1969	Acute/chronic (7 days)	Rat brain	Acute: increase in deaminated catechols; decreased NE following dose of labeled NE Chronic: labeled NE decreased; decrease in NM and deaminated O-methylated metabolites
	Stern et al. 1969	Chronic (48 hours)	Rat brain	Decrease in elimination rate constant of NE
	Cameron and Smith 1980	Chronic (7 and 14 days)	Rat brain	7 days: increase in ^{3}H-NE uptake 14 days: ^{3}H-NE uptake at baseline or below
	Wolf et al. 1989	Chronic (7 days)	Chick embryo neurons	^{3}H-NE uptake increased
	Wilk et al. 1972	Chronic (3 weeks)	Human CSF	MHPG increased in two manic patients; decreased during drug treatment
	Shopsin et al. 1973	Chronic (several weeks)	Human CSF	MHPG increased in two manic patients; decreased during drug treatment
	Schildkraut 1974	Chronic (5 months)	Human urine	Decrease in urinary sum[a] excretion rate, when compared with manic period
	Beckmann et al. 1975	Chronic (3–4 weeks)	Human urine	Decrease in urinary MHPG excretion rate in clinical responders
	Linnoila et al. 1983	Chronic (6 weeks)	Human urine	Decrease in urinary excretion rates of MHPG, VMA, and sum[a]

continued

Table 5–1. Effects of psychotropic drugs on norepinephrine and its metabolites *(continued)*

Drug	Reference	Treatment duration	Tissue	Norepinephrine effects
	Swann et al. 1987	Chronic (18 days)	Human CSF and urine	Decrease in MHPG in CSF; decrease in urinary NE excretion rate
	Rudorfer et al. 1985	Chronic (7 days)	Human plasma and urine	No effect on plasma NE; no effect on urinary excretion rate of sum[a]
	Manji et al. 1991	Chronic (14 days)	Human plasma and urine	Increase in NE and NM urinary excretion rates
Stimulants				
Amphetamine	Bareggi et al. 1978a	Acute	Rat brain	Increase in MHPG-SO_4 immediately; then decrease in later MHPG-SO_4 accumulation following probenecid
	Calderini et al. 1975	Acute	Rat brain	Decrease in NE, with increase of MHPG-SO_4
	Stolk et al. 1974	Acute	Rat brain	Increase in labeled NM following labeled NE injection
	Roffman et al. 1978	Acute	Rat brain	Increase in MHPG-SO_4
	Cassens et al. 1979	Chronic (24 hours and 10 days)	Rat brain	24 hours: increase in MHPG-SO_4 10 days: no change in MHPG-SO_4
	Angrist et al. 1972	Chronic (several days)	Human CSF and urine	MHPG in CSF decreased during amphetamine; urinary MHPG excretion did not change consistently
	Schildkraut et al. 1971	Acute (amphetamine withdrawal)	Human urine	Urinary MHPG excretion increased during amphetamine use then decreased during withdrawal
	Fawcett et al. 1972	Chronic (3–4 days)	Human urine	Increase in urinary MHPG excretion rate during amphetamine in tricyclic antidepressant responders

	Beckmann et al. 1976	Chronic (2 days)	Human urine	Decrease in urinary MHPG excretion rate
	Shekim et al. 1977, 1979a, 1979b	Chronic	Human urine (2 weeks)	Decrease in urinary excretion rates of MHPG and NM
	Brown et al. 1981	Chronic (8 days)	Human urine	Decrease in urinary MHPG excretion rate
	Zametkin et al. 1985a, 1985b	Chronic (2 weeks)	Human urine and plasma	Decrease in urinary excretion rates of MHPG, VMA, NE, and NM; plasma NE and MHPG unchanged
	Elia et al. 1990	Chronic (3 weeks)	Human urine	Decrease in urinary excretion rates of MHPG, VMA, and sum[a]
Methylphenidate	Zametkin et al. 1985b	Chronic (2 weeks)	Human urine	Increase in urinary excretion rates of NE and NM
	Elia et al. 1990	Chronic (3 weeks)	Human urine and plasma	No change in urinary excretion rates; increase in plasma NE concentration
	Yu-cun and Yu-feng 1984	Chronic (3 weeks)	Human urine	Decrease in urinary MHPG excretion rate
Pemoline	Zametkin et al. 1986	Chronic (4 weeks)	Human urine	No change in urinary excretion rates of MHPG, NE, NM, or VMA
Neuroleptics				
Haloperidol	Berridge and Sharman 1973	Acute	Rat brain	Increase in MHPG-SO_4
	Keller et al. 1973	Acute	Rat brain	Increase in MHPG-SO_4
	Nose and Takemoto 1975	Acute	Mouse and rat brain	Increase in MHPG-SO_4
	Ader et al. 1980	Acute/chronic (15 days)	Rat brain	Acute: increase in MHPG-SO_4 Chronic: no change in MHPG-SO_4
	Singhal et al. 1983	Acute/chronic (21 days)	Rat brain	Acute: increase in MHPG-SO_4 Chronic: decrease in MHPG-SO_4
	Kaneko et al. 1992	Chronic (53 days)	Human plasma	Plasma MHPG decreased in schizophrenic patients compared with baseline

continued

Table 5–1. Effects of psychotropic drugs on norepinephrine and its metabolites *(continued)*

Drug	Reference	Treatment duration	Tissue	Norepinephrine effects
Clozapine	Berridge and Sharman 1973	Acute	Rat brain	Increase in MHPG-SO_4
	Keller et al. 1973	Acute	Rat brain	Increase in MHPG-SO_4
	Ader et al. 1980	Acute/chronic (15 days)	Rat brain	Acute: increase in MHPG-SO_4 Chronic: no change in MHPG-SO_4
Thioridazine	Keller et al. 1973	Acute	Rat brain	Increase in MHPG-SO_4
Chlorpromazine	Alfredsson et al. 1977	Acute	Rat brain	Increase in MHPG-SO_4
	Ader et al. 1980	Acute	Rat brain	Increase in MHPG-SO_4
	Singhal et al. 1983	Acute/chronic (21 days)	Rat brain	Acute: increase in MHPG-SO_4 Chronic: decrease in MHPG-SO_4
Pimozide	Keller et al. 1973	Acute	Rat brain	Slight increase in MHPG-SO_4
Fluphenazine	Ader et al. 1980	Acute	Rat brain	Slight increase in MHPG-SO_4
Remoxipride	Magnusson and Fowler 1989	Acute	Rat brain	Slight increase in total MHPG

Note. MHPG = 3-methoxy-4-hydroxyphenylglycol; SO_4 = sulfate; NM = normetanephrine; NE = norepinephrine; dopa = dihydroxyphenylalanine; DHPG = 3,4-dihydroxyphenylglycol; CSF = cerebrospinal fluid; VMA = vanillylmandelic acid; AUC = area under the curve; ^{3}H-NE = ^{3}H-labeled norepinephrine.

[a]The sum of NE plus its major metabolites (NE, NM, VMA, MHPG).

studied, and used as an index of overall NE activity. In humans the majority of MHPG present in the CSF is unconjugated, whereas in rats the majority of MHPG is present as the sulfate ester (Maas and Landis 1976; Schanberg et al. 1968; Sjöquist et al. 1975). Unconjugated MHPG can easily travel across the blood-brain barrier via passive diffusion; conjugated MHPG is transported via acid transport, which can be inhibited by probenecid.

Early investigators (Fawcett et al. 1972) hypothesized that MHPG could be used as a unique index of CNS NE activity because 25%–30% of urinary MHPG had its origin in the CNS, whereas most, if not all, of vanillylmandelic acid (VMA) or normetanephrine (NM) originated from metabolism of NE from peripheral pools. However, Blombery et al. (1980) and Mårdh et al. (1981) concluded that MHPG produced centrally or peripherally are equally likely to undergo further oxidation to VMA.

In an effort to obtain more direct information concerning the effects of psychotropic agents on the CNS NE system, many investigators have performed lumbar punctures and assayed the amount of MHPG in the CSF. However, in drug-free depressed patients, Linnoila et al. (1982c) and Roy et al. (1988) found that the 24-hour urinary excretion rates of NE, NM, MHPG, and VMA are highly intercorrelated. This suggests that the formation of all of the NE metabolites are dependent on overall NE synthesis and release. Since the NE system functions in a parallel fashion (Elam et al. 1984) in both the periphery and central compartments, CSF sampling may not offer any unique advantage over less invasive urine collections.

A measurement of NE, NM, MHPG, and VMA can account for almost 90% of NE; therefore, under steady-state conditions, a 24-hour collection of urine yields an overall estimate of total body NE turnover. Plasma or CSF measures reflect sympathetic activity at a single time point, whereas a 24-hour urine measurement allows an integrated measure that is less subject to any acute fluctuations in NE function. Urine collections also offer the advantage of requiring a less invasive methodology than do either plasma or CSF. Since many psychotropic drugs have the capacity to alter NE metabolism, measuring only one NE metabolite offers an incomplete picture of the effect of that agent on overall NE functioning. The alteration of the fraction of NE metabolized to different NE metabolites may offer additional information regarding the pharmacological actions of the drug.

Although the studies reviewed below have sought to determine the effects of psychotropic agents on the NE system, there are other factors that can serve as unwanted sources of variance. Interassay variability can often be assumed to be an important source of variance, especially in some of the earlier studies. Many of the older studies used less specific and sensitive methodology, whereas the more recent studies use methods such as high-performance liquid chromatography with electrochemical detection or mass fragmentography, which have resulted in great advances in analytical techniques. Other major sources of variance may be environmental factors such as diet or an insufficient drug-free period prior to obtaining baseline values. Muscettola et al. (1977) found that the low monoamine diet had a significant effect on urinary MHPG measurements in depressed patients but not in control subjects. Charney et al. (1982) found that patients who abruptly discontinued tricyclic antidepressants underwent a "rebound" withdrawal over a 2-week period of time, during which the MHPG excretion was transiently elevated.

The currently available studies in both animals and humans on the effects of different classes of psychotropic agents on the NE system are reviewed in this chapter.

Antidepressants

Since many antidepressants have direct effects on noradrenergic reuptake or metabolism, it is understandable that the majority of studies to determine the effects of psychotropic drugs on the NE system have been devoted to antidepressants. Both animal and human studies have been performed, using both acute and chronic dosing strategies.

Tricyclic Antidepressants

Animal studies have been conducted to determine the effects of single doses of desipramine, imipramine, amitriptyline, and nortriptyline on the NE system. Studies investigating desipramine or imipramine have generally found significant effects. In rats, following a single dose of desipramine, Roffman et al. (1977), Tang et al. (1978, 1979), and Bareggi et al. (1978b) found decreased concentrations of MHPG-SO_4 in brain tissue. Acute administration of desipramine also caused a significant increase in endogenous NM in specific areas of the rat brain, the hypothalamus and the cortex (Racagni et al. 1982). After giving rats a single dose of desipramine, Nielsen and Braestrup (1977) and Rosloff and Davis (1978) injected the rats with tracer doses of tritiated tyrosine and found a decrease in conversion to NE in the brain. Similar results have been noted following single doses of imipramine. Roffman et al. (1977) found that MHPG-SO_4 in rat brain was decreased when measured 6 hours following a single imipramine injection. Schildkraut et al. (1970) injected rats with a single dose of imipramine, followed 6 hours later by an intraventricular injection of tritiated NE. They found an increase in tritiated NE at 270 minutes, which they concluded was due to a decrease in NE clearance from the brain. In a more recent study, the effects of an acute dose of imipramine or desipramine on NE synthesis were determined in rats (Moret and Briley 1992). An aromatic acid decarboxylase inhibitor was injected 1 hour following a single dose of antidepressant. Animals were sacrificed 30 minutes later, and the dihydroxyphenylalanine (dopa) concentration was assayed in the frontoparietal cortex of dissected brains. Both imipramine and desipramine caused a decrease in dopa accumulation in comparison with control animals (30% and 45%, respectively), reflecting an acute decrease in the NE synthesis pathway. Single doses of amitriptyline and nortriptyline have not been associated with such dramatic effects on the NE system. Neither amitriptyline (Roffman et al. 1977; Tang et al. 1978) nor nortriptyline (Roffman et al. 1977) caused statistically significant decreases or increases in the endogenous brain concentrations of MHPG-SO_4 in rats. In mice (Schubert et al. 1970), a single dose of amitriptyline caused no reduction in the accumulation of ^{14}C-labeled NE, following a constant rate infusion of ^{14}C-labeled tyrosine, whereas both desipramine and nortriptyline caused significant reductions in ^{14}C-labeled NE accumulation. The authors concluded that this effect was due to a decrease in the synthesis and turnover of NE.

Although the above studies collectively suggest that an acute dose of desipramine or imipramine will cause a decrease in NE turnover, results following chronic antidepressant dosing are more contradictory. When rats were injected with tritiated tyrosine following 20 days of desipramine, the labeled DHPG-SO_4 (3,4-dihydroxyphenylglycol-sulfate), MHPG (free and conjugated), and NE concentrations were decreased 2 hours following the final desipramine dose (Nielsen and Braestrup 1977). Also, Racagni et al. (1982) found a decrease in NM brain content following 18 days of treatment with either mianserin or desipramine, in contrast to the increase they found using an acute dose. Karoum et al. (1984) administered desipramine to rats for 2 weeks and measured MHPG and NE content in different areas of the brain. Treatment resulted in a decrease in the hypothalamic MHPG concentration, but no change in NE content. Schildkraut et al. (1970) also found a lower concentration of endogenous NE following 3 weeks of treatment with imipramine. After injection of tritiated NE into the cisterna magna, these same authors found increased concentrations of labeled NM, but decreased deaminated *O*-methylated metabolites (MHPG and VMA). However, Roffman et al. (1977) found significant increases in the brain concentrations of endogenous MHPG-SO_4 following 2 weeks of imipramine ($P < .01$) or desipramine ($P < .001$), but no change following nortriptyline or amitriptyline. Also, 4 days of desipramine administration resulted in a decrease in the endogenous MHPG concentration measured in rat brain, but when desipramine was given for 21 days, an increase in total brain MHPG was found (Tang et al. 1978).

Although there are some discrepancies in these studies, both the acute and chronic animal studies suggest that tricyclic antidepressants are associated with an overall decrease in synthesis and metabolism of NE. Additionally, both acute (Nyback et al. 1975) and chronic (Svensson and Usdin 1978) imipramine are associated with a decrease in noradrenergic neuronal firing rate. Chronic treatment (11 days) with imipramine is associated with a significant decrease ($P < .001$) in the average neuronal firing rate in the locus coeruleus, probably secondary to feedback inhibition mediated by α_2-autoceptors (Svensson and Usdin 1978).

Most studies in humans have focused on CSF or urinary MHPG as a reflection of central NE activity, but several studies have collected urine, CSF, and plasma MHPG, or NE, VMA, and NM, in addition to MHPG, which allows a more comprehensive analysis. The effects of desipramine, imipramine, amitriptyline, nortriptyline, and clomipramine treatment on the NE system have been investigated, primarily in depressed patients.

Potter and colleagues published a series of reports comparing desipramine with the serotonin (5-HT) reuptake inhibitor zimelidine. Twelve depressed patients received zimelidine and desipramine in a double-blind crossover trial, following a 2- to- 4-week placebo period. Urine and CSF were obtained during both the placebo and the third or fourth week of drug treatment. Desipramine caused a significant reduction in both the CSF concentration ($P < .001$) (Potter et al. 1985) and urinary excretion rate of MHPG ($P < .01$) (Potter et al. 1981). Urine collected during the pretreatment and treatment periods was also assayed for NM, NE, and VMA, in addition to MHPG (Linnoila et al. 1982a; Potter et al. 1984). The NM excretion rate was not significantly affected, but the rates of NE ($P < .05$), MHPG ($P < .01$), and VMA ($P < .01$) were all significantly decreased by desipramine, as was the sum of NE plus its metabolites ($P < .01$).

Other investigators have also studied desipramine effects on NE metabolism. Dahl et al. (1982) compared the biochemical effects of desipramine with femoxetine (a specific 5-HT reuptake blocker) and found a slight, but nonsignificant, decrease in CSF MHPG in seven patients who reccived desipramine for 6 weeks. Donnelly et al. (1986) measured urinary excretion rates of MHPG, VMA, and NE in 17 hyperactive boys who received desipramine for a 2-week treatment period. In agreement with Linnoila et al. (1982a), they found significant decreases in the urinary excretion rates of NE ($P < .01$), MHPG ($P < .01$), and VMA ($P < .05$). They also measured supine and standing plasma NE, MHPG, and VMA and found a significant decrease in supine plasma MHPG ($P < .0001$), accompanied by a significant increase in standing NE ($P < .05$).

Imipramine treatment has also generally been associated with a decrease in NE turnover. In one of the earliest studies, Schildkraut et al. (1964) assayed 24-hour urine collections for VMA in 17 depressed patients, both before and following a minimum of 3 weeks of treatment with imipramine or phenelzine. They found a significant ($P < .05$) reduction in 24-hour VMA excretion in the six patients who received imipramine. Roccatagliata et al. (1981) found a significant decrease in CSF MHPG ($P < .05$) in seven depressed patients who received imipramine and exhibited a good clinical response, but a nonsignificant change in MHPG in the five patients who did not show a clinical response. However, this study included a small population, and patients were not placed on a low monoamine diet.

Several groups have compared biochemical response to imipramine and amitriptyline. Beckmann and Goodwin (1975) found significant decreases in urinary MHPG excretion ($P < .001$ in patients who received either drug) in depressed patients treated with 2–3 weeks of either imipramine ($n = 7$) or amitriptyline ($n = 8$). Bowden et al. (1985) analyzed results from 66 depressed patients who participated in a large collaborative study.

Gut data was not obtained from one of the subjects. The patients received either amitriptyline ($n = 34$) or imipramine ($n = 31$), and lumbar punctures were performed on day 11 of the drug-free baseline period and after 20 days of antidepressant treatment. The MHPG values were significantly lower ($P < .02$) during treatment in the patients who received either drug.

The effects of amitriptyline on the noradrenergic system have been investigated in three additional studies. Sacchetti et al. (1976) conducted their study in five depressed subjects who remained drug free for at least 4 weeks before treatment with amitriptyline. Daily 24-hour urine collections were conducted over a 20-day period; during the last 15 days all patients received amitriptyline. Two patients exhibited an overall decrease of MHPG; the other three showed an increase. This study was hampered by the small population, a short treatment period, and lack of dietary control. Modai et al. (1979) studied 15 elderly bipolar, depressed persons who received amitriptyline for a 5-week period. Urine collected prior to treatment (after a minimum 2-week drug-free period) and following 5 weeks was assayed for MHPG. In contrast to most other studies, these investigators found a significant increase ($P < .05$) in the urinary excretion of MHPG during amitriptyline treatment. In the third study (Robinson et al. 1983) plasma NE, epinephrine, and DHPG were measured in depressed patients who received either amitriptyline or phenelzine for 6 weeks. Although no significant changes occurred in plasma epinephrine or NE levels, a significant decline in plasma DHPG was noted during treatment with either amitriptyline or phenelzine ($P < .0001$), but the decrease was significantly greater in those treated with phenelzine.

The effects of clomipramine and nortriptyline on the NE system have also been evaluated. Bertilsson et al. (1974) compared CSF monoamine metabolite concentrations in 11 depressed patients who received clomipramine with 7 who received nortriptyline. Lumbar punctures were performed at two different time points: following an initial washout period of 4–7 days and following 3 weeks of antidepressant treatment. Both treatments were associated with a significant decrease in the CSF MHPG concentrations when compared with the pretreatment values. Both of these findings were replicated in further studies by these same investigators. Clomipramine caused a decrease in CSF MHPG concentrations in 14 depressed patients (Åsberg et al. 1977), in 27 depressed patients (Träskman et al. 1979), and in an additional group of 12 depressed inpatients

(Mårtensson et al. 1991). Thorén et al. (1980) compared clomipramine ($n = 8$) with nortriptyline ($n = 8$) in patients suffering from obsessive-compulsive disorder and found that a significant reduction in MHPG concentration in CSF was caused by both clomipramine ($P < .001$) and nortriptyline ($P < .01$) following 3 weeks of drug treatment.

It is not surprising that tricyclic antidepressants cause a decrease in MHPG values, since it has been known for some time that they block the uptake of NE into the neuron. Following tricyclic antidepressant treatment, less NE will be metabolized to MHPG since the majority of monoamine oxidase (MAO) is located intraneuronally. Available evidence suggests that NE is preferentially metabolized by catechol-*O*-methyltransferase to NM during tricyclic antidepressant treatment. Studies measuring NM, VMA, NE, and MHPG excretion rates show that the NM excretion rate is decreased less than the other metabolites. However, animal studies that show a decrease in NE neuronal firing in the locus coeruleus, and the human studies that reveal a decrease in total body NE turnover, indicate that an overall decrease in NE turnover occurs during treatment with tricyclic antidepressants.

Specific 5-HT Reuptake Inhibitors

It would be expected that drugs that primarily or exclusively block 5-HT, rather than NE, reuptake should affect 5-HT, but not NE, neurotransmission. Investigators have compared specific 5-HT reuptake inhibitors with specific NE reuptake inhibitors. Experiments conducted in rats (Moret and Briley 1992) have revealed that neither acutely nor chronically (21 days) administered citalopram affected dopa accumulation in the fronto-parietal cortex following aromatic acid decarboxylase inhibition. These results suggest that citalopram does not alter NE synthesis at the level of aromatic acid decarboxylase.

Although 5-HT reuptake inhibitors may not alter NE synthesis, they do appear to affect NE function. Siwers et al. (1977) studied six depressed patients who received zimelidine after a psychotropic drug-free period of at least 3 weeks. Both zimelidine and its metabolite, norzimelidine, are specific 5-HT reuptake inhibitors. Lumbar punctures conducted prior to and following at least 20 days of zimelidine treatment revealed that zimelidine caused a slight but statistically significant ($P < .05$) reduction in CSF

MHPG. This same group of investigators published a follow-up study (Bertilsson et al. 1980) in 12 depressed patients who received zimelidine following 4–7 days of placebo. Lumbar punctures were performed at the end of the placebo period and following 3 weeks of zimelidine treatment. The uptake of neurotransmitters in this group of patients was tested using plasma from the patients while they received zimelidine, which showed an inhibition of 5-HT reuptake (about 52% of baseline) but only a minor effect on NE reuptake (about 83% of baseline). Not surprisingly, the CSF concentration of 5-hydroxyindoleacetic acid (5-HIAA), the major 5-HT metabolite, was decreased by 29%; however, somewhat surprisingly, a significant decrease in the concentration of MHPG was also seen (11%, $P < .01$). Wålinder et al. (1981) also measured the CSF concentration of MHPG following 3 weeks of zimelidine treatment in 10 depressed patients and found a significant decrease ($P < .05$).

The four reports by Potter and colleagues (Linnoila et al. 1982a; Potter et al. 1981, 1984, 1985) comparing zimelidine with other antidepressants have been noted previously in this review. Potter et al. (1981) treated 12 depressed patients with both desipramine and zimelidine using a double-blind, placebo-controlled crossover design. CSF was obtained in 8 patients who participated in both arms of the crossover comparison during the third or fourth week of drug treatment. MHPG levels in the CSF decreased significantly during desipramine (33%, $P < .05$), and although the CSF MHPG concentration decreased during zimelidine treatment, the change was not statistically significant. However, the 24-hour urine collections revealed a significant decrease in the MHPG urinary excretion rate during both desipramine and zimelidine ($P < .01$ for both). A later report (Potter et al. 1985) noted a significant decrease in the CSF concentration of MHPG ($P < .02$) when all 12 of the patients who received zimelidine were included in the analysis. An in-depth analysis of the urinary excretion of NE, NM, and VMA, in addition to MHPG, in 11 of these patients was later published (Potter et al. 1984). Zimelidine caused a significant decrease in the excretion rates of all metabolites except NM, as well as a significant decrease ($P < .05$) in the sum of NE plus its metabolites. Unlike desipramine, zimelidine did not cause a significant increase in the NM/NE ratio or a decrease in the MHPG/NE ratio.

A study by Åberg-Wistedt et al. (1985) compared zimelidine with alaproclate, also a selective 5-HT reuptake inhibitor. Studies conducted in rats show that alaproclate differs from zimelidine by preferentially blocking 5-HT reuptake in the hippocampus and hypothalamus, and exhibiting a lower potency in other areas of the brain, whereas zimelidine shows equivalent 5-HT reuptake blocking potency in most areas of the brain. Fourteen depressed patients received zimelidine and 10 received alaproclate for 3 weeks; lumbar punctures were performed prior to and following treatment. In agreement with the previous studies, zimelidine caused a significant drop in the concentration of MHPG in the CSF ($P < .001$) as well as a large decrease in the 5-HIAA concentration, whereas alaproclate did not cause significant changes in either 5-HIAA or MHPG concentrations.

Studies have also been conducted using 5-HT reuptake inhibitors that are more specific than zimelidine, such as femoxetine, fluoxetine, and citalopram. Dahl et al. (1982) compared femoxetine with desipramine in a double-blind study of depressed outpatients. Patients received either medication for 6 weeks, and CSF samples were obtained prior to treatment and at the end of the treatment period. Although the MHPG concentration in CSF decreased in the 7 patients receiving desipramine and 10 patients receiving femoxetine, the difference was not statistically significant. However, the results of this study were probably affected by the small sample size and by the length of the drug-free period (which was unspecified). Bjerkenstedt et al. (1985) gave three different doses of citalopram (5 mg, 25 mg, and 50 mg) to 26 depressed patients, using a randomized, double-blind study design. Patients were free of antidepressant medication for at least 7 days prior to the initiation of the study. Medication dose was held constant for 4 weeks, but lumbar punctures were performed prior to and following 2 weeks of treatment. Even though the treatment duration was very short, an analysis of variance showed a significant decrease in the CSF concentration of MHPG ($P < .05$) during drug treatment. Mårtensson et al. (1989) studied changes in MHPG concentration in CSF following fluoxetine in 13 patients. Patients did not receive any psychotropic medication for 1 month prior to the study. Lumbar punctures were performed prior to treatment and following 3–6 weeks of treatment. A significant decrease in the MHPG concentration in CSF was noted using the analysis of variance for repeated measures ($P < .01$); MHPG was decreased at 3 weeks (36.5 ± 6.8 from baseline 41.0 ± 5.7 nmol/L) but even more significant was the decrease following 6 weeks of treatment (33.6 ± 4.2 nmol/L).

Investigators have, somewhat unexpectedly, found that even very specific 5-HT reuptake inhibitors have caused a significant decrease in MHPG concentrations in CSF or urinary MHPG excretion rate. Additionally, zimelidine causes a significant decrease in the total body NE turnover. Interestingly, unlike desipramine, zimelidine caused a significant increase in the VMA/NE, which seems to suggest that zimelidine causes an overall decrease in NE neuronal firing, rather than altering the metabolic pathway by blocking NE reuptake or degradation.

Monoamine Oxidase Inhibitors (MAOIs)

Much of the early evidence linking affective disorders to dysfunctions in the noradrenergic system was obtained by study of the MAOIs. As with tricyclic antidepressants, animal studies were conducted to determine the biochemical effects of MAOIs on the CNS noradrenergic system. Schildkraut (1970) measured tritiated VMA, DHPG, and MHPG following a single intracisternally administered injection of tritiated NE in rats that were pretreated with a single dose of tranylcypromine or pargyline. Following pargyline, tritiated NM was increased, and MHPG, DHPG, and VMA were decreased. A similar pattern was noted following tranylcypromine, accompanied by a decrease in tritiated NE uptake, suggesting an additional weak NE reuptake blocking effect. Yang and Neff (1974) injected rats with single doses of the enzyme-specific MAOIs, clorgyline (MAO-A) or deprenyl (MAO-B), then measured endogenous NE, dopamine, and 5-HT in brain tissue 2–4 hours later. Clorgyline caused significant increases in endogenous levels of all three neurotransmitters, but deprenyl increased only dopamine concentrations. Increases in NE following an acute dose of clorgyline may be responsible for the significant decrease (80%) in accumulation of dopa in rat frontoparietal cortex following aromatic acid decarboxylase inhibition, in comparison with the control group (Moret and Briley 1992). Braestrup et al. (1975) also injected rats with deprenyl or clorgyline 6 hours prior to the determination of endogenous homovanillic acid, 3,4-dihydroxyphenylacetic acid, and MHPG concentrations in brain tissue. Clorgyline caused a fall in the concentration of MHPG, but deprenyl caused a fall in MHPG only at high dose, confirming its MAO-B specificity.

Early studies in humans were performed using MAO-nonspecific enzyme inhibitors. One of the earliest investigations of MAOI effects on NE metabolism was conducted by Sjoerdsma et al. (1959). Following 5–10 days of treatment with an investigational MAOI, pheniprazine (1-phenyl-2-hydrazinopropane), the investigators found no change in the urinary excretion of NE, which suggested to the authors the presence of an efficient alternate metabolic pathway for NE. Studnitz (1959) gave iproniazid to six healthy humans for 3–4 days and found a significant decrease in urinary excretion rate of VMA. Schildkraut et al. (1964) also found that 3 weeks of treatment with phenelzine caused a significant decrease in the 24-hour urinary excretion of VMA in six hospitalized depressed patients. Beckmann and Murphy (1977) treated 12 depressed patients with 60 mg/day of phenelzine for a 4-week period. They found a significant decrease ($P < .001$) in the urinary excretion rate of MHPG, which did not differ between responders and nonresponders. Thirty-six depressed inpatients who underwent a mean 15.6-day psychotropic drug-free period were placed on a low monoamine diet, and 24-hour urine samples were obtained prior to and following 4 weeks of treatment with phenelzine (Sharma et al. 1990). The urinary MHPG decreased significantly ($P = .008$) during phenelzine treatment, but the decreases in urinary MHPG were not significantly correlated with either the pretreatment urinary MHPG or the decrease in Hamilton Scale for Depression scores. Plasma NE, epinephrine, and DHPG concentrations were sampled in depressed patients following 6 weeks of phenelzine (Robinson et al. 1983). The authors found significant decreases in both DHPG and NE, but not epinephrine. There was a weak, but significant, correlation between the percentage of platelet MAO inhibition and the percentage of decrease in DHPG at 4 weeks ($r = 0.54$, $P < .005$) and at 6 weeks ($r = 0.37$, $P < .05$).

Researchers at the National Institute of Mental Health have conducted a series of studies investigating the effects of MAO-A or MAO-B specific inhibitors on the NE system. Either clorgyline or pargyline were administered to 18 depressed patients for a 4-week period, and changes in the NE metabolite profile in both CSF and plasma were reported by Murphy et al. (1979). The MHPG concentration in CSF was significantly decreased for both drugs ($P < .01$) but VMA and NE concentrations were not altered by either drug. Additionally, clorgyline, but not pargyline, caused a significant decrease in the standing ($P < .01$) and resting ($P < .05$) plasma NE concentrations. Pickar et al. (1981) administered clorgyline (30 mg/day), pargyline (30 mg/day), and deprenyl (MAO-B inhibitor, 10 mg/day) to depressed patients for a 4-week treatment period. Both clorgyline and pargyline produced significant reductions in plasma MHPG concentration ($P < .005$). Deprenyl also caused a slight decrease in plasma MHPG (17%), but the change was not significant, probably due to the small dose, since at the higher doses deprenyl will lose its MAO-B selectivity. In the final study, 11 depressed patients received deprenyl for 3 weeks at varying doses (Sunderland et al. 1985), which resulted in a dose-dependent decrease in plasma MHPG during deprenyl treatment (25% with 10 mg/day, 43% with 30 mg/day, and 68% with 60 mg/day).

Additional studies have been conducted using MAO-A inhibitors. Following 4–6 weeks of clorgyline treatment in four depressed bipolar patients, Linnoila et al. (1982b) found a significant decrease in MHPG and VMA urinary excretion rates, accompanied by an increase in NM excretion rate. The 24-hour excretion rate of the sum of NE and its metabolites was also decreased by 55%. Potter et al. (1984, 1985) included the four patients from the above study in two later reports comparing the biochemical effects of clorgyline with those of zimelidine and desipramine. Further analysis of the urinary excretion rates (Potter et al. 1984) revealed a significant decrease in the sum of the excretion rates of NE and its metabolites ($P < .01$) during clorgyline when compared with placebo. The four patients in the earlier reports were also combined with six additional patients in a report comparing zimelidine or desipramine with clorgyline (Potter et al. 1985). Lumbar punctures were performed prior to treatment and following 5–8 weeks of clorgyline therapy, and CSF was assayed for MHPG. Clorgyline caused a very significant decrease in the CSF concentration of MHPG ($P < .001$), more pronounced than with either zimelidine or desipramine. In a study that included 14 hyperactive boys, Zametkin et al. (1985a) compared MAOIs (clorgyline or tranylcypromine) with *d*-amphetamine using a randomized, double-blind, crossover design. Drug treatment followed a 2-week placebo period and lasted for 4 weeks, followed by a 2-week placebo washout before crossover to the alternate drug treatment. Plasma NE and MHPG concentrations and urinary excretion rates of MHPG, NE, NM, and VMA were obtained prior to treatment and at the end of each treatment period. The values obtained during tranylcypromine and clorgyline treatment were combined for analysis. During MAOI treatment, there were significant reductions in excretion rates of urinary MHPG and VMA ($P < .001$); the NM excretion rate showed a significant increase ($P < .01$). Additionally, plasma MHPG ($P < .001$) and NE ($P < .05$) concentrations were significantly decreased. In agreement with these results, Ross et al. (1985) found a significantly lower supine plasma NE concentration in 10 patients who had received 6–10 weeks of clorgyline ($P < .005$).

Dajas et al. (1984) treated 26 depressed patients with moclobemide, a reversible MAO-A inhibitor, following a 5- to 7-day washout period. Patients were not allowed bananas, cocoa, or coffee on the morning prior to the urine collection. During pretreatment and following 7 and 14 days of treatment, a 1-hour urine collection was obtained after the patient had remained supine for 60 minutes. Urinary NE excretion decreased significantly (*P* value not reported) in patients following 14 days of treatment with moclobemide, whether or not they responded to therapy. This study suffers from several methodological flaws: 1) a 1-hour urine collection is likely to reflect changes in renal clearance or circadian variation in NE system activity, whereas a 24-hour collection would adjust for these variations; and 2) both the drug-free period and the treatment periods were very short. A more recent study (Berlin et al. 1990) examined DHPG plasma concentrations and urinary NM following 1 week of treatment with moclobemide and toloxatone (another reversible MAO-A inhibitor) in 12 healthy volunteers. On the eighth treatment day, blood samples were drawn every 2 hours, and urine was collected over 24 hours. Both drugs caused significant decreases in the DHPG area under the plasma concentration versus time curve, although the decrease was much larger with moclobemide. Additionally, moclobemide, but not toloxatone, significantly increased urinary NM ($P < .001$).

Together, these studies indicate that nonspecific MAOIs change the urinary excretion rates of NE and its metabolites. As expected from agents that inhibit the deamination pathway for NE, the aminated to deaminated metabolite ratio (NM/[MHPG + VMA]) increased. Nonreversible MAO-A inhibitors exhibit a profile that is very similar to the nonspecific inhibitors. The MAO-B inhibitors pargyline and deprenyl also show a dose-dependent pattern of change in NE metabolism reminiscent of the nonspecific inhibitors. In addition to alterating in NE metabolic pathways, reversible MAO-A inhibitors also appear to cause a decrease in total body NE turnover.

Atypical Antidepressants

The effects of buproprion on the NE system have been extensively studied by one group of researchers. Buproprion differs from other antidepressants both structurally and pharmacologically; it has weak dopamine-reuptake blocking activity but it does not block reuptake of either NE of 5-HT, nor does it have significant effects on the histaminergic or cholinergic neurotransmitter systems. Ten depressed patients received a low monoamine diet and a 3-week placebo washout period, followed by buproprion administration (Golden et al. 1988). Plasma and CSF samples and 24-hour urine collections were obtained from the patients, prior to treatment and following 4–6 weeks of drug treatment. Plasma NE was obtained during the supine position and after standing for 5 minutes. Unlike desipramine (Ross et al. 1983) or phenelzine (Robinson et al. 1983), buproprion caused no change in plasma NE during this orthostatic challenge. The total body NE turnover (sum of the excretion rates of NE plus its metabolites) decreased significantly ($P < .05$) during treatment, but there was no change in the NE/(MHPG + VMA) ratio. Although the CSF concentration of MHPG decreased during buproprion (89% of baseline), the change was not statistically significant, possibly due to the small sample size ($n = 6$).

Summary

Although not all studies agree, the weight of evidence strongly suggests that all of the antidepressant agents reviewed above cause a decrease in noradrenergic activity. Animal studies reveal a decrease in MHPG formation, a decrease in labeled tyrosine incorporation into NE, and a decrease in the neuronal firing rate of the locus coeruleus, following both single and long-term dosing. Studies conducted in humans show, almost universally, that chronic dosing of these antidepressants results in a decrease in the CSF concentration of MHPG, and even more significantly a decrease in total body NE turnover. Despite their differing pharmacological actions, these antidepressants cause a decrease in total body turnover of NE, even though some of the agents have no direct pharmacological action on the NE system.

Lithium

Subsequent to observations that antidepressants appeared to alter the function of the noradrenergic system, it was hypothesized that lithium, which has efficacy in the treatment of both mania and depression, might alter NE neurotransmission. Schildkraut and colleagues conducted two studies in rats to determine both the acute and the longer-term effects of lithium on the CNS NE system. In the first study (Schildkraut et al. 1966), tritiated NE was injected into the cisterna magna, followed by lithium chloride at 1, 2, and 3 hours. The animals receiving lithium showed no difference in overall brain concentration of NE; however, they did exhibit higher brain concentrations of tritiated deaminated catechols (3,4-dihydroxyphenylglycol and 3,4-dihydroxymandelic acid, $P < .0001$) and lower NM ($P < .02$) than control animals. In the second study, Schildkraut et al. (1969) again administered a single dose of lithium, and subsequently found increased tritiated deaminated catechols ($P < .05$), slightly (but nonsignificantly) decreased tritiated NM, and significantly decreased tritiated NE ($P < .05$). An additional group of animals received lithium for 7 days, followed by an injection of tritiated NE given 3.5 hours following the final lithium dose. In lithium-treated animals that were sacrificed 150 minutes following the labeled NE dose, the levels of tritiated NE were significantly decreased ($P < .01$), as were those of tritiated NM and deaminated *O*-methylated metabolites (MHPG and VMA). The authors hypothesized that NE deamination may be increased in both acute and chronic lithium treatment, resulting in lower NM concentrations following both acute and chronic treatment. However, this does not explain the apparent decrease in overall NE turnover during chronic lithium treatment. Alternatively, acid transport inhibition would result in increases in deaminated catechols following acute treatment, although it would not explain the decrease in NE concentrations.

Stern et al. (1969) administered lithium to rats for a 48-hour period prior to the injection of either tritiated NE or a tyrosine hydroxylase inhibitor (α-methyl tyrosine) and found that lithium increased the elimination rate constant of NE from the rat brain. The authors concluded that lithium caused an increase in the NE turnover rate, which contrasted with the results of Schildkraut et al. (1966), who concluded that acute lithium did not change the NE turnover rate. Cameron and Smith (1980) investigated both short- and long-term effects of lithium on ^{3}H-labeled NE uptake in slices from rat brain. Seven days of lithium treatment resulted in a significant increase in NE uptake in all brain regions (brain stem, hypothalamus, parietal cortex, and caudate nucleus). However, following 14 days of treatment,

^{3}H-NE uptake decreased to control (or below) levels. Another study, conducted in chick embryo neurons, also found that ^{3}H-NE uptake significantly increased ($P < .001$) following 7 days of treatment with lithium (Wolf et al. 1989). The effects of a longer period of lithium treatment were not investigated in this study. Thus, although the animal studies do not yield a consensus of opinion concerning the effects of lithium on the NE system, it appears that lithium may acutely increase NE uptake and turnover, but this effect may be attenuated over time.

The earliest evaluations in humans of the effects of lithium on the NE system were case reports. Wilk et al. (1972) found that elevated CSF concentrations of MHPG in two manic patients decreased into the normal range following 3 weeks of lithium treatment. Shopsin et al. (1973) found that elevated MHPG concentrations in the CSF in two manic patients decreased to a more normal range during treatment with lithium and that the decrease in CSF MHPG level appeared to correlate with clinical improvement. Schildkraut (1974) reported evidence of an increase in NE turnover in a single manic patient who exhibited a transient increase in urinary MHPG excretion over the first 10 days of lithium, followed by a gradual decrease over the next 5 months. The urinary excretion of the sum of NE and its metabolites (NE, NM, MHPG, and VMA) decreased significantly ($P < .05$) when compared with the first week of lithium treatment. Beckmann et al. (1975) conducted the first actual study of lithium effects on the NE system. In 10 hospitalized patients, 24-hour urine collections for MHPG were conducted prior to and following 3–4 weeks of lithium treatment. The urinary MHPG excretion rate increased in some of the patients and decreased in others, although there was a nonsignificant trend toward a decrease in the urinary excretion rate of MHPG with lithium in nonresponders and an increase in responders. The authors concluded that the change in MHPG excretion in responders was due to a change in state rather than a biochemical effect of lithium treatment.

Two exhaustive evaluations of the effects of lithium on the NE system presented evidence of a lithium-induced decrease in NE turnover. Linnoila et al. (1983) reported the results of a double-blind study containing eight depressed patients, five of whom received lithium. All patients were placed on a low monoamine diet and, following a 5- to 8-week placebo period, they received lithium carbonate for 6 weeks. During the last 2 weeks of the placebo period and during the fifth or sixth week of lithium treatment, 24-hour urine collections were obtained. Lithium treatment caused decreases in the urinary excretion rates of MHPG ($P < .05$), VMA ($P < .001$), and the sum of NE plus its major metabolites ($P < .001$). The findings from 14 patients who received lithium, following a 15-day washout period, were reported by Swann et al (1987). Nine patients who were manic during pretreatment exhibited elevated MHPG concentrations in the CSF, which decreased during lithium treatment. CSF and 24-hour urine collections were obtained prior to treatment and following 18 days of lithium carbonate treatment. During lithium treatment, the CSF MHPG concentration in patients who had been manic decreased and was no longer elevated in comparison with healthy or depressed subjects. However, significant decreases in CSF concentrations of MHPG occurred in both the lithium responders and non-responders ($P < .05$, overall), suggesting a biochemical effect related to lithium treatment rather than a state effect. These investigators also found that lithium caused a significant decrease in the urinary NE excretion rate ($P < .05$), but no change in any of the other NE metabolite excretion rates (MHPG, VMA, and NM), although the ratio of NE and total NE metabolites decreased during lithium treatment.

Rudorfer et al. (1985) administered lithium to 12 healthy volunteers for 7 days and found no change in plasma NE concentration following an orthostatic challenge during lithium treatment. Nor did lithium result in a significant decrease in the sum NE and its metabolites in the urine, although there was a slight decrease. The short duration of this study probably affected the results. In a follow-up study (Manji et al. 1991), 11 healthy volunteers received 2 weeks of lithium treatment. Following 1 week of lithium, the 24-hour urinary outputs of NM and NE did not show any significant changes from baseline. However, at 2 weeks both of these measures were significantly increased ($P < .05$ for both) when compared with baseline. Additionally, plasma NE showed a trend toward an increase after 2 weeks of treatment. The discrepancies between the results of these two studies and the findings of others might be attributed to the use of healthy volunteers rather than depressed or manic patients.

Unfortunately, the results of studies investigating the effects of lithium on the NE system do not yield a consensus of opinion. The authors of studies conducted with rats have concluded that lithium may cause an increase of NE uptake, followed by a decrease to control (or below) levels. The findings from human studies are also contradictory. Beckmann et al. (1975)

found a tendency for urinary MHPG excretion to increase in lithium responders and concluded that this change was due to a change in state rather than the direct pharmacological action of lithium. However, Swann et al. (1987) found that the concentration of MHPG in CSF decreased regardless of whether the patient responded to lithium, and they concluded that a reduction in CSF MHPG concentration was a biochemical, rather than state, effect. Overall, the more recent studies in depressed or manic patients seem to indicate that lithium treatment is associated with a decrease in total body NE turnover. Although this might be a biochemical effect, the results of studies conducted by Rudorfer et al. (1985) and Manji et al. (1991) in healthy volunteers argue against this. It is possible that lithium effects on NE are state dependent, decreasing NE release in manic or depressed patients, but increasing NE release in healthy persons.

Stimulants

Numerous studies of the effects of amphetamines (especially *d*-amphetamine) on the NE system have been published. The findings from investigations in rats have been generally consistent. Single doses of amphetamine have usually resulted in transient increases, followed by slight decreases in the MHPG concentration in the brain. Bareggi et al. (1978a) found an increase ($P < .01$) in MHPG-SO_4, accompanied by a slight decrease in NE concentration ($P < .01$), in rat brain at 2 hours following a single 5 mg/kg amphetamine injection. However, 4 hours following this dose of amphetamine, there was a decrease ($P < .01$) in the accumulation of MHPG-SO_4 in rats that were pretreated with probenecid. Calderini et al. (1975) administered acute doses of *d*-amphetamine or *d*- or *l*-fenfluramine to rats and found a decrease in the brain concentration of NE following the 7.5 and 15 mg/kg doses of *d*-amphetamine at 60 and 240 minutes, and at 120 minutes following a 3.75 mg/kg dose. This was accompanied by an increase in MHPG-SO_4 at 120 minutes following the 15 mg/kg dose. *d*-Fenfluramine was less potent than *l*-fenfluramine, and *l*-fenfluramine caused an increase in brain NE concentration at 1 and 2 hours following the injection of a high dose (15 mg/kg). This NE increase was accompanied by an increase in MHPG-SO_4 at 60 and 120 minutes. Stolk et al. (1974) found that *d*-amphetamine caused an increase in tritiated NM 2 hours following injection of tritiated NE, but no change in the deaminated-*O*-methylated, or deaminated catecholamine fractions. Roffman et al. (1978) studied specific areas of the brain following both *d*- and *l*-amphetamine and found an increase ($P <$ .005) in the whole brain MHPG-SO_4, especially prominent in the hypothalamus, 2 hours following 2.5 and 5.0 mg/kg doses of *d*-amphetamine administration. Following 24 hours of amphetamine during which their patients were given escalating doses, Cassens et al. (1979) also noted an acute, but nonsignificant, increase in MHPG-SO_4 measured at 2 hours following the final amphetamine dose. The MHPG-SO_4 then decreased below the control concentration ($P < .05$) at 48 hours, but returned to baseline at 72 hours following the final dose of amphetamine. However, the investigators found no change in MHPG-SO_4 brain concentrations in rats who received a fixed dose (5 mg/kg, twice daily) over a 10-day period.

An early investigation in four nonschizophrenic amphetamine-abusing persons was conducted by Angrist et al. (1972). Following a 10-day, drug-free period, hourly doses of racemic amphetamine were given. Lumbar punctures were performed before and near the end of the experiment, and urine collections were obtained during the period of amphetamine ingestion. Although urinary VMA excretion did not change, MHPG excretion increased in one subject, did not change in another subject, and decreased in the last subject (urine was lost in the fourth subject). The concentration of MHPG in CSF decreased during amphetamine in three of four subjects. Schildkraut et al. (1971) studied four amphetamine abusers during withdrawal. Urinary MHPG excretion increased during amphetamine administration, then declined during withdrawal. The lowest point was reached between 24 to 60 hours following amphetamine discontinuation; thereafter MHPG excretion returned to the normal range.

Fawcett et al. (1972) used a *d*-amphetamine challenge to study the relationship between the urinary MHPG response to amphetamine and clinical response to tricyclic antidepressants. They treated 12 patients with one or two, 3- to 4-day trials of *d*-amphetamine, followed by imipramine or desipramine, and found that the patients who responded to tricyclic antidepressant treatment had an increase in urinary excretion of MHPG during the *d*-amphetamine challenge. Beckmann et al. (1976) also investigated the interactions between amphetamine and lithium with catecholamine metabolism in eight depressed patients who had remained drug free for 2 weeks. Lithium was administered for 10 days; *d*- and *l*-amphetamine were given for 2 days each; and, finally, amphetamine plus lithium were given (2 days each, *l*- and *d*-isomers). Throughout the study, 24-hour urine collections for MHPG were obtained. When *d*-amphetamine was administered alone, the urinary MHPG decreased ($P < .05$), but *l*-amphetamine had no effect. When lithium and *d*-amphetamine were given concomitantly, the expected decrease in urinary MHPG was attenuated.

The majority of studies evaluating the effects of stimulants on the NE system have been performed in children with attention deficit hyperactivity disorder. Shekim et al. conducted three studies of *d*-amphetamine effects in hyperactive children. In the first study (Shekim et al. 1977), 7 hyperactive and 12 control children underwent a 2-week drug-free period prior to treatment, then the hyperactive children received 2 weeks of *d*-amphetamine treatment (0.5 mg/kg/day). Urine was collected before treatment and after 2 weeks of *d*-amphetamine treatment. During *d*-amphetamine administration, the urinary excretion rates of MHPG and NM decreased significantly ($P < .01$). In the second two studies, 15 hyperactive children (Shekim et al. 1979b) and 23 hyperactive children (Shekim et al. 1979a) underwent a 2-week drug-free period, followed by a 2-week period of *d*-amphetamine administration. In both studies, the investigators noted a significant decrease in the urinary excretion rate of MHPG, which was more pronounced in those who exhibited a clinical response.

A study conducted by Brown et al. (1981) confirmed the above results: 24-hour urine collections were obtained from eight hyperactive boys during placebo treatment and following 3 and 8 days of *d*-amphetamine treatment (0.74 ± 0.98 mg/kg/day). The urinary MHPG excretion rate decreased significantly, when compared with pretreatment values, by the eighth day collection ($P < .05$).

Researchers at the National Institute of Mental Health have published a series of studies comparing the effects of *d*-amphetamine with those of methylphenidate, pemoline, or MAOIs. In the first study (Zametkin et al. 1985b), the urinary excretion rates of NE and metabolites in 10 hyperactive boys who received methylphenidate (0.8 ± 0.12 mg/kg/day) were compared with those in another group of 12 hyperactive boys who received *d*-amphetamine for 2 weeks. *d*-Amphetamine caused a slight (but not statistically significant) decrease in urinary NE excretion, as well as a significant decrease in MHPG ($P < .05$) excretion, whereas during methylphenidate treatment, the NE and NM excretion rates both increased significantly ($P < .02$). Methylphenidate also caused a slight (but not significant) increase in the MHPG excretion rate. Neither drug caused a significant change in the urinary excretion of the sum of NE plus its metabolites.

In the second study (Zametkin et al. 1985a), *d*-amphetamine caused significant decreases in MHPG, VMA, NE, and NM urinary excretion rates; the MAOIs caused significant decreases in urinary MHPG and VMA, but an increase in urinary NM excretion rates. Plasma NE and MHPG were not altered following 4 weeks of *d*-amphetamine. In contrast to *d*-amphetamine, MAOI therapy was associated with decreases in plasma NE ($P < .05$) and MHPG ($P < .001$) concentrations. Zametkin et al. (1986) hypothesized that a drug-induced change in NE urinary excretion should accompany treatment with any drug that causes an improvement in attention deficit disorder. However, results following 4 weeks of pemoline treatment (2.9 mg/kg/day) in 11 hyperactive boys did not agree with this hypothesis. A significant clinical improvement was found, but no significant changes in the urinary excretion rates of any of the NE metabolites (NE, NM, MHPG, VMA) were found. Although the MHPG and NM excretion rates decreased 12.9% and 45.0%, respectively, the sample population was too small for these differences to achieve statistical significance.

In the most recent study, Elia et al. (1990) treated 31 hyperactive boys with *d*-amphetamine and methylphenidate using a double-blind, crossover design. Patients received both drugs for a 3-week period, with a 3-week washout separating the two treatment periods. *d*-Amphetamine caused a decrease in the urinary excretion rates of MHPG ($P < .0001$), VMA ($P < .0025$), and the sum of NE and its metabolites ($P < .0005$). Methylphenidate caused no change in any of the urinary excretion rates, but it did cause an increase in the plasma NE concentration.

Unlike Elia et al. (1990), Yu-cun and Yu-feng (1984) found that 3 weeks of methylphenidate resulted in a significant decrease in urinary MHPG-SO_4 in 38 children with minimal brain dysfunction ($P < .05$). They also found that the children who showed a better clinical response exhibited larger decreases in the urinary MHPG-SO_4 excretion rate. In human urine, only 6% of MHPG is eliminated in the unconjugated form, so an increase in the excretion rate of MHPG-SO_4 is analogous to an increase in the overall excretion of MHPG.

The effects of amphetamine on NE turnover are compatible with its pharmacological actions. Single doses in animals generally cause increases in MHPG and NM in the brain, accompanied by a decrease in NE concentrations. The NE-releasing properties possessed by amphetamine are compatible with these findings. In humans, short-term dosing of amphetamine has been associated with an increase, a decrease, or no change in MHPG urinary excretion. However, chronic amphetamine administration is associated with a decrease in urinary excretion of NE metabolites. Methylphenidate, conversely, appeared to increase the NE, NM, and MHPG urinary excretion rates in one study, whereas it caused a significant decrease in MHPG excretion in another study. Pemoline had no significant effects on any of the NE metabolites measured in one small study. Although the effects of amphetamine on the NE system have been extensively studied, the results of studies with the other stimulants—methylphenidate and pemoline—indicate that more research is needed before any definitive statements can be made concerning their effects on the NE system.

Neuroleptics

Most of the studies determining the effects of neuroleptics on the NE system have been conducted in animals. Studies in both rats and mice have found increases in the total brain MHPG-SO_4 concentration following single doses of haloperidol (Ader et al. 1980; Berridge and Sharman 1973; Keller et al. 1973; Nose and Takemoto 1975; Singhal et al. 1983), clozapine (Ader et al. 1980; Berridge and Sharman 1973; Keller et al. 1973), thioridazine (Keller et al. 1973), and chlorpromazine (Ader et al. 1980; Alfredsson et al. 1977; Singhal et al. 1983). However, pimozide (Keller et al. 1973), fluphenazine (Ader et al. 1980), and remoxipride (Magnusson and Fowler 1989) caused only slight increases in brain MHPG concentration, and chlorpromazine metabolites did not affect brain MHPG concentrations (Alfredsson et al. 1977). Many investigators have hypothesized that the increase in brain MHPG is due to an α_2 autoceptor blockade. If this is the case, then cyclic adenosine monophosphate formation should be increased, but Ader et al. found that the formation of cyclic adenosine monophosphate was not altered by neuroleptics. Long-term neuroleptic administration appears to result in an alteration of effects on the NE system that are seen following acute administration. Following 15 days of daily clozapine or haloperidol dosing, Ader et al. found that the forebrain MHPG concentration in rats returned to normal. Following 21 days of treatment with either haloperidol or chlorpromazine, Singhal et al. found significant decreases in MHPG-SO_4 in rat cerebral cortex. In a single study in humans, 19 schizophrenic patients exhibited decreased plasma MHPG concentrations from baseline following an average of 53 days of treatment with haloperidol (Kaneko et al. 1992). Unfortunately, it is not possible to determine whether this effect was due to the medication or to an improvement in schizophrenic symptoms. At baseline the patients in this study exhibited elevated MHPG plasma concentrations when compared with control subjects.

Summary

The data presented in this chapter suggest that drugs primarily affecting mood alter the functioning of the noradrenergic system. Although the effects of lithium are still inconclusive, all other agents that possess antidepressant effects (MAOIs, tricyclic antidepressants, atypical antidepressants) cause an overall decrease in the total body NE turnover. This may be a specific property of drugs having antidepressant efficacy, since individually these drugs effectively treat other disorders, but collectively they are only all effective in the treatment of depression. Although this finding certainly does not prove that a dysfunction of the NE system is the cause of depression, it does imply that alterations in NE functioning may be associated with the improvement of a depressive episode. It has become apparent that neurotransmitter systems interact extensively such that, under normal circumstances, a homeostasis is maintained. It is also apparent that since central neurotransmission is integrated, it is virtually impossible to perturb one neurotransmitter system without affecting others. As an example, the 5-HT selective reuptake inhibitors decrease total body NE turnover without directly affecting the reuptake, metabolism, or synthesis of NE. Psychotropic drugs that are effective in the treatment of depression exhibit a variety of pharmacological effects; collectively, these actions may amount to a kind of "reset" button.

When evaluating the effects of any psychotropic agent on NE function, considerable attention should be devoted to the duration of treatment, since acute treatment effects often differ from chronic treatment effects. The specific study parameter chosen as a measure of NE function is also of importance; an acute single time-point measurement may reflect variability due to such factors as stress, diet, or activity. This feature can be utilized to reveal the relative capacity of a system, such as has been done with various challenge paradigms. Determinations of NE or NM are thought to reflect presynaptic release of NE, whereas measurements of the ratios of NE or NM to MHPG or VMA will reflect alterations of neuronal reuptake or degradation of NE. If information regarding the overall tone of the NE system is desired, then the most useful measurement would be an integrated measurement over time, such as the 24-hour urine collection for NE and its metabolites. Overall sympathetic tone is a reflection of the steady state or equilibrium situation; therefore, it is not surprising that the concentrations of NE and its metabolites show a strong correlation when they are measured in urine, CSF, or plasma. Consequently, any of these body fluids could be used to determine overall sympathetic tone, but 24-hour urine collections are less invasive and are less likely to reflect random variability. When the effects of an acute perturbation are sought, such as in the case of a challenge paradigm, a body fluid that is capable of reflecting acute changes in the NE system (e.g., plasma) would probably be the most appropriate choice.

The interaction that occurs between biochemical effects and disease state is another important consideration. It would appear that the actions of lithium on the NE system may vary depending on whether the population being studied is healthy, manic, or depressed. It is quite possible that other psychotropic agents may also exhibit state-dependent biochemical effects.

Unfortunately, since many of these studies were conducted 15 or more years ago, interpretation of the results is confounded by the use of older and, in many cases, less accurate techniques. Although a considerable amount of data has been obtained from studies in animals and humans, currently all our information concerning the effects of drugs on the CNS NE system is obtained indirectly. In the future, newer imaging techniques may eventually allow a closer and more direct examination of the CNS effects.

References

Åberg-Wistedt A, Alvariza M, Bertilsson L, et al: Alaproclate a novel antidepressant? a biochemical and clinical comparison with zimelidine. Acta Psychiatr Scand 71:256–268, 1985

Ader J-P, Sebens JB, Korf J: Central levels of noradrenaline, 3-methoxy-4-hydroxyphenylethyleneglycol and cyclic AMP in the rat after activation of the locus coeruleus neurons: influence of single and repeated neuroleptic treatment. Psychopharmacology (Berl) 70:239–245, 1980

Alfredsson G, Wiesel F-A, Skett P: Levels of chlorpromazline and its active metabolites in rat brain and the relationship to central monoamine metabolism and prolactin secretion. Psychopharmacology (Berl) 53:13–18, 1977

Angrist B, Shopsin B, Gershon S: Metabolites of monoamines in urine and cerebrospinal fluid, after large dose amphetamine administration. Psychopharmacologia 26:1–9, 1972

Åsberg M, Ringberger V, Sjöquist F, et al: Monoamine metabolites in cerebrospinal fluid and serotonin uptake during treatment with chlorimipramine. Clin Pharmacol Ther 21:201–207, 1977

Bareggi SR, Markey K, Paoletti R: Effects of amphetamine, electrical stimulation and stress on endogenous MOPEG-SO_4 levels in rat brain. Pharmacological Research Communications 10:65–73, 1978a

Bareggi SR, Markey K, Genovese E: Effects of single and multiple doses of desipramine (DMI) on endogenous levels of 3-methoxy-4-hydroxyphenylglycol-sulfate (MOPEG-SO_4) in rat brain. Eur J Pharmacol 50:301–306, 1978b

Beckmann H, Goodwin FK: Antidepressant response to tricyclics and urinary MHPG in unipolar patients, clinical response to imipramine or amitriptyline. Arch Gen Psychiatry 32:17–21, 1975

Beckmann H, Murphy DL: Phenelzine in depressed patients, effects on urinary MHPG excretion in relation to clinical response. Neuropsychobiology 3:49–55, 1977

Beckmann H, St Laurent J, Goodwin FK: The effect of lithium on urinary MHPG in unipolar and bipolar depressed patients. Psychopharmacologia 42:277–282, 1975

Beckmann H, van Kammen DP, Goodwin FK, et al: Urinary excretion of 3-methoxy-4-hydroxyphenylglycol in depressed patients: modifications by amphetamine and lithium. Biol Psychiatry 11:377–387, 1976

Berlin I, Zimmer R, Thiede H-M, et al: Comparison of the monoamine oxidase inhibiting properties of two reversible and selective monoamine oxidase-A inhibitors moclobemide and toloxatone, and assessment of their effect on psychometric performance in healthy subjects. Br J Clin Pharmacol 30:805–816, 1990

Berridge TL, Sharman DF: The effect of tranquillizing drugs on the concentration of the sulphate ester of 4-hydroxy-3-methoxyphenylethane-1,2,diol in rat brain. Br J Pharmacol 50:156–158, 1973

Bertilsson L, Åsberg M, Thorén P: Differential effect of chlorimipramine and nortriptyline on cerebrospinal fluid metabolites of serotonin and noradrenaline in depression. Eur J Clin Pharmacol 7:365–368, 1974

Bertilsson L, Tuck JR, Siwers B: Biochemical effects of zimelidine in man. Eur J Clin Pharmacol 18:483–487, 1980

Bjerkenstedt L, Edman G, Flyckt L, et al: Clinical and biochemical effects of citalopram, a selective 5-HT reuptake inhibitor: a dose-response study in depressed patients. Psychopharmacology (Berl) 87:253–259, 1985

Blombery PA, Kopin IJ, Gordon EK, et al: Conversion of MHPG to vanillylmandelic acid: implications for the importance of urinary MHPG. Arch Gen Psychiatry 37:1095–1098, 1980

Bowden CL, Koslow SH, Hanin I, et al: Effects of amitriptyline and imipramine on brain amine neurotransmitter metabolites in cerebrospinal fluid. Clin Pharmacol Ther 37:316–324, 1985

Braestrup C, Andersen H, Randrup A: The monoamine oxidase B inhibitor deprenyl potentiates phenylethylamine behavior in rats without inhibition of catecholamine metabolite formation. Eur J Pharmacol 34:181–187, 1975

Brown GL, Ebert MH, Hunt RD, et al: Urinary 3-methoxy-4-hydroxyphenylglycol and homovanillic acid response to *d*-amphetamine in hyperactive children. Biol Psychiatry 16:779–787, 1981

Calderini G, Morselli PL, Garattini S: Effect of amphetamine and fenfluramine on brain noradrenaline and MOPEG-SO_4. Eur J Pharmacol 34:345–350, 1975

Cameron OG, Smith CB: Comparison of acute and chronic lithium treatment of ^{3}H-norepinephrine uptake by rat brain slices. Psychopharmacology (Berl) 67:81–85, 1980

Cassens G, Kuruc A, Orsulak PJ, et al: Amphetamine withdrawal: effects on brain levels of MHPG-SO_4 in the rat. Communications in Psychopharmacology (Berl) 3:217–223, 1979

Charney DS, Heninger GR, Sternberg DE, et al; Abrupt discontinuation of tricyclic antidepressant drugs: evidence for noradrenergic hyperactivity. Br J Psychiatry 141:377–386, 1982

Dahl L-E, Lundin L, Le Fevre P, et al: Antidepressant effect of femoxetine and desipramine and relationship to the concentration of amine metabolites in cerebrospinal fluid: a double-blind evaluation. Acta Psychiatr Scand 66:9–17, 1982

Dajas F, Lista A, Barbeito L: High urinary norepinephrine excretion in major depressive disorders: effects of a new type of MAO inhibitor (moclobemide, RO 11-1163). Acta Psychiatr Scand 70:432–437, 1984

Donnelly M, Zametkin AJ, Rapoport JL, et al: Treatment of childhood hyperactivity with desipramine: plasma drug concentration, cardiovascular effects, plasma and urinary catecholamine levels, and clinical response. Clin Pharmacol Ther 39:72–81, 1986

Elam M, Yao T, Svensson TH, et al: Regulation of locus coeruleus neurons and splanchnic, sympathetic nerves by cardiovascular afferents. Brain Res 290:281–287, 1984

Elia J, Borcherding BG, Potter WZ, et al: Stimulant drug treatment of hyperactivity: biochemical correlates. Clin Pharmacol Ther 48:57–66, 1990

Fawcett J, Maas JW, Dekirmenjian H: Depression and MHPG excretion, response to dextroamphetamine and tricyclic antidepressants. Arch Gen Psychiatry 26:246–251, 1972

Golden RN, Rudorfer MV, Sherer MA, et al: Buproprion in depression, I: biochemical effects and clinical response. Arch Gen Psychiatry 45:139–143, 1988

Kaneko M, Honda Y, Kanno T, et al: Plasma free 3-methoxy-4-hydroxyphenylglycol in acute schizophrenics before and after treatment. Neuropsychobiology 25:126–129, 1992

Karoum F, Korpi ER, Linnoila M, et al: Reduced metabolism and turnover rates of rat brain dopamine, norepinephrine, and serotonin by chronic desipramine and zimelidine treatments. Eur J Pharmacol 100:137–144, 1984

Keller HH, Bartholini G, Pletscher A: Increase of 3-methoxy-4-hydroxyphenylethylene glycol in rat brain by neuroleptic drugs. Eur J Pharmacol 23:183–186, 1973

Linnoila M, Karoum F, Calil HM, et al: Alteration of norepinephrine metabolism with desipramine and zimelidine in depressed patients. Arch Gen Psychiatry 39:1025–1028, 1982a

Linnoila M, Karoum F, Potter WZ: Effect of low-dose clorgyline on 24-hour urinary monoamine excretion in patients with rapidly cycling bipolar affective disorder. Arch Gen Psychiatry 39:513–516, 1982b

Linnoila M, Karoum F, Potter WZ: High correlation of norepinephrine and its major metabolite excretion rates. Arch Gen Psychiatry 39:521–523, 1982c

Linnoila M, Karoum F, Rosenthal N, et al: Electroconvulsive treatment and lithium carbonate: their effects on norepinephrine metabolism in patients with primary, major depressions. Arch Gen Psychiatry 40:677–680, 1983

Maas JW, Landis H: The occurrence of free vs conjugated MHPG in non-human and human primate brain. Psychopharmacology Communications 2:403–410, 1976

Magnusson O, Fowler CJ: Comparison of the effects of the novel antipsychotic agent remoxipride on dopamine and noradrenaline turnover in the rat brain. Pharmacol Toxicol 65:295–298, 1989

Manji HK, Hsiao JK, Risby ED, et al: The mechanisms of action of lithium, I: effects on serotoninergic and noradrenergic systems in normal subjects. Arch Gen Psychiatry 48:505–512, 1991

Mårdh G, Sjöquist B, Änggård E: Norepinephrine metabolism in man using deuterium labelling: the conversion of 4-hydroxy-3-methoxyphenylglycol to 4-hydroxy-3-methoxymandelic acid. J Neurochem 36:1181–1185, 1981

Mårtensson B, Nyberg S, Toresson G, et al: Fluoxetine treatment of depression, clinical effects, drug concentrations and monoamine metabolites and N-terminally extended substance P in cerebrospinal fluid. Acta Psychiatr Scand 79:586–596, 1989

Mårtensson B, Wägner A, Beck O, et al: Effects of clomipramine treatment on cerebrospinal fluid monoamine metabolites and platelet ^{3}H-imipramine binding and serotonin uptake and concentration in major depressive disorder. Acta Psychiatr Scand 83:125–133, 1991

Modai I, Apter A, Golomb M, et al: Response to amitriptyline and urinary MHPG in bipolar depressive patients. Neuropsychobiology 5:181–184, 1979

Moret C, Briley M: Effect of antidepressant drugs on monoamine synthesis in brain in vivo. Neuropharmacology 7:679–684, 1992

Murphy DL, Lipper S, Campbell IC, et al: Comparative studies of MAO-A and MAO-B inhibitors in man, in Monoamine Oxidase: Structure, Function, and Altered Functions. Edited by Singer TP, VonKorff RW, Murphy DL. New York, Academic Press, 1979, pp 457–475

Muscettola G, Wehr T, Goodwin FK: Effect of diet on urinary MHPG excretion in depressed patients and normal control subjects. Am J Psychiatry 134:914–916, 1977

Nielsen M, Braestrup C: Chronic treatment with desipramine caused a sustained decrease of 3,4-dihydroxyphenylglycol-sulphate and total 3-methoxy-4-hydroxyphenylglycol in the rat brain. Naunyn Schmiedebergs Arch Pharmacol 300:87–92, 1977

Nose T, Takemoto H: The effect of penfluridol and some psychotropic drugs on monoamine metabolism in central nervous system. Eur J Pharmacol 31:351–359, 1975

Nyback HV, Walters JR, Aghajanian GK, et al: Tricyclic antidepressants: effects on the firing rate of brain noradrenergic neurons. Eur J Pharmacol 32:302–312, 1975

Pickar D, Cohen RM, Jimerson DC, et al: Tyramine infusions and selective monoamine oxidase inhibitor treatment, II: interrelationships among pressor sensitivity changes, platelet MAO inhibition, and plasma MHPG reduction. Psychopharmacology (Berl) 74:8–12, 1981

Potter WZ, Calil HM, Extein P, et al: Specific norepinephrine and serotonin uptake inhibitors in man: a cross over study with pharmacokinetic, biochemical, neuroendocrine and behavioral parameters. Acta Psychiatr Scand Suppl 290:152–165, 1981

Potter WZ, Karoum F, Linnoila M: Common mechanism of action of biochemically "specific" antidepressants. Prog Neuropsychopharmacol Biol Psychiatry 8:153–161, 1984

Potter WZ, Scheinin M, Golden RN, et al: Selective antidepressants and cerebrospinal fluid, lack of specificity on norepinephrine and serotonin metabolites. Arch Gen Psychiatry 42:1171–1177, 1985

Racagni G, Mocchetti I, De Angelis L, et al: *In vivo* effects of typical and atypical antidepressants on brain noradrenergic system: behavioral correlations, in Typical and Atypical Antidepressants: Molecular Mechanisms. Edited by Costa E, Racagni G. New York, Raven, 1982, pp 185–197

Robinson DS, Johnson GA, Nies A, et al: Plasma levels of catecholamines and dihydroxyphenylglycol during antidepressant drug treatment. J Clin Psychopharmacol 3:282–287, 1983

Roccatagliata G, Albano C, Abbruzzese G: CSF-MHPG in depressive syndromes: basal values and imipramine-induced modifications. Neuropsychobiology 7:169–171, 1981

Roffman M, Kling MA, Cassens G, et al: The effects of acute and chronic administration of tricyclic antidepressants on MHPG-SO_4 in rat brain. Communications in Psychopharmacology (Berl) 1:195–206, 1977

Roffman M, Cassens G, Schildkraut JJ: Effects of the *d*- and *l*-isomers of amphetamine on the levels of 3-methoxy-4-hydroxyphenylglycol sulfate in whole rat brain and rat brain regions. Biochem Pharmacol 27:1774–1777, 1978

Rosloff BN, Davis JM: Decrease in brain NE turnover after chronic DMI treatment: no effect with iprindole. Psychopharmacology (Berl) 56:335–341, 1978

Ross RJ, Zavadil AP, Calil HM, et al: Effects of desmethylimipramine on plasma norepinephrine, pulse, and blood pressure. Clin Pharmacol Ther 33:429–437, 1983

Ross RJ, Scheinin M, Lesieur P, et al: The effect of clorgyline on noradrenergic function. Psychopharmacology (Berl) 85:227–230, 1985

Roy A, Guthrie S, Karoum F, et al: High intercorrelations among urinary outputs of norepinephrine and its major metabolites, a replication in depressed patients and controls. Arch Gen Psychiatry 45:158–161, 1988

Rudorfer MV, Karoum F, Ross RJ, et al: Differences in lithium effects in depressed and healthy subjects. Clin Pharmacol Ther 37:66–71, 1985

Sacchetti E, Smeraldi E, Cagnasso M, et al: MHPG, amitriptyline and affective disorders: a longitudinal study. International Pharmacopsychiatry 11:157–162, 1976

Schanberg SM, Breese GR, Schildkraut JJ, et al: 3-Methoxy-4-hydroxyphenylglycol sulfate in brain and cerebrospinal fluid. Biochem Pharmacol 17:2006–2008, 1968

Schildkraut JJ: Tranylcypromine: effects on norepinephrine metabolism in rat brain. Am J Psychiatry 126:925–931, 1970

Schildkraut JJ: The effects of lithium on norepinephrine turnover and metabolism: basic and clinical studies. J Nerv Ment Dis 158:348–360, 1974

Schildkraut JJ, Klerman GL, Hammond R, et al: Excretion of 3-methoxy-4-hydroxymandelic acid (VMA) in depressed patients treated with antidepressant drugs. J Psychiatr Res 2:257–266, 1964

Schildkraut JJ, Schanberg SM, Kopin IJ: The effects of lithium ion on H^3-norepinephrine metabolism in brain. Life Sci 5:1479–1483, 1966

Schildkraut JJ, Logue MA, Dodge GA: The effects of lithium salts on the turnover and metabolism of norepinephrine in rat brain. Psychopharmacologia 14:135–141, 1969

Schildkraut JJ, Winokur A, Applegate CW: Norepinephrine turnover and metabolism in rat brain after long-term administration of imipramine. Science 168:867–869, 1970

Schildkraut JJ, Watson R, Draskoczy PR, et al: Amphetamine withdrawal: depression and MHPG excretion. Lancet 2:485–486, 1971

Schubert J, Nybäck H, Sedvall G: Effect of antidepressant drugs on accumulation and disappearance of monoamines formed *in vivo* from labelled precursors in mouse brain. J Pharm Pharmacol 11:136–139, 1970

Sharma RP, Janicak PG, Javaid JI, et al: Platelet MAO inhibition, urinary MHPG, and leukocyte β-adrenergic receptors in depressed patients treated with phenelzine. Am J Psychiatry 147:1318–1321, 1990

Shekim WO, Dekirmenjian H, Chapel JL: Urinary catecholamine metabolites in hyperkinetic boys treated with *d*-amphetamine. Am J Psychiatry 134:1276–1279, 1977

Shekim WO, Dekirmenjian H, Chapel JL, et al: Norepinephrine metabolism and clinical response to dextroamphetamine in hyperactive boys. J Pediatr 95:389–394, 1979a

Shekim WO, Dekirmenjian H, Chapel JL: Urinary MHPG excretion in minimal brain dysfunction and its modification by *d*-amphetamine. Am J Psychiatry 136:667–671, 1979b

Shopsin B, Wilk S, Gershon S, et al: Cerebrospinal fluid MHPG: an assessment of norepinephrine metabolism in affective disorders. Arch Gen Psychiatry 28:230–233, 1973

Singhal RL, Rastogi RB, Lapierre YD: Diazepam potentiates the effect of neuroleptics on behavioural activity as well as dopamine and norepinephrine turnover: do benzodiazepines have antipsychotic potency? J Neural Transm 56:127–138, 1983

Siwers B, Ringberger V, Tuck JR, et al: Initial clinical trial based on biochemical methodology of zimelidine (a serotonin uptake inhibitor) in depressed patients. Clin Pharmacol Ther 21:194–200, 1977

Sjoerdsma A, Lovenberg W, Oates JA, et al: Alterations in the pattern of amine excretion in man produced by a monoamine oxidase inhibitor. Science 130:225, 1959

Sjöquist B, Lindström B, Änggård E: Mass fragmentographic determination of 4-hydroxy-3-methoxyphenylglycol (HMPG) in urine, cerebrospinal fluid, plasma and tissues using a deuterium-labelled internal standard. J Chromatogr 105:309–316, 1975

Stern DN, Fieve RR, Neff NH, et al: The effect of lithium chloride administration on brain and heart norepinephrine turnover rates. Psychopharmacologia 14:315–322, 1969

Stolk JM, Barchas JD, Goldstein M, et al: A comparison of psychotomimetic drug effects on rat brain norepinephrine metabolism. J Pharmacol Exp Ther 189:42–50, 1974

Studnitz WV: Effect of marsilid on excretion of 3-methoxy-4-hydroxymandelic acid in man. Scand J Clin Lab Invest 11:224–225, 1959

Sunderland T, Mueller EA, Cohen RM, et al: Tyramine pressor sensitivity changes during deprenyl treatment. Psychopharmacology (Berl) 86:432–437, 1985

Svensson TH, Usdin T: Feedback inhibition of brain noradrenaline neurons by tricyclic antidepressants: α-receptor mediation. Science 202:1089–1091, 1978

Swann AC, Koslow SH, Katz MM, et al: Lithium carbonate treatment of mania, cerebrospinal fluid and urinary monoamine metabolites and treatment outcome. Arch Gen Psychiatry 44:345–354, 1987

Tang SW, Helmeste DM, Stancer HC: The effect of acute and chronic desipramine and amitriptyline treatment on rat brain total 3-methoxy-4-hydroxyphenylglycol. Naunyn Schmiedebergs Arch Pharmacol 305:207–211, 1978

Tang SW, Helmeste DM, Stancer HC: Interaction of antidepressants with clonidine on rat brain total 3-methoxy-4-hydroxyphenylglycol. Can J Physiol Pharmacol 57:435–437, 1979

Thorén P, Äsberg M, Bertilsson L, et al: Clomipramine treatment of obsessive-compulsive disorder, II: biochemical aspects. Arch Gen Psychiatry 37:1289–1294, 1980

Träskman L, Äsberg M, Bertilsson L, et al: Plasma levels of chlorimipramine and its demethyl metabolite during treatment of depression. Clin Pharmacol Ther 26:600–610, 1979

Wålinder J, Carlsson A, Persson R: 5-HT reuptake inhibitors plus tryptophan in endogenous depression. Acta Psychiatr Scand Suppl 290:179–190, 1981

Wilk S, Shopsin B, Gershon S, et al: Cerebrospinal fluid levels of MHPG in affective disorders. Nature 235:440–441, 1972

Wolf MA, Louis J-C, Vincendon G: Antidepressant effect of lithium: a neurochemical study on neuronal uptake of norepinephrine and serotonin. Prog Neuropsychopharmacol Biol Psychiatry 13:765–773, 1989

Yang H-YT, Neff NH: The monoamine oxidases of brain: selective inhibition with drugs and the consequences for the metabolism of the biogenic amines. J Pharmacol Exp Ther 189:733–740, 1974

Yu-cun S, Yu-feng W: Urinary 3-methoxy-4-hydroxyphenylglycol sulfate excretion in seventy-three schoolchildren with minimal brain dysfunction syndrome. Biol Psychiatry 19:861–870, 1984

Zametkin AJ, Karoum F, Linnoila M, et al: Stimulants, urinary catecholamines, and indoleamines in hyperactivity: a comparison of methylphenidate and dextroamphetamine. Arch Gen Psychiatry 42:251–255, 1985a

Zametkin A, Rapoport JL, Murphy DL, et al: Treatment of hyperactive children with monoamine oxidase inhibitors, II: plasma and urinary monoamine findings after treatment. Arch Gen Psychiatry 42:969–973, 1985b

Zametkin AJ, Linnoila M, Karoum F, et al: Pemoline and urinary excretion of catecholamines and indoleamines in children with attention deficit disorder. Am J Psychiatry 143:359–362, 1986

6

Animal Models of Stress, Adrenergic Dysfunction, and Cardiac Disease

BENJAMIN H. NATELSON, M.D.
QIANG CHANG, B.S.
JOHN E. OTTENWELLER, PH.D.
ROBERT S. CONWAY, PH.D.
WALTER N. TAPP, PH.D.

The early work of Selye (1946) tied a knot between stress and disease that has captured the attention and interest of physicians for decades. Although stress is known to produce disease, a major unanswered question has to do with the severity of the disease produced. Our early thinking on this question was heavily influenced by the body of work tying stress to

This work was supported by VA Medical Research funds.

gastric pathology. That literature made it clear that stress reliably produced disease but of a pathologically trivial nature. This was the case because stress-induced gastric disease was characterized by lesions that did not penetrate the muscularis mucosae and thus were erosions rather than the more serious ulcers.

An exception to that was the famous "executive monkey" experiment (Brady et al. 1958). In that experiment, monkeys were taught to avoid shock and were followed over time; they then became sick and died with severe ulcer disease. That study was a bellwether in psychosomatic medicine because it reported that psychological factors inherent in stress had potentially lethal consequences. To evaluate the model further, Natelson et al. (1977) set out to duplicate it. They designed a paradigm consisting of progressively more demanding operant schedules characterized by shock avoidance and conflict. Importantly, they found that the frequency of gastric lesions correlated with stressor intensity; regardless of either stressor intensity or lesion frequency, however, lesion severity was also relatively mild. In fact, lesions often disappeared despite the continuation of stress.

These findings suggested to Natelson et al. that a critical medical consideration was missing from the psychosomatic equation. That factor was the presence of organ vulnerability. With this idea, Natelson (1983) reviewed the clinical literature. Two important quotes sum up all the studies. First, "exposure to stressors alone is almost never a sufficient explanation for the onset of illness in ordinary human experience" (Rabkin and Struening 1976, p. 1018). Second, assuming a relatively constant physical environment, "exposure to cultural change, social change, and change in interpersonal relations may lead to a significant change in health if . . . a person has preexisting illness or susceptibility to illness, and . . . perceives the change as important" (Hinkle 1974, p. 42). Thus the clinical literature supported our own hypothesis.

Now it was time to begin to test it experimentally. Because our review (Natelson 1983) had indicated that the link between stress and serious disease was strongest for heart disease, we decided to do our experiments in animals with heart disease. For our experimental animal, we chose the cardiomyopathic hamster (CMH). The CMH inherits its disease as an autosomal dominant trait. The disease has four histological and clinical phases during the life of the hamster. During the "prenecrotic" phase, the hamsters show no clinical or histological evidence of disease. The "necrotic" phase begins at about 45 days of age for male CMHs, and by 60 days lesions are found in 100% of hamsters. This phase continues until the hamsters are 80–90 days old, at which time healing has begun and only mild, scattered new lesions are seen. The final two phases are the "hypertrophic" and the "terminal" phases. These manifest themselves by progressive cardiac hypertrophy and dilatation, resulting in a prolonged period of compensated heart failure followed by a short period of uncompensated heart failure leading to death. In our hands (Ottenweller et al. 1987), the hypertrophic phase begins at about 6 months of age, and the hamsters succumb to congestive heart failure at a median age of 13 months.

Adrenergic dysfunction seems to be associated with the disease process exhibited by these animals. Sympathetic activity is higher in the CMH heart than in the healthy one since norepinephrine synthesis and turnover is elevated in all stages (Angelakos et al. 1971). During the necrotic phase, myocardial content of norepinephrine increases concomitantly with progression of necrosis. Moreover, epinephrine administration rapidly accelerates the disease process, producing cardiac necrosis and congestive heart failure (Bajusz et al. 1969). An increase in both α_1- and β-adrenergic receptor density is observed in the CMH hearts even before the development of overt congestive heart failure (Karliner et al. 1981).

The cause of this experimental cardiomyopathy appears to be small vessel vasospasm. The vasospastic etiology of the cardiomyopathy was discovered using silicone rubber infusion of the coronaries in these animals (Factor et al. 1982). The vasospastic process is so severe that it produces microinfarcts such that by the end of the necrotic phase of the disease, the heart has multiple foci of myocytolytic necrosis and fibrosis that resemble in part human congestive cardiomyopathies (Factor and Sonnenblick 1985). Because the hamster's disease is not confined to smooth muscle but is found also in striated muscle, the logistics of having to study the beating heart can be bypassed with studies of a striated muscle preparation. Data from this preparation strongly suggest a role for the adrenergic nervous system in precipitating small vessel vasospasm.

We have done this using the very thin cremaster muscle (Conway et al. 1987). In the necrotic phase, active myocytolytic necrosis occurs in this muscle as well as in the heart of the myopathic hamster. During the limited time window of the necrotic phase, the myopathic arterioles become hyperactive to norepinephrine and constrict to 35%–50% of their resting internal diameter over a narrow range of norepinephrine. In contrast, the control hamster and the CMH during the hypertrophic phase—that is, only a few months older than during the necrotic phase—show no significant differences in arteriolar responsiveness to norepinephrine and little active necrosis. Administration of the α-blocker prazosin prevents abnormal microvascular constriction and stops the development of cardiomyopathic changes in the myocardium (Sonnenblick et al. 1985). These results suggest that CMH disease is heavily dependent on abnormal vasomotion, which is influenced by adrenergic effects, especially in the necrotic phase.

Importantly, this animal model is a useful tool to investigate a human disorder of vasomotion called *syndrome X*. Syndrome X is the diagnosis given to patients with typical exertional angina but normal coronary angiograms. The pathogenesis of this disorder is thought to reflect dynamic abnormalities of small coronary arteries that result in either less vasodilator reserve or vasoconstriction (Cannon et al. 1985). The prearteriolar vessels, too small to be seen by angiography, undergo abnormal vasoconstriction and produce anginal symptoms of coronary ischemia (Epstein and Cannon 1986).

Some evidence exists to suggest that syndrome X is a disease of the adrenergic nervous system, as seems to be the case for the CMH. Patients with syndrome X have increased heart rates during Holter monitoring and elevated plasma catecholamine levels, both of which suggest adrenergic hyperactivity (Galassi et al. 1991). A marked circadian variation in the occurrence of ischemic episodes in patients with syndrome X and a diminution in the frequency of these attacks with β-blockade (Bugiardini et al. 1989) further support a role for the adrenergic nervous system in this disease process. Moreover, cold pressor test produces coronary α-adrenergic hyperreactivity in a subset of syndrome X patients having evidence of frank myocardial dysfunction in the form of abnormal rest electrocardiograms (Montorsi et al. 1991). Finally, nifedipine's therapeutic efficacy in this condition is inversely proportional to the patient's sympathetic response to exercise (Montorsi et al. 1990). Unfortunately, the data for a role of α-adrenergic factors are not clear-cut in that treatment with α-adrenergic blocking agents does not block the ischemic response to exercise in these patients (Galassi et al. 1989).

Although initial beliefs about prognosis were that this syndrome was always benign, a 4-year follow-up study found left ventricular performance in patients with syndrome X to be worse than that of control subjects (Opherk et al. 1989). Some patients with syndrome X eventually develop congestive heart failure (Opherk et al. 1989). Importantly, one pathological study has found the sort of microvascular histological changes noted in the CMH to occur in humans—thus providing evidence that spasm of the coronary microcirculation is involved in the genesis of congestive cardiomyopathy in humans as well as in hamsters (Toussaint et al. 1987). Data such as these emphasize the value of the CMH as an animal model of this disease.

Because of this, the CMH model seemed an excellent choice to test our hypothesis that cardiac vulnerability was necessary for stress to produce potentially lethal effects. It provided a spectrum of vulnerability ranging from a healthy appearing animal with no disease, to a healthy appearing animal with covert disease, to an obviously sick animal with overt evidence of congestive heart failure.

We initially tested our hypothesis with hamsters at three stages of their disease (Tapp and Natelson 1988): 1) at 4 months, when active vasospasm had ended but failure had not yet begun; 2) at 6 months, when cardiac dilatation was just beginning; and 3) at 10 months, when the hamsters were in overt heart failure as evidenced by poor grooming and apparent subcutaneous edema. Healthy hamsters of the same ages were used as control subjects.

Animals were given two stress sessions a day for 4 consecutive days; the stressors consisted of foot shock for the experimental animal alone or with another animal for 30-minute sessions or 2-hour sessions of either prone or supine cold immobilization. Before the hamsters were sacrificed, 60 hours were allowed to pass after the last stress session to allow *transient* pathophysiological effects to pass. Figure 6–1 shows the data for heart weight. As heart failure develops, the heart dilates and hypertrophies (i.e., gets heavier). Thus unstressed control CMHs showed increasing heart weight over the course of this experiment. In other work using larger numbers of animals, we have detected the first signs of cardiac enlargement to occur in 6-month-old hamsters (Ottenweller et al. 1987).

Healthy hamsters were unaffected by the stress paradigm. The stressed CMHs on the other hand showed a multiplicative effect of stress. The 6-month-old stressed hamsters developed the same cardiac size as did unstressed 10-month-old hamsters; that is, stress appeared to have accelerated the spontaneous course of heart failure. Also, the 6-month-old stressed CMHs had significantly heavier hearts than did stressed healthy hamsters. But the 10-month CMH that was already in overt failure had significantly more dire consequences from exposure to stress than the younger animal. Their hearts became extremely dilated, and half of the animals in this group succumbed during the 7-day experimental period.

These different effects of stress can be handled parsimoniously in a single model where the pathological effects of stress are considered to be the result of a multiplicative relationship between stress and organic vulnerability. For purposes of conceptual shorthand (rather than as a quantitative statement or rigorous formalism), we would write

Pathological effects ≈ stress × organic vulnerability

Using this formulation, we can see that a constant magnitude of stress multiplied by a small vulnerability will produce small, possibly undetect-

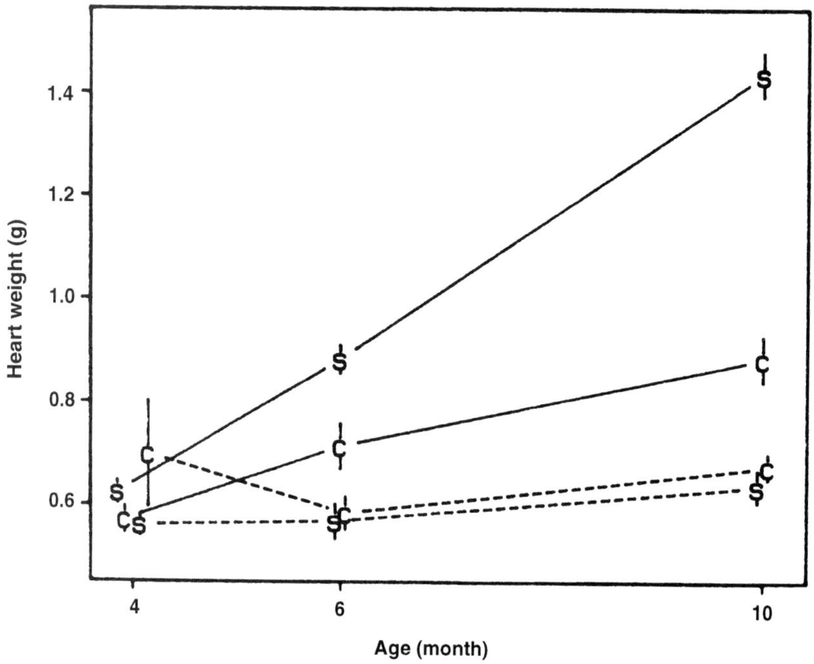

Figure 6–1. Mean heart weights (± SEM) for healthy control hamsters (dashed lines) and for cardiomyopathic hamsters (solid lines). Hamsters were randomly assigned to control conditions (*C*s) or to stress conditions (*S*s). Healthy hamsters were not affected by the stress regimen; for the cardiomyopathic hamsters, the effect of stressor application was multiplied by the hamster's state of health.

able, pathological effects as in the case of the healthy hamsters and the 4-month-old CMHs. But when organic vulnerability is high, as in 10-month-old CMHs with overt heart failure, the effects of the same magnitude of stress are multiplied to produce serious, potentially lethal illness. If similar data as these hold for people, physicians may not need to be overly concerned about the effects of stress in comparatively healthy patients. However, they should probably be progressively more concerned about stress and stress management in patients with known disease or in elderly patients who are more likely to be harboring covert disease; in such vulnerable individuals, stress could have serious, potentially lethal consequences.

Although we have not as yet begun to research the mechanism by which stress interacts with heart disease to produce this increased risk in our animals with covert or overt heart disease, we assume that adrenergic mechanisms are operative. Stress is well known to produce release of adrenal epinephrine as well as activation of the sympathetic nervous system (Natelson et al. 1987). These responses are likely to lead to increased total systemic vascular resistance and heart rate and thus to increased cardiac work. The effect of these pathophysiological changes would be to increase the load on the heart. Inevitably, increased load will lead to changes in the Starling curves indicative of diminished myocardial force. In unpublished studies (B. H. Natelson, W. N. Tapp, May–July 1985), we have found evidence for similar changes occurring immediately after stress in healthy animals. However, when stress stops, the homeostatic process allows the healthy heart to return to its normal configuration. Why the failing heart cannot show similar resiliency is a major question for further research.

To prepare for our next experiments, we decided to move from our multiple stressor paradigm to a more standardized single stressor. We did this because the multiple stress protocol did not allow us to determine which stressors were producing the pathophysiological effect. Also, we reasoned that the probability of seeing habituation would be low if a relatively intense stressor was used. That reasoning was based on the literature as well as our own work on habituation of the stress response (Natelson et al. 1988; Pitman et al. 1990). Of the stressors we had employed, data existed to indicate that supine cold immobilization produced the most profound pathophysiological changes (Paré and Glavin 1986). Therefore, we decided to employ this stressor in 2-hour daily sessions over 5 consecutive days. Since we already had a good idea what stress would do to older CMHs, we decided to explore its effects in younger hamsters—those in the lesion-forming period of their lives. Our specific reason for deciding to do this was that the development of the myocytolytic lesions appeared to be sensitive to sympathetic inputs to the heart; that is, sympatholytics decreased cardiac pathology (Factor and Cho 1983), and sympathomimetics increased it, sometimes so greatly as to produce lethal outcomes (Lossnitzer et al. 1978). Since stress also activates the sympathetic nervous system, we reasoned that stress could have lethal consequences in hamsters at this stage of their disease.

Our initial experiment supported that hypothesis. We monitored survival during the stress period and for 17 days thereafter (Tapp et al. 1989b). In comparison with age-matched healthy hamsters that did not succumb to stress, five of nine stressed CMHs did succumb—a significant difference; none of the nonstressed control hamsters, either healthy or cardiomyopathic, succumbed during this experiment. Stressor intensity was found to be a critical variable in these younger hamsters too. In contrast to the lethal effects of supine cold immobilization in CMHs, prone cold immobilization—done in a way that allowed small amounts of movement—produced no fatalities (Tapp et al. 1989b).

We thought about these results in the context of our original experiment. We were impressed that stress was often lethal in these young hamsters whereas it had never been so in the slightly older hamsters, whose active disease process had slowed or even stopped and in which the process of heart failure was just beginning (i.e., the 4- and 6-month-old CMHs). Two relatively trivial explanations might have been the reason for the discrepancy. First, the experiments were done at different times; second, stressor intensities might have differed across the two experiments. Daily supine cold immobilization was probably a more intense stressor than the multiple stressor paradigm that we had originally used (Tapp and Natelson 1988). Furthermore, as is clear from both our original study and the one done on younger CMHs, stressor intensity was an important variable in influencing experimental outcome. If, however, these trivial reasons could be eliminated, that would indicate that real differences in susceptibility to stress existed in the hamsters despite their being only a few months apart in age. This would suggest that the presence of small vessel spasm carried a different risk for the animal when confronted with stress than in its absence despite the beginning of congestive heart failure.

To assess this issue further, we did several experiments on CMHs that were 2.5 and 5 months old. The first experiment used 2 hours per day of supine cold immobilization. Although no significant difference in the overall survival curves was found, the patterns of the curves for the two groups were somewhat dissimilar, suggesting that the groups were different. In the 5-month-old group, a number of hamsters succumbed in the first few days of the stress paradigm, but then no further deaths occurred. In contrast, in the younger group, a sigmoid-shaped survival curve was seen, with hamsters dying over a longer time span than was the case in the older group.

This difference in the pattern of the survival curves plus the fact that 75% of the younger hamsters succumbed whereas only 61% of the older ones led us to believe that perhaps the intensity of the stressor might have obscured differences between the pathological processes leading to death in the two groups. If the stressor were less intense, we reasoned that perhaps the early mortality in the older group would not occur. In the next experiment, stressor intensity was manipulated simply by decreasing the duration of time that the hamsters were exposed to cold. They were immobilized for the same 2-hour period as in the prior experiment but were placed in the cold for only one of those hours.

Figure 6–2 shows that our reasoning was sound. A statistically significant difference in survival curves was seen: 85% of the younger hamsters succumbed to stress whereas only 46% of the older ones did. Thus this sequence of experiments indicated that the enhanced susceptibility to stress-related mortality exhibited by younger CMHs as compared with hamsters only a few months older is a real phenomenon and cannot be attributed to methodological differences in our earlier experiments.

One possible explanation for the different survival curves found for the two groups of hamsters during the current studies is that different pathological processes may have been responsible for the deaths in each group. Because younger hamsters have more active coronary vasospasm, we reasoned that this process might have been involved in their death. Because older hamsters have beginning signs of congestive heart failure, we reasoned that heart failure might be the pathogenetic mechanism responsible for the lethal effects of stress in these animals. To support the second of these hypotheses, we would expect to find evidence of heart failure in the older hamsters that died from stress but not in the younger hamsters.

Figure 6–3 depicts data supporting this hypothesis. Hearts from the experiment where the survival curves were the same were prepared histologically for microscopic examination along with hearts from unstressed younger and older hamsters. As a basis for comparison, we provided our associate, an anatomical pathologist, with slides of healthy hamsters hearts. We asked him to review unidentified slides looking for ventricular dilatation, defined as enlarged ventricular cavity accompanied by relatively thin ventricular wall diameter. The figure shows that younger hamsters that succumbed to stress showed no evidence of cardiac dilatation, whereas the older hamsters that succumbed to stress all showed dilatation. Thus these data support our belief that the differences in survival between the two groups reflect differences in the nature of the heart disease in the groups. Despite the fact that both groups of hamsters were at increased risk from stress, they differed to a major degree in that active microvascular spasm was occurring in the younger hamsters (Conway et al. 1987). Finding that these hamsters were also at increased risk from succumbing to stress means that the presence of an active vasospastic process within coronary vessels

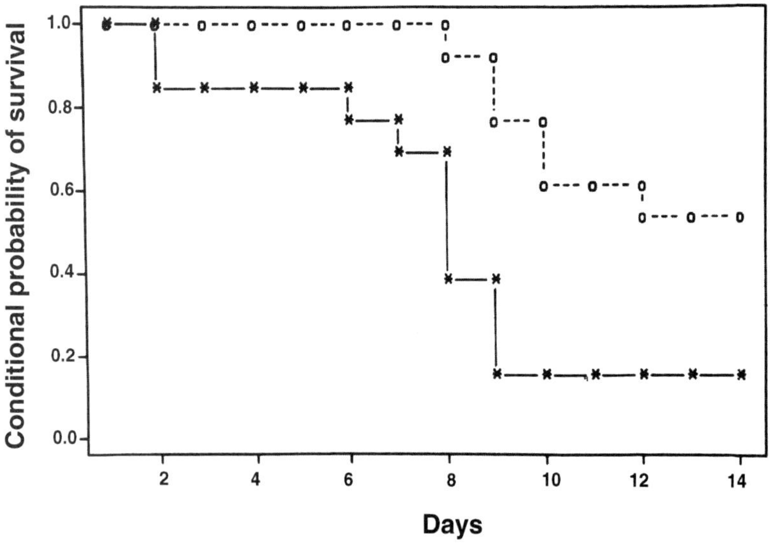

Figure 6–2. Conditional probability of survival for older cardiomyopathic hamsters (circles and dashed line) and for younger cardiomyopathic hamsters (asterisks and solid line) during the 5-day period during which the stressor was repeatedly administered. A significant difference in survival is seen favoring the older hamsters.

may be an additional risk factor in determining the medical consequences of stress in CMHs independent of congestive heart failure.

Finding that stress may activate different pathological mechanisms to produce lethal effects may be important in the practice of medicine. Our earlier work has alerted medical practitioners to the fact that either covert or overt congestive heart failure increases the risk that serious medical consequences will follow stress (Tapp and Natelson 1988). The work reported here shows that stress dramatically shortens the life of hamsters with active coronary vasospasm. As noted previously, one set of cardiac diseases that occur in humans is also due to abnormalities of coronary vasomotion in vessels that are free of arteriosclerotic disease (Cannon and Epstein 1988; Selzer et al. 1976). These diseases affect both the large and small coronary

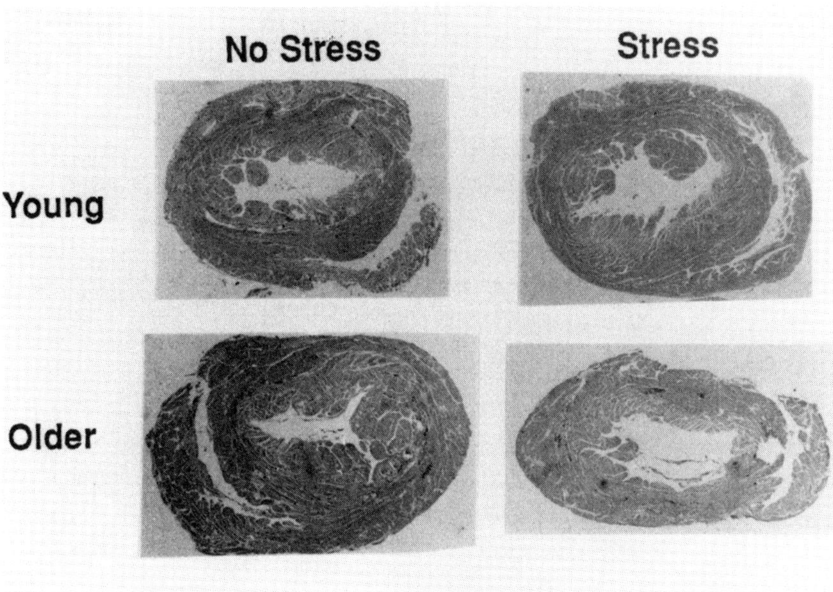

Figure 6–3. This composite shows representative examples of the cardiac ventricles from each of the four groups of cardiomyopathic hamsters studied. Each cross-section was evaluated in a blind design by a cardiac pathologist. The pathologist rated each section for severity of ventricular dilation (i.e., enlarged ventricular cavity accompanied by relatively thin ventricular wall diameter). Only the hearts from the stressed older hamsters were found to be dilated.

vessels and produce chest pain and heart attacks with their attendant disabilities and premature death. Since coronary vasospasm is known to play a role in the genesis of myocardial infarction (Maseri et al. 1978), coronary vasospasm may be the cause of death in the 10% of sudden death victims who have normal hearts and normal coronary arteries. An important question is to determine how psychosocial factors influence coronary vasomotion in patients with disordered coronary vessels. The current focus of researchers trying to understand these processes is to study the coronary vessel wall (e.g., Shepherd and Vanhoutte 1985). We had previously hypothesized that neurobehavioral mechanisms might also be responsible for some of these human diseases (Natelson 1987). The data reported here support that thesis and encourage physician-scientists and practitioners to consider psychological problems in the patient's environment in evaluating and treating patients with chest pain and normal coronary vessels. Moreover, our finding that hamsters with vasospasm are at increased risk of unexpected death when compared with the same hamsters without vasospasm suggests that this animal model can be used to study and understand better the relations among stress, vasomotion, and sudden death in the absence of coronary arteriosclerosis.

But if psychological aspects of stress have lethal consequences in the young CMH during its period of microvascular spasm, it should be possible to interfere with this process to some degree by treating the animal with tranquilizers. An initial experiment with alprazolam (Tapp et al. 1989b) showed this to be the case: 14 of 20 saline-injected, stressed controls succumbed, whereas only 3 of the 20 drug-treated, stressed group died.

To test whether the protective effect of alprazolam was a property shared by other benzodiazepine drugs, we decided to compare its effect with that of the prototypic benzodiazepine, diazepam (Tapp et al. 1989a). To allow such a comparison, we developed doses of subcutaneous alprazolam and diazepam that produced the same endpoint of behavioral tranquilization. Our operational definition of hamster tranquilization was the amount of each drug that would allow us to position the animal in the supine position with its forelimbs spread apart without its making vigorous attempts at escape. Because alprazolam is a short-acting drug that is given several times per day in its clinical use, we elected to give our animals twice-daily injections and to begin this treatment regimen a week prior to the onset of stress. We elected to inject the hamsters late in their inactive period and then 4 hours later when their active period began. Diazepam injections were given on the same schedule. The second injection of each drug or of vehicle was set to occur just before the time that CMHs would undergo supine cold immobilization. We chose to stress the hamsters in their active period because of evidence indicating the pathophysiological effects of stress are greater at this time than during an animal's rest period (Ader 1964).

At the time of injections, a record was made as to whether the hamster vocalized or not. During the stress period, rectal temperatures were taken before and at the end of each stress session.

Concerning vocalizations, statistical evaluation indicated differences in vocalization frequency among all vehicle-treated hamsters (no vocalizations were heard during the injection period in 15 of 95 evaluations), all alprazolam-treated hamsters (no vocalizations were detected during the injection period in 41 of 106 evaluations), and all diazepam-treated hamsters (no vocalizations were heard during the injection protocol in 72 of 91 evaluations). The analysis suggested that vehicle-treated hamsters were aroused most by the injection regimen, alprazolam-treated hamsters less so, and diazepam-treated hamsters even less.

Based on those data, one might expect the diazepam-treated hamsters to resist stress even better than alprazolam-treated hamsters. Figure 6–4 shows that this was not the case. The hatched bars show the data for the stressed hamsters and the open bars for the unstressed control hamsters. Again vehicle-treated hamsters showed the highest incidence of succumbing to stress (7 of 11), whereas alprazolam-treated hamsters showed significant protection (2 of 12 succumbed). Diazepam treatment, however, did not protect the hamsters from stress (7 of 12 succumbed).

Differences in the effects of the two benzodiazepines on thermoregulation did not appear to contribute to these differences in survival. Pre- and poststress temperatures on the first day of stress were not different across the three treatment groups. Temperatures fell from a prestress average of 37.2°C to 33.9 ± 0.7°C (SEM) for vehicle-treated hamsters, 31.7 ± 1.1°C for alprazolam-treated hamsters, and 32.2 ± 0.3°C for diazepam-treated hamsters. However, the animal's ability to protect its core temperature from the stressor predicted whether or not it would survive, but this effect was independent of whether or not the hamster was treated with a benzodiazepine. Figure 6–5 shows core temperatures before and after the last stress session. The figure shows the data from the survivors' temperature on their fifth stress session and from the nonsurvivors' records on the day prior to their deaths. Animals that succumbed to stress had lower basal temperatures than survivors and protected their body temperature much more poorly than survivors. Thus, although ability to maintain core temperature was a good predictor of the hamster's ability to resist the lethal effects of stress, thermoregulatory responses did not explain the differential effect of the drugs on survival.

If drug-induced differences in maintaining thermal homeostasis cannot be invoked to explain the differences in survival, what other explanations could exist? One possibility is that alprazolam reduces stress-induced activation of central nervous system circuits more than diazepam. A second possibility is that the two benzodiazepines may have varying effects on blocking cardiovascular pathophysiological processes. Vogel's group has shown that alprazolam decreases isoproterenol-induced myocardial damage in the rat (Berkowitz et al. 1988). Other work by this group showed that alprazolam, but not diazepam, diminished stress-induced epinephrine in-

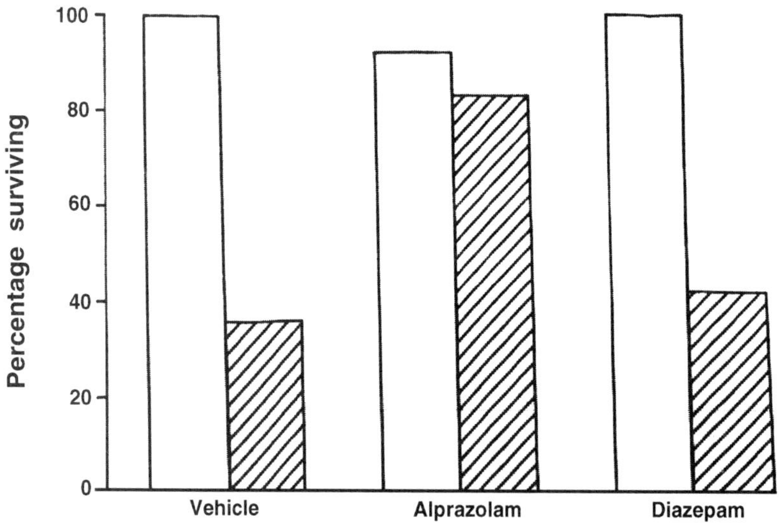

Figure 6–4. Percentage of cardiomyopathic hamsters surviving in the unstressed groups (open bars) and stressed groups (hatched bars). Note that alprazolam selectively protects stressed cardiomyopathic hamsters from the lethal consequences of stress.

creases (Vogel et al. 1984). Thus our working hypothesis is that alprazolam reduced epinephrine release in our stressed hamsters and that the loss of this additional sympathomimetic load allowed the hamster to survive the stress protocol.

In summary, our work makes the point that the presence of organic cardiac disease greatly increases a subject's risk of suffering serious cardiac consequences from stress. In our heart failure model, more severe heart failure increased the risk of repeated presentation of a single stressor in a multiplicative way. But when hamsters of similar age—some with and others without coronary microvascular spasm—were stressed, hamsters with the active vasospastic disease process succumbed more frequently than strain mates with fewer vasospastic episodes. These experiments document the importance of stress under conditions of cardiac vulnerability. Moreover, we are currently working on the hypothesis that an adrenergic mechanism is responsible for both aspects of CMH disease.

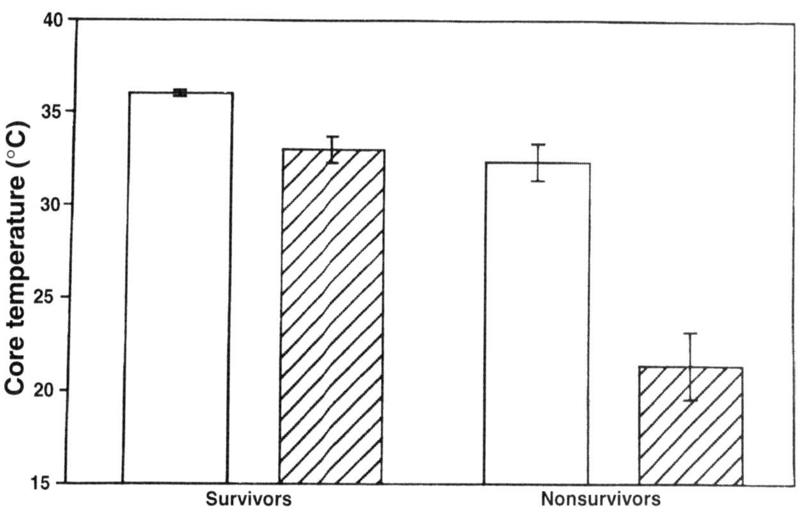

Figure 6–5. Mean core temperatures (°C) (± SEM) for hamsters that survived and did not survive stress. Data from survivors are depicted before (open bars) and after (hatched bars) the fifth stress session. Data from hamsters that succumbed to stress are depicted before (open bars) and after (hatched bars) the stress session on the day before they died. Note that nonsurvivors did not thermoregulate as well as did survivors. This effect was independent of drug treatment.

Our work to date has focused on the effects of cardiac vulnerability on the cardiac consequences of stress. We expect that these findings probably can be generalized to other systems. For example, were we to return to our earlier work on peptic ulcer disease, we expect that we might be able to produce frank peptic ulcers if the stomachs of the experimental subjects were made vulnerable by a treatment with an agent such as aspirin. Work on generalizing our finding will be important in determining the significance of our work. But at least as far as heart disease is concerned, our data suggest that physicians need not be too concerned by complaints of stress in young healthy individuals; instead, they should increase their concern about such complaints in elderly individuals who may be harboring covert heart disease and even more so in patients with known heart disease. In addition, our animal experiments raise the possibility of a role for alprazolam in the care of patients with heart disease during periods of stress.

References

Ader R: Gastric erosions in the rat: effects of immobilization at different points in the activity cycle. Science 145:406–407, 1964

Angelakos ET, Carballo LC, Daniels JB, et al: Adrenergic neurohumors in the heart of hamsters with hereditary myopathy during cardiac hypertrophy and failure, in Recent Advances in Studies on Cardiac Structure and Metabolism. Edited by Bajusz E, Rona G. Baltimore, MD, University Park Press, 1971, pp 262–278

Bajusz E, Homburger F, Baker JR, et al: Dissociation of factors influencing myocardial degeneration and generalized cardiocirculatory failure. Ann N Y Acad Sci 156:396–420, 1969

Berkowitz K, Conahan S, Vogel WH: Alprazolam decreases isoproterenol induced myocardial damage in the rat. Cardiovasc Res 22:414–416, 1988

Brady JV, Porter RW, Conrad DG, et al: Avoidance behavior and the development of gastroduodenal ulcers. J Exp Anal Behav 1:69–72, 1958

Bugiardini R, Borghi A, Biagetti L, et al: Comparison of verapamil versus propranolol therapy in syndrome X. Am J Cardiol 63:286–290, 1989

Cannon RO, Epstein SE: "Microvascular angina" as a cause of chest pain with angiographically normal coronary arteries. Am J Cardiol 61:1328–1343, 1988

Cannon RO, Leon MB, Watson RM, et al: Chest pain and "normal" coronary arteries: role of small coronary arteries. Am J Cardiol 55:50B–60B, 1985

Conway RS, Factor SM, Sonnenblick EH, et al: Microvascular reactivity of the myopathic Syrian hamster cremaster muscle. Cardiovasc Res 21:796–803, 1987

Epstein SE, Cannon RO: Site of increased resistance to coronary flow in patients with angina pectoris and normal epicardial coronary arteries. J Am Coll Cardiol 8:459–461, 1986

Factor SM, Cho S: Alpha adrenergic blockade of the cardiomyopathic Syrian hamster: further evidence for the microvascular etiology of micronecrosis. Federation Proceedings 42:920, 1983

Factor SM, Sonnenblick EH: The pathogenesis of clinical and experimental congestive cardiomyopathy: recent concepts. Prog Cardiovasc Dis 27:395–420, 1985

Factor SM, Minase T, Cho S, et al: Microvascular spasm in the cardiomyopathic Syrian hamster: a preventable cause of focal myocardial necrosis. Circulation 66:342–354, 1982

Galassi AR, Kaski JC, Pupita G, et al: Lack of evidence for α-adrenergic receptor-mediated mechanisms in the genesis of ischemia in syndrome X. Am J Cardiol 64:264–269, 1989

Galassi AR, Kaski JC, Crea F, et al: Heart rate response during exercise testing and ambulatory ECG monitoring in patients with syndrome X. Am Heart J 122:458–463, 1991

Hinkle LE: The effect of exposure to culture change, social change, and changes in interpersonal relationships on health, in Stressful Life Events. Edited by Dohrenwend BS, Dohrenwend BP. New York, Wiley, 1973, pp 9–44

Karliner JS, Alabaster C, Stephens H, et al: Enhanced noradrenaline response in cardiomyopathic hamsters: possible relation to changes in adrenoceptors studied by radioligand binding. Cardiovasc Res 15:296–304, 1981

Lossnitzer K, Mohr W, Konrad A, et al: Hereditary cardiomyopathy in the Syrian golden hamster: influence of verapamil as calcium antagonist, in Cardiomyopathy and Myocardial Biopsy. Edited by Kaltenbach M, Loogen F, Olfen EGJ. New York, Springer-Verlag, 1978, pp 27–37

Maseri A, L'Abbate A, Baroldi G, et al: Coronary vasospasm as a possible cause of myocardial infarction: a conclusion derived from the study of "preinfarction" angina. N Engl J Med 299:1271–1277, 1978

Montorsi P, Cozzi S, Loaldi A, et al: Acute coronary vasomotor effects of nifedipine and therapeutic correlates in syndrome X. Am J Cardiol 66:302–307, 1990

Montorsi P, Fabbiocchi F, Loaldi A, et al: Coronary adrenergic hyperreactivity in patients with syndrome X and abnormal electrocardiogram at rest. Am J Cardiol 68:1698–1703, 1991

Natelson BH: Stress, predisposition and the onset of serious disease: implications about psychosomatic etiology. Neurosci Biobehav Rev 7:511–527, 1983

Natelson BH: Coronary vasospasm: a neuro-behavioral event? Med Hypotheses 24:369–374, 1987

Natelson BH, Dubois A, Sodetz FJ: Effect of multiple stress procedures on monkey gastro-duodenal mucosa, serum gastrin and hydrogen ion kinetics. American Journal of Digestive Diseases 22:888–897, 1977

Natelson BH, Creighton D, McCarty R, et al: Adrenal indices of stress in laboratory rats. Physiol Behav 39:117–125, 1987

Natelson BH, Ottenweller JE, Pitman DE, et al: Effect of stressor intensity on habituation of the adrenocortical stress response. Physiol Behav 43:41–46, 1988

Opherk D, Schuler G, Wetterauer K, et al: Four-year follow-up study in patients with angina pectoris and normal coronary arteriograms ("syndrome X"). Circulation 80:1610–1616, 1989

Ottenweller JE, Tapp WN, Chen TS, et al: Cardiovascular aging in Syrian hamsters: similarities between normal aging and disease. Exp Aging Res 13:73–84, 1987

Paré WP, Glavin GB.: Restraint stress in biomedical research. Neurosci Biobehav Rev 10:339–370, 1986

Pitman DL, Ottenweller JE, Natelson BH: The effect of stressor intensity on habituation and sensitization of glucocorticoid responses in rats. Behav Neurosci 104:28–36, 1990

Rabkin JG, Struening EL: Life events, stress and illness. Science 194:1013–1020, 1976

Selye H: The general adaptation syndrome and the diseases of adaptation. J Clin Endocrinol 6:117–230, 1946

Selzer A, Langston M, Ruggeroli C, et al: Clinical syndrome of variant angina with normal coronary arteriogram. N Engl J Med 295:1343–1347, 1976

Shepherd JT, Vanhoutte PM: Spasm of the coronary arteries: causes and consequences (the scientist's viewpoint). Mayo Clin Proc 60:33–46, 1985

Sonnenblick EH, Fein F, Capasso JM, et al: Microvascular spasm as a cause of cardiomyopathies and the calcium-blocking agent verapamil as potential primary therapy. Am J Cardiol 55:179B–184B, 1985

Tapp WN, Natelson BH: Consequences of stress: a multiplicative function of health status. FASEB J 2:2268–2271, 1988

Tapp WN, Natelson BH, Grover E, et al: Alprazolam but not diazepam protects hamsters with heart disease from the medical consequences of stress. Pharmacol Biochem Behav 33:633–636, 1989a

Tapp WN, Natelson BH, Creighton D, et al: Alprazolam reduces stress-induced mortality in cardiomyopathic hamsters. Pharmacol Biochem Behav 32:331–336, 1989b

Toussaint M, Pfister DD, Da Lage C: Congestive cardiomyopathy and spasm of coronary microvascularisation in man. Ann Pathol 7:223–226, 1987

Vogel WH, Miller J, DeTurck KH, et al: Effects of psychoactive drugs on plasma catecholamines during stress in rats. Neuropharmacology 23:1105–1108, 1984

7

Techniques of Adrenergic Functional Assessment

RICHARD A. NELESEN, PH.D.
MICHAEL G. ZIEGLER, M.D.

In vivo evaluation of the autonomic nervous system, and in particular the sympathetic nervous system, has been the goal of investigators in many fields. Through research on reactivity to challenges, patterns have emerged that relate changes of adrenergic function to stress, anxiety, and cardiovascular disease. Many techniques are available to evaluate the contribution of adrenergic activity to changes in cardiovascular homeostasis.

As illustrated in Figure 7–1, sympathetic nervous system evaluation can be performed at a number of different levels.

Direct recording of sympathetic nerve activity to skin and muscle can

This work was supported in part by Grants HL36005, HL40102, HL07261, and RR00827 from the National Institutes of Health.

quantitate electrical activity to nerve terminals. These nerve terminals release norepinephrine (NE), and the adrenal medulla releases epinephrine. Plasma catecholamine levels reflect the escape of neurotransmitter to the circulation. NE and epinephrine mediate adrenergic function α- and β-adrenergic receptors. Stimulation of these receptors regulates changes in their number and affinity. Drugs also stimulate these receptors, and drug sensitivity can provide an in vivo measure of receptor function. Receptor levels can be estimated on platelets, lymphocytes, and fat cells. The receptors mediate cellular responses via second messengers, such as cyclic adenosine monophosphate (cAMP). Stimulated levels of cAMP are important in the assessment of adrenergic receptor function. This cascade of events results in end-organ responses such as changes in blood pressure, heart rate, cardiac output, vascular resistance, and perspiration. Instrumenta-

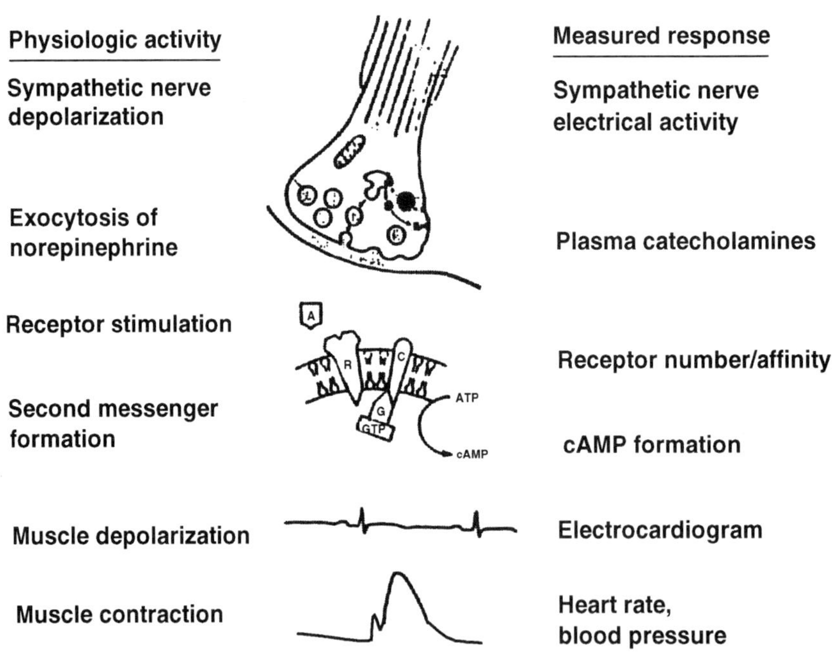

Figure 7–1. The various levels of evaluation of adrenergic function. ATP = adenosine triphosphate; cAMP = cyclic adenosine monophosphate.

tion is available that measures these responses.

In this chapter, we review in vivo techniques available to evaluate adrenergic function at each of these steps.

Techniques That Evaluate End-Organ Response

Early reactivity research focused on the end-organ responses of sympathetic nervous system activity, such as increases in blood pressure and heart rate. Inferences about adrenergic drive to the heart and vasculature have been made based on changes in blood pressure and heart rate to various challenges. These challenges, which elicit increases in systolic blood pressure (SBP), presumably result in a concomitant increase in cardiac output without a similar change in diastolic blood pressure (DBP). This leads one to infer that these challenges are mediated mainly by β-adrenergic stimulus to the heart than by α-adrenergic stimulus to the vasculature. The opposite may be inferred if there is a preferential increase in DBP, since this is strongly influenced by changes in vascular resistance.

Blood Pressure and Heart Rate

β-Agonists, such as isoproterenol, stimulate the heart to beat more quickly and forcefully and decrease peripheral resistance. This lowers DBP while increasing SBP. α-Agonists, such as phenylephrine, raise both SBP and DBP. Epinephrine is a powerful β-agonist; NE is more potent at α- than β-receptors. However, when challenges increase both SBP and DBP, it becomes difficult to make inferences as to the underlying adrenergic influence on the myocardium and vasculature.

There are a number of noninvasive techniques available to elucidate the underlying mechanisms of heart rate and blood pressure responses. Heart rate changes differently to various behavioral challenges with different attentional demands (Lacey 1967). Tasks requiring the rejection of external sensory demands (i.e., cognitive tasks) result in decreased heart rate, and tasks that required sensory intake (i.e., reaction time tasks) increase heart rate. These patterns are influenced by pharmacological manipulations (see later for further discussion). The heart rate acceleration to sensory intake tasks is inhibited by β-adrenergic blocking agents, but these drugs have no effect on the heart rate deceleration of sensory rejection tasks. Changes in heart rate do not always represent adrenergic drive to the myocardium, as a large component in heart rate variability is determined by vagal tone (Berne and Levy 1981). Because of this problem, a number of different measures have been examined to determine adrenergic tone of the myocardium under challenge. These techniques include measures such as T-wave amplitude and various measures collectively known as systolic time intervals.

T-Wave Amplitude

The T-wave is the electrical signal that accompanies myocardial electrical repolarization after each heart beat. T-wave amplitude is a noninvasive measure that requires only the simple instrumentation of a reliable electrocardiogram (ECG). It may be an index of myocardial sympathetic nervous system tone (Furedy and Heslegrave 1983). Pharmacological studies (Biberman et al. 1971; Hartwell et al. 1942) indicate that T-wave amplitude is decreased by drugs that increase sympathetic tone; in contrast, T-wave amplitude is increased by drugs that diminish adrenergic tone. The mechanism that appears to be responsible for T-wave amplitude attenuation with increased sympathetic activity is the nonuniform, asynchronous shortening of the various refractory periods in the myocardium (Han and Moe 1964; Han et al. 1964). Thus, alterations in the amplitude of the T-wave may be interpreted in terms of improved synchrony or greater rate of repolarization of individual segments of the ventricle (Randall 1977), unlike the improved synchrony of contractions during depolarization that occurs with increased sympathetic nervous system activity (Obrist 1981). Specifically, since the spread of repolarization is from the ventricular epicardium to the endocardium, which is opposite to the depolarization sequence, the reduction in the longer endocardial refractory periods has been suggested as the underlying mechanism of T-wave amplitude attenuation in response to adrenergic stimulation (Biberman et al. 1971). The ultimate result of adrenergic stimulation is to invert the T-wave polarity from positive to negative (Furedy and Heslegrave 1983). As noted previously, T-wave amplitude requires only minimal equipment; it is measured as the amount of change relative to the ECG baseline determined during the isoelectric P-Q interval. This method is relatively simple to obtain and quantitate, but has not become a popular measure of adrenergic drive on the heart.

Systolic Time Intervals

A group of measures collectively called systolic time intervals also responds to cardiac adrenergic stimuli. These measures include the cardiac preejection period (PEP), left ventricular ejection time (LVET), electromechanical systole (QT/QS2), the Heather Index, cardiac acceleration index, and pulse transit time. These measures are noninvasive, but require the use of other instrumentation, such as phonocardiography (PCG) and impedance cardiography (ICG). These measures can be confirmed through other techniques, such as cardiac catheterization or echocardiography. Figure 7–2 depicts these events as determined from the simultaneous recording of ECG, PCG, and ICG.

Cardiac adrenergic drive alters the relationship between electrical and mechanical systole (Larsen et al. 1986). β-Adrenergic drive accelerates both the rate and force of myocardial contractility (Winegrad 1982). The ratios of PEP/LVET and QT/QS2 are two indices of cardiovascular function that may reflect sympathetic drive on the myocardium. Each component index is based on systolic time interval, PEP, LVET, electrical systole (QT), and mechanical systole (QS2). These measurements are derived from the ECG, PCG, and/or ICG recordings. The PEP interval begins with the onset of QT at the ECG Q-wave and ends at the onset of left ventricular ejection. The PCG measures the first heart sound (S1), which begins prior to ejection and aortic opening. S1 is a poor measure of the onset of left ventricular ejection but the B-wave of the ICG occurs near the time of aortic opening and has been used to signify the end of PEP and the onset of ventricular ejection (Lewis et al. 1977). PEP is an important period in assessing contractility since the length of the PEP relates to the velocity of myocardial contraction. Both PEP and LVET are related to heart rate, so the ratio of PEP to LVET has been used as an index of contractility because it does not require corrections for heart rate (Lewis et al. 1977). Indeed, PEP/LVET ratio correlates highly with left ventricular ejection fraction and the rate of ventricular pressure change (Ahmed et al. 1972). A drawback of the use of this ratio is that both PEP and LVET are highly sensitive to both preload and afterload; therefore, during situations in which large increases in peripheral resistance occur, the sensitivity of the ratio may be compromised (Lewis et al. 1977). When there is an increase in β-adrenergic drive, the relative shortening of the PEP interval is greater than the change in LVET, and the ratio decreases. Thus, PEP/LVET is inversely related to β-adrenergic drive on the myocardium in the absence of preload or afterload.

Both QT and QS2 are inversely related to heart rate (Boudoulos et al. 1981). Since the regressions, like PEP and LVET, are nearly all parallel, the ratio of QT/QS2 can be used without correction for heart rate. Increased β-adrenergic drive shortens the QS2 more than the QT interval (Boudoulos et al. 1981; Johnson et al. 1981). Thus, the ratio is positively related to

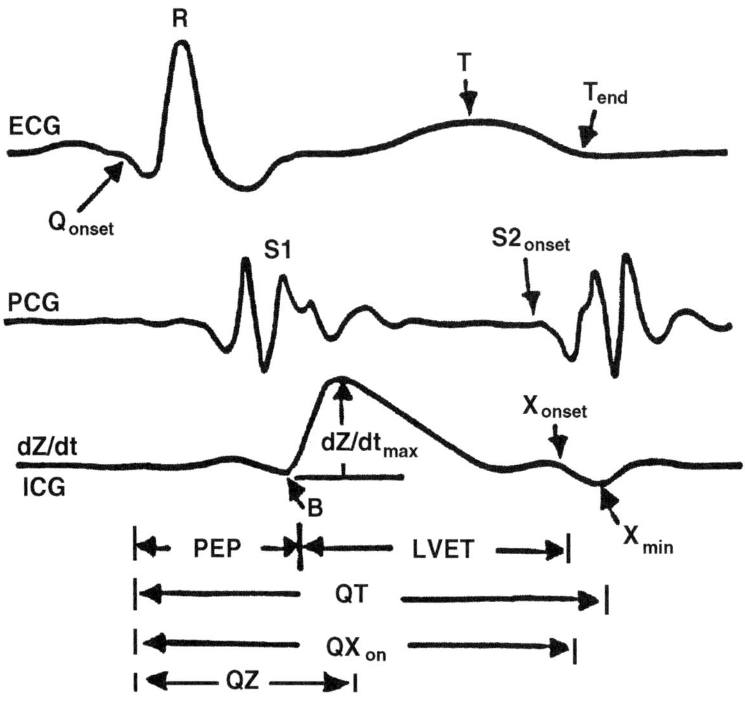

Figure 7–2. Cardiovascular waveforms, as derived from electrocardiogram (ECG), phonocardiogram (PCG), and impedance cardiogram (ICG), defining the relevant signal events. dZ/dt = first time derivative of the change (dZ) in mean thoracic impedance; dZ/dt_{max} = maximum height of the dZ/dt wave; PEP = preejection period; LVET = left ventricular ejection time; QT = electrical systole; QX_{on} = onset of the X-wave; QZ = interval from Q to dZ/dt_{max}.

sympathetic myocardial tone. It relies on mechanical events that may be less sensitive to changes in preload or afterload. The second time point in the QS2 interval relies on PCG detection of the second heart sound (S2), so it is difficult to detect during speech, which totally obliterates this PCG signal. The onset of the ICG X-wave (X_{on}), however, correlates temporally with S2 and can be used instead of S2 in this ratio (QT/QX_{on}) (Nagel et al. 1989).

PEP/LVET and QT/QX_{on} are two indices of central sympathetic cardiovascular drive. A third useful measure of cardiac adrenergic drive is the Heather Index of myocardial contractility. The Heather Index is the ratio of dz/dt_{max} to the QZ interval (interval from Q to dZ/dt_{max}). This index uses information reflecting the maximum velocity of left ventricular ejection (dz/dt_{max}) and the period required to generate the contractile force (QZ) to open the aortic valve and reach the maximum velocity. This measure appears to reflect left ventricular contractility (Heather 1969). Another measure used to infer the adrenergic drive to the myocardium includes the acceleration index, which reflects the acceleration of the rapid ventricular ejection phase following the aortic opening. This is defined as the ratio of dz/dt_{max} to the B to dz/dt_{max} interval (Hurwitz et al. 1993). These measures have been suggested to reflect left ventricular contractility.

ICG also determines stroke volume, and cardiac output is stroke volume times heart rate. Vascular resistance can be calculated from cardiac output and blood pressure. These two indices further elucidate the effects of adrenergic stimulation. The use of ICG and blood pressure have shown that behavioral challenges do indeed produce different patterns of cardiovascular performance suggestive of β-adrenergic stimulation, α-adrenergic stimulation, or a mixed pattern of both types of stimulation (Allen et al. 1987; Hurwitz et al., in press; Sherwood et al. 1986).

Thus, in situations where it is impractical to use more invasive methods, the combination of ECG, ICG, and blood pressure determination provide information as to end-organ responses to sympathetic nervous system activity. Systolic time interval measures also provide clues as to when increases in the heart rate are likely to be related to sympathetic nervous system activity rather than vagal withdrawal. They also provide an estimate of the inotropic forces elicited by adrenergic activation. Noninvasive indices of cardiovascular performance provide estimates of the end-organ response to sympathetic nervous system activity, but do not evaluate the sympathetic nervous system itself. To do this, more direct methods of evaluation are needed, such as measurement of the catecholamines or nerve activity.

Catecholamines

NE, epinephrine, and dopamine are catecholamines released from nerves and the adrenal medulla into blood, cerebrospinal fluid, and urine. NE is primarily released from sympathetic nerves, with some release from the adrenal medulla. Its major effects follow release from sympathetic nerves across a synapse to adjacent adrenergic receptors. A small fraction of this NE spills over into the bloodstream. Plasma levels of NE correlate with sympathetic nervous system activity. Epinephrine is released from the adrenal medulla into the bloodstream, where it acts as a hormone, stimulating both α- and β-adrenergic receptors. Blood levels of epinephrine provide a measure of adrenomedullary stimulation. Small epinephrine changes have measurable effects on end-organ responses. Plasma, cerebrospinal fluid, and urinary catecholamine levels are a potentially useful guide to sympathetic nervous system activity. These values, however, have been tainted by the volatility of the catecholamines themselves and errors of inconsistent assay techniques. Problems associated with the stress of collecting samples initially produced higher values than were actually present in the resting organism.

Catecholamine Assay Techniques

A number of catecholamine assay techniques have been developed with a goal of sufficient sensitivity and selectivity to measure plasma catecholamine levels. Early techniques, such as fluorometry proved sufficiently sensitive for the high catecholamine levels in urine, but were inadequate for plasma. As a result, other methods were developed. The most widely used techniques today are the radioenzymatic methods and high-performance liquid chromatography (HPLC) with electrochemical detection.

PNMT radioenzymatic assay. In the adrenal, phenylethanolamine-*N*-methyltransferase (PNMT) converts NE into epinephrine by transfer of a methyl group from *S*-adenosylmethionine (SAM) to the primary amine of

NE (Henry et al. 1980; Lake et al. 1976). The enzyme can be partially purified from bovine adrenal glands. In the test tube, radiolabeled ^{3}H-SAM is used as a methyl donor to convert NE to ^{3}H-epinephrine, which can be isolated and counted by scintillation spectroscopy. The enzyme is fairly specific for β-hydroxylated phenylethanolamines and does not appreciably label dopamine or further metabolize epinephrine. The PNMT technique is the most reliable assay available for measuring plasma NE levels and can accommodate large numbers of samples easily (Lake et al. 1976). Some laboratories have experienced difficulty in adequately purifying PNMT, therefore the technique has been used less widely than the catechol-*O*-methyltransferase (COMT) method.

The PNMT assay has several other advantages over the COMT-based radioenzymatic technique. The PNMT assay is more specific for NE than the COMT technique because PNMT has much greater substrate specificity than COMT. PNMT is not inhibited by aluminum so catecholamines can be preconcentrated on alumina prior to the assay, and the assay can be made extremely sensitive. The same preconcentrating step eliminates inhibiting substances and allows the assay to be standardized for tissue or plasma samples. Henry et al.'s (1980) techniques for purifying radiolabeled epinephrine from other radiolabeled contaminants without using chromatography makes the PNMT method one of the most rapid NE assay systems available.

COMT assay. COMT is an enzyme that will *O*-methylate almost any molecule with a catechol group by transferring a methyl group from SAM. In its simplest form, the catecholamines in unextracted plasma are incubated in a buffer solution with COMT, magnesium, and ^{3}H-SAM. The derived ^{3}H-NE and ^{3}H-epinephrine metabolites are then oxidized to vanillin, which is extracted into toluene and counted by liquid scintillation spectroscopy.

COMT converts NE to normetanephrine, epinephrine to metanephrine, and dopamine to methoxytryamine. A more specific form of the assay separates these metabolites by thin-layer chromatography to measure the individual catecholamines. Many drugs, such as isoproterenol, dobutamine, and methyldopa can interfere with the assay, but when the assay is combined with the appropriate separation techniques, these catechol drugs can also be measured (Durrett and Ziegler 1980). COMT is strongly inhibited by calcium, which must be chelated out of solution by EGTA to bind inhibiting calcium. Plasma proteins, endogenous SAM, some compounds in urine, aluminum, and ascorbic acid also interfere with the enzymatic activity of COMT and may provide spurious assay values (Durrett and Ziegler 1980). Kennedy and Ziegler (1990) have developed a preextraction concentration technique that eliminates these interfering compounds and improves the sensitivity of the assay. In this technique, samples are extracted into heptane with diphenylborate, then into dilute acetic acid. This extraction procedure has an efficiency of 78% for NE. It excludes calcium and other COMT inhibitors present in plasma, urine, and tissue. This procedure increased the sensitivity of the assay by a factor of 10.

After chromatographic separation, NE and epinephrine can be measured more reliably than dopamine because of high blank values in dopamine measurement. Since samples are separated by thin-layer chromatography, very high levels of one substance can contaminate an adjacent chromatographic band, artificially increasing levels of the adjacent substance. The COMT assay is also complex and subject to error from standardization and pipetting. On the other hand, the COMT assay has several advantages. It can be performed with 50 μl of plasma or 0.5 μl of urine and can measure several catechol drugs and drug metabolites (Durrett and Ziegler 1980). This assay with the preextraction step is sensitive enough to measure basal levels of plasma epinephrine (Kennedy and Ziegler 1990).

HPLC assay. HPLC separates catecholamines and their metabolites into sharp peaks. Many papers define chromatographic conditions for separation of the catecholamines and their detection. Good separation of the catecholamines provides specificity; the choice of detector is important for sensitivity. Purification of biological samples is often required prior to HPLC analysis to improve the sensitivity and specificity. A number of different methods have evolved to extract the catecholamines, including organic solvent extraction, protein precipitation, cation exchange, boric acid affinity, and alumina extraction. The most commonly used of these techniques are alumina extraction for plasma and cation exchange for urine. HPLC columns have been improved in the last decade, and 3- to 5-μm spherical packing materials generate high speed and efficiency. A number of different column types are available, but the reverse phase ion-pair chromatography is the most widely used. This technique involves adding an ion pairing agent (e.g., heptane sulfonic acid) to the mobile phase. Ion-exchange chromatography is making a resurgence as it reverses the position of epinephrine and NE, giving a higher signal for epinephrine (Mefford 1987).

After separation of the catechol compounds, they are quantitated by a detector. Urine has enough catecholamines to be detected by ultraviolet light absorption. Fluorometric detection is far more sensitive, but calls for either pre- or postcolumn derivatization (Abe and Konaka 1988; Imai et al. 1977; Mori 1986). The most widely used method of postseparation detection is electrochemical detection, which takes advantage of the oxidation potentials of the catecholamines and provides satisfactory selectivity and sensitivity (Goto 1986; Stulik and Pacakova 1986). Two types of electrodes are currently is use: the thin-layer flow cell (amperometric) and the flow through cell (coulometric). Both have their advantages; the former being inexpensive to maintain, and the latter offering a better signal-to-noise ratio and the ability to link detectors in series. The coulometric method enables the analyst to take advantage of reversible catecholamine oxidation, thus eliminating interfering nonreversible compounds (Minegeshi and Ishizaki 1984). Recently developed pulsed detectors that allow the analyst to change applied potential and gain provide an assay range from a few picograms to tens of nanograms of catecholamine.

Other techniques that have been used to quantitate catecholamines include fluorometry, radioimmunoassay, gas chromatography, and capillary zone electrophoresis. Fluorometry is an old technique that gave a wide range of values for plasma catecholamines from very high (Hjemdahl 1984) to very low (Miura et al. 1977). Radioimmunoassay techniques have not worked well for the catecholamines, because antibodies obtained cross-react with compounds similar in structure (Grota and Brown 1986). Gas chromatography has been used for the measurement of the catecholamines since the 1960s. Two techniques have been used: electron capture detection, which lacks specificity, and mass spectroscopy. The latter can be very accurate and provides a reference standard against which other techniques can be compared (Ziegler et al. 1976). This technique is so time consuming and expensive that it is not suitable for routine catecholamine analysis.

Hjemdahl (1984) compared a number of laboratories and the four commonly used methods of catecholamine analysis (i.e., COMT, PNMT, HPLC, and fluorometric techniques). He found all techniques had problems with accuracy and reproducibility. The development of better extraction methods and more sensitive and reliable detection techniques has probably increased the reliability of the radioenzymatic method and HPLC with electrochemical detection since that study was published. The choice of an assay system will depend on individual needs. Those researchers who need to work with small volumes of sample will probably be served best by the COMT assay; those needing flexibility, HPLC; and those who are interested only in NE, the PNMT assay. All of these assays are difficult, so it is recommended that the advice of a laboratory with experience in a particular technique is sought before attempting it.

Sampling, Degradation, and Storage of Catecholamines

Plasma catecholamine levels vary with age, posture, time of day, medication, disease state, mental status, and stress (see Table 7–1). Catecholamine levels also vary with sampling site. Blood sampling from the antecubital vein is convenient, but the sympathetic innervation of the hand and forearm contribute substantially to NE in antecubital venous blood (Chang et al. 1986). The hand and forearm also clear a large fraction of NE from their arterial blood supply (Folkow 1982). Extraction of

Table 7–1. Factors that can influence human plasma catecholamine levels

Environment of the subject
  Hospitalized versus outpatient
  Familiarity with the clinic setting
Posture
Age
Tobacco smoking
Diet
  Carbohydrate intake
  Sodium intake
  Caffeine consumption
Medications
  Diuretics
  Vasodilators
  β-Adrenergic antagonists
  α-Adrenergic antagonists
  Reserpine
	Antidepressants
	Marijuana
	Central nervous system stimulants
	Anorexiants
	Dopamine
	Neuroleptics
	Time of day
	Physical conditioning
	Various illnesses
	  Depression
	  Mania
	  Thyroid disease
	  Congestive heart failure
	  Hypertension
	Blood sample processing and assay methodology

NE varies inversely with forearm blood flow over a range of 10%–90% (Chang et al. 1986; Goldstein et al. 1987). However, arterial blood sampling does not obviate artifacts due to sampling site, since sympathetically mediated NE release can be underestimated by arterial sampling when cardiac output is increased. High cardiac output enhances the delivery of catecholamines to extraction sites and increases the rate of NE clearance from the bloodstream. During mental challenge, hemodynamic changes have been closely linked to changes in NE release into blood, weakly linked to changes in arterial NE, and completely unrelated to changes in antecubital venous NE (Goldstein et al. 1987).

Plasma NE levels are the result of the rate of NE secretion into the blood and the rate of NE clearance. Plasma NE has a brief half-life of 1–4 minutes. The rate of NE clearance is affected by cardiac output, drugs, and disease. High plasma NE levels can reflect increased NE release, impaired NE clearance, or both.

Catecholamine stability also depends on sampling site. Catecholamines in cerebrospinal fluid decay in minutes at room temperature. In contrast, catecholamines in plasma can be stable at room temperature for several hours (Weir et al. 1986). Blood samples collected into heparin or EDTA and plasma separated by refrigerated centrifugation, and stored at –70°C appear to be stable for at least 1 year (Kagedal and Goldstein 1988; Ziegler 1989). 6 M hydrochloric acid is an effective preservative for urine catecholamines (Kagedal and Goldstein 1988), but could also hydrolyze conjugated catecholamines and result in artifactually high urine catecholamine levels.

Urinary Catecholamines

Urinary catecholamines might provide a measure of adrenergic function over a prolonged time period without the use of intravenous catheters. One must bear in mind, however, that most urinary catecholamines are not derived from plasma. As noted previously, urinary dopamine levels are very high, whereas plasma levels are quite low; the apparent source of this urinary dopamine is circulating dihydroxyphenylacetic acid (dopa). Urinary epinephrine levels are also quite high and do not appear to change despite adrenalectomy in humans (von Euler et al. 1961), or adrenal demedullation or denervation in rats (Leduc 1961; Ziegler et al. 1989). It appears that epinephrine is synthesized in the kidney through the enzyme *N*-methyltransferase (Ziegler et al. 1989), which is distinct from adrenal PNMT.

Despite these alternate sources, urinary catecholamines do appear to reflect changes in sympathetic function in relation to various challenges. The use of urinary catecholamine levels for long- and short-term measurement requires different strategies of collection. Urinary catecholamine measurements reflect all of the catecholamine-stimulating activities that have occurred since the subject last voided.

Long-term sample collection (e.g., 24 hours) will reflect the basic stress of that period, but a short-term increase will be diluted by lower levels of other less stressful periods. Short-term sample collection can be used to show the effect of a challenge and usually represents periods of several minutes to a few hours. The largest problem with this technique is the subject's ability to void when requested.

Persons living near the Three Mile Island nuclear accident near Harrisburg, Pennsylvania, in March 1979, had chronic elevations of urinary NE and epinephrine relative to persons who lived in a control area. Furthermore, these levels increased during active periods of the cleanup procedure (Baum et al. 1981). Frankenhauser and colleagues showed that women in nontraditional roles excrete more urinary epinephrine than women in traditional roles (Rauste-von Wright et al. 1981). They examined occupational settings and found that lack of control over one's work pace results in higher urinary catecholamine levels than self-paced work despite the workload (Frankenhauser and Johansson 1976).

Urinary catecholamines are derived from plasma catecholamines, renal nerves, and extraneuronal renal catecholamine synthesis. They can reflect sympathetic nervous activity in many circumstances.

Catecholamine Responses to Challenge

Plasma levels of NE and epinephrine increase minutes after physical or psychological stresses. One of the simplest tests to evaluate sympathetic nervous system function is to stand from a supine posture. The baroreflexes double plasma NE levels in 5 minutes (Lake et al. 1984). NE increases are proportional to the vertical component of gravity; thus, seated subjects' NE levels are about 60% greater than while supine. Most, and perhaps all, stress increases NE release. The rate of NE release is greater with increasing age, but some tissues in older subjects respond less to NE. Undergoing venipuncture or change in posture can increase NE release, so the circumstances of blood sampling are important. Standing doubles NE levels, and the pain of venipuncture can cause a further increase.

Exercise is a potent stimulus to NE release, even isometric exercise with one arm. Isometric exercise near the limits of a person's arm strength is more potent than prolonged moderate exercise in raising NE levels. Isotonic exercise gives a fairly predictable sympathetic response in control subjects at submaximal exertion. Physically fit subjects have relatively low NE levels at moderate workloads, but can attain much higher NE levels than sedentary subjects during maximal exercise. When subjects exercise at 20%–80% of their maximal capacity, plasma NE levels increase exponentially with the level of work. Above 80% maximum effort, NE levels increase even more sharply so that at maximal exertion NE levels may reach 50 times basal.

Physiological and psychological stresses activate catecholamine release. When the experimental interventions include unfamiliar stimuli, catecholamine responses to psychological stimuli are added to the responses to physiological stimuli (Rose 1980). Similarly, catecholamine responses to such physiological stimuli as sodium deprivation (Nichols et al. 1980), upright posture (Henry et al. 1980), exercise (Watson et al. 1980), and exposure to cold (Jessen 1980) all increase plasma NE. Extremes of these stimuli also increase cortisol. Physiological and psychological factors must be carefully controlled since catecholamine responses to these challenges differ. Control subjects were given physiological stimulation consisting of immersion of one hand in cold water (cold pressor test) or the psychological stimulus of an arithmetic test. SBP increased during the cold pressor test, during mental arithmetic, and during a combination of the two tests. The plasma epinephrine response, however, was different. The psychological stimulus produced a larger increase in plasma epinephrine (100 pg/ml) than the physiological stimulus (25 pg/ml), whereas the physiological stressor increased NE (75 pg/ml) more than epinephrine (50 pg/ml) (LeBlanc et al. 1979).

Dimsdale and Moss (1980) observed different NE and epinephrine responses during isotonic exercise and public speaking. During public speaking, epinephrine levels increased twofold, whereas exercise increased NE levels threefold with little change in epinephrine levels. They suggested that psychological stimuli induced an adrenomedullary epinephrine release, whereas physiological stimuli primarily produced release of neuronal NE.

Ward et al. (1983) also examined the effects of physiological and psychological stimuli on catecholamine responses. They collected blood samples every 2 minutes as eight male subjects underwent venipuncture, isometric handgrip, knee bends (20 performed over 2 minutes), two blood pressure measurements, mental arithmetic, and a cold pressor stress. Mental arithmetic was a cognitive stressor that involved no movement. Isometric handgrip and knee bends were active physical stressors. The cold pressor test and venipuncture were passive painful stressors. The blood pressure readings were included to examine whether catecholamine release was so sensitive that this simple noninvasive procedure would produce significant increases in NE and epinephrine levels. The psychological stressor (mental arithmetic) caused the greatest increase in epinephrine, with venipuncture, knee bends, and cold pressor producing similar increases in epinephrine levels. Blood pressure and handgrip did not significantly increase epinephrine levels. On the other hand, the cold pressor test increased NE levels most, followed by venipuncture, mental arithmetic, knee bends, and handgrip. Blood pressure determination did not significantly change NE levels. Physical stresses increased NE levels, whereas psychological stressors produced a greater change in epinephrine levels.

Subjects adapt rapidly to laboratory stimuli, but in real-life situations human subjects may have chronically elevated sympathetic neuronal responses. Timio and Gentili (1976) made serial measurements of urinary catecholamine on 32 healthy men during piecework with payment by results, during work on an assembly line, and during work under ordinary conditions (for salary and away from an assembly line). NE and epinephrine levels were significantly higher for piecework and assembly-line workers compared with salaried workers under ordinary conditions. In a further study, urinary-free NE and epinephrine were higher throughout 4 days of piecework, with lower levels during fixed-wage days away from the assembly line. These differences persisted on retesting after 6 months of alternating work regimens. Thus, prolonged exposure and habituation did not decrease the catecholamine response to piecework and assembly-line work.

The plasma catecholamines (NE and epinephrine) are a reliable, albeit invasive, measure of the sympathetic nervous system response to stress. They are released proportionate to the severity of stress, and the amount of change in NE or epinephrine appears to differentiate the nature of the stressful stimulus. The plasma catecholamine response is related to the heart rate and blood pressure responses observed during stressful situations.

Receptor Assays

Receptor binding techniques measure the functional link between neuroendocrine signals and the end-organ responses they stimulate. Receptors determine the magnitude of the target organ's response, so receptor assessment is an important addition to assessment of adrenergic function. NE and epinephrine initiate physiological events by binding to adrenergic receptors. This mediates changes in heart rate, contractile force, vascular constriction and dilation, and blood pressure responses. Adrenergic receptors are classified into α_1, α_2, β_1, and β_2 based on the efficacy of agonists and antagonists in eliciting responses and on receptor binding affinities. α-Adrenergic receptors are more responsive to epinephrine and NE than isoproterenol, and stimulation results in vasoconstriction. β-Adrenergic receptors are more sensitive to isoproterenol (Weiner 1985) and mediate vasodilation, renin release, and increased heart rate.

Radioligand binding techniques can measure receptor density and receptor affinity for drugs and hormones. Tissues are incubated with a radiolabeled drug or hormone, and the unbound radioligand is removed by filtration. The bound radioactivity is used to calculate receptor indices. A number of theoretical and practical issues affect the binding experiments. Details are addressed elsewhere (Motulsky and Insel 1987; Robertson et al. 1989; Theode-Reynolds et al. 1986).

Although radioligands are chosen to be specific for a given receptor, there is usually some nonspecific binding to the membrane protein. The amount of nonspecific binding is determined by incubating excess concentrations of a nonlabeled ligand with the tissue and radioligand. The nonlabeled ligand will bind all of the specific receptors and leave only nonspecific sites for the radioligand. Subtracting the radioactivity from the nonspecific sites in the presence of unlabeled drug from that obtained without unlabeled drug (total binding) yields specific binding (Robertson et al. 1989).

Mathematical analysis of binding at various concentrations of radioligand yields receptor number and affinity. These measures are expressed as: B_{max}, a measure of the number of receptors; and Kd, the disassociation constant, a measure of the affinity of the radioligand for the receptor. The Kd represents the concentration of radioligand required to bind half of the receptors. The higher the disassociation constant, the lower the affinity of the receptor for ligand. Receptor affinity is a property of the receptor binding to a particular ligand and is not directly associated with receptor number (Mills and Dimsdale 1988).

β-Receptors mediate their effects via cAMP. The rate of cAMP production in response to the β-agonist isoproterenol provides another measure of β-receptors. The α_1- and β_1-receptors are most important in the cardiovascular system, but tissues containing these receptors are not easily accessible in humans. However, β_2-receptors are found on lymphocytes, and α_2-receptors are found on platelets.

The relationships among the adrenergic receptors, circulating catecholamine levels, and cardiovascular measures are important. It has been well documented that there is an association between levels of circulating NE and epinephrine with blood pressure and heart rate (Cryer 1980; Weiner 1985; Ziegler and Lake 1984). In the extreme cases of pheochromocytoma, where NE and epinephrine levels are very high, or autonomic neuropathy, where NE and epinephrine levels are very low, there is clear down-regulation and up-regulation of both α- and β-receptors. Most other human studies have produced less apparent receptor alterations in response to physiological changes in catecholamine levels. Exposure to β-adrenergic agonists decreases β-adrenergic receptors, whereas an exposure to antagonists increases β-receptors on lymphocytes (Aarons et al. 1983; Fraser et al. 1987) and other tissues (Tjujimoto et al. 1984). Furthermore, there is a positive correlation between lymphocyte β-adrenoceptor status with cardiac and lung β-adrenoceptor number (Aarons and Molinoff 1982; Brodde et al. 1986) and cardiac isoproterenol sensitivity (Fraser et al. 1987).

There are indications that adrenergic receptors can be influenced by sodium intake (Ambrosioni et al. 1982), hypertension (Manuck and Proietti 1982), age (Bertel et al. 1980; Matthews and Stoney 1988), fitness (O'Rourke and Kennedy 1988), and depression and anxiety (Cameron et al.

1990; Dimsdale 1988). Age, hypertension, and sodium intake interact to affect adrenergic receptor responsiveness (Dimsdale et al. 1987, 1988). It is clear that very high catecholamine levels downregulate α- and β-receptors, and very low catecholamine levels upregulate adrenergic receptors. Elderly persons and blacks have low β-receptor responses, whereas age and race have relatively little effect on α-receptors. There have been many studies of receptor levels in personality and psychiatric disorders, but results have often been contradictory.

Direct Measurement of Sympathetic Nerve Activity

Microneurography uses microelectrodes inserted percutaneously into peripheral sympathetic nerves for recording of single or multiple action potentials. This technique can record from any nerve that is accessible percutaneously. These nerves include the radial, median, tibial, and sural nerves, but the peroneal nerve is the most commonly recorded nerve because of its easy access. The peroneal nerve provides a stable recording site with minimal artifacts and has a high frequency of relatively pure nerve fascicles to muscle or skin.

Muscle sympathetic nerve activity (MSNA) can be obtained through any of the above-mentioned nerves. Stimulation of the nerve results in twitching of the innervated muscle without paresthesia (Anderson and Mark 1989). Both cardiopulmonary and carotid baroreceptors have a strong influence on MSNA. Inhibition of cardiopulmonary baroreceptors by lower-body negative pressure increased MSNA (Sundloff and Wallin 1978), whereas stimulation of the carotid baroreceptor by neck suction produced a profound but transient reduction in MSNA (Wallin and Eckberg 1982).

Static isometric handgrip increased MSNA, which was maintained after release of the handgrip (Mark et al. 1985). Although a rhythmic handgrip did not have a prolonged effect (Victor et al. 1987), rhythmic contraction of large muscle groups (arm cycling) was associated with increases in MSNA at higher workloads (Victor et al. 1987). These increases lagged behind increases in heart rate. MSNA also increased during a cold pressor test (Victor et al. 1985). Simultaneous recording from two muscle nerves (radial and peroneal) revealed similar patterns of sympathetic bursts at rest (Sundloff and Wallin 1977). However, Anderson et al. (1987) found that mental arithmetic increased MSNA activity in the peroneal nerve, but not in the radial nerve.

MSNA correlates with plasma NE levels (Morlin et al. 1983; Wallin et al. 1981), suggesting that adrenergic terminals in skeletal muscle provide a major source of plasma NE (Wallin 1984).

Percutaneous electrodes can also record skin sympathetic nerve activity (SSNA). Stimulation of the electrode produces paresthesia but not muscle twitch. MSNA is primarily vasoconstrictor; SSNA is vasoconstrictor, sudomotor, and piloerector (Hagbarth 1979). SSNA does not vary with fluctuations in blood pressure (Delius et al. 1972; Hagbarth et al. 1972), indicating little baroreceptor influence.

Body cooling increases SSNA, producing vasoconstriction. Subsequent body warming initially decreases SSNA, and then a progressive increase as perspiration begins (Wallin 1981). SSNA increases during mental challenge and decreases during relaxation. Both pleasant and unpleasant novel stimuli increase SSNA when first presented, but SSNA decreases after several repetitions (Hagbarth et al. 1972). These changes correlate with blood flow to the innervated cutaneous tissue (Wallin et al. 1974).

Agonists and Antagonists in Assessing Adrenergic Function

Physiological responses to the catecholamines are mediated by cells containing catecholamine receptors. Cellular sensitivity to the catecholamines can be inferred by infusing agonist drugs. NE has a moderate affinity for β_1-receptors and little effect at β_2-receptors. In contrast, epinephrine is a powerful agonist at both β_1- and β_2-receptors. A comparison of responses to stress before and after a β-blockade helps one infer if the stimulus is primarily in response to NE or epinephrine. Pharmacological agents have provided insight into the role of the sympathetic nervous system. α- and β-receptors were discovered through the effects of adrenergic antagonists and agonists.

Andren (1985) examined the effects of propranolol (a nonselective β-antagonist) and metoprolol (a β_1-selective antagonist) on mental arithmetic; cold pressor test; and noise on blood pressure, cardiac output and vascular resistance (as determined from ICG), and NE and epinephrine. The cold pressor and noise both caused vasoconstriction through increases in vascular resistance and no change in cardiac output. β-Blockade (either selective or nonselective) brought about no change in the hemodynamic responses to the cold pressor test. Mental arithmetic caused an increase in blood pressure, resulting from an increase in heart rate and cardiac output and from decreases in vascular resistance. Nonselective β-blockade caused a reverse in the hemodynamic pattern during mental arithmetic, and β_1-selective blockade caused less vasoconstriction, indicating that the cardiovascular changes following mental arithmetic were due to epinephrine. Epinephrine stimulates both α- and β-adrenoceptors, leading to β_2-mediated vasodilation. If the vascular β_2-receptors are blocked, epinephrine stimulation of α-receptors increases DBP. Selective β_1-blocking agents, such as metoprolol have little effect on vascular β_2-receptors.

In a study examining the effect of α-agonists and antagonists specific for α_1- and α_2-receptors, Jie et al. (1985) found that both α_1- and α_2-receptor stimulation result in vasoconstriction and that NE and epinephrine interact with both receptor subtypes. The cold pressor test normally reduces forearm blood flow; α_1-adrenoceptor blockade with prazosin given intra-arterially to the forearm increased forearm blood flow during the cold pressor test. Drug strategies such as this permit one to determine which catecholamine and what type of receptor mediates a physiological response.

Conclusion

In this chapter, we focused on a number of techniques to evaluate adrenergic function in vivo. The methods of end-organ response evaluation are easy to obtain, both in the laboratory and in the field. These techniques provide basic information as to sympathetic function, but are of limited value for detailed evaluation of adrenergic activation.

Measurement of epinephrine and NE reflect adrenergic function more specifically. Plasma levels of these catecholamines can be used to estimate changes in adrenergic activity over relatively short periods, whereas urinary measures can evaluate the effects of longer stressful events. Coupling catecholamine levels with the measures of adrenergic receptors provides a more specific evaluation of adrenergic activation. The most specific measurement of adrenergic activity is through direct recording of sympathetic nerve activity. This technique, however, is limited in its application, as it requires the subject to remain relatively quiet and in the laboratory during the recording.

Although all of these techniques for assessing adrenergic function provide specific information, the coupling of catecholamine measurement with end-organ responses will provide the investigator with the most flexible and reliable assessment of adrenergic function.

References

Aarons RD, Molinoff PB: Changes in the density of beta-adrenergic receptors in rat lymphocytes, heart and lung after chronic treatment with propranolol. J Pharmacol Exp Ther 221:439–443, 1982

Aarons RD, Nies, Gerber JG, et al: Decreased beta adrenergic receptor density on human lymphocytes after chronic treatment with agonists. J Pharmacol Exp Ther 224:1–6, 1983

Abe K, Konaka R: Specific determination of 3-methoxy-4-hydroxyphenylethylene glycol in urine by liquid chromatography with post-column reaction. Clin Chem 34:87–90, 1988

Ahmed SS, Levinson GE, Schwarz CJ, et al: Systolic time intervals as measures of the contractile state of the left ventricular myocardium in man. Circulation 52:559–571, 1972

Allen MT, Obrist PA, Sherwood A, et al: Evaluation of myocardial and peripheral vascular responses during reaction time, mental arithmetic and cold pressor tasks. Psychophysiology 24:648–655, 1987

Ambrosioni E, Costa FV, Borghi C, et al: Effects of moderate salt restriction on intra lymphocytic sodium and pressor response to stress in borderline hypertension. Hypertension 4:789–794, 1982

Anderson EA, Mark AL: Microneurographic measurement of sympathetic nerve activity in humans, in Handbook of Research Methods in Cardiovascular Behavioral Medicine. Edited by Schneiderman N, Weiss S, Kaufmann PG. New York, Plenum, 1989, pp 107–115

Anderson EA, Wallin BG, Mark AL: Dissociation of sympathetic nerve activity to arm and leg during mental stress. Hypertension 9:III-114–III-119, 1987

Andren L: The effect of β-blockade on stress induced vasoconstriction, in Adrenergic Blood Pressure Regulation. Edited by Birkenhager WH, Folkow B, Struyker Boudier HAJ. Amsterdam, Excerpta Medica, 1985, pp 117–131

Baum A, Gatchel RS, Fleming R, et al: Chronic and acute stress associated with Three Mile Island accident and decontamination: preliminary findings of longitudinal study. Technical report submitted to the US Nuclear Regulatory Commission, Washington, DC, 1981

Berne RM, Levy MN: Cardiovascular Physiology. St Louis, MO, CV Mosby, 1981

Bertel O, Fritz FR, Kiowski W, et al: Decreased beta-adrenoreceptor responsiveness to age, blood pressure and plasma catecholamines in patients with essential hypertension. Hypertension 2:130–138, 1980

Biberman L, Sarma RN, Surawicz B: T-wave abnormalities during hyperventilation and isoproterenol infusion. Am Heart J 81:166–174, 1971

Boudoulos H, Geleris P, Lewis RP, et al: Linear relationships between electrical systole and heart rate. Chest 80:613–617, 1981

Brodde OE, Kretch R, Ikazano K, et al: Human beta-adreno-receptors: relation of myocardial and lymphocyte beta-adrenoreceptor density. Science 231:1584–1585, 1986

Cameron OG, Smith CB, Lee MA, et al: Adrenergic status in anxiety disorders: platelet alpha$_2$-adrenergic receptor binding, blood pressure, pulse and plasma catecholamines in panic and generalized anxiety disorder patients and in normal subjects. Biol Psychiatry 28:3–20, 1990

Chang PC, van der Krogt JA, Vermeij P, et al: Norepinephrine removal and release in the forearm of healthy subjects. Hypertension 8:801–809, 1986

Cryer PE: Physiology and pathophysiology of the sympatho-adrenal neuroendocrine system. N Engl J Med 303:436–444, 1980

Delius W, Hagbarth KE, Hongell A, et al: Manoeuvers affecting sympathetic outflow in human skin nerves. Acta Physiol Scand 84:177–186, 1972

Dimsdale JE: The effect of depression on cardiovascular reactivity, in Stress and Coping Across Development. Edited by Field TM, McCabe PM, Schneiderman N. Hillsdale, NJ, Lawrence Erlbaum Associates, 1988, pp 132–166

Dimsdale JE, Moss J: Plasma catecholamines in stress and exercise. JAMA 243:340–342, 1980

Dimsdale JE, Graham R, Ziegler MG, et al: Age, race, diagnosis and sodium effects on the pressor response to infused norepinephrine. Hypertension 10:564–569, 1987

Dimsdale JE, Ziegler MG, Graham R: Salt, race and diagnosis affects isoproterenol sensitivity. Clin Exp Hypertens [A] 10:747–756, 1988

Durrett LR, Ziegler MG: A sensitive radioenzymatic assay for catechol drugs. J Neurosci Res 5:587–598, 1980

Folkow B: Physiological aspects of primary hypertension. Physiol Rev 62:347–504, 1982

Frankenhauser M, Johansson G: Task demand as reflected in catecholamine excretion and heart rate. Journal of Human Stress 2:15–23, 1976

Fraser JA, Nadeau JHJ, Robertson D, et al: Regulation of human leucocyte beta receptors by endogenous catecholamines: relationship of leucocyte beta receptor density to the cardiac sensitivity to isoproterenol. J Clin Invest 67:1777–1784, 1987

Furedy JJ, Heslegrave RS: A consideration of recent criticisms of the T-wave amplitude index of myocardial sympathetic activity. Psychophysiology 20:204–211, 1983

Goldstein DS, Eisenhofer G, Sax FL, et al: Plasma norepinephrine pharmacokinetics during mental challenge. Psychosom Med 49:591–605, 1987

Goto M: Dual-electrode electrochemical detection of catecholamines and related compounds analyzed by micro HPLC, in Quantitative Analysis of Catecholamines and Related Compounds. Edited by Krstulovic AM. New York, Wiley, 1986, pp 187–208

Grota LJ, Brown GM: Immunoassay of catecholamines and other phenethyiamine derivatives, in Quantitative Analysis of Catecholamines and Related Compounds. Edited by Krstulovic AM. New York, Wiley, 1986, pp 282–301

Hagbarth KE: Exteroceptive, proprioceptive and sympathetic activity recorded with micro electrodes from human peripheral nerves. Mayo Clin Proc 54:353–365, 1979

Hagbarth KE, Hallin RG, Hongell A, et al: General characteristics of sympathetic activity in human skin nerves. Acta Physiol Scand 84:164–176, 1972

Han J, Moe GK: Non-uniform recovery of excitability in ventricular muscle. Circ Res 14:44–60, 1964

Han J, de Jalon J, Moe GK: Adrenergic effects on ventricular vulnerability. Circ Res 14:516–524, 1964

Hartwell AS, Burrett JB, Braybield A, et al: The effect of exercise and four commonly used drugs on the normal human electrocardiogram with particular reference to T-wave changes. J Clin Invest 21:409–417, 1942

Heather LW: A comparison of cardiac output values by the impedance cardiograph and dye dilution techniques in cardiac patients, in Development and Evaluation of an Impedance Cardiographic System to Measure Cardiac Output and Other Cardiac Parameters (CR 101965). Edited by Kublick WG, Witsoe DA, Patterson RP. Houston, TX, NASA Publications, 1969, pp 22–27

Henry DP, Luft FC, Weinberger MH, et al: Norepinephrine in urine and plasma following provocative maneuvers in normal and hypertensive subjects. Hypertension 2:20–28, 1980

Hjemdahl P: Inter-laboratory comparison of plasma catecholamine determinations using several different assays. Acta Physiol Scand 527:43–54, 1984

Hurwitz BE, Nelesen RA, Saab PG, et al: Dynamic cardiovascular processes supporting blood pressure elevation to three behavioral stressors. Biol Psychology 36:75–95, 1993

Imai K, Tsukamoto M, Tamura Z: L C of rat-brain dopamine and norepinephrine. J Chromatogr 137:357–362, 1977

Jessen K: An assessment of human regulatory nonshivering thermogenesis. Scandinavian Society of Anesthesiology 24:138–143, 1980

Jie K, van Brummelen P, Vereij P, et al: Alpha- and β_2-adrenoceptor mediated vasoconstriction in the forearm: differences between normotensive and hypertensive subjects. J Hypertens 3:S89–S91, 1985

Johnson BF, Meeran MF, Frank A, et al: Systolic time intervals in measurement of inotropic response to drugs. Br Heart J 46:513–521, 1981

Kagedal B, Goldstein DS: Catecholamines and their metabolites. J Chromatogr 429:177–233, 1988

Kennedy B, Ziegler MG: A more sensitive and specific radio-enzymatic assay for catecholamines. Life Sci 47:2143–2154, 1990

Lacey JI: Somatic response patterning and stress: some revisions of activation theory, in Psychological Stress. Edited by Appley MH, Trumbel R. New York, Appleton-Century-Crofts, 1967, pp 10–18

Lake CR, Ziegler MG, Kopin IJ: Use of norepinephrine for evaluation of sympathetic neuronal function of man. Life Sci 18:1315–1326, 1976

Lake CR, Chernow B, Feuerstein G, et al: The sympathetic nervous system in man: its evaluation and the measurement of plasma NE, in Norepinephrine. Edited by Ziegler MG, Lake CR. Baltimore, MD, Williams & Wilkins, 1984, pp 1–26

Larsen PB, Schneiderman N, Pasin RD: Physiological bases of cardiovascular psychophysiology, in Psychophysiology: Systems Processes and Applications. Edited by Coles MGH, Donchin E, Porges SW. New York, Guilford, 1986, pp 336–348

LeBlanc J, Cote J, Jobin, et al: Plasma catecholamine and cardiovascular responses to cold and mental activity. J Appl Physiol 47:1207–1211, 1979

Leduc J: Excretion of catecholamines in rats exposed to cold. Acta Physiol Scand 51:94–96, 1961

Lewis RP, Rittigeris SE, Forester WF, et al: A critical review of systolic time intervals. Circulation 56:146–158, 1977

Manuck SB, Proietti JM: Parental hypertension and cardiovascular response to cognitive and isometric challenge. Psychophysiology 19:481–489, 1982

Mark AL, Victor RG, Nerhed C, et al: Microneurographic studies of the mechanisms of sympathetic nerve responses to static exercise in humans. Circ Res 57:461–469, 1985

Matthews KA, Stoney CM: Influence of sex and age on cardiovascular response during stress. Psychosom Med 50:46–56, 1988

Mefford IN: Lauroyl sarcosine: a weak cation-exchange reagent for on column modification of reversed-phase material. J Chromatogr 393:441–446, 1987

Mills PJ, Dimsdale JE: The promise of receptor studies in psychophysiologic research. Psychosom Med 50:555–566, 1988

Minegeshi A, Ishizaki T: Rapid and simple method for the simultaneous determination of 3,4 dihydroxy phenylacetic acid, 5-hydroxy indole-3-acetic acid and 4 hydroxy-3-methoxy phenylacetic acid in human plasma by high-performance liquid chromatography with electrochemical detection. J Chromatogr 308:55–63, 1984

Miura Y, Campese V, DeQuattro V, et al: Plasma catecholamines via an improved fluorimetric assay: comparison with an enzymatic method. J Lab Clin Med 89:421–427, 1977

Mori K: Automated HPLC analysis with fluorimetric detection: the HI method, in Quantitative Analysis of Catecholamines and Related Compounds. Edited by Krstulovic AM. New York, Wiley, 1986, pp 340–356

Morlin C, Wallin BG, Eriksson BM: Muscle sympathetic activity and plasma noradrenaline in normotensive and hypertensive man. Acta Physiol Scand 119:117–121, 1983

Motulsky HJ, Insel PA: In vitro methods for studying human adrenergic receptors: methods and applications, in Adrenergic Receptors in Man. Edited by Insel PA. New York, Marcel Dekker, 1987, pp 139–160

Nagel JH, Shyu LY, Reddy SP, et al: New signal processing techniques for improved precision of non-invasive impedance cardiography. Ann Biomed Eng 17:517–534, 1989

Nichols MG, Kiowski W, Zweifler AJ, et al: Plasma norepinephrine variations with dietary sodium intake. Hypertension 2:27–32, 1980

Obrist PA: Cardiovascular Psychophysiology: A Perspective. New York, Plenum, 1981

O'Rourke DF, Kennedy EE: Cardiovascular responsivity to physical and mental stressors: relationship to fitness and age. Psychosom Med 50:206, 1988

Randall WC: Sympathetic control of the heart, in Neural Regulation of the Heart. Edited by Randall WC. New York, Oxford University Press, 1977, pp 56–70

Rauste-von Wright M, von Wright J, Frankenhauser M: Relations between sex-related psychological characteristics during adolescence and catecholamine excretion during achievement stress. Psychophysiology 18:362–370, 1981

Robertson D, Parfynonova Y, Menshikov M, et al: Receptors, in Handbook of Research Methods in Cardiovascular Behavioral Medicine. Edited by Schneiderman N, Weiss S, Kaufmann PG. New York, Plenum, 1989, pp 221–236

Rose RM: Endocrine response to stressful psychological events. Psychiatr Clin North Am 3:25–27, 1980

Sherwood A, Allen MT, Obrist PA, et al: Evaluation of beta-adrenergic influences on cardiovascular and metabolic adjustments to physical and psychological stress. Psychophysiology 23:89–104, 1986

Stulik K, Pacakova V: Electrochemical detection of catecholamines and related compounds, in Quantitative Analysis of Catecholamines and Related Compounds. Edited by Krstulovic AM. New York, Wiley, 1986

Sundloff G, Wallin BG: The variability of muscle sympathetic nerve activity in resting recumbent man. J Physiol (Lond) 272:399–414, 1977

Sundloff G, Wallin BG: Effect of lower body negative pressure on human muscle nerve sympathetic activity. J Physiol (Lond) 278:525–532, 1978

Theode-Reynolds KR, Motulsky JH, Feldman RD: Temperature-dependent binding of hydrophilic beta-adrenergic receptor ligands to intact human lymphocytes. Life Sci 39:1325–1334, 1986

Timio M, Gentili S: Adrenosympathetic over activity under conditions of work stress. British Journal of Preventative Social Medicine 30:262–265, 1976

Tjujimoto G, Manger WM, Hoffman BB: Desensitization of beta-adrenergic receptors by pheochromocytoma. Endocrinology 114:1272–78, 1984

Victor R, Leimbach W, Wallin G, et al: Microneurographic evidence for increased cerebral sympathetic neural drive during the cold pressor test. J Am Coll Cardiol 5:II-415, 1985

Victor RG, Seals DR, Mark AL: Differential control of heart rate and sympathetic nerve activity during dynamic exercise: insight from direct intraneural recordings in humans. J Clin Invest 79:508–516, 1987

von Euler US, Ikkos D, Luft R: Adrenaline excretion during resting conditions and after insulin in adrenalectomized human subjects. Acta Endocrinol (Copenh) 38:441, 1961

Wallin BG: Sympathetic nerve activity underlying electrodermal and cardiovascular reactions in man. Psychophysiology 18:470–476, 1981

Wallin BG: Sympathetic activity in human extremity nerves and its relationship to plasma norepinephrine, in Norepinephrine. Edited by Ziegler MG, Lake CR. Baltimore, MD, Williams & Wilkins, 1984, pp 431–438

Wallin BG, Eckberg DL: Sympathetic transients by abrupt alterations of carotid baroreceptor activity in humans. Am J Physiol 242:H185–H190, 1982

Wallin BG, Delius W, Hagbarth KE: Regional control of sympathetic outflow in human skin and muscle nerves, in Central Rhythmic and Regulation. Edited by Umbach W, Koepchen HP. Stuttgart, Hippokrats Verlag, 1974, pp 266–277

Wallin BG, Sundloff G, Eriksson BM, et al: Plasma noradrenaline correlates to sympathetic muscle nerve activity in normotensive man. Acta Physiol Scand 111:69–73, 1981

Ward MM, Mefford IN, Parker SD, et al: Epinephrine and norepinephrine responses in continuously collected human plasma to a series of stressors. Psychosom Med 45:471–486, 1983

Watson RDS, Littler WA, Eriksson BM: Changes in plasma noradrenaline and adrenaline during isometric exercise. Clin Exp Pharmacol Physiol 7:399–402, 1980

Weiner N: Norepinephrine, epinephrine and the sympathomimetic amines, in Goodman and Gillman's The Pharmacological Basis of Therapeutics, 6th Edition. Edited by Gillman AG, Goodman LS, Rall TW, et al. New York, MacMillan, 1985, pp 145–180

Weir TB, Smitt CC, Round JN, et al: Stability of catecholamines in whole blood plasma and platelets. Clin Chem 32:882–883, 1986

Winegrad S: Calcium release from cardiac sarcoplasmic reticulum. Annu Rev Physiol 44:451–462, 1982

Ziegler MG: Catecholamine measurement in behavioral research, in Handbook of Research Methods in Cardiovascular Behavioral Medicine. Edited by Schneiderman N, Weiss S, Kaufmann PG. New York, Plenum, 1989, pp 167–184

Ziegler MG, Lake CR: Norepinephrine. Baltimore, MD, Williams & Wilkins, 1984

Ziegler MG, Lake CR, Foppen FH, et al: Norepinephrine in cerebrospinal fluid. Brain Res 108:436–440, 1976

Ziegler MG, Kennedy B, Elayan H: Rat renal epinephrine synthesis. J Clin Invest 84:1130–1133, 1989

Section II

Psychosomatic Considerations

8

Stress and Science

DAVID S. GOLDSTEIN, M.D., PH.D.

Scientific consideration of the role of adrenergic function in health and disease should begin with a theory about what stress is. In this chapter, I present a few fundamental concepts (Bernard's *milieu interieur* and Cannon's "homeostasis"), analyze the stress theory of Selye, propose new definitions of stress and distress, view stress in the perspective of evolution, and provide concepts for understanding the psychobiological roles of catecholaminergic systems.

Bernard and the *Milieu Interieur*

About 150 years ago, Claude Bernard (1813–1878) concentrated on the fundamental physiological issue of the time: How does the living body work? Do living beings share a vital essence that is beyond understanding by physical or chemical laws? Or can all bodily processes be explained in terms of laws based on observation and experimentation? These were the opposing views of the vitalist and determinist schools.

Bernard was the foremost proponent of the determinist view.

Bernard's creed was that by scientific observation, one could grasp the laws governing and so predict the activities of body systems. The guiding principle of what he called "scientific determinism" was that if the experimental conditions were identical, then the experimental results would be the same—that is, the results would be determined.

Bernard's theory of the *milieu interieur,* explaining the actions of body processes to achieve the goal of constancy of the internal environment, and Darwin's theory of evolution, explaining the emergence of physical and behavioral characteristics according to natural selection, were approximately contemporary. Bernard's and Darwin's theories proposed analogous laws according to which individuals and species survive and evolve. The present conception of stress continues this tradition, in viewing stress responses as having evolved because they subserve adaptive goals.

The apparent purposiveness of stress responses does not imply either that they are experienced consciously or that they result from an exercise of will, just as the apparent purposiveness of a thermostat does not imply that it "wants" to maintain a constant room temperature. The crucial distinction between the theories of Bernard and Darwin, compared with those espoused by teleological vitalists, has been the ability to test the derivative ideas by observation and experiment.

Bernard's most enduring contribution to scientific medicine has been his theory of the *milieu interieur,* the internal environment, the "inner sea" bathing our cells. Bernard's theory answers the question, Why do body systems function as they do? by answering: to maintain a stable internal environment. He suggested that the body maintains a constant internal environment by myriad, continual, compensatory reactions, tending to restore a state of equilibrium, thereby allowing independence from the external environment. He taught that "it is the fixity of the *'milieu interieur'* which is the condition of free and independent life. All the vital mechanisms, varied as they are, have only one object, that of preserving constant the conditions of life in the internal environment" (Bernard 1878; qtd. in Olmstead and Olmstead 1952, p. 224).

Though this view must seem straightforward or even simple-minded today, it is new in the history of medicine. Because it imputes a goal, or purpose, to bodily processes, it is essentially teleological. As will be seen, it also is the basis for the present conception of the "what" and "why" of stress.

During the era of ascendance of the germ theory of disease at the end of the 1800s, the importance of the equilibrium of the organism was relatively ignored. By now, however, in the era of immunology and molecular genetics, Bernard seems vindicated, with many diseases thought to result from deleterious interactions between heredity and environment, or from genetic flaws outright.

Bernard's idea of the *milieu interieur* is the first and main guiding principle for the present views about interrelationships among stress, catecholaminergic systems, and disease: All life forms possess means to preserve their integrity, by compensatory activation of systems to maintain the constancy of the internal environment.

Cannon and Homeostasis

Walter B. Cannon (1871–1945) extended Bernard's concepts by showing that the maintenance of a stable internal environment depends crucially on catecholamines, the chemical messengers of what has been termed the *sympathoadrenal system.*

The existence of this system can be traced to primordial challenges all higher organisms have faced as a direct consequence of their complex structure. Whereas single-celled organisms normally are surrounded by the nutrient medium they need, higher organisms must use internal systems—especially the cardiovascular system—to maintain cellular temperature, take in and distribute oxygen, ingest and deliver fuel and water, and deliver by-products of cellular metabolism to excretion sites. The evolution of increasingly complex (and increasingly energy-requiring) systems to deliver fuel and remove waste has paralleled the evolution of the complex organization that enhances resistance to the vicissitudes of the external environment. This theme is discussed later in the section about stress in evolutionary perspective. The price of independence and insulation from the external environment is the maintenance of internal systems to deliver fuel, remove waste, provide water with the correct tonicity and electrolyte contents, and optimize temperature for cellular processes.

Efficient body functioning depends on continuous monitoring and coordination of blood flows to the various organs, allowing appropriate levels and redistribution of the cardiac output to meet internal and external demands, yet maintaining blood flow to the vital organs—the heart, lungs, and especially the brain. This monitoring depends on perceptions of the outside world and sensations from the inside, and the coordination depends mainly on the sympathoadrenal nervous system.

Adjustments in sympathoadrenal activity accompany virtually every human action and reaction and are the main means for acute regulation of cardiovascular function—from the subtle, unconscious shifts of the distribution of blood volume that occur each time we stand, to the desperate distress syndrome of shock. Because of the prominence of the sympathoadrenal system in cardiovascular regulation and the obviousness of cardiovascular changes during stress responses, I emphasize in this chapter these aspects of adrenergic function and dysfunction.

The body's effector systems determine the patterns of physiological and chemical responses attending stress. The sympathoadrenal system and the pituitary-adrenocortical system are the two main stress effector systems of the body. There are several others as well, and new ones are continually being proposed.

Although Cannon considered the sympathoadrenal system to function as a unit during emergency reactions, abundant evidence supports a distinction between its neuronal and hormonal components.

The sympathoneural system, the neuronal component of the sympathoadrenal system, consists of nerve networks of postganglionic neurons. Sympathetic nerves and nerve terminals densely enmesh arterioles and form lattice-like networks in the myocardium and glands such as the salivary and sweat glands.

Stimulation of sympathetic nerves almost instantaneously constricts arterioles and therefore increases regional resistance to blood flow. The increased vascular resistance shifts the blood to other regions with less resistance. Different patterns of sympathoneural activation in the kidneys, the gut (splanchnic) region, the heart, the skin, and skeletal muscle therefore result in rapid changes in the distribution of blood volume. Diffuse sympathetic stimulation invariably increases the total arteriolar resistance to blood flow, so that blood pressure increases. Stimulation of sympathetic nerves to the heart increases the force and rate of contraction, amplifying the pressor effect. Stimulation of nerves to the salivary glands increases salivation, to the eye dilates the pupil, to sweat glands increases sweating, to hair follicles causes goose bumps (piloerection), to the kidneys inhibits excretion of sodium, and to skeletal muscle causes trembling.

The chemical transmitter at sympathetic nerve endings is norepinephrine, or noradrenaline. Norepinephrine in the periphery satisfies the three main criteria for defining a neurotransmitter: It is a chemical that is released from nerve terminals by electrical action potentials, interacts with specific receptors on nearby structures, and in so doing produces specific physiological responses.

The sympathoneural system is responsible primarily for adaptive adjustments in blood flow distribution, cardiac function, and glandular activity. These adjustments, which often are not sensed, are mediated mainly by release of norepinephrine and occupation of the adrenoceptors on the smooth muscle or secretory cells. Because the sympathoneural system is a nerve network, the structure of the system allows the possibility that different stressors can elicit different patterns of sympathoneural outflows in the various vascular beds. The resulting differential vasoconstriction can in turn produce the alterations in distribution of blood flow. Examples of situations associated with predominantly sympathoneural activation include orthostasis, mild exercise, postprandial hemodynamic changes, mild changes in environmental temperature, and performance of nondistressing locomotor tasks.

The adrenomedullary component of the sympathoadrenal system includes chromaffin cells of the adrenal medulla. These cells receive mainly preganglionic input from cell bodies of the intermediolateral columns of the spinal cord. The adrenomedullary cells secrete catecholamines—in humans mainly epinephrine, or adrenaline—directly into the bloodstream. The adrenomedullary component of the sympathoadrenal system therefore is hormonal.

Epinephrine affects the function of most body systems. Thus, exogenously administered epinephrine rapidly increases the rate and force of cardiac contraction; increases myocardial cell automaticity; dilates bronchioles and increases the rate of breathing; redistributes blood volume toward the heart, brain, and skeletal muscle and away from the skin, kidneys, and gut; enhances the aggregability of platelets; relaxes smooth muscle of the uterus and gut; increases blood glucose by a variety of means including glycogenolysis; dilates pupils; increases activity of the renin-angiotensin-aldosterone system; decreases serum potassium concentrations; increases the metabolic rate; and produces psychological effects such as increased alertness, decreased fatigue, and intensification of emotions.

All these effects can be viewed as adaptive during several types of emergency reactions, including traumatic hemorrhage, hypoglycemia, asphyxiation, circulatory collapse, and emotional distress, when the individual senses an overall threat to well-being or survival.

Homeostasis

Cannon coined the term *homeostasis* to describe the coordinated physiological processes that maintain most of the steady states in the organism (Cannon 1939). According to Cannon, the most important process determining the phylogenetic development of physiological and even psychological patterns of response has been essentially the same process that has determined the development of anatomical patterns: natural selection. Cannon suggested that rapid activation of homeostatic systems preserves the internal milieu by producing compensatory and anticipatory adjustments that enhance the likelihood of survival of the organism. He explained the emotional responses of rage and fear—which he called "fight or flight" responses—in these terms.

One would predict from this view that in cardiovascular disorders, alterations in sympathetic activity reflect an attempt at compensation to maintain homeostasis, rather than reflect a primary etiologic abnormality. This is the essence of the problem in elucidating the role of the sympathetic system in cardiovascular disorders.

Consideration of the large constellation of effects of sympathetic stimulation leads to two questions. First, depending on the specific threat, would not one or more of these effects be irrelevant to homeostasis? Cannon recognized this potential limitation, and his explanation is discussed later in this chapter. Second, during emergency situations such as "fight-or-flight" encounters, what is homeostasis? Values for many parameters would be regulated at new levels, in essence redefining homeostasis, at least temporarily. This Cannon did not recognize. The present theory holds that emotional distress alters homeostatic function.

From the work and ideas of Cannon, we glean the following principles. First, homeostatic systems maintain most of the steady states in the organism. Second, the sympathoadrenal system regulates cardiovascular performance to maintain homeostasis, especially during emergency reactions.

The theories of Bernard and Cannon did not incorporate stress per se as a scientific concept, nor the relationship between stress and disease. Hans Selye concentrated on these issues.

Selye and the General Adaptation Syndrome

Hans Selye (1907–1982) popularized stress as a scientific and medical concept. He elaborated a theory of stress as a condition shared by all organisms in their interaction with the environment: "Stress is the nonspecific response of the body to any demand upon it" (Selye 1974, p. 14).

The Doctrine of Nonspecificity

Selye emphasized from the start that stress and stress responses are nonspecific (Selye 1936); that is, they can be brought on regardless of the stimulus. After specific responses are removed from consideration, a nonspecific syndrome remains:

> In addition to their specific actions, all agents to which we are exposed also produce a nonspecific increase in the need to perform adaptive functions and thereby to reestablish normalcy. This is independent of the specific activity that caused the rise in requirements. The nonspecific demand for activity as such is the essence of stress. (Selye 1974, pp. 14–15)

The General Adaptation Syndrome

Although nonspecific with respect to the inciting agents, the stress response itself was viewed to consist of a stereotyped pattern. The stereotyped pathophysiological pattern to which Selye referred consisted of three components: enlargement of the adrenal glands, involution of the thymus gland (associated with atrophy of lymph nodes and inhibition of inflammatory responses), and peptic ulceration of the stomach. A decrease in the circulating number of eosinophils was added later to this triad. All these responses are able to be elicited by exogenously administrated glucocorticoids.

The stress response was also thought to have characteristic stages in a triphasic "general adaptation syndrome." The first stage was a rapid "alarm" reaction, including hypothermia, hypotension, neural depression, decreased muscle tone, hemoconcentration, altered capillary and red blood cell permeability, catabolism, hyperkalemia, acidosis, leukopenia followed by leukocytosis, hyperglycemia followed by hypoglycemia, and gastric erosions (Selye 1950). The alarm stage was thought to consist of two periods: shock and countershock, the latter due to responses of the adrenal cortex (releasing "corticoids") and responses of the adrenal medulla (releasing "adrenalines"). The countershock responses were thought to reverse many of the above-mentioned abnormalities and to induce hypertension, hyperglycemia, increased blood volume, alkalosis, diuresis, and often hyperthermia (Selye 1950). Morphologically there would be enlargement of the adrenal cortex and acute involution of the thymico-lymphatic apparatus due to the release of anti-inflammatory corticoids.

After an undefined length of time, a longer-lasting stage of "resistance" would follow the alarm stage. The resistance stage was thought to be characterized by increased resistance to the particular stressor agent but decreased resistance to other stimuli (Selye 1950). During this stage, most of the above morphological and biochemical changes would disappear. The stage of resistance was dominated by a balanced effect of "syntoxic" and "catatoxic" hormones of the adrenal cortex. The adrenal corticosteroids (cortisol in humans, corticosterone in rats) were thought to be syntoxic in that they would help the body to put up with aggressors by acting as "tissue tranquilizers," inhibiting defensive reactions such as immune and inflammatory responses. Catatoxic agents were thought to be proinflammatory, destroying aggressors by destructive enzymatic attack. The catatoxic hormones of the body were never identified conclusively.

The stage of resistance would end when stores of "adaptation energy" were spent. This would usher in the third stage, "exhaustion," with a resurgence of pituitary-adrenocortical hyperactivity, attendant gastrointestinal ulcers and immunological failure, and eventual death of the organism.

Eustress, Distress, and Diseases of Adaptation

According to Selye's theory, stress is not necessarily deleterious. He coined the term *eustress* to refer to stress that was not harmful and possibly was helpful to the body. *Distress* referred to damaging or unpleasant stress (Selye 1974). Excessive, repeated, or inappropriate stress responses were viewed as maladaptive, and Selye coined the phrase "diseases of adaptation" to refer to situations where the general adaptation syndrome is "derailed" (Selye 1956, p. 47). The contributions of stress to the diseases of adaptation were suggested mainly on the basis of effects of large doses of glucocorticoids or mineralocorticoids. If abnormal (hyper-, hypo-, or dysadaptive) responses did not directly cause these diseases, then they were thought to predispose the individual to develop these diseases, based on tendencies called "conditioning factors."

Selye proposed an immense list of diseases of adaptation. Hyperfunctional and dysfunctional conditions included Cushing's disease, adrenal tumors, chromaffinomas, renal artery stenosis, hypertension, periarteritis nodosa, nephrosclerosis, nephritis, rheumatic and inflammatory diseases, gouty arthritis, peptic ulceration, eclampsia, diabetes, allergic and hypersensitivity disorders, and psychosomatic disorders. Hypofunctional conditions included Addison's disease, Waterhouse-Friderichsen syndrome, cancer, and diseases of resistance in general (Selye 1950, 1974). The most severely affected targets were thought to be the cardiovascular system, the joints, and metabolism.

Critique of Selye's Stress Theory

Selye's stress theory became and remains popular. Among other things, the theory provides a ready explanation for how any distressing experience can lead to virtually any disease state. By 1982, Selye's International Institute of Stress had catalogued more than 200,000 publications about various aspects of the stress concept (Selye 1982). Although Selye's stress theory has certainly provoked thought and research, the theory has proven to be deficient in several important respects.

Circularities. Circularities vitiate Selye's theory. The first circularity is that the theory views stress either as a response pattern or as a nonobservable *condition* (Selye 1956) that leads invariably to but can be detected only by the general adaptation syndrome. The occurrence of the syndrome would be the only means by which the existence of the condition could be deduced, and only the condition could produce the syndrome.

To explain how different patients could evince different diseases of adaptation, Selye hypothesized that "conditioning factors" selectively enhance or inhibit particular stress effects. The conditioning factors could be internal (e.g., hereditary predispositions, aging, sex) or external (e.g., drugs, nutritional factors). Conditioning could explain any deviation from the predicted pattern of stress-induced lesions; however, the presence of conditioning could be detected only by this deviation.

Distress was defined as stress that was unpleasant or harmful to the body (Selye 1974). The only means to determine whether a particular stress was a distress or "eustress" would be to observe whether there was observable tissue damage or shortened survival.

Adaptation energy was defined as that which was consumed during adaptive work. When that energy was depleted, the state of exhaustion would lead to death. The measure of the success of the adaptive response to a particular stressor was not the ability to mitigate the intensity or noxiousness of the stressor but its effects on pathology and survival. If these were affected negatively, then the response was viewed as maladaptive. If there were no pathological consequences to a stress response, then the active and passive defenses were viewed as in balance; if there were pathological consequences, then this was viewed as due to an imbalance and maladaptation. The only way to determine whether such an imbalance had occurred was by the pathological effects.

Selye claimed that all he knew about adaptation energy was that constant exposure to any stressor would exhaust the adaptation energy (Selye 1956). He considered the human life span to be determined primarily by the amount of available adaptation energy. Indeed, Selye suggested that there was a fundamental, specific element of life, a unit of reactivity to biological stimuli, that could be used to define life. He called this functional unit of life a "reacton" (Selye 1956, p. 233). The idea of adaptation energy not only constitutes a circularity in Selye's theory, it also smacks of vitalism.

Inconsistencies. As noted above, Selye (1956) defined stress as the nonspecific *response* of the body to any demand imposed on it. Elsewhere, he claimed that this was an operational definition and that stress was the "sum of all the wear and tear in the body caused by life at any one time" (p. 222), or that stress was a nonobservable *condition* that led invariably to and therefore was detectable and measurable by the general adaptation syndrome.

The stage of resistance was thought to be characterized by decreased responsiveness to the stressor to which the organism had been exposed repeatedly, but increased responsiveness to other stressors. Although several studies have confirmed this observation (e.g., Nisenbaum and Abercrombie

1992; Nisenbaum et al. 1991), the phenomenon of "dishabituation" or "stress-switch hyper-responsiveness" argues against the doctrine of non-specificity.

The doctrine of nonspecificity. Cannon was one of the first critics of Selye's stress theory. Cannon's critique centered on the doctrine of non-specificity and the implicit assumption that a stereotyped response pattern can be adaptive, regardless of the character of the stressor. How could the general adaptation syndrome be viewed as adaptive in the sense of preserving the internal environment effectively in the face of widely differing challenges? Since a nonspecific stress response would not have provided an advantage in natural selection, a stereotyped stress response would not have evolved.

As research has uncovered more and more systems that participate in stress responses and seem to serve distinct homeostatic needs, it becomes less and less clear whether, after removing these specific reactions from consideration, any truly nonspecific reactions would be left.

Mason (1971, 1975) also disputed Selye's doctrine of nonspecificity, noting that in response to different stressors, activity of the pituitary-adrenocortical system can increase, decrease, or remain unaffected (Mason 1971). This implies that the triad of the general adaptation syndrome cannot be a stereotyped indicator of the stress. Mason suggested that the basis for similarities in neuroendocrine responses to different physical stressors is not the nonspecificity of responses of the body to any demand but that all the stressors used in Selye's research produce the experience of anxiety or emotional distress, resulting in similar neuroendocrine manifestations. A central nervous system source of distress responses would help to explain one of the main admitted failures of stress research testing Selye's theory: the inability to identify carriers of the "alarm signals" from peripheral sites (Selye 1982).

The emphasis on inflammation and the pituitary-adrenocortical axis. Selye overemphasized analogies to inflammation and the role of the pituitary-adrenocortical system in the stress response. By dwelling on only one effector system, Selye virtually defined the presence or absence of physiological stress by the presence or absence of pituitary-adrenocortical activation. Other neuroendocrine systems—such as the sympathoadrenal system, the vasopressin system, the renin-angiotensin-aldosterone system, the endogenous opioid system, the system regulating growth hormone secretion, and the parasympathetic nervous system—participate in different stress responses, as discussed elsewhere (Natelson et al., Chapter 6, this volume). Because Selye never incorporated these other systems adequately into his theory, he did not consider possible adaptive patterning of responses of these systems. Moreover, since pituitary-adrenocortical function can vary as part of specific homeostatic responses, alterations in pituitary-adrenocortical function could not specifically reflect the occurrence or intensity of the nonspecific element.

Selye overgeneralized from effects of pharmacological doses of adrenocortical steroids to the occurrence of "diseases of adaptation," thought to result from maladaptive stimulation of endogenous release of those steroids. For example, although Selye was correct that deoxycorticosterone combined with high salt intake produces hypertension in uninephrectomized rats, and that clinical diseases such as hyperaldosteronism produce salt-sensitive hypertension, he erred in implying that excessive release of mineralocorticoids into the bloodstream characterizes, causes, or even contributes to more common forms of hypertension. His suggestion that adrenalectomy or hypophysectomy could be beneficial in hypertensive patients in general (Selye 1956) has been discarded.

Had Selye obtained data about activities of other stress systems besides the pituitary-adrenocortical system, he probably would not have proposed the stages of the general adaptation syndrome. For example, repetition of immobilization stress does not result in habituation of plasma catecholamine responses, and no stage of exhaustion is encountered (Kvetnansky and Mikulaj 1970).

Neuroendocrine evidence has suggested that stimuli that are perceived as novel may selectively activate the pituitary-adrenocortical system (Frankenhaeuser 1975; Herd 1986). This leads to the hypothesis that the stage of resistance in Selye's general adaptation syndrome may actually reflect habituation of the pituitary-adrenocortical response during repetitive exposure to noxious but no longer novel stimuli. Since hypercortisolemia characterizes a proportion of patients with depression (Gold et al. 1988), the occurrence of pituitary-adrenocortical activation during clinical "giving up" (Henry et al. 1986) may indicate a phenomenon corresponding to the stage of exhaustion.

Two conclusions from Selye's theory and experiments do seem justified, and they are incorporated in the present theory. First, stress is an unobservable condition that leads to adaptive responses. Second, adrenal release of glucocorticoids often accompanies stress responses.

A Definition of Stress

Selye appeared to be closer to a valid definition when he asserted that stress is a "condition" or a "state" than when he asserted that stress is a nonspecific, stereotyped pattern of response. In the former sense, stress can be thought of as a generic form of what in psychology is called an "intervening variable." This is the starting point for the present definition.

Stress as an Intervening Variable

An intervening variable is a theoretical construct in psychology that links a stimulus and a behavioral response.

Consider the experience of hunger. If one were to deprive an individual of food, one would make that individual hungry, but the stimulus, the withdrawal of food, would not itself be the hunger. If one were to observe an individual speeding to a buffet table and gorging on the food, one would infer that the individual must be hungry, but the behavior, although perhaps indicating the intensity of the hunger, would not be the hunger itself. The hunger would not be directly observable. It would be experiential. From the point of view of the observer, it would be an intervening variable, in this case in the category of motivational states.

Similarly, consider the emotion of anger. Certain stimulus situations are likely to elicit this emotion, and several behaviors or measurements of hemodynamic or biochemical parameters may measure its intensity, but the anger itself, although experienced by the individual, would not be observed directly.

One can conceptualize many intervening variables, including all emotional feelings and motivational states.

The present theory views stress as a type of intervening variable condition, with several characteristics, as described later. Weiner (1991) similarly uses the term *stressful experience.* According to the present view, however, both the stimulus and response may be physiological or psychological. Indeed, in some cases, such as conditioned avoidance responses attended by autonomic activation, this distinction becomes blurred.

Advocates from different schools of psychology have argued about whether emotions and motivational states are necessary scientifically. Because they are experiential, emotions and motivational states are not amenable to direct examination. Given two theories, one with and one without an untestable component, the law of parsimony would favor the simpler one. The behaviorist school posits that the study of the rules governing relationships between stimuli and responses is a sufficient and appropriate basis for experimental psychology.

Nevertheless, a term such as *hunger* is easier to conceptualize than is "the collection of central neural and neuroendocrine processes, initiated by food deprivation, that produces a characteristic experience and elicits effector behaviors such as the search for and ingestion of food." Viewed strictly, the experiential aspect may be unscientific because its existence may not be testable by observation or experimentation; therefore, the intervening variable, hunger, may not explain most parsimoniously the relationship between food deprivation and the search for and ingestion of food. Since all persons have shared the experience of hunger at some time, however, there is a sense of reality to that experience that seems to transcend a simple stimulus-response relationship.

The same arguments can be applied to intervening variables in general and therefore to stress. One may question the scientific necessity of stress as an intervening variable. For example, it can be argued that adding the word *stress* to hypoglycemia or hemorrhage is unnecessary in studying effects of these manipulations on responses of neuroendocrine effector systems. Unlike emotional and motivational intervening variables, however, stress—at least according to the present conception—does not necessarily imply any particular conscious experience. It is possible that all the intervening neurophysiological and neurochemical mechanisms for a stress response may eventually be elucidated. The term therefore seems useful as a conceptual abbreviation. One must bear in mind that any aspect of stress that requires conscious experience may not be amenable to scientific testing. The present definition therefore does not assume that stress is experienced consciously. What an *unconscious* experience is seems more of a philosophical issue than a scientific issue. To put it simply, the present view is based on stress as a condition, which is not the same thing as an experience:

> Stress is a condition where expectations, whether genetically programmed, established by prior learning, or deduced from circumstances, do not match the current or anticipated perceptions of the internal or external environment, and this discrepancy between what is observed or sensed and what is expected or programmed elicits patterned, compensatory responses.

The following discussion elaborates on the main elements of this definition.

Homeostats

Not all intervening variables are stresses. The main determinant of whether an intervening variable is a stress is the effort at a *compensatory* response.

Whether stress occurs depends on the organism's sensing something that leads to a compensatory reaction. It matters little whether the reaction is rational or successful or even adaptive in fact for the reaction to indicate stress.

In the occurrence of stress, what is sensed is a disruption or a threat of disruption of homeostasis. This sensation requires a comparative process, where information made available to the brain is compared with set points for responding. Consider the analogy of a thermostat. Feedback about temperature reaches a thermostat that is set for a certain temperature. A sufficiently large discrepancy between the measured temperature and the set temperature closes a switch in the thermostat and ignites the furnace. Sufficient reduction of the discrepancy shuts down the furnace. This keeps the room temperature within a certain range, with the average temperature corresponding to the thermostat setting.

The body has many such homeostatic comparators—they can generically be called *homeostats.* Afferent information to a homeostat is compared with a set point for responding, determined by a regulator. The homeostat uses one or more effectors to change values for the controlled variable. The loop is closed by monitoring changes in the levels of the controlled variable, via one or more monitored variables. At least one homeostat regulates each stress response.

A physiological stress system requires a sensor that can monitor the controlled variable faithfully. The organism responds not to the perturbation itself, nor even to alterations in values of the variable the system is designed to control, but to sensed changes of values for the variable that the sensor actually monitors. A key to understanding the function of any stress system is to identify what the relationship is between what is "supposed" to be regulated and what actually is sensed. From the analogy of the thermostat, the "purpose" of the device is to maintain room temperature, the controlled variable. The temperature itself does not determine whether the thermostat turns on the furnace; rather, when the temperature in the thermostat changes, two metal ribboned strips expand differentially, bending the ribbon and allowing it to close an electrical contact. The extent of bending of the ribbon is the monitored variable. Of course, in this case, the monitored variable reflects the controlled variable with very high fidelity. But suppose one of the metal bands corroded so that the ribbon bent less for a given change in temperature, or suppose that the two bands began to separate. The system would continue to regulate the monitored variable, the extent of bending, but the room temperature would drift to a new level.

In the case of the human arterial baroreceptor reflex, stretch receptors in the walls of major arteries such as the carotid arteries sense pulse-related distortion of the vessel wall. Unlike the situation with the room thermostat, for physiological systems one must make an educated guess at what the controlled variable is (i.e., what the homeostatic system regulates). As the name baroreceptor implies, cardiovascular researchers have thought that the system regulates mean arterial pressure, although pulse pressure and pulse rate also affect baroreceptor afferent activity (Gero and Gerova 1967). The afferent information to the brain leads to reflexive responses, including sympathetic inhibition, that, among other things, relaxes blood vessels and thereby counters the initial perturbation of blood pressure.

Under normal circumstances, the extent of receptor stretching, the monitored variable, should faithfully indicate the extent of change in arterial pressure, the presumed controlled variable. But what happens in arteriosclerosis, when the thickened and rigidified arterial walls decrease the extent of pulse-related distortion? Even if the baroreceptors themselves functioned normally, the decreased distortion would attenuate the amount of baroreceptor stimulation. This would decrease the afferent baroreflex information, tending to increase sympathetic outflow and therefore blood pressure. In other words, all other things being the same, arteriosclerosis would reset the function of the barostatic system in such a manner that blood pressure and sympathetic outflow would increase. In fact, casting the carotid sinus baroreceptors does increase blood pressure (Kezdi 1967). Since in many populations arterial rigidity increases with normal aging, this phenomenon may help to explain why blood pressure and sympathetic nerve activity tend to increase with age.

Another key to understanding the function of a homeostatic system is to determine which homeostats mediate the compensatory response. The neuroendocrine response to water deprivation illustrates this point. The response pattern in this setting does not result from the deprivation itself but from changes in at least *two* monitored variables, serum osmolality and blood volume, each of which is regulated by a homeostat. Osmolality is sensed in the hypothalamus, and the main known effector the osmostat uses is vasopressin. Effective circulating blood volume is monitored not by volume sensors per se but by "low pressure" stretch receptors (i.e., low pressure baroreceptors) in the cardiac atria and to a lesser extent in the ventricular myocardium and possibly other cardiovascular regions. We can assume for now that the osmoreceptors and volume receptors actually monitor osmolality and blood volume faithfully. Several systems regulate effective circulating blood volume, including the sympathoneural system (via renal and cardiac sympathetic nerves), atrial natriuretic factor, the renin-angiotensin-aldosterone system, and vasopressin. Although alterations in the activities of several effectors determine the stress response pattern, for the present purpose, alterations in activity of only one effector, the vasopressin system, are considered. Vasopressin is an effector for *both* the osmostat and the "volustat." Since water deprivation tends to increase osmolality and to decrease blood volume, and since in this setting the osmostat and volustat both would direct increases in vasopressin release, vasopressin levels increase markedly in response to water deprivation (e.g., Seckl et al. 1986; Shore et al. 1988).

What happens to vasopressin levels when the perturbation tends to increase *both* osmolality and blood volume concurrently? The osmostat would stimulate vasopressin release, whereas the volustat would inhibit vasopressin release. Depending on the gain of the two homeostats, and on the dynamics of the monitored variables, vasopressin levels could increase, decrease, or not change at all. Thus, in baboons chronically ingesting hypertonic saline (Turkkan and Goldstein 1991), and in dogs on a high-salt diet with constant water intake (Krieger et al. 1990), circulating vasopressin levels do not change, despite the osmolar load. In humans, ingestion of oral hypertonic saline after dehydration transiently decreases plasma vasopressin levels from the high levels that result from dehydration alone (Seckl et al. 1986). Finally, Quillen and Cowley (1983) reported that hypervolemia shifts to the right the curve relating vasopressin levels to osmolality. In other words, hypervolemia increases the set point of the osmostat. These results demonstrate that to predict the response of an effector to a perturbation, one must take into account the effects of the perturbation on all homeostats that use the effector.

A tremendous array of homeostatic systems detect perturbations of monitored variables. Arterial baroreceptors regulate blood pressure acutely by using the sympathetic and parasympathetic effectors. Chemoreceptors in the macula densa of the kidney regulate balance of sodium and potassium by using renin-angiotensin-aldosterone effector. Chemoreceptors in the carotid sinus region and in the brain stem regulate arterial oxygen, pH, and carbon dioxide levels by reflexive stimulation of phrenic and sympathetic nerve activities. Nociceptors in the skin send information about pain via spinothalamic tracts and elicit sympathoneural, adrenomedullary, and pituitary-adrenocortical activation. Gastrointestinal distention by food stimulates local increases in blood flow, regional sympathoneural inhibition, and vagally mediated parasympathetic activation. Glucose and temperature sensors, probably located in the hypothalamus, elicit different efferent patterns, with hypoglycemia stimulating growth hormone release and corticotropin release from the anterior pituitary gland, cortisol release from the adrenal cortex, glucagon release from the pancreas, and epinephrine release from the adrenal medulla; release of insulin from the pancreas decreases. Effectors for temperature regulation include cholinergic and noradrenergic receptors in the skin.

According to the current definition, even a simple homeostatic reflex such as the arterial baroreflex may reflect stress, in that a perceived discrepancy between a set point for a monitored variable and information about the actual level of that variable elicits compensatory responses to decrease the discrepancy. Is stress, defined in this way, too general then to be meaningful, since it would encompass virtually all physiological and psychological adjustments? As discussed at the beginning of this chapter, the value of the definition will depend on the ability to generate hypotheses that are testable by observation and experiment. One such prediction, about the homeostatic bases for vasopressin responses to water deprivation, was discussed earlier. Homeostatic systems seem to operate according to several principles, presented later; from these principles, testable hypotheses can be derived.

Given the multiplicity of homeostatic systems, predictions about patterns of responses to stressors that potentially affect many homeostats, each of which may have particular gains and time constants for several effectors, would require rather sophisticated computer modeling. This can (Bogen 1989) and should be done; the only alternative would be to regress to Selye's dubious doctrine of nonspecificity.

Principles of Homeostat Operation

All homeostatic systems operate according to a few principles. Despite the simplicity of these principles, they help to resolve persistently controversial issues in the area of stress, autonomic function, and clinical pathology.

Monitored versus controlled variables. In the analogy with a thermostat as a homeostatic system, values for the monitored variable (the extent of bending of the bimetallic ribbon) faithfully indicate values for the controlled variable (room temperature), and the distinction between the two seems semantic. In physiological systems, however, the distinction between monitored and controlled variables is crucial.

The monitored variable is a tangible, measurable entity such as the extent of stretching or the concentration of hydrogen ions. In contrast, the controlled variable, regulation of which is the "goal" of the homeostatic system, is not sensed directly. Controlled variables in physiological systems are analogous to motivations, emotions, and other presumed drives in psychological systems. They are intervening variables. One can never prove the existence of intervening variables, since these are theoretical constructs. They have value in science partly because they help us to comprehend, to "explain" phenomena, but mainly because they help us to predict phenomena. Controlled variables are inductions. One should feel free to discard a controlled variable if empirical observations disagree with predictions based on regulation of the controlled variable.

Substantial progress has been made in identifying the physical location and regulation of brain centers mediating baroreflex regulation of the circulation. The central nervous pathways have been mapped out, and simple baroreflex arcs involving the nucleus of the solitary tract have been identified. These results have helped to identify the site where and the mechanisms by which the baroreflex homeostat operates; they cannot identify with surety what and why the barostat *is*. One may hope that the locations and regulation of psychological homeostats—"psychostats"—will also be revealed using new functional testing such as positron-emission tomographic scanning to detect regional metabolic activity and neurotransmitter turnover.

Negative feedback. One of the principles by which homeostatic systems operate is regulation by negative feedback. Increases in values of the monitored variable result in changes in effector activity that oppose and thereby "buffer" the changes in that variable. Thus, hypercortisolemia inhibits pituitary-adrenocortical system activity; stimulation of arterial baroreceptor afferents inhibits sympathoneural activity; increased cardiopulmonary filling inhibits vasopressin release; increases in the filtered load of sodium inhibit renin-angiotensin-aldosterone system activity; and atrial decompression inhibits atriopeptin release. This feedback regulation can be modulated at several levels and therefore can be quite complex.

Positive feedback loops can occur in clinical medicine, but when they do, they always signify an unstable situation that poses a threat. For example, when an insulin-dependent diabetic patient contracts a bacterial infection, septicemia can rapidly ensue. The patient can present with both hypotension and diabetic ketoacidosis. The fall in blood pressure disinhibits sympathetic outflow reflexively (Chernow 1988), and endotoxin increases circulating catecholamine levels (Gullichsen et al. 1989; Murray et al. 1989). Catecholamines counter insulin effects in several ways (Garber et al. 1976; Shah et al. 1984). This worsens the ketoacidosis.

Mechanisms of cardiac decompensation in heart failure probably involve several positive feedback loops. Ventricular chamber enlargement normally augments contractility and performance, according to Starling's law of the heart (Braunwald et al. 1968; Frye and Braunwald 1960). In patients with heart failure and cardiomegaly, for a given increment in chamber size, the increment in ventricular performance decreases; that is, the slope of the performance-volume relationship flattens. If the slope were to become negative, then with further enlargement of the ventricular chamber, ventricular performance would decrease rather than increase. This would result rapidly in myocardial depression and fatal pulmonary edema.

Most patients with heart failure die before reaching the descending portion of the Starling curve. A major reason for this is the likely operation of other positive feedback loops, in which the sympathetic nervous system figures prominently. The maintenance of ventricular performance in compensated heart failure depends not only on the Starling mechanism but also on sympathetic neural outflow (Gaffney and Braunwald 1963). In patients with heart failure, myocardial sympathetic outflow, as indicated by the spillover rate of norepinephrine into the cardiac venous drainage, increases markedly (Hasking et al. 1986). High myocardial levels of catecholamines such as norepinephrine predispose to the development of arrhythmias. Any arrhythmia in this setting would rapidly decrease cardiac output, reflexively evoking further increases in cardiac sympathetically mediated release of norepinephrine, which would further increase the likelihood of developing a fatal ventricular arrhythmia.

A second type of sympathetic positive feedback loop in heart failure would result from the frequent concurrence of heart failure and coronary ischemia. Heart failure recruits cardiac sympathetic outflow, which increases myocardial oxygen consumption by increasing cardiac rate and contractility. This would worsen the imbalance between oxygen supply and demand and precipitate a lethal myocardial infarction or ventricular arrhythmia. Hypoxic ischemia also can evoke norepinephrine release by local effects at the sympathetic terminals (Schomig et al. 1987).

A third type of sympathetic positive feedback loop in heart failure would be due to simultaneous increases in sympathetic outflow to the heart and to other body regions. These changes increase cardiac filling and work, due to direct and indirect sodium-retaining effects of renal sympathetic stimulation (Hasking et al. 1986) and to peripheral vasoconstriction. The increases in cardiac afterload could impair ventricular performance more than the increases in cardiac filling would augment ventricular performance, because the heart would be operating at a flattened portion of the Starling curve. The net worsening of ventricular performance would stimulate sympathetic outflow further. Cardiac overfilling also leads to pulmonary edema, producing hypoxemia, acidosis, and respiratory distress, all of which would exaggerate the sympathetic response. The combination of these factors would produce rapid cardiac decompensation and death.

Multiple effectors. Homeostatic systems generally use more than one effector. For example, activation of the body's glucostat rapidly increases glucagon and growth hormone levels, augments activities of the adrenomedullary and pituitary-adrenocortical systems, and suppresses insulin release. Unloading of cardiopulmonary baroreceptor (deactivating the "volustat") increases skeletal sympathoneural outflow and renal release of renin, whereas cardiac release of atriopeptin declines.

This redundancy has several consequences. One is that disabling an effector should compensatorily activate the others. Disabling an effector temporarily releases the monitored and controlled variables from regulation; afferent activity changes, and, assuming no change in the homeostat settings, this would lead to reflexive alterations in the output of other available effectors. This would enable partial or even complete eventual attainment of the homeostatic "goal." The efficiency of the other effectors would determine whether the compensatory activation would actually normalize values for the controlled variable.

One may hypothesize that since multiple effectors enable a degree of control of physiological variables by compensatory activation when one effector malfunctions, natural selection would have favored the evolution of systems including multiple effectors.

Because of compensatory activation of alternative effectors, studies using blockade or destruction of a single effector to assess the contribution of the effector to a monitored variable will underestimate that contribution, or even fail to detect it, if the homeostat uses effectors other than the one blocked or destroyed. The multiplicity of effectors regulating blood pressure helps to explain inconsistent results of studies using sympatholytic procedures or adrenoceptor blockade to evaluate the sympathetic contribution to high blood pressure. Elucidating such a contribution requires either monitoring or control of the activities of the other effectors.

A second and conceptually crucial consequence of multiple effectors is the potential for patterning of the effector responses. Patterning of neuroendocrine, physiological, and behavioral effectors would increase the likelihood of adaptiveness to the particular challenge to homeostasis, providing another basis for natural selection to favor the evolution of systems with multiple effectors.

Experimental evidence supporting this patterning (e.g., de Boer et al. 1990a, 1990b; Henry 1992; Rappaport et al. 1982; Young et al. 1984) refutes Selye's doctrine of nonspecificity. This patterning is important not only for stress theory but also for elucidating the role of catecholaminergic systems in mediating the relationships between stress and disease.

Effector sharing. Conversely, several homeostats can regulate the activity of an effector system. I have already discussed sharing of the vasopressin effector by the osmostat and volustat. Other examples of effector sharing abound in physiology. Both arterial hypotension and decreased cardiopulmonary filling stimulate sympathoneural activity; both decreased cardiopulmonary filling and increased serum osmolality stimulate vasopressin release; both hypoglycemia and emotional distress stimulate adrenomedullary activity; both exercise and hypoglycemia stimulate growth hormone secretion; and both a decreased sodium concentration in the glomerular filtrate and decreased renal perfusion pressure stimulate renin-angiotensin-aldosterone system activity.

One should avoid drawing inferences about the functioning of any single homeostat from measurements of the activity of any single effector because sharing of the activity of an effector by more than one homeostat increases the likelihood of drawing a false negative conclusion about the relationship between the state of activity of the effector and the extent of activation of any homeostat using that effector. For example, debate has raged for years about whether borderline hypertensive patients have increased sympathetic nerve activity. Many homeostats share the sympathoneural effector. When the activity of one of them, the low pressure barostatic system, is monitored by measurements of central venous pressure, then excessive skeletal muscle sympathetic activity *for a given central venous pressure* seems glaringly obvious (Rea and Hamdan 1990); in the absence of this monitoring, the frequency distributions of nerve activities in hypertensive and normotensive groups overlap hopelessly.

Virtually every monitored variable is subject to influences by more than one effector. The combination of multiple effectors and effector sharing can easily explain the notoriously poor correlations between values for activity of a single effector and of a single monitored variable—such as the correlations between sympathetic nerve activity and blood pressure in "resting" subjects. Blood pressure is only one of many factors determining sympathetic activity (effector sharing); many systems besides the sympathetic nerve system contribute to blood pressure regulation (multiple effectors).

Inadequate consideration of these two principles of operation of homeostatic systems will produce Type I errors.

Release from homeostatic restraint. Blockade of afferent information or interference with a homeostat increases the variability of the monitored variable. Thus, baroreceptor deafferentation increases the variability of blood pressure (Cowley et al. 1973), as does destroying bilaterally the likely brain-stem site of the barostat (Nathan and Reis 1977).

A more difficult issue is whether release of a monitored variable from homeostatic restraint increases the tendency for the variable to drift to a new level—a level that may be pathophysiological. In particular, researchers have debated for many years whether baroreceptor "debuffering" increases "resting" blood pressure; that is, whether debuffering produces a form of neurogenic hypertension.

Part of the problem is the meaning of the term *resting*. Many interacting systems determine blood pressure levels, and the activities of these systems never cease in life. *Resting* denotes a state where activities of these systems are maintained at certain levels and the organism seems inactive and not in distress. Without overt anthropomorphism, the term is circular and therefore meaningless scientifically.

Even if one were to avoid use of the term *resting* and control the external environmental conditions rigorously, organisms with impaired baroreflexes would have excessive classically conditioned sympathetic responses to seemingly neutral cues (Nathan et al. 1978). Moreover, debuffering may decrease thresholds for arousal, vigilance, or rage (Bartorelli et al. 1960; Lacey 1967; Lacey and Lacey 1970); the accompanying sympathetic activation would then increase blood pressure.

Natural selection must have favored the development of systems to support the delivery of blood to the brain and heart during emergency reactions. As Cannon (1929, 1939) demonstrated, prominent among these systems is the sympathoadrenal system, activation of which increases blood pressure and cardiac output. If this system is an emergency system, then most of its activity should be inhibited under more normal circumstances. Release of the effector from baroreceptor restraint would bias the blood pressure upward. (In terms of blood pressure regulation, the weaker influence of the other main arterial baroreflex effector, the parasympathetic system, can be ignored here.) Conversely, baroreflexes should prevent hypotension more effectively than buffer hypertension. Although the literature on the latter point is inconsistent, at least some supportive evidence has been reported (Rea and Eckberg 1987).

For all these reasons, one would expect that interference with homeostatic regulation of blood pressure should increase the variability of blood pressure, augment hypertensive responses to various stressors, and produce acute and perhaps chronic hypertension. In rabbits (Berthelot et al. 1982) and baboons (Shade et al. 1990), denervation of the carotid sinus and aortic baroreceptors does produce sustained increases in mean arterial pressure, blood pressure lability, and increased plasma levels of norepinephrine. In humans, baroreceptor debuffering due to surgical denervation (Ahn et al. 1989) or local anesthetic blockade (Fagius et al. 1985; Guz et al. 1966) increases blood pressure and indices of sympathetic nervous activity acutely; however, in the case reported by Aksamit et al. (1987), irradiation-induced arterial baroreceptor denervation was associated with episodes of paroxysmal hypertension and very high plasma norepinephrine levels but not with sustained hypertension.

Resetting. Finally, homeostatic systems can reset, via altered regulation of the homeostat. In essence, this resetting redefines homeostasis. For some stress responses, such as during exercise, short-term changes in homeostatic settings clearly enhance the long-term well-being and survival of the organism. A main premise of present stress theory is that as homeostatic systems evolved, so did mechanisms to regulate them.

Nevertheless, when superimposed on a substrate of cardiovascular pathology, such as heart failure or coronary artery disease, homeostatic resetting can produce harmful or even fatal effects. This is because, during many stresses (e.g., exercise, cold exposure), the net hemodynamic effects of the resetting are increased cardiac work or afterload, resulting from global or patterned increases in sympathetic outflow. Thus although homeostat resetting during stress generally benefits the individual, the resetting could also worsen a largely independent cardiac pathological state.

Challenges to homeostasis may or may not require homeostat resetting in order for the individual to respond appropriately. For example, during performance of the Valsalva maneuver, as the straining decreases cardiac filling and output, sympathetic vasomotor outflow increases, and the resulting vasoconstriction increases diastolic and therefore mean arterialpressure to approximately the baseline value.

In contrast, distress—to be defined shortly—*always* resets homeostats. Homeostat resetting during distress can be viewed as the basis for many psychosomatic theories of disease. For example, resetting of arterial baroreflex function due to repeated "defense reactions" remains a viable mechanistic hypothesis for the pathogenesis of hypertension (Folkow 1982, 1987, 1988).

Homeostatic resetting in pathological conditions can produce unexpected clinical consequences. For instance, patients with chronic obstructive pulmonary disease often have decreased responsiveness of the chemoreceptor homeostatic response to hypercarbia. They depend on hypoxemia to regulate ventilation (hypoxic drive). Administering pure oxygen in this setting removes the hypoxic drive, causing hypoventilation, accumulation of carbon dioxide, narcosis, and possibly respiratory arrest and death.

The therapeutic efficacy of nitroglycerin in patients with angina pectoris may depend on resetting of baroreflexes. Administration of nitroglycerin to healthy volunteer subjects invariably increases pulse rate reflexively as stroke volume falls (Goldstein 1983). Since systolic blood pressure falls, whereas heart rate increases, the pressure-rate product, an index of myocardial oxygen consumption, remains largely unchanged. In patients with angina pectoris, nitroglycerin reduces the imbalance between myocardial oxygen demand and coronary arterial oxygen supply, relieving the chest discomfort. One explanation for this beneficial effect is that, as a result of barostat resetting (in turn probably arising from arterial rigidification), the nitroglycerin-induced decrease in cardiac filling does not reflexively increase cardiac rate compensatorily, so that myocardial oxygen consumption decreases.

Distress

The present theory distinguishes stress from distress. Distress is a form of stress with four additional features: consciousness, aversiveness, observable signs, and pituitary-adrenocortical and adrenomedullary activation.

Consciousness. According to the present conception, the existence of stress does not require consciousness. Selye (1956) would have agreed, since he claimed that stress reactions can occur in anesthetized animals, in lower animals without nervous systems or undergoing mechanical damage to denervated limbs, and even in cells cultured outside the body.

In contrast, the experience of *dis*tress does require consciousness, because distress occurs when there is not only a challenge to homeostasis, but also a perception by the organism that homeostatic mechanisms may be overwhelmed. The latter requires a conscious interpretation of sensory information.

This is a more generalized statement of the concept of psychological stress as a consequence of a perceived inability to cope (Lazarus 1966). The sense of an inability to cope or of a lack of controllability is basic to psychological theories about feelings associated with distress (Krantz and Lazar 1987; Lazarus 1966, 1991). It is suggested here that distress is experienced when the organism perceives the inadequacy of compensatory adjustments to either a psychological or physiological stressor.

There is little reason to separate psychological and physiological models of stress, despite independent development of the two research traditions (Krantz and Lazar 1987). The present stress model is analogous to the psychological stress model proposed by Lazarus (1966). Both share key main elements: appraisal processes based on afferent information, perception of threats to homeostasis, elicitation of compensatory response patterns, and continuous reappraisal of the success of the homeostatic effort.

Fundamental questions remain about whether and how psychological stress can bring on physical disease in otherwise healthy individuals, and about the extent to which physiological stress contributes to psychopathology. The existence, location, regulation, and pathology of psychostats associated with the experience of distress are also poorly understood.

Aversiveness. Distress is an experience where the organism tends to avoid situations that may produce the same experience in the future—that is, distress is negatively reinforcing. It motivates escape and avoidance learning. The experience of distress would be expected also to increase vigilance behavior and enhance long-term memory of the distressing event—obviously adaptive neurological adjustments.

Animals can react instinctively not only to a stressor but to symbolic substitutes that partly resemble the natural stimulus. For example, monkeys that have never seen a snake become visibly upset on exposure to a snake;

rabbits that have never seen a hawk freeze when a shadow in the shape of a hawk goes by; and stickleback fish instinctively attack any red object in their territory. In other species, the risk of the errors of rigid instinct—inappropriate responses to symbolic cues—is minimized by the plasticity afforded by learning.

Even "primitive" animals have the capacity to learn to withdraw or escape from noxious stimuli (Kandel 1983) or, conversely, to habituate after prolonged or repeated exposure to a stimulus. These forms of learning mirror each other, the former reflecting a sensitization and the latter a desensitization. The fact that primitive animals have these capabilities indicates the remarkably durable survival advantage of learning.

Classical (or Pavlovian) conditioning represents an important refinement of these responses (Pavlov 1927). Habituation and sensitization exemplify nonassociative learning, where the organism learns about a single stimulus. In contrast, classical conditioning (and operant conditioning, to be discussed shortly) involves learning associations among stimuli.

Specifically, in classical conditioning, repeated pairing of a neutral stimulus (e.g., a bell ringing) with an unconditioned stimulus that elicits an instinctive unconditioned response (e.g., salivation at the presentation of meat powder to a hungry dog; limb withdrawal after a local shock) results eventually in the elicitation of the unconditioned response (or components of it) by the previously neutral conditioned stimulus. That is, the conditioned stimulus now elicits a conditioned response. For example, depending on the unconditioned stimulus with which the ringing has been paired, the dog salivates when the bell rings or withdraws its leg.

Although Pavlov believed that the acquisition of conditioned reflexes requires cerebral cortices, the "crowning achievement in the nervous development of the animal kingdom" (Pavlov 1927, p. 1), even invertebrates such as the sea snail, Aplysia, have the capacity to learn by classical conditioning (Kandel 1983; Kandel and Tauc 1965).

Most classical conditioning experiments have involved an external unconditioned stimulus, such as an electric shock to the skin. This does not imply that the unconditioned stimulus must be external. For example, in rats, hyperglycemia can be learned by classical conditioning after repeated pairing of a previously neutral cue with injections of insulin (Siegel 1972). Pavlov (1927) himself demonstrated that nausea and vomiting can be learned by classical conditioning after repeated pairing of a conditioned stimulus (approach of the experimenter) with an internal unconditioned stimulus (injected morphine). Moreover, although classical conditioning experiments often have focused on presumably involuntary alimentary, glandular, or cardiovascular responses mediated by smooth muscle, Pavlov emphasized that conditioned reflexes include distinct external motor and internal secretory components (Pavlov 1927).

Instrumental, or operant, conditioning appears to represent a more advanced form of learning, requiring a cerebral cortex (Thornton and Van-Toller 1973). In instrumental conditioning, the likelihood of a behavior increases when the behavior leads to positive reinforcement (reward) and decreases when the behavior leads to negative reinforcement (punishment). Conversely—and circularly—reinforcement has been defined as an event that strengthens the response it follows (Rubin and McNeil 1987). The conditioning is "operant" in that the individual's behavior operates on the environment, determining the occurrence of reinforcement. The conditioning is "instrumental" in that the learning is a means to an end, with the occurrence of reinforcement contingent on the behavior. Operant conditioning therefore differs from Pavlovian conditioning, where the individual's behavior is irrelevant to the occurrence of reinforcement. Both forms of conditioning require the memory of an association between reinforcement and behavior; in Pavlovian conditioning, behavior (the unconditioned response and conditioned response) is consequent to the reinforcement (the unconditioned stimulus), whereas in operant conditioning, reinforcement is consequent to the behavior.

The circular definition of reinforcement in the definition of operant conditioning appears to be the price psychological theory has paid for discarding intervening variables such as "drives" from the mechanism of instrumental learning. According to the drive-reduction theory (Hull 1943), an individual acquires a behavior because the behavior reduces a drive. For example, a hungry rat learns to press a bar to receive food pellets, since the reinforcement, food, reduces the drive, hunger. Drive-reduction theory encountered difficult criticism that is beyond the scope of this presentation.

Although it had been thought that only voluntary behaviors, mediated by skeletal muscle, can be acquired by operant conditioning, biofeedback experiments in the 1960s and 1970s supported the hypothesis that visceral, glandular, and cardiovascular responses can also be modified by operant conditioning (Kamiya et al. 1971; Miller 1969).

Avoidance learning is a form of operant conditioning where the individual learns to avoid negative reinforcement by producing behaviors that decrease the likelihood of that reinforcement. According to the drive-reduction theory, the punishment increases the experience of fear or anxiety, and learning the avoidance behavior reduces the drive. This is where the present view of distress enters the picture. In the morphine experiment described by Pavlov, the conditioned response, salivation and nausea, probably would have been accompanied by other classically conditioned autonomic responses, including cardiovascular changes, if these parameters had been measured. In terms of the cardiovascular system, however, any such conditioned responses would be overshadowed by two other processes occurring at the same time.

According to the present conception, if an organism experiences distress consistently in a given situation, the subsequent perception of reexposure to the situation elicits distress as a classically conditioned response. Thus, in the morphine experiment described by Pavlov, when the experimenter entered the room, the dog would become "restless," not as a conditioned response to morphine injection but as a conditioned response to the actual experimenter. In anthropomorphic terms, the dog would recognize that every time the experimenter entered the room, suffering would follow, and this would elicit a feeling of anxiety, independent of the specific effects of morphine. One factor contributing to the cardiovascular response in this setting would be to the involuntary neuroendocrine activation evoked by whatever emotional state led to the appearance of restlessness.

The second additional process would depend on the aversive quality of the distress. Encountering an aversive conditioned response can lead to two behavioral responses. One is a rapid, involuntary, conditioned withdrawal response. The response may be so rapid, the individual may not have time to consider the significance of the conditioned stimulus. The sheer rapidity of withdrawal (or escape) responses helps to explain why they have endured through evolution. A slower behavioral response is voluntary avoidance. The basis for learning the avoidance behavior is the reinforcement provided by reducing the distress. Whenever avoidance behavior occurs, supportive neuroendocrine changes accompany it.

Concrete examples may help here. Suppose there are two signal lights in an animal's cage, one red and the other white. When the red light is on, there is a likelihood that an electric shock to the tail will follow. When the white light is on, then there is no chance of electric shock. When the red light is on, pulling a lever in the box decreases the likelihood but does not completely prevent the occurrence of the shock. What does the animal eventually learn to do when the red light turns on, and what would the cardiovascular consequences be?

First, by classical conditioning, the red light would acquire the properties of an aversive conditioned stimulus. The cardiovascular reaction to this conditioned stimulus would mimic that to the unconditioned stimulus, shock. Second, skeletal muscle in the tail would contract as a conditioned withdrawal response. Third, the animal would experience distress, which would be associated with patterned neuroendocrine changes producing cardiovascular effects. Fourth, by operant avoidance conditioning, the animal would learn to pull the lever, reducing its distress. Patterned neuroendocrine and cardiovascular changes would accompany the operantly conditioned skeletal muscle contractions.

This example illustrates sharing of the sympathoneural effector by several homeostats, including those related to skeletal muscle contraction and to the experience of distress, and suggests that situations evoking distress typically involve a complex interplay of classically and operantly conditioned behaviors.

Communication. A third characteristic of distress, in addition to being conscious and aversive, is that it evokes external messages, signs that other members of the species can interpret as indicating the emotional state and intent of the organism.

Darwin (1872/1965) emphasized that the various outward manifestations of emotion provide important means of intraspecific communication with survival value for the species, and that physiological arousal, by intensifying emotions, amplifies further the physiological stress responses that accompany those emotions.

The latter can be viewed as a psychological positive feedback loop. From the previous discussion, psychological positive feedback loops must be inherently unstable, and they are. For example, fear, if not unchecked by some interposed negative feedback, degenerates to panic, which can be self-destructive. The worst recorded episode of mass death due to panic occurred June 6, 1941, when about 4,000 people died as a result of panic in an air raid shelter in Chungking (Zhong qing), China (McFarlan 1992).

Analogously, anger can escalate to murderous rampage or frenzy and grief to suicide.

Perceptions of signs of distress by other members of the species can elicit involuntary, instinctive responses. Lorenz (1952) described the behavior of two fighting wolves after one submitted and extended its neck to the victor.

> Every second you expect violence and await with bated breath the moment when the winner's teeth will rip the jugular vein of the loser. But your fears are groundless, for it will not happen. . . . A dog or wolf that offers its neck to its adversary in this way will never be bitten seriously. . . . Since the fight is stopped so suddenly by this action, the victor frequently finds himself straddling his vanquished foe in anything but a comfortable position. So to remain, with his muzzle applied to the neck of the "under-dog" soon becomes tedious for the champion and, seeing that he cannot bite anyway, he soon withdraws. (pp. 188–189)

This principle applies analogously in humans, with whom the fiercest combat usually ends abruptly when one side shows universally understood signs of surrender and submission.

The communication value of external signs of distress helps to explain the continued elaboration of observable components of distress responses. Although true "fight-or-flight" reactions are rare, even in the wild, all social animals continually elicit and instinctively comprehend psychosocial cues (Sapolsky 1990).

Alterations in sympathoadrenal activation often produce these external signs. For example, sympathetic nerves to cutaneous blood vessels of the face mediate facial blushing (Drummond and Lance 1987; Metz et al. 1978). Whereas blushing in modern society connotes embarrassment or humiliation, its evolutionary significance remains obscure. If two people confront each other, one appearing pale, mumbling, with palms exposed, and the other appearing flushed, shouting, with fists clenched, the former is communicating signs of capitulating, and the latter is communicating signs of prevailing. In evolution, these signs may have been by-products of genetically determined neurocirculatory adjustments supporting fleeing and fighting. In modern society, they serve signal functions of their own.

Neuroendocrine activation due to homeostat resetting. A fourth characteristic of distress is neuroendocrine activation—in particular, enhanced adrenomedullary release of catecholamines and adrenocortical release of steroids. Increased adrenomedullary release of epinephrine is one of the most sensitive neurochemical indices of this activation.

The experience of emotional distress probably requires both physiological arousal and appropriate cognitions by the organism (Schachter and Singer 1962). Although several emotions include distressing elements (e.g., grief, fear, anxiety, panic, guilt), other emotions do not (e.g., libido, joy). Even positive emotional experiences can be associated with adrenomedullary activation (Frankenhaeuser 1975). According to the present concept, adrenomedullary activation does not imply the experience of distress, since such an experience also requires appropriate cognitions, and the activation can accompany even nondistressing emotions; however, in organisms with a functional sympathoadrenal system, the absence of adrenomedullary activation excludes the experience of distress.

Cannon viewed the neuronal and hormonal components of the sympathetic system as acting in concert to preserve homeostasis. The foregoing discussion has generally assumed the validity of this notion; however, describing the pattern of neuroendocrine activation attending distress requires a distinction to be drawn between the neuronal and adrenomedullary components of the sympathoadrenal system. According to the present conception, it is specifically the adrenomedullary hormonal component that characterizes distress.

Reversible resetting of many homeostats during distress produces the neurocirculatory and neuroendocrine activation, including the activation of the hypothalamic-pituitary-adrenocortical and adrenomedullary systems. Consistent with this view, stimulation of the "defense area" of the hypothalamus or elicitation of the "defense reaction" decreases the sensitivity of the cardiac limb of the arterial baroreflex (Koizumi and Kollai 1981; Lundin et al. 1983, 1984; Sleight et al. 1978). Indeed, the defining characterization of distress in terms of hypothalamic-pituitary-adrenocortical and adrenomedullary activation may be restated as follows: Distress always alters the functioning of homeostatic systems, with two of the most prominent changes in activities of effector systems being those in activities of the hypothalamic-pituitary-adrenocortical and adrenomedullary systems.

If distress invariably resets homeostats, why does it do so? The resetting must have afforded survival advantages in natural selection. The many physiological, neurochemical, and psychological adjustments attending distress may reflect preprogrammed strategies. In animals, these strategies include a capacity for learning, which, in turn, enables simulations of future events. These processes would have resulted in progressively more rapid expression of response patterns whose compensatoriness was confirmed over millions of years and became embedded genetically. The evolution of means to express these responses even *in anticipation* of distress-evoking situations indicates that the survival advantage of anticipatory changes outweighed the potentially deleterious effects of redefining homeostasis temporarily.

What are these adjustments, the products of these resettings, during distress? Hemodynamically, blood flow is diverted to the vital organs—especially the brain—and pumped at an increased rate to skeletal muscle, to deliver glucose and oxygen and to remove metabolic waste; and platelet aggregability increases (enhancing the ability to plug vascular holes). Biochemically, glucose is released into the bloodstream and gluconeogenesis spurred. Behaviorally, hyperventilation maximizes oxygenation, removes the metabolic by-product carbon dioxide, and increases pH and decreases serum potassium (countering acidosis and hyperkalemia that would result from tissue trauma). The organism becomes alert and vigilant. Memory processes change to facilitate long-term recall of the distressing event and associative escape and avoidance learning. Sweating increases, fostering correction of temperature by evaporation as metabolic processes accelerate. The individual largely instinctively communicates, by its appearance and actions, information about its internal state.

With repetition of exposure to a situation evoking distress, the organism habituates, in the sense that effector components whose expression depends on novelty or unpredictability recede; concurrently, however, the organism becomes hyperreactive to other potential stressors (Abercrombie et al. 1992; Kvetnansky et al. 1984; McCarty et al. 1992). The latter phenomenon of stressor-switch hyperresponsiveness has been called dishabituation (McCarty et al. 1992). These findings seem reminiscent of events in the stage of resistance, the second phase of Selye's general adaptation syndrome. In the stage of resistance, however, all the components of the unitary stress response would diminish, not only those components dependent on novelty or unpredictability. As noted above, Selye's theory does not explain dishabitation because the occurrence of dishabitation implies stressor specificity.

In conclusion, distress includes experiential, neuroendocrine, and behavioral components. The experiential component is not directly observed but to some extent can be inferred from self-report in humans and from the occurrence of avoidance and escape learning in laboratory animals. Distress is a negative feeling that usually includes a sense of failure to cope or loss of control—a perception that homeostatic mechanisms are being or will be overwhelmed. Neuroendocrine manifestations of distress include adrenomedullary and pituitary-adrenocortical activation, due to reversible resetting of homeostats. Behavioral manifestations include vocalizations and movements and autonomically mediated cutaneous and neuromuscular signs that have definite meanings for other members of the species, with both the expressions by the distressed individual and the perceptions of the distress by others being to a large extent inherited.

Distress Versus Selye's Concept of Stress

How does the present view of distress differ from Selye's concept of stress? First, Selye's theory emphasizes that the stress syndrome is nonspecific, independent of the stressor. According to the present conception, the elicitation of distress responses depends importantly on the character, intensity, and meaning of the stressor as perceived by the organism and on the organism's perceived ability to cope with it.

Second, distress responses—as all stress responses—are viewed here as compensatory in origin and purpose, attempting in some way to mitigate the effects of a stressor. This applies not only to neuroendocrine activation (such as the glucose counterregulatory actions of pituitary-adrenocortical and adrenomedullary stimulation during insulin-induced hypoglycemia) but also to psychological activation (such as conditioned aversive and instrumental avoidance learning). In this sense, the present conception agrees with that of Gray (1991):

> Fear or anxiety that is associated with real or perceived threat constitutes psychological stress. This "stress" or so-called defense response is composed of a set of relatively well-defined biological changes. . . . These changes help to promote the readiness and execution of behaviors that will ultimately increase the probability of survival of the organism. (p. 39)

Gray, however, does not distinguish stress as an experiential entity from stress as a stereotyped biological pattern, just as Selye sometimes defined stress as a condition and sometimes as a response pattern.

Third, Selye's general adaptation syndrome consisted of a series of well-defined stages: alarm, resistance, and exhaustion. Experimental results have not consistently confirmed the habituation of and never have demonstrated the exhaustion of adrenomedullary responses after chronically repeated episodes of distress (Kvetnansky and Mikulaj 1970; McCarty et al. 1992); the present theory does not incorporate these stages.

Fourth, the present theory assumes that distress is always consciously experienced. Neither adrenomedullary nor pituitary-adrenocortical activation implies the experience of distress. Since Selye never developed a clear-cut distinction between stress (which could be conscious or unconscious) and distress, his writings do not reveal whether stress and distress are distinguishable in terms of an experiential requirement.

Fifth, and most relevant to the subject matter of this book, the present theory does not assume that distress causes physical or psychiatric disease. Selye (1974) characterized distress as unpleasant or harmful, without separating these two very different characteristics. He never incorporated the relationship between distress and disease explicitly in his theory; distress as a concept appears to have been added long after he proposed maladaptation from excessive or inappropriate stress responses or the operation of "conditioning factors." Rather than concluding that the relationship between stress and disease was a matter for research, he simply defined a new term *distress* to denote stress that causes disease.

Writers about stress (Asterita 1985; Beamish et al. 1985; Berger 1982; Eliot 1974) generally have not recognized this paradox of Selye's theory and have either assumed the linkage between stress and psychiatric and physical diseases or else have adopted the term *distress* without explaining its meaning specifically

It has been proposed that the basis for stress-related pathology is the repeated experience of stress responses that, due to the constraints of modern society, cannot be expressed outright. Folkow (1987), for example, has suggested that chronic, frequent repetition of even mild "defense reactions" can produce high blood pressure. This type of notion is one of the founding—and largely unproven—ideas of psychosomatic medicine.

Weiner (1991) seems to have adopted Selye's circular argument that stress is harmful when it causes disease: "The linear relation between stressful experience and disease, which Selye described, occurs when the experience is injurious, unavoidable, or incontrollable. In most other instances, however, when it is not, little or no organ damage occurs" (p. II-2). Another layer of circularity is added with the assertion that stressful experience probably leads to disease "only in predisposed, high-risk populations or when subclinical disease is also present" (Weiner 1991, p. II-2).

The present conception implies no necessary link between stress or distress and the long-term development of disease. This does not imply that acute stress, by stimulation of neuroendocrine activation, cannot induce severe and even fatal pathology. For example, a large plasma epinephrine response during exercise in a patient with coronary heart disease could augment myocardial oxygen consumption, platelet aggregability, and ventricular irritability, resulting in angina pectoris, myocardial infarction, or sudden death. Nor does this imply that the assertion that chronic emotional distress leads to pathological consequences is unreasonable. Rather, the issue is a matter for experimental testing using longitudinal controlled studies.

Several other modern theories of stress are analogous to the present one. These theories also discard Selye's notions about the nonspecificity and stereotyped nature of the stress response, in favor of adaptive, compensatory response patterns elicited by challenges to homeostasis as those challenges are perceived by the organism.

The psychological stress model proposed by Lazarus (1966, 1991) has already been cited; the present theory extends this model to physiological stress and neuroendocrine and circulatory responses.

Levine and Ursin (1991) provided a definition of stress that includes three elements: stimulus input, a central processing system, and response output (with biological and psychological processes viewed as integral parts of a general homeostatic principle). According to Weiner (1991), stressors are selective pressures from the physical and social environments that threaten or challenge the organism and that elicit compensatory response patterns. Weiner has also asserted that the concept of homeostasis has by now lost its usefulness because there are no truly steady states in living organisms and that the body's "fundamental operating modes" are oscillatory. Weiner has argued that illness or disease occurs when the stability of a system's usual operating mode is lost.

Both Levine and Ursin (1991) and Weiner (1991) have used the terms *distress* and *stress* largely interchangeably. According to the present conception, the distinction is more than semantic. Distress is viewed here as a form of stress with additional, defining features—adrenomedullary and pituitary-adrenocortical activation; autonomic and behavioral communicated signs that are largely instinctive; and a conscious, aversive experience related to the perception of a failure of homeostatic mechanisms.

A second distinction of the present theory from others is the emphasis on homeostat resetting that often accompanies stress and always accompanies distress; such resetting is not a component of the above stress theories. According to the present conception, this resetting is determined by higher brain centers as part of intrinsically coordinated behavioral, endocrine, and physiological patterns. Thus, according to the present view, homeostasis is not static at all; instead, homeostasis is continually being redefined. There are no "steady states," not only because some effector systems are inherently oscillatory, but because homeostatic settings are continually being readjusted.

Just as homeostatic systems have evolved, so have means to regulate them and coordinate them during stress, and resetting of homeostats is an option for a satisfactory compensatory response. A major task for future research is to identify central neural processes that are the bases for these readjustments and their results—stress response patterns.

Stress in Evolutionary Perspective

Darwin and Ethology

Charles Darwin (1809–1882) introduced the notion that emotional expressions and their autonomic accompaniments evolved. Darwin suggested, based on numerous interviews, readings, and observations, that not only physical characteristics but also emotional behaviors are inherited. In fact, he used the latter to support his theory of the "descent of man."

Darwin also recognized that the autonomic concomitants of emotion have been advantageous not only in preparing the organism for emergency responses, but also for providing the basis for visible signs that serve important, universally understood communication functions within the species. Research about primates in the wild (Sapolsky 1990) has supported the suggestion that in day-to-day life, most of the activation of our "stress" systems is not to mitigate threats to physiological or metabolic homeostasis but to accompany instinctive social behaviors (e.g., dominance displays, sexual pursuit, and submissive escape).

Darwin was acquainted with the work of Claude Bernard but had only a rudimentary understanding of the role of the sympathoadrenal system in producing most of the observable signs of emotional distress. Darwin did recognize, however, that the origin of these signs was in the nervous system, and he understood the complexity of the interaction between the brain and heart during emotion.

The current theory of stress and distress is based on three principles that Darwin introduced. First, autonomic changes that accompany emotional experiences and their behavioral manifestations evolved because they were advantageous in natural selection. Second, one advantage of autonomic arousal is rapid or anticipatory adjustments in cardiovascular function. Third, another advantage is to provide a universally understood means of intraspecific communication.

Stress responses are at least partly inherited. Several laboratory animal models have genetic differences in activities of the major stress systems. Spontaneously hypertensive and borderline hypertensive rats of the Okamoto strain have excessive responses of renal sympathetic nerve activity or plasma catecholamine levels during various forms of psychological stress (Lundin and Thoren 1982; McCarty and Kopin 1978). Lewis-strain rats have a genetic deficiency of corticotropin responses after injection of substances evoking inflammation (Sternberg et al. 1989). Brattleboro-strain rats have a genetic absence of hypothalamic production of vasopressin. Blizard et al. (1980) reported different plasma catecholamine levels in rat strains selectively bred for differences in response to stress.

Clinically, studies of identical twins have indicated that levels of various neurochemicals, including catecholamines, have relatively high heritability (Luft et al. 1987). Twins tend to have similar cardiovascular stress responses (Carmelli et al. 1985). Lacey and Lacey (1958) elaborated a principle of autonomic response-stereotypy, in which individuals respond with a consistent physiological activation pattern across stressors. Whether response-stereotypy has a genetic component has not been explored.

If stress responses are to a large extent inherited, then they should have evolved, because they would have been subject to the same selective pressures that determined the evolution of physical characteristics—in particular, the survival advantage of adaptation by the individual.

The ethologist Richard Dawkins (1989) has presented a comprehensive theory about the evolution of behavior. The following propositions summarize his "selfish gene" theory.

First, the fundamental unit of selection is not the species, nor, strictly speaking, the individual, but the gene, the unit of heredity. A gene is any portion of chromosomal material that potentially lasts for enough generations to serve as a unit of natural selection. A gene is not a single physical piece of DNA in a chromosome but all replicas of that piece, distributed throughout the world. Groups of genes can be selected as units.

Second, "survival of the fittest" means survival of the stable. Selection has favored genes that cooperate within a body, so that it is usually convenient to consider the body as an integral agent for propagating genes. A body is a "survival machine" for preserving and propagating genes.

Third, in animals, behavior, mediated by muscles, evolved because it enabled rapid movement. Generating complex, timed movements requires coordination by a computer, the brain, the unit of which is the neuron. The genes control behavior indirectly, like a computer programmer. They set up the machine but cannot control it, because of time-lag problems. Genes work by controlling protein synthesis, which is powerful but slow. They provide strategies, that is, preprogrammed behavioral policies. An evolutionarily stable strategy is one that, if most members of a population adopt it, cannot be bettered by an alternative strategy.

Finally, genes predict behavioral strategies by incorporating a capacity for learning, which requires memory and feedback. Learning can occur by trial and error; simulations represent vicarious trial-and-error predictions of future events. The evolution of the capacity to simulate has culminated in subjective consciousness.

Applying the selfish gene theory, natural selection clearly would favor the evolution of stress responses. Stress responses are the body's means for preserving a stable internal environment. They do so by continual adjustments in activities of homeostatic systems, based on genetically determined algorithms, learning by combinations of classical and operant conditioning, and conscious simulations of likely future events. The mystery, and the challenge for stress research, is not why stress responses evolved, but how they work. Ascending the phylogenetic scale, organisms have depended increasingly on complex, energy-requiring systems for life. For example, reef corals, relatively simple animals, live a notoriously fragile existence, surviving only in unusual, suitable niches within a small range of salinity, turbulence, and temperature. They benefit from the lack of energy expenditure to maintain their cellular temperature; however, they cannot live in water colder than 18°C.

More versatile cold-blooded reptiles and other lower vertebrates survive a larger range of external temperature by using the sun and shade instinctively (Attenborough 1979; Brattstrom 1965; Swanson 1987) and probably by adjusting cardiovascular system performance, but this requires more organization and more energy.

The bigeye tuna, which has muscles adapted to warm water, possesses a circulatory heat exchanger, where a network of vessels containing blood warmed by exercising muscles wraps around vessels containing chilled blood due to their closeness to the fish's outer surface. The heat exchangers regulate the temperature of the blood before it enters the heart. When the fish ascend to warm water, the exchanger shuts off, accelerating the ability of the fish to warm the blood before they descend again (Holland et al. 1992). The rapidity of the shutdown in the heat exchanger suggests a form of neurocirculatory regulation.

Warm-blooded mammals require and use much more fuel than do reptiles or fish, expending a large proportion of the metabolic fuel simply to maintain body temperature. As in reptiles, mammals possess not only physiological but also instinctive behavioral means to regulate temperature. The importance of this inheritance in adapting to external environmental changes was demonstrated in a news story about llamas that had been transported to and kept on a Caribbean island while awaiting transfer to the United States. Having evolved in a cold, mountainous, treeless climate, they did not instinctively seek shade from the sun. When their coats prevented evaporative heat loss, many succumbed to heat stroke before they could be rescued (The New York Times, April 21, 1989, p. 1).

Most humans spend their lives wrapped in protective clothing or in environments where the external temperature itself is regulated.

Other stress responses are also phylogenetically very old. Withdrawal or escape responses to physical or chemical manipulation are observable even in relatively simple organisms such as Aplysia (Kandel 1983; Kandel and Tauc 1965). Mechanisms for preserving body water and essential nutrients—oxygen and glucose—must also have developed early in evolution.

The remarkably large repertoire of glucose "counterregulatory" systems indicates the importance of appropriate adjustments in blood glucose levels in different situations, including exercise, emotion, and the postprandial state. So many behaviors use up glucose that a variety of means to maintain glucose levels evolved with those behaviors. Since all cells require glucose, these situations call for generalized metabolic responses, elicited by hormones reaching all tissues of the body. Cellular glucoprivation poses a generalized threat, where no patterned neural response or readjustment of blood volume distribution can suffice. Glucopenia, therefore, evokes marked hormonal responses, including large, rapid increases in circulating epinephrine levels, due to stimulation of adrenomedullary secretion.

Generalized threats such as glucopenia often elicit emotional distress. In insulin-induced hypoglycemia, the patient's anxious, pale, and sweaty appearance is an important clinical sign. Drugs that block effects of epinephrine mask this protective response and can render the patient and those caring for the patient unaware of the life threat.

Agonistic behavior, such as attack and dominance displays, also includes strong genetically determined components and has evolved. Aggression facilitates predation, reproduction, social organization, and the division of ecological space (Lorenz 1963). Aggression is a double-edged sword, and since unchecked aggression would threaten the survival of the species, instinctive means to limit it have also evolved.

A more recent challenge, especially relevant to human evolution, has been the assumption of an upright posture, or orthostasis. Here, patterned neural discharges result in constriction of arterioles in the legs and gut, increased pulse rate, and increased total peripheral resistance to blood flow, all of which tend to redistribute the blood volume toward the heart and brain. Pumping of calf skeletal muscle contributes to this redistribution. Ordinarily, the concerted neural responses, mediated mainly by the sympathetic nervous system, are so rapid and effective, the stressor is countered without it being perceived consciously—although it is sensed internally.

The ability of snakes to maintain blood pressure during orthostasis exemplifies the interplay of physical characteristics, physiology, and behavior in the dependence of orthostatic stress responses on the ecological heritage of the species (Lillywhite 1988). In climbing snakes, the heart is located relatively close to the head, and the thin tail minimizes pooling of blood during climbing. In contrast, in sea snakes, the heart is located in the center of the body. When sea snakes are exposed experimentally to orthostatic stress in a tilting tube, their blood pressure falls, whereas climbing snakes in the same situation maintain their blood pressure, both because of the above physical features and because of active muscle pumping behavior.

In humans, powerful homeostatic mechanisms have evolved to maintain cardiac filling during orthostasis. The importance of these mechanisms in human well-being is illustrated by several clinical autonomic disorders and the effects of weightlessness in space flight. Weightlessness is, of course, a novel stress in human evolution. One manifestation of "space sickness" has resulted from the lack of evolution of physiological and neuroendocrine mechanisms to counter the absence of gravitational force. During exposure to zero gravity, the absence of orthostatic pooling of blood in the legs increases venous return to the heart. This leads to patterned activation or inhibition of the several primitive systems that maintain cardiac filling. Prominent among the adjustments are increased excretion of water and salt, probably at least partly due to decreased renal sympathetic activity; decreased activity of the renin-angiotensin-aldosterone system; increased atrial natriuretic factor release by the heart; and decreased vasopressin release by the pituitary gland. A new state of equilibrium occurs, and sympathoneural activity, as indicated by norepinephrine levels, returns to about baseline during prolonged space flight (Davydova et al. 1989; Kvetnansky et al. 1988). When the astronauts return to earth, however, orthostatic pooling of blood combines with the relatively low blood volume to evoke symptomatic orthostatic hypotension despite increased sympathetic activity.

Many other internal or external challenges to homeostasis can be listed. If they operate long enough, stress responses to counter them evolve because the evolution of means to cope with these challenges are advantageous in natural selection.

Although stress response patterns often include activation of the sympathoneural and adrenomedullary systems, this is in concert with other systems, such as the pituitary-adrenocortical system and the parasympathetic nervous system. Several examples of these patterns are introduced later.

As stress homeostatic systems evolved, and as systems to regulate them in patterns evolved, so did the effector systems themselves. Peripheral catecholaminergic systems seem to have evolved in sequence, with dopamine the most primitive catecholamine. For example, sea anemones contain dopa but do not contain catecholamines or the enzyme catalyzing the conversion of dopa to catecholamines (Carlberg 1983). In the great majority of invertebrates, dopamine is the dominant catecholamine; concentrations of norepinephrine, when detected, are less than those of dopamine, and epinephrine is absent (Welsh 1972). In organs or plasma of amphibia, epinephrine is the dominant catecholamine; in reptiles, either norepinephrine or epinephrine may predominate (Stinner and Ely 1993); and in mammals, norepinephrine predominates (Holzbauer and Sharman 1972). Hart et al. (1989) reported that of 31 animal groups, the lamprey had the highest basal circulating concentrations of dopamine and epinephrine; sharks and domestic animals had the highest concentrations of norepinephrine; and the eel and lumpfish had the lowest concentrations of norepinephrine.

Catecholaminergic effector systems include an autocrine-paracrine dopaminergic system in the kidneys; a hormonal system based largely on adrenomedullary secretion of epinephrine; and a nerve network, the sympathetic nervous system, where norepinephrine functions as the main neurotransmitter. Autocrine-paracrine systems can theoretically meet the needs of simple organisms that lack mechanisms of neurocirculatory regulation; more sophisticated organisms can use hormones to coordinate organ functions. In the most complicated organisms, including humans, autonomic nerve networks enable complex central neural regulation of regional resistances to blood flow and of glandular secretion. The regulation of nerve networks by the brain also fosters a close association between learned skeletal muscle movements and autonomic outflows to smooth muscle that support those movements. These principles may help to comprehend the evolution of catecholaminergic systems.

The sympathetic nervous system is immature at birth (Friedman et al. 1968). If ontogeny recapitulates phylogeny, this immaturity further indicates relatively late development of the sympathetic nerve network in evolution. Maturation of the sympathetic nervous system after birth allows a degree of plasticity of sympathoneural development based on experiences of the individual early in life. In contrast, the adrenal medulla forms in utero in mammals, and low and unchanging basal tissue concentrations of epinephrine persist in sympathetically innervated organs throughout the life span (Friedman et al. 1968).

One apparent consequence of this differential evolution of catecholaminergic systems is that metabolic stressors evoke preferentially adrenomedullary hormonal responses, whereas stressors involving alterations in skeletal muscle contraction (e.g., locomotion) evoke preferentially sympathoneural responses.

What became of the autocrine-paracrine catecholaminergic systems? They still appear to operate, at least in the kidneys and adrenal glands. In the kidneys, locally formed dopamine contributes to regulation of sodium balance (Carey et al. 1990; Siragy et al. 1989). In the adrenal cortex, locally formed dopamine inhibits aldosterone secretion. In both organs, however, more powerful hormonal and neuronal systems now predominate.

In the rabbit myocardium, nests of chromaffin cells have been observed at all ages studied, even in the fetus (Friedman et al. 1968). The source and significance of these cells remain mysterious. Perhaps these cells are also remnants of a primitive autocrine-paracrine system. One may speculate that other cells dispersed throughout the body and with largely unknown functions, such as "APUD" (amino precursor uptake and decarboxylation) cells, "SIF" (small intensely fluorescent) cells in sympathetic ganglia, enterochromaffin cells of the gut, and mast cells in the lungs, represent the remnants of this primordial system.

Dopamine concentrations in plasma are remarkably low, and concentrations of the inactive conjugate, dopamine sulfate, are about 50-fold greater than those of unconjugated dopamine. When dopamine is injected into humans, dopamine sulfate levels rapidly increase (Ratge et al. 1991). The source and physiological meaning of conjugated dopamine are unknown. Since the conjugating mechanism maintains circulating dopamine concentrations at levels far lower than required to produce physiological effects, one may speculate that mechanisms evolved to inactivate dopamine escaping from organs into the bloodstream—that is, that the high levels of dopamine sulfate reflect overriding of a primitive autocrine-paracrine catecholaminergic system. In the gut, sulfation may enable a "gut-blood barrier" for ingested catecholamines (Cuche et al. 1990).

Envelopment of the adrenal medulla by the adrenal cortex is a remarkable feature in mammals. The two components have entirely different embryological sources (Langman 1969). The adrenal medullary cells are derived from ectodermal sympathetic neuroblasts. They migrate anteriorly to penetrate the fetal adrenal, which is derived from mesoderm. The sympathetic cells cluster to form the adrenal medulla. Since blood flows in a corticomedullary direction in the adrenal gland, this arrangement results in very high local concentrations of adrenal steroids in the blood passing through the adrenal medulla. Glucocorticoids contribute importantly to regulation of adrenomedullary epinephrine synthesis. Since angiotensin II is formed in the adrenal cortex, and since angiotensin II stimulates catecholamine secretion in cultured adrenomedullary cells (Zimlichman et al. 1987), high local concentrations of angiotensin II in blood traversing the adrenal medulla may contribute to regulation of adrenomedullary secretion. The architectural arrangement of the adrenal cortex and medulla therefore fosters the coordination of activities of major hormonal stress effector systems.

Primitive Specificity

A major thesis of this chapter is that activities of stress effector systems are coordinated in relatively specific patterns, including neuroendocrine patterns. These patterns serve different homeostatic needs and are produced by the actions of different homeostats. One may speculate that the survival advantages afforded by redundant effectors fostered the evolution of stress systems in which each homeostat would drive more than one effector. The availability of multiple effectors would, in turn, have enabled the expression of patterns of effector activation, and the survival advantages afforded by adaptive coordinated responses could have fostered the evolution of response patterns with a degree of "primitive specificity" for different stressors.

Stress responses involving specific neuroendocrine, physiological, and behavioral patterns occur commonly in daily life. Examples are orthostasis, temperature regulation, exercise, and the postprandial state. Even emotional or psychological stress responses have a degree of specificity. The distinctive patterning of responses of these systems subserves different adaptive needs.

For each stress, neuroendocrine and physiological changes are coupled with behavioral changes. For example, the regulation of total body water in humans depends on an interplay between behavior (the search for water and drinking), an internal experience or feeling (thirst), and the elicitation of a neurohormonal response pattern (in this case dominated by vasopressin, the antidiuretic hormone; and to a lesser extent angiotensin, a potent dipsogen, or stimulator of drinking). Activation of homeostats almost always produces not only neuroendocrine and physiological effects but also behavioral responses, and the pattern of elicited responses is compensatory in decreasing the perturbation as sensed by the homeostats. Of course, any stressor, if prolonged or extreme, can elicit distress and thereby activation of many effectors.

Because of traditional boundaries among physiologists, endocrinologists, and psychologists, the interactions among these changes in producing integrated patterns of response are incompletely understood. Thus, studies about vasopressin and activity of the renin-angiotensin-aldosterone system during blood volume depletion rarely have paid attention to mechanisms of thirst and salt hunger. This situation developed partly because it used to be thought that the voluntary, conscious responses of the organism to the external environment—the province of psychology—were mediated by acts of skeletal muscle; whereas involuntary, unconscious responses of smooth muscle and glands to maintain the internal environment—the province of physiology and endocrinology—were regulated by the autonomic and endocrine systems. The development of the field of neuroendocrinology has questioned the validity of this distinction, since conclusive evidence has accrued that the external environment, via the central nervous system—in particular the hypothalamus—affects autonomic and endocrine activity. Biofeedback and conditioning studies have also suggested that modifications of some autonomic functions can be learned (Goldstein et al. 1977; Miller 1969; Siegel 1972). Conversely, the state of physiological arousal affects behavior by, among other ways, intensifying emotions (Schachter and Singer 1962) and communicating external signs of internal states. Finally, recent findings involving circumventricular organs, sites in the brain that are devoid of a blood-brain barrier, have introduced the possibility that hormones can reach central neural sites, eliciting various behaviors and altering autonomic and neuroendocrine activity.

Folkow (1988) has argued that limbic-hypothalamic patterns, designed to afford protection to the individual and species during exposure to adverse environmental conditions, are always expressed as a triad, with closely linked situation-specific somatomotor behavior, visceromotor autonomic changes, and hormonal changes. One may even speculate that patterns of behavioral, experiential, neuroendocrine, and autonomic activities during stress have become so tightly intertwined during the course of evolution that they are now regulated as units. As noted earlier in the discussion of Dawkins' selfish gene theory, groups of genes can be selected as a unit; it may be that the behavioral, neuroendocrine, autonomic, and perhaps experiential components of stress responses are regulated and expressed as units, providing a basis for "primitive specificity" of stress responses.

The view that stress responses can be differentiated biochemically contrasts with those of both Cannon and Selye. Largely ignoring the roles of other systems, Cannon asserted that the same sympathoadrenal response would meet most or all important threats to the internal environment. Differential regulation of the sympathoneural and adrenomedullary components of the sympathoadrenal system during different forms of stress (Young et al. 1984) constitutes one of the main pieces of evidence in favor of the notion of primitive specificity.

Different emotional behaviors can have very different cardiovascular concomitants and so presumably different autonomic determinants. To give one example, responses of renal and iliac blood flow velocity, blood pressure, and heart rate during various behaviors were measured in baboons by Smith et al. (1988). The behaviors included self-grooming, grooming another animal, yawning, sitting, walking, orienting, attack, and sight of a conspecific of the opposite sex. Attack toward the observer was associated with very marked decreases in renal flow and increases in heart rate, whereas iliac flow was increased, and arterial pressure decreased. Exposure to a conspecific of the opposite sex increased heart rate, decreased renal flow, increased iliac flow, and increased arterial pressure. Startle was associated with increased iliac flow, decreased renal flow, increased heart rate, and little change in mean arterial pressure. Orienting was associated with decreased renal flow, increased iliac flow, decreased arterial pressure, and increased heart rate. Exposure of a monitored male baboon to a male conspecific produced profound, prolonged decreases in renal flow and increases in heart, with small and variable increases in blood pressure and iliac flow. Before overt attack by a dominant male, renal and mesenteric flow already were decreased and heart rate and blood pressure increased (Astley et al. 1991).

The patterns of cardiovascular response in these different behavioral situations therefore are diverse. Because of the importance of dominance in the daily life of the baboon and determining access to food, territory, attention, and sexual partners, it is not surprising that challenge to dominance produces especially large cardiovascular stress responses, in which decreased renal blood flow and increased heart rate are prominent features. Laboratory psychological stressors in humans are similarly associated with sympathetically mediated decreases in renal blood flow and increases in heart rate associated with regional increases in release of norepinephrine. Smith et al. (1988) concluded that different behaviors seem to have thier own cardiovascular signature. By analyses of relevant environmental changes, video recordings of behavior, and telemetered cardiovascular changes, their group hopes to develop a code relating behavioral with cardiovascular dynamics (Astley et al. 1991). One would hope further that the code would incorporate the neuroendocrine responses that link these dynamics, and the homeostats that determine them all.

Summary and Conclusion

Claude Bernard suggested that the body maintains a constant internal environment by myriad, continual compensatory reactions, tending to restore a state of equilibrium, thereby allowing independence from the external environment. He wrote that "it is the fixity of the *'milieu interieur'* which is the condition of free and independent life. All the vital mechanisms, varied as they are, have only one object, that of preserving constant the conditions of life in the internal environment" (Bernard 1878).

Cannon extended Bernard's concept by showing that the maintenance of a stable internal environment depends importantly on circulatory regulation by the sympathoadrenal system. Cannon coined the term *homeostasis* and proposed that homeostatic systems maintain most of the steady states in the organism. Adjustments in sympathoadrenal activity accompany virtually every human action and reaction and are the main means for acute regulation of cardiovascular function.

Hans Selye popularized stress as a scientific and medical concept. His view was that stress and stress responses are nonspecific; that is, they can be brought on regardless of the stimulus. The stress response itself was viewed by Selye as consisting of a stereotyped pattern, with enlargement of the adrenal glands, involution of the thymus (associated with atrophy of lymph nodes and inhibition of inflammatory responses), and peptic ulceration—all now known to be effects of high circulating levels of adrenal corticosteroids. The stress response was also thought to have characteristic stages: alarm, resistance, and exhaustion. Several major problems with Selye's stress theory, including the circularities, inconsistencies, and the doctrine of nonspecificity, have led to its abandonment by many researchers. Nevertheless, two general points from his theory and experiments seem justified: Stress is an unobservable condition that leads to adaptive responses; and adrenal release of glucocorticoids and catecholamines often accompanies stress responses.

According to the present conception, stress is a type of intervening variable. It is a condition where expectations—whether genetically programmed, established by prior learning, or deduced from circumstances—do not match the current or anticipated perceptions of the internal or external environment, and this discrepancy between what is observed or sensed and what is expected or programmed elicits patterned compensatory responses.

Distress is a consciously experienced form of stress, characterized by specific behavioral and autonomic communicated signs, pituitary-adrenocortical and adrenomedullary activation, and a negative feeling that motivates escape behavior or avoidance learning.

The body has many homeostatic comparators—they can be generically called "homeostats." All seem to share several characteristics. Homeostatic systems are subject to negative feedback regulation. They usually include more than one effector, allowing for a degree of compensatory activation of effectors to achieve homeostatic goals and also enabling the elaboration of relatively specific patterns of effector responses. Homeostats can share an effector system, obscuring relationships between effector system activity and values for any single physiological parameter. These characteristics of homeostatic systems explain why, in the absence of rigorous controls, plasma levels of norepinephrine, the sympathetic neurotransmitter, rarely correlate positively with mean arterial pressure across subjects, even though the sympathoneural system is an important pressure-regulating effector system.

Homeostatic systems are prone to resetting during stress. Simple homeostatic reflexes such as the baroreflex can be subserved unconsciously by centers in the medulla oblongata of the brain stem, but the functional characteristics of these reflexes can be readjusted by higher centers. Distress always resets homeostats.

Disruption of a homeostatic system increases the variability of the monitored variable and tends to result in drifting of the absolute value to a new level. Dysfunction of circulatory homeostatic reflexes therefore would be expected to contribute to cardiovascular instability and possibly to stress-related pathology.

The notion that emotional expressions and their autonomic accompaniments evolved did not originate with Cannon but with Charles Darwin. The process that has determined the phylogenetic development of physiological and even psychological patterns of response has been essentially the same as the process that has determined the development of anatomical patterns—natural selection in evolution. As homeostatic systems evolved, so did mechanisms regulating them. These mechanisms enable homeostasis to be redefined continually in life.

Activities of the body's stress systems are regulated in primitively specific patterns. Many of these patterns, which are at least partly inherited, can be understood teleologically on the basis of preservation of the internal environment and natural selection in evolution. The physiological, biochemical, experiential, and behavioral components usually are closely linked.

The scientific value of the homeostatic theory of stress and distress will depend not so much on the ability to explain phenomena but on the ability to generate hypotheses that observation and experiment can test.

References

Abercrombie ED, Nisenbaum LK, Zigmond MJ: Impact of acute and chronic stress on the release and synthesis of norepinephrine in brain: microdialysis studies in behaving animals, in Stress: Neuroendocrine and Molecular Approaches. Edited by Kvetnansky R, McCarty R, Axelrod J. New York, Gordon & Breach Science Publishers, 1992, pp 29–42

Ahn SS, Marcus DR, Moore WS: Post-carotid endarterectomy hypertension: association with elevated cranial norepinephrine. J Vasc Surg 9:351–360, 1989

Aksamit TR, Floras JS, Victor RG, et al: Paroxysmal hypertension due to sinoaortic baroreceptor denervation in humans. Hypertension 9:309–314, 1987

Asterita MF: The Physiology of Stress. New York, Human Sciences Press, 1985

Astley CA, Smith OA, Ray RD, et al: Integrating behavior and cardiovascular responses: the code. Am J Physiol 261:R172–R181, 1991

Attenborough D: Life on Earth. Boston, MA, Little, Brown, 1979

Bartorelli C, Bizzi E, Libretti A, et al: Inhibitory control of sinocarotid pressoceptive afferents on hypothalamic autonomic activity and sham rage behavior. Arch Ital Biol 98:308–326, 1960

Beamish RE, Singal PK, Dhalla NS (eds): Stress and Heart Disease. Boston, MA, Martinus Nijhoof Publishing, 1985

Berger FM: How to be happy in spite of life's vicissitudes, in Life Stress: A Companion to the Life Sciences, Vol 3. Edited by Day SB. New York, Van Nostrand Reinhold, 1982, pp 104–109

Bernard C: Leçons sur les Phénomenones de la Vie Communs aux Animaux et aux Végtaux. Paris, France, Balliere et fils, 1878

Berthelot A, Hamilton CA, Petty MA, Reid JL: Central and peripheral α-adrenoceptor number and responsiveness after sinoaortic denervation in the rabbit. J Cardiovasc Pharmacol 4:567–574, 1982

Blizard DA, Liang B, Emmel DK: Blood pressure, heart rate, and plasma catecholamines under resting conditions in rat strains selectively bred for differences in response to stress. Behav Neural Biol 29:487–492, 1980

Bogen DK: Simulation software for the Macintosh. Science 246:138–142, 1989

Brattstrom BH: Body temperature in reptiles. American Midland Naturalist 73:376–422, 1965

Braunwald E, Ross J, Sonnenblick EH: Mechanisms of Contraction of the Normal and Failing Heart. Boston, MA, Little, Brown, 1968

Cannon WB: Bodily Changes in Pain, Hunger, Fear and Rage. New York, Appleton, 1929

Cannon WB: The Wisdom of the Body. New York, WW Norton, 1939

Carey RM, Siragy HM, Felder RA: Physiological modulation of renal function by the renal dopaminergic system. J Auton Pharmacol 10 (suppl 1):S47–S51, 1990

Carlberg M: Evidence of dopa in the nerves of sea anemones. J Neural Transm 57:75–84, 1983

Carmelli D, Chesney MA, Ward MM, et al: Twin similarity in cardiovascular stress response. Health Psychol 4:413–423, 1985

Chernow B: Catecholamines and circulatory shock, in Progress in Catecholamine Research. Edited by Belmaker RH, Sandler M, Dahlstrom A. New York, Alan R Liss, 1988, pp 475–476

Cowley AW, Liard JF, Guyton AC: Role of the baroreceptor reflex in daily control of arterial blood pressure and other variables in dogs. Circ Res 32:564–576, 1973

Cuche J-L, Brochier P, Klioua N, et al: Conjugated catecholamines in human plasma: where are they coming from? J Lab Clin Med 116:681–686, 1990

Darwin C: The Expression of the Emotions in Man and Animals (1872). Chicago, IL, The University of Chicago Press, 1965

Davydova NA, Kvetnansky R, Usakov AC: Activity of sympatho-adrenal system of cosmonauts during prolonged space flights on station "Saljut-7," in Stress: Neurochemical and Humoral Mechanisms. Edited by Van Loon GR, Kvetnansky R, McCarty R, et al. New York, Gordon & Breach Science Publishers, 1989, pp 999–1013

Dawkins R: The Selfish Gene. New York, Oxford University Press, 1989

de Boer SF, De Beun R, Slangen JL, et al: Dynamics of plasma catecholamine and corticosterone concentrations during reinforced and extinguished behavior in rats. Physiol Behav 47:691–698, 1990a

de Boer SF, Slangen JL, Van der Gugten J: Plasma catecholamine and corticosterone levels during active and passive shock-prod avoidance behavior in rats: effects of chlordiazepoxide. Physiol Behav 47:1089–1098, 1990b

Drummond PD, Lance JW: Facial flushing and sweating mediated by the sympathetic nervous system. Brain 110:793–803, 1987

Eliot RS (ed): Stress and the Heart. Mount Kisco, NY, Futura Publishing, 1974

Fagius J, Wallin BG, Sundlof G, et al: Sympathetic outflow in man after anaesthesia of the glossopharyngeal and vagus nerves. Brain 108:423–438, 1985

Folkow B: Physiological aspects of primary hypertension. Physiol Rev 62:347–504, 1982

Folkow B: Physiology of behavior and blood pressure regulation in animals, in Handbook of Hypertension, Vol 9: Behavioral Factors in Hypertension. Edited by Julius S, Bassett DR. New York, Elsevier, 1987, pp 1–18

Folkow B: Stress, hypothalamic function and neuroendocrine consequences. Acta Med Scand Suppl 723:61–69, 1988

Frankenhaeuser M: Experimental approaches to the study of catecholamines and emotion, in Emotions: Their Parameters and Measurement. Edited by Levi L. New York, Raven, 1975, pp 209–234

Friedman WF, Pool PE, Jacobowitz D, et al: Sympathetic innervation of the developing rat heart: biochemical and histochemical comparisons of fetal, neonatal, and adult myocardium. Circ Res 23:25–32, 1968

Frye RL, Braunwald E: Studies on Starling's law of the heart, I: the circulatory response to acute hypervolemia and its modification by ganglion blockade. J Clin Invest 39:1043–1050, 1960

Gaffney TE, Braunwald E: Importance of the adrenergic nervous system in the support of circulatory function in patients with congestive heart failure. Am J Med 34:320–324, 1963

Garber AJ, Cryer PE, Santiago JV, et al: The role of adrenergic mechanisms in the substrate and hormonal response to insulin-induced hypoglycemia in man. J Clin Invest 58:7–15, 1976

Gero J, Gerova M: Significance of the individual parameters of pulsating pressure in stimulation of baroreceptors, in Baroreceptors and Hypertension. Edited by Kezdi P. Oxford, Pergamon, 1967, pp 17–30

Gold PW, Goodwin FK, Chrousos GP: Clinical and biochemical manifestations of depression: relation to the neurobiology of stress. N Engl J Med 319:348–353, 413–420, 1988

Goldstein DS: Arterial baroreflex sensitivity, plasma catecholamines, and pressor responsiveness in essential hypertension. Circulation 68:234–240, 1983

Goldstein DS, Ross RS, Brady JV: Biofeedback heart rate training during exercise. Biofeedback Self Regul 2:127–138, 1977

Gray TS: Limbic pathways and neurotransmitters as mediators of autonomic and neuroendocrine responses to stress, in Stress: Neurobiology and Neuroendocrinology. Edited by Brown MR, Koob GF, Rivier C. New York, Marcel Dekker, 1991, pp 73–89

Gullichsen E, Nelimarkka O, Scheinin M, et al: Plasma levels of atrial natriuretic factor and catecholamines in endotoxin shock in dogs. Acta Chir Scand 155:567–571, 1989

Guz A, Noble MIM, Widdicombe JG, et al: The role of vagal and glossopharyngeal afferent nerves in respiratory sensation, control of breathing and arterial pressure regulation in conscious man. Clin Sci 30:161–170, 1966

Hart BB, Stanford GG, Ziegler MG, et al: Catecholamines: study of interspecies variation. Crit Care Med 17:1203–1222, 1989

Hasking GJ, Esler MD, Jennings GL, et al: Norepinephrine spillover to plasma in patients with congestive heart failure: evidence of increased overall and cardiorenal sympathetic nervous activity. Circulation 73:615–621, 1986

Henry JP: Biological basis of the stress response. Integrative Physiology and Behavioral Science 27:66–83, 1992

Henry JP, Stephens PM, Ely DL: Psychosocial hypertension and the defense and defeat reactions. J Hypertens 4:687–697, 1986

Herd JA: Neuroendocrine mechanisms in coronary heart disease, in Handbook of Stress, Reactivity, and Cardiovascular Disease. Edited by Matthews KA, Weiss SM, Detre T, et al. New York, Wiley, 1986, pp 49–70

Holland KN, Brill RW, Chang RK, et al: Physiological and behavioural thermoregulation in bigeye tuna (*Thunnus obesus*). Nature 358:410–412, 1992

Holzbauer M, Sharman DF: The distribution of catecholamines in vertebrates, in Handbook of Experimental Pharmacology, Vol 33. Edited by Eichler O, Farah A, Herken H, et al. New York, Springer-Verlag, 1972, pp 110–185

Hull CL: Principles of Behavior. New York, Appleton, 1943

Kamiya J, Barber TX, DiCara LV, et al: Biofeedback and Self-Control. New York, Aldine Atherton, 1971

Kandel ER: From metapsychology to molecular biology: explorations into the nature of anxiety. Am J Psychiatry 140:1277–1293, 1983

Kandel ER, Tauc L: Mechanism of heterosynaptic facilitation in the giant cell of the abdominal ganglion of Aplysia depilans. J Physiol (Lond) 181:28–47, 1965

Kezdi P (ed): Baroreceptors and Hypertension. Oxford, Pergamon, 1967

Koizumi K, Kollai M: Control of reciprocal and non-reciprocal action of vagal and sympathetic efferents: study of centrally induced reactions. J Auton Nerv Syst 3:483–501, 1981

Krantz DS, Lazar JD: The stress concept: issues and measurement, in Behavioral Factors in Hypertension. Edited by Julius S, Bassett DR. New York, Elsevier, 1987, pp 43–58

Krieger JE, Liard J-F, Cowley AW Jr: Hemodynamics, fluid volume, and hormonal responses to chronic high-salt intake in dogs. Am J Physiol 259:H1629–H1636, 1990

Kvetnansky R, Mikulaj L: Adrenal and urinary catecholamines in rat during adaptation to repeated immobilization stress. Endocrinology 87:738–743, 1970

Kvetnansky R, Nemeth S, Vigas M, et al: Plasma catecholamines in rats during adaptation to intermittent exposure to different stressors, in Stress: The Role of Catecholamines and Other Neurotransmitters. Edited by Usdin E, Kvetnansky R, Axelrod J. New York, Gordon & Breach Publishers, 1984, pp 537–562

Kvetnansky R, Davydova NA, Noskov VB, et al: Plasma and urine catecholamine levels in cosmonauts during long-term stay on space station Salyet-7. Acta Astronautica 17:181–186, 1988

Lacey JI: Somatic response patterning and stress: some revisions of activation theory, in Psychological Stress. Edited by Appley MH, Trumble R. New York, Appleton-Century-Crofts, 1967, pp 14–42

Lacey JI, Lacey BC: Verification and extension of the principle of autonomic response-stereotypy. Am J Psychol 71:50–73, 1958

Lacey JI, Lacey BC: Some autonomic-central nervous system interrelationships, in Physiological Correlates of Emotion. Edited by Black P. New York, Academic Press, 1970, pp 205–227

Langman J: Medical Embryology. Baltimore, MD, Williams & Wilkins, 1969

Lazarus RS: Psychological Stress and the Coping Process. New York, McGraw-Hill, 1966

Lazarus RS: Emotion and Adaptation. New York, Oxford University Press, 1991

Levine S, Ursin H: What is stress? in Stress: Neurobiology and Neuroendocrinology. Edited by Brown MR, Koob GF, Rivier C. New York, Marcel Dekker, 1991, pp 3–21

Lillywhite HB: Snakes, blood circulation and gravity. Sci Am 259:92–98, 1988

Lorenz K: King Solomon's Ring. New York, Thomas Y Crowell, 1952

Lorenz K: On Aggression. New York, Bantam Books, 1963

Luft FC, Miller JZ, Weinberger MH, et al: Influence of genetic variance on sodium sensitivity of blood pressure. Klin Wochenschr 65:101–109, 1987

Lundin S, Thoren P: Renal function and sympathetic activity during mental stress in normotensive and spontaneously hypertensive rats. Acta Physiol Scand 115:115–124, 1982

Lundin S, Rickstein S-E, Thoren P: Interaction between mental stress and baroreceptor control of heart rate and sympathetic activity in conscious spontaneously hypertensive (SHR) and normotensive (WKY) rats. J Hypertens 1 (suppl 2):68–70, 1983

Lundin S, Rickstein S-E, Thoren P: Interaction between "mental" stress and baroreceptor reflexes concerning effects on heart rate, mean arterial pressure and renal sympathetic activity in conscious spontaneously hypertensive rats. Acta Physiol Scand 120:273–281, 1984

Mason JW: A re-evaluation of the concept of "non-specificity" in stress theory. J Psychiatr Res 8:323–333, 1971

Mason JW: A historical view of the stress field, II. Journal of Human Stress 1:6–12, 22–36, 1975

McCarty R, Kopin IJ: Alterations in plasma catecholamines and behavior during acute stress in spontaneously hypertensive and Wistar-Kyoto normotensive rats. Life Sci 22:997–1005, 1978

McCarty R, Konarska M, Stewart RE: Adaptation to stress: a learned response? in Stress: Neuroendocrine and Molecular Approaches. Edited by Kvetnansky R, McCarty R, Axelrod J. New York, Gordon & Breach Science Publishers, 1992, pp 521–535

McFarlan D (ed): The Guinness Book of World Records. New York, Bantam Books, 1992

Metz SA, Halter JB, Porte D Jr, et al: Autonomic epilepsy: clonidine blockade of paroxysmal catecholamine release and flushing. Ann Intern Med 88:189–193, 1978

Miller NE: Learning of visceral and glandular responses. Science 163:434–445, 1969

Murray MJ, Offord KP, Yaksh TL: Physiologic and plasma hormone correlates of survival in endotoxic dogs: effects of opiate antagonists. Crit Care Med 17:39–47, 1989

Nathan MA, Reis DJ: Chronic labile hypertension produced by lesions of the nucleus tractus solitarii in the cat. Circ Res 40:72–81, 1977

Nathan MA, Tucker LW, Serverini WH, et al: Enhancement of conditioned arterial pressure responses in cats after brainstem lesions. Science 201:71–73, 1978

The New York Times, April 21, 1989, p. 1

Nisenbaum LK, Abercrombie ED: Enhanced tyrosine hydroxylation in hippocampus of chronically stressed rats upon exposure to a novel stressor. J Neurochem 58:276–281, 1992

Nisenbaum LK, Zigmond MJ, Sved AF, et al: Prior exposure to chronic stress results in enhanced synthesis and release of hippocampal norepinephrine in response to a novel stressor. J Neurosci 11:1478–1484, 1991

Olmstead JMD, Olmstead EH: Claude Bernard and the Experimental Method in Medicine. New York, Henry Schuman, 1952

Pavlov IP: Conditioned Reflexes: An Investigation of the Physiological Activity of the Cerebral Cortex. London, Oxford University Press, 1927

Quillen EW Jr, Cowley AW Jr: Influence of volume changes on osmolality-vasopressin relationships in conscious dogs. Am J Physiol 244:H73–H79, 1983

Rappaport EB, Young JB, Landsberg L: Effects of 2-deoxy-D-glucose on the cardiac sympathetic nerves and the adrenal medulla in the rat: further evidence for a dissociation of sympathetic nervous system and adrenal medullary responses. Endocrinology 110:650–656, 1982

Ratge D, Kohse KP, Steegmuller U, et al: Distribution of free and conjugated catecholamines between plasma, platelets and erythrocytes: different effects of intravenous and oral catecholamine administrations. J Pharmacol Exp Ther 257:232–238, 1991

Rea RF, Eckberg DL: Carotid baroreceptor-muscle sympathetic relation in humans. Am J Physiol 253:R929–R934, 1987

Rea RF, Hamdan M: Baroreflex control of muscle sympathetic nerve activity in borderline hypertension. Circulation 82:856–862, 1990

Rubin Z, McNeil EB: Psychology: Being Human. New York, Harper & Row, 1987

Sapolsky RM: Stress in the wild. Sci Am 262:116–123, 1990

Schachter S, Singer J: Cognitive, social, and physiological determinants of emotional state. Psychol Rev 69:379–399, 1962

Schomig A, Fischer S, Kunz T, et al: Nonexocytotic release of endogenous noradrenaline in the ischemic and anoxic rat heart: mechanism and metabolic requirements. Circ Res 60:194–205, 1987

Seckl JR, Williams TDM, Lightman SL: Oral hypertonic saline causes transient fall of vasopressin in humans. Am J Physiol 251:R214–R217, 1986

Selye H: A syndrome produced by diverse nocuous agents. Nature 138:32, 1936

Selye H: The Physiology and Pathology of Exposure to Stress: A Treatise Based on the Concepts of the General-Adaptation Syndrome and the Diseases of Adaptation. Montreal, Canada, Acta, 1950

Selye H: The Stress of Life. New York, McGraw-Hill, 1956

Selye H: Stress Without Distress. New York, New American Library, 1974

Selye H: Stress: Eustress, distress, and human perspectives, in Life Stress: A Companion to the Life Sciences, Vol 3. Edited by Day SB. New York, Van Nostrand Reinhold, 1982, pp 3–13

Shade RE, Bishop VS, Haywood JR, et al: Cardiovascular and neuroendocrine responses to baroreceptor denervation in baboons. Am J Physiol 258:R930–R938, 1990

Shah SD, Tse TF, Clutter WE, et al: The human sympathochromaffin system. Am J Physiol 247:E380–E384, 1984

Shore AC, Markandu ND, Sagnella GA, et al: Endocrine and renal response to water loading and water restriction in normal man. Clin Sci 75:171–177, 1988

Siegel S: Conditioning of insulin effects. Journal of Comparative and Physiological Psychology 78:233–241, 1972

Siragy HM, Felder RA, Howell NL, et al: Evidence that intrarenal dopamine acts as a paracrine substance at the renal tubule. Am J Physiol 257:F469–F477, 1989
Sleight P, Fox P, Lopez R, et al: The effect of mental arithmetic on blood pressure variability and baroreflex sensitivity in man. Clinical Science and Molecular Medicine 55 (suppl 4):381S–382S, 1978
Smith OA, Golanov E, Astley CA, et al: Behavior and the cardiovascular system, in Neurocardiology. Edited by Kulbertus HE, Franck G. Mount Kisco, NY, Futura Publishing, 1988, pp 279–289
Sternberg EM, Hill JM, Chrousos GP, et al: Inflammatory mediatory-induced hypothalamic-pituitary-adrenal axis activation is defective in streptococcal cell wall arthritis-susceptible Lewis rats. Proc Natl Acad Sci U S A 86:2374–2378, 1989
Stinner JN, Ely DL: Blood pressure during routine activity, stress, and feeding in black racer snakes (*Coluber constrictor*). Am J Physiol 264:R79–R84, 1993
Swanson LW: The hypothalamus, in Handbook of Chemical Neuroanatomy, Vol 5: Integrated Systems of the CNS, Part I: Hypothalamus, Hippocampus, Amygdala, Retina. Edited by Bjorklund A, Hokfelt T, Swanson LW, et al. New York, Elsevier, 1987, pp 1–124
Thornton EW, Van-Toller C: Operant conditioning of heart-rate changes in the functionally decorticate curarised rat. Physiol Behav 10:983–988, 1973
Turkkan JS, Goldstein DS: Stress and sodium hypertension in baboons: neuroendocrine and pharmacotherapeutic assessments. J Hypertens 9:969–975, 1991
Weiner H: Behavioral biology of stress and psychosomatic medicine, in Stress: Neurobiology and Neuroendocrinology. Edited by Brown MR, Koob GF, Rivier C. New York, Marcel Dekker, 1991, pp 23–51
Welsh JH: Catecholamines in the invertebrates, in Handbook of Experimental Pharmacology, Vol 33. Edited by Eichler O, Farah A, Herken H, et al. New York, Springer-Verlag, 1972, pp 79–109
Young JB, Rosa RM, Landsberg L: Dissociation of sympathetic nervous system and adrenal medullary responses. Am J Physiol 247:E35–E40, 1984
Zimlichman R, Goldstein DS, Zimlichman S, et al: Angiotensin II increases cytosolic calcium and stimulates catecholamine release in cultured bovine adrenomedullary cells. Cell Calcium 8:315–325, 1987

9

The Symptoms of Adrenergic Activation

OLIVER G. CAMERON, M.D., PH.D.

> Wherever there is a heart and an intellect, the diseases of the physical frame are tinged with the peculiarities of these.
>
> —Nathaniel Hawthorne, *The Scarlet Letter*

The ultimate goal of medicine is the correct diagnosis and the effective treatment of disease. Diagnosis and treatment are based, if possible, on an understanding of the pathophysiology and, ideally, etiology of each specific illness. Nevertheless, a disease is typically first defined as a discrete entity not based on an understanding of pathophysiology or etiology but rather on recognition that a specific combination of symptoms and signs tend to cluster (i.e., a syndrome).

Symptoms are usually considered manifestations of illness and, as such, are thought to be of secondary interest, useful only insofar as they lead to a correct diagnosis. If *symptom* is defined more broadly, however, as any subjective sensation reported by an individual (not just as an indicator of the presence of some illness), then the physiology of symptoms per se becomes of interest as well. (For the remainder of this exposition, the terms *symptom* and *subjective sensation* are used synonymously).

Symptoms per se are of interest for two reasons. First, individual symptoms do not covary with specific disease entities—a given symptom can occur in association with a variety of diseases, and the presence of a given disease only rarely means that any given specific symptom will invariably be observed (Kroenke 1992). Second, a wide variety of subjective sensations occur, in all individuals and under many circumstances, that do not reflect pathological functioning; examples include the awareness of heart action after exercise, the occurrence of an itch, the lump in the throat one sometimes feels when about to cry—many more examples could be offered. From where do these sensations come? What is their physiology? How does an individual sense or become aware of his or her heartbeat, for example? Furthermore, at least one psychiatric disorder—somatization disorder—is defined by DSM-IV (American Psychiatric Association 1994) criteria, which requires the presence of a "history of many physical complaints," which must include four pain symptoms, two gastrointestinal symptoms, one sexual symptom, and one pseudoneurological symptom, and also requires that "after appropriate investigation, each of the [preceding] symptoms . . . cannot be fully explained by a known general medical condition or the direct effects of a substance . . ." and that "when there is a general medical condition, the physical complaints or resulting . . . impairment are in excess of what would be expected from the history, physical examination, or laboratory findings" (pp. 449–450). In other words, at least one syndrome is defined precisely by the presence but unexplainability of symptoms (Kellner 1991).

The purpose of this chapter is to explore these questions within the circumscribed context of symptoms that are likely to be mediated at least in part by adrenergic mechanisms. The reader should bear in mind that the considerations being reviewed apply not just to the adrenergically mediated symptoms discussed in this chapter, but to symptom production as a generic fact of psychobiology. It should also be noted that, although levels of the catecholamines epinephrine and norepinephrine are commonly studied markers of adrenergic function in blood, dopamine is present in blood concentrations of the order of magnitude of epinephrine, and it also has a variety of physiological effects if administered in pharmacological doses (Hoffman and Lefkowitz 1990). The question of whether or not dopamine per se in any concentration seen under physiological conditions produces subjective sensations apparently has been given minimal consideration.

History

The physiology of suprarenal extracts, the earliest of which were undoubtedly impure mixtures of catecholamine and other substances, has been studied since the end of the 19th century. The active substance in the extracts was named "epinephrine" in 1899, and first synthesized early in this century (Hoffman and Lefkowitz 1990). (It should be noted that *epinephrine* and *adrenaline* are synonyms, as are *norepinephrine* and *noradrenaline.*) Studies of the effects of infusions of adrenergic substances into humans were first reported around 1920 (Redmond 1979; Van Toller 1979). Controversy concerning whether the physiological effects of sympathomimetic substances, which typically do not cross the blood-brain barrier (MacKenzie et al. 1976; Oleson et al. 1978; Oldendorff 1971), actually mimic fear and related emotions or rather produce "cold, as-if" emotional feelings involving only the physical sensations of emotion was very prominent earlier in this century and has not been fully resolved to this date. The theory that peripheral physical sensations are the primary source of emotional experiences is usually credited to James and Lange (Fehr and Stern 1970; Lyons 1980; Mayes 1979). The major source of confusion in this arena appears to involve the question of whether the experimental production of physical sensations similar to symptoms experienced under other circumstances (e.g., in individuals with anxiety disorders) can induce the full emotional reaction (e.g., cause the anxious person to react both physically *and psychologically*) (Lyons 1980; Van Toller 1979).

In this chapter, I review research addressing the relationship between symptoms and adrenergic function in several areas, including 1) cardiorespiratory adrenergic symptoms, 2) adrenergic sensations and hypoglycemia, 3) symptoms associated with systemic administration of adrenergic substances, 4) stress and adrenergic symptoms, and 5) central nervous system changes and adrenergic symptoms. I conclude with a discussion of the results and implications.

Cardiorespiratory Symptoms

It is generally assumed that those symptoms that are mediated adrenergically, and more generally autonomically, are well understood and well described. Awareness of augmented heart action, changes in respiration, tremor, and increased sweating are the sensations most commonly attributed to adrenergic activation (although sweating is sympathetic but mediated by cholinergic neurotransmission). However, these data seem to be based primarily on results of studies of systemic infusions of adrenergic agonist (and antagonist) substances; sensations associated with endogenous adrenergic activation are less well described and understood.

The sensation perhaps most widely assumed to be associated with adrenergic activation is awareness of heartbeat. There have been many studies, reported mainly in the psychophysiology and biofeedback literatures, of the ability of humans to discriminate heart rate. Several studies have reported that subjects can learn to discriminate heart rate (Bergman and Johnson 1971; Davis et al. 1986; Grigg and Ashton 1982; Jones et al. 1987). Nevertheless, despite higher heart rates, type A individuals (Essau and Jamieson 1987), healthy individuals after exercise (Cacioppo et al. 1987; Jones and Hollandsworth 1981), and postinfarct patients (Jones et al. 1985) were no more accurate than control groups. Some studies have identified factors important in perception of heartbeat; for example, lean body weight (Rouse et al. 1988), cerebral laterality (Montgomery and Jones 1984), higher basal heart rate (Ludwick-Rosenthal and Nuefeld 1985), and male gender (Katkin 1985) all appear to favor enhanced performance. Finally, although there has been some research on this question (Droste et al. 1988), it is not clear why only some individuals feel cardiac pain in association with cardiac ischemia (Barsky et al. 1990; Deanfield et al. 1983), and why only some individuals at only some times are aware of cardiac dysrhythmias such as premature ventricular contractions (Brodsky et al. 1977; Sobotka et al. 1981).

In addition to studies of discrimination of heart action, a general relationship has been reported between somatic symptoms and mood (Persson and Sjobert 1987). Two studies reported a specific relationship between heart rate and mood (Johnston and Anastasiades 1990; Roth 1976). There is also a small amount of data indicating that blood pressure, as well as heartbeat, can be discriminated (Greenstadt et al. 1986; Pennebaker et al.

1982), and a suggestion that respiratory symptoms are associated with psychological symptoms and functioning (Dales et al. 1989).

It is not clear to what extent adrenergic mechanisms are involved in the ability to discriminate heart rate, blood pressure, respiratory changes, and cardiac pain. These physiological functions are influenced by other neurotransmitters and receptors as well as adrenergic mechanisms. For example, it will be important to explore the role of vagus nerve activity in the discrimination process. The research to date has addressed itself in depth to determining the ability of individuals to make such discriminations, including the development of optimal training procedures, but has paid less attention to the physiology that mediates this ability.

Hypoglycemia

Studies of the ability of diabetic subjects to discriminate their own blood glucose levels have reported a range of accuracies across subjects and across studies (Diamond et al. 1989; Moses and Bradley 1985; Pennebaker et al. 1981). Changes in physical symptoms, cognitive functioning, and mood have been associated with fluctuations in glucose level (Diamond et al. 1989; Gonder-Frederick et al. 1989; Moses and Bradley 1985; Pennebaker et al. 1981; Taylor and Rachman 1988). Adrenergic changes are specifically associated with reductions in blood glucose level (hypoglycemia) (Marks and Rose 1981; Service 1983). A number of studies have evaluated the relationship between hypoglycemia and symptom report.

The relationship among hypoglycemia, adrenergic symptoms, and increases in epinephrine levels has been recognized since the 1950s or earlier, although it was clear even then that epinephrine release did not account for all of the observed symptoms (French and Kilpatrick 1955). More recently, it was demonstrated in healthy individuals that the glucose level at which release of hypoglycemic counterregulatory hormones occurred (including epinephrine and norepinephrine) is higher (i.e., *less* hypoglycemic) than the glucose level associated with occurrence of symptoms (Schwartz et al. 1987), as one would expect if a significant increase in levels of adrenergic substances is necessary to produce the symptoms observed. The threshold for hormone release was in the high 60s (mg/dl of plasma glucose); the threshold for symptoms was in the lower 50s. Symptoms occurred at a plasma epinephrine level of approximately 300 pg/ml (normal basal levels are below 50 pg/ml); norepinephrine levels were not significantly increased at the glucose level associated with symptoms. In another study, healthy adults first reported awareness of hypoglycemia at a glucose level in the mid- to high 50s, with a concomitant increase in plasma epinephrine to 100–200 pg/ml (Cameron et al. 1988). Awareness was more closely associated with the magnitude of the increase in epinephrine level than with the decrease in glucose level; other counterregulatory hormones, including norepinephrine, appeared to play little role in hypoglycemia awareness.

The results with healthy individuals indicate that awareness of hypoglycemia is determined, not exclusively but mainly, by the symptoms produced by the counterregulatory increase in epinephrine (Cameron et al. 1988; French and Kilpatrick 1955; Schwartz et al. 1987). This issue has been further explored in persons with a common complication of diabetes mellitus, autonomic neuropathy, which is associated with a reduction in the counterregulatory increase in epinephrine (Hoeldtke et al. 1982; Marks and Rose 1981; Service 1983). Studies have indicated that autonomic neuropathy and the reduction in counterregulatory epinephrine increase is associated with unawareness of hypoglycemia, whereas diabetic patients with normal counterregulatory epinephrine release have awareness at glucose levels comparable to those in healthy individuals (Heller et al. 1987; Hoeldtke et al. 1982; Pauli et al. 1991). One investigative group has suggested that a decrease in β-adrenergic receptor sensitivity as indicated by a reduced heart rate response to isoproterenol administration, in addition to a decrease in counterregulatory epinephrine increase, occurs in some diabetic patients and is associated with hypoglycemia unawareness (Berlin et al. 1987, 1988). The mechanism underlying this desensitization is not clear, since the epinephrine levels were actually lower in the group of unaware diabetic patients; that is, desensitization was not apparently due to elevations of circulating catecholamines. Because unaware patients had more severe diabetes for a longer duration, receptor damage secondary to chronic exposure to elevated plasma glucose levels is a possibility that should be tested.

Other hypoglycemia-related syndromes are insulinoma and postprandial hypoglycemia. Insulinoma is a rare insulin-secreting tumor that produces hypoglycemia and the symptoms associated with it (Service et al.

1976). Postprandial hypoglycemia is a syndrome suggestive of hypoglycemia that occurs several hours after eating and is thought to be caused by excessive insulin secretion in response to the meal (Permutt 1976). Significant hypoglycemia is actually observed in only a minority of individuals complaining of such symptoms (Palardy et al. 1989), and the authors of at least one study have concluded that in the majority of such individuals, these symptoms are attributable to "emotional disturbances" (Johnson et al. 1980).

The significant involvement of adrenergic symptoms, counterregulatory epinephrine increase, and possibly β-adrenergic receptor responses in awareness of hypoglycemia suggests that β-adrenergic blockade should impair hypoglycemia awareness. Although it has been reported that β-adrenergic blocking drugs can produce such impairment (Blum and Ruoff 1983; Marks and Rose 1981; Service 1983), some authors have indicated that hypoglycemia awareness is not lost in diabetic patients in response to treatment with β-adrenergic blocking drugs, even if the tachycardiac reaction is diminished (Blohme et al. 1981; Clausen-Sjobom et al. 1987; Deacon et al. 1977). Furthermore, a study in healthy adults (Cameron 1989) demonstrated that acute administration of propranolol, a nonselective β-adrenergic blocking drug, reduced the adrenergic symptoms associated with hypoglycemia but did not impair awareness or change the blood glucose level at which awareness first occurred; it was suggested that "subjects use adrenergic symptoms when they occur in making the choice of active substance [insulin-induced decreases in blood glucose] or inactive substance [saline], but can use other symptoms with equal accuracy when adrenergic symptoms are not available" (p. 170).

Systemic Administration of Adrenergic Substances

Many studies of the systemic administration of adrenergic substances to humans have been reported, starting when partially purified adrenergic substances first became available early in this century. Ever since these initial studies, it has been reported frequently that administration of these substances, especially epinephrine, is associated with reports of "fear" or "anxiety," whereas administration of norepinephrine produces few symptoms (for reviews, see Frankenhaeuser 1971; Redmond 1979; Van Toller 1979). Several studies have been reported in patients with specific anxiety disorders; they are reviewed by Abelson and Cameron (Chapter 16, this volume).

Systemic administration is usually defined as exogenous administration, given for either experimental or medicinal purposes. Pheochromocytomas (catecholamine-secreting tumors), however, can also be considered as sources of systemic catecholamine administration, albeit from an endogenous source. Although pheochromocytomas have usually been considered as productive of "anxiety," research has raised questions about this association (Starkman et al. 1985, 1990). In these studies, it was found that patients with pheochromocytomas rarely reported a sufficient number of anxiety symptoms to merit a diagnosis of any specifically defined anxiety disorder. Furthermore, epinephrine levels were correlated with reported levels of somatic symptoms but were not associated with psychological anxiety, and norepinephrine levels were not correlated with either somatic or anxiety symptoms in patients with pheochromocytomas. Thus, these studies support the association of epinephrine but not norepinephrine with somatic sensations, but do not support the claim that increases in either catecholamine produce anxiety symptoms. It should be noted, however, that receptor desensitization might modify symptom profiles in those individuals who are exposed chronically to high catecholamine levels, such as patients with pheochromocytomas (Krause et al. 1988), who differ from those of individuals who receive only one or a few experimentally produced increases in systemic catecholamine levels (i.e., reduce symptomatic response sensitivity to a given plasma catecholamine level).

Although earlier research had indicated that increases in systemic epinephrine produce symptoms, the required level of plasma epinephrine and the nature of the symptoms have been systematically investigated only recently, in part because reliable assays of catecholamines at the concentrations present in plasma became available only about 15 years ago. Using a drug discrimination paradigm (Stolerman et al. 1989a, 1989b), it was demonstrated that plasma epinephrine levels of between 250 and 300 pg/ml (normal is typically less than 50) were required to produce awareness of epinephrine-induced sensations (Cameron et al. 1990). The symptoms that were significantly correlated with epinephrine increase were breathing changes, fast heartbeat, trembling or shaking, hard or irregular heartbeat, and tingling sensations; a subjective awareness of a change in heart action was the most robustly associated symptom.

Some research has been reported concerning symptoms associated with administration of other adrenergic substances. For example, α_2-adrenoreceptor agents such as clonidine and yohimbine have been used experimentally in studies of the physiology of pathological anxiety states, as has the β-adrenoreceptor agonist isoproterenol. Under the assumption that anxiety symptoms are mediated at least in part by adrenergic mechanisms, β-adrenoreceptor antagonists have also been studied as potential therapeutic agents in anxious individuals (Cole 1984; Noyes 1985). These agents have also been administered to other patient groups, such as persons with hypertension or other cardiovascular disease, but questions about specific symptomatology per se have rarely been researched in these groups except insofar as they constitute unwanted side effects, often in "overdose" situations. Furthermore, since most of these drugs enter the central nervous system, it is usually not clear whether symptoms are due to peripheral or central effects, or to some combination.

Stress

Scores of studies have reported that a number of hormones, most notably catecholamines and glucocorticoids, change during stress (Donovan 1988; Lake and Ziegler 1985; Rose 1980; Van Toller 1979; Ziegler and Lake 1984). The simultaneous occurrence of increases in catecholamines and symptoms similar to those produced by administration of adrenergic substances have led many investigators to attribute the symptoms of stress to systemic catecholamine increases. As noted earlier (Cameron et al. 1988, 1990), however, norepinephrine appears to produce few symptoms, and epinephrine produces subjective sensations only when plasma levels exceed those typically observed in stressful circumstances. It therefore seems likely that the adrenergic symptoms associated with stress are mediated not by increases in systemic catecholamine levels due to diffuse sympathetic discharge, but rather by changes in the activity of the sympathetic innervation to the organ in question (e.g., the heart) (Esler et al. 1989; Lader and Tyrer 1972; Lewis et al. 1984).

Symptoms and Central Nervous System Adrenergic Mechanisms

In addition to the adrenergic mechanisms in the peripheral nervous system, there are several well-described adrenergic nuclei and tracts in the central nervous system. Norepinephrine and dopamine are the primary catecholaminergic neurotransmitters for these tracts, but epinephrine has also been identified as a neurotransmitter in the central nervous system (Cooper et al. 1991; Heal and Marsden 1990; Hoffman and Lefkowitz 1990; Lake and Ziegler 1985; Mason 1984; Meltzer 1987; Ziegler and Lake 1984). Dopamine seems to be present in several discrete independent tracts. In contrast, more than half of the norepinephrine in the brain originates from one nucleus, the locus coeruleus, although there are other noradrenergic nuclei in the mammalian brain. These adrenergic systems have synaptic input from a variety of other neurotransmitters.

The adrenergic neurotransmitters norepinephrine and epinephrine interact with adrenergic receptors of two major classes, designated α and β, with at least two subclasses (labeled 1 and 2) in each major class (Hoffman and Lefkowitz 1990; Meltzer 1987; Smith and Hollingsworth, Chapter 4, this volume); there are also at least two types of dopaminergic receptors. Adrenergic receptors are distributed widely throughout the central (and peripheral) nervous system, both postsynaptically and as presynaptic autoreceptors on adrenergic neurons.

The relationship between the physiologies of the central adrenergic systems and the peripheral adrenergic system is not well understood. Central nervous system efferents involving many different neurotransmitters affect peripheral sympathetic activity, and peripheral afferents influence the central nervous system at various levels (see Loewy and Spyer 1990). There is pharmacological evidence of an association between functioning of these two systems (Maas and Leckman 1983). To the extent that central adrenergic mechanisms influence peripheral adrenergic functioning, central activity may be involved in the production of peripheral adrenergic symptomatology.

In addition to the potential importance of the central nervous system for peripheral adrenergic symptom production, physiological changes originating in the periphery may influence the central nervous system physiologically in ways that are associated with subjective awareness. This could be due either to sympathetic afferents influencing central activity, or to effects of circulating substances in the periphery entering the central nervous system across the blood-brain barrier.

The literature concerning the central nervous system representation of autonomic function is too large to review here (see Cechetto and Saper 1990). Autonomic afferents seem to have their major cortical representation in the insular cortex, and, thus, it seems likely that this region of the cerebral cortex is highly involved in awareness of interoceptive visceral sensations.

As noted above, the catecholamines themselves do not cross the blood-brain barrier in physiologically significant amounts. Many adrenergically active drugs, however, such as the α_2-adrenoreceptor partial agonist clonidine and the antagonist yohimbine, do enter the brain and are physiologically and subjectively active in opposite directions (Abelson and Cameron, Chapter 16, this volume). Most of the available β-adrenoreceptor antagonists—propranolol, for example—also enter the brain and appear to produce subjective symptoms in many individuals (Kielholz 1977). It is not clear, however, whether the symptoms produced by these drugs are due to peripheral or central nervous system effects (or both).

One method of studying adrenergic symptoms is to induce changes in adrenergic functioning with drugs known to be specific for a particular neurotransmitter receptor population. The subjective effects in response to the administration of drugs such as α_2-adrenoreceptor agonists and antagonists and the β-adrenoreceptor antagonists can be characterized quite precisely with drug discrimination procedures (Stolerman et al. 1989a, 1989b).

The relationship between subjective and discriminable effects of several drugs in humans has been evaluated (Schuster and Johanson 1988). Results of these studies indicate that studies of discriminability and subjective changes, as might be expected, are in substantial agreement, leading to the inference that results of studies of drug discriminability in animals can be used to predict symptom profiles in humans. Unfortunately, and perhaps surprisingly, animal studies of discriminability of the endogenous catecholamines (epinephrine, norepinephrine, and dopamine) are extremely limited (Cook et al. 1960; Schuster and Brady 1964). Although some studies indicate that individuals are aware of increases in epinephrine but less so of norepinephrine, only one study has used a discrimination procedure to evaluate epinephrine symptomatology discriminability in humans (Cameron et al. 1990, see above), and no study has been done with norepinephrine. Animal studies with agonists or antagonists for adrenergic receptors are beyond the scope of this chapter; references are available for the interested reader (see Stolerman et al. 1989a, 1989b). In addition to the use of drugs, which are likely to have peripheral as well as central effects (unless administered directly into the brain), brain electrical stimulation studies can be used to focus on an anatomically specific, and potentially neurotransmitter-specific, brain region (e.g., Skelton et al. 1982).

A number of medical and psychiatric disorders are associated with at least presumptive evidence of abnormalities of central nervous system adrenergic functioning, including (at least) hypertension and other cardiovascular diseases; parkinsonism and other extrapyramidal, basal ganglia movement disorders; cognitive and learning disorders; sleep disorders; and a variety of psychiatric disorders, including depression, anxiety, schizophrenia, eating disorders, and addictive disorders. Adrenergic changes associated with the aging process and rhythmic functions may also be implicated in a variety of disease states. It is not clear, however, in any of these disorders if the "adrenergic" symptoms are actually due to adrenergic changes rather than to changes in other neurotransmitter systems, and if any adrenergically mediated symptoms are due to central or peripheral nervous system changes, or to some combination of the two.

Discussion and Conclusion

In this chapter it has been argued that adrenergically, as well as non-adrenergically, mediated subjective sensations or symptoms are likely to be of major importance in themselves to the understanding of human psychobiology and should not be viewed as merely one of the manifestations of a series of medical and psychiatric illnesses. Additionally, it has been pointed out that which symptoms are actually "adrenergic" is not necessarily as clear as usually assumed. Furthermore, despite the fact that the James-Lange theory in its original form is undoubtedly very oversimplified, the relationship between somatic sensations and the experience of fear and other emotional states remains poorly understood. Finally, the mechanisms mediating these subjective sensations have not been clarified. For example, the magnitude of increase in plasma epinephrine associated with mental or psychological stress is insufficient to produce the symptoms usually ascribed to adrenergic activation, whereas norepinephrine appears to play an even less important role in symptom production, and the role of the other catecholamine, dopamine, is completely unexplored. In addition, the increase in the inotropic and chronotropic actions of the heart, which sometimes are associated with subjective awareness, might be due to withdrawal of vagal tone or some as-yet-unknown mechanism rather than due to increased adrenergic input. Finally, the relationship between the occurrence of adrenergic symptoms and the arousal status (Neiss 1988) of the organism is unclear.

The major question to be answered to understand the relationship between adrenergic activation and the occurrence of symptoms is: What is the physiological mechanism by which awareness occurs? Anatomically and physiologically, what and where are the "interoceptors" (Chernigovskiy 1967; Razran 1961; Reed et al. 1990), or, for example, as asked and answered affirmatively in an editorial, is the heart a sensory organ (Abboud 1989)? How and where are these "interoceptors" connected to the central nervous system? There are some additional questions raised by the research reviewed above. What accounts for the individual differences observed in the awareness of visceral functioning? For example, why are some but not all cardiac dysrhythmias and myocardial infarctions "silent"? How does the individual use an apparent hierarchy of symptoms in becoming aware of a change in visceral function? For example, in the absence of β-adrenergic blockade, adrenergic symptoms are strongly associated with awareness of hypoglycemia, whereas individuals demonstrate an undiminished capacity to sense hypoglycemia in the face of β-adrenergic blockade despite the fact that adrenergic symptoms are much less prominent. Are the attentional processes important?

Some information is available concerning possible mechanisms of subjective awareness of visceral symptoms. First, there have been sensory receptors identified in visceral organs, including, of course, pain receptors, but also others such as the mechanoreceptors in the heart (Abboud 1989; Stark et al. 1991). Second, the autonomic nervous system contains many afferent fibers, such as the predominance of such fibers in the vagus nerve (Loewy and Spyer 1990). Third, anatomical pathways in the central nervous system for these afferent inputs have been identified (Loewy and Spyer 1990). Fourth, these pathways reach the cerebral cortex where awareness is most likely mediated; the visceral organs are represented somatotopically in the insular cortex (Cechetto and Saper 1990). Finally, research in animals has indicated that stimulation of visceral organs can activate the major central noradrenergic nucleus, the locus coeruleus (Svensson 1987). What is now needed are studies relating this anatomy and physiology to symptomatology. For example, would there be a predictable relationship in humans (and animals) among physiological activation of visceral organs such as the heart, activation of specific regions of the insular cortex, and subjective awareness of this activation? Conversely, would activation that did not produce symptoms also not activate the insula?

References

Abboud FM: Ventricular syncope: is the heart a sensory organ? N Engl J Med 320:390–392, 1989

American Psychiatric Association: Diagnostic and Statistical Manual of Mental Disorders, 4th Edition. Washington, DC, American Psychiatric Association, 1994

Barsky AJ, Hochstrasser B, Coles A, et al: Silent myocardial ischemia: is the person of the event silent? JAMA 264:1132–1135, 1990

Bergman JS, Johnson HJ: The effects of instructional set and autonomic perception on cardiac control. Psychophysiology 8:180–190, 1971

Berlin I, Grimaldi A, Payan C, et al: Hypoglycemic symptoms and decreased β-adrenergic sensitivity in insulin-dependent diabetic patients. Diabetes Care 10:742–747, 1987

Berlin I, Grimaldi A, Landault C, et al: Lack of hypoglycemic symptoms and decreased β-adrenergic sensitivity in insulin-dependent diabetic patients. J Clin Endocrinol Metab 66:273–278, 1988

Blohme G, Lager I, Lonnroth P, et al: Hypoglycemic symptoms in insulin-dependent diabetics: a prospective study of the influence of beta-blockade. Diabete Metab 7:235–238, 1981

Blum M, Ruoff M: Response to hypoglycemia masked by propranolol. N Engl J Med 309:112–113, 1983

Brodsky M, Wu O, Denes P, et al: Arrhythmias documented by 24-hour continuous electrocardiographic monitoring in 50 male medical students without apparent heart disease. Am J Cardiol 39:390–395, 1977

Cacioppo JT, Tassinary LG, Stonebraker TB, et al: Self-report and cardiovascular measures of arousal: fractionation during residual arousal. Biol Psychol 25:135–151, 1987

Cameron OG: Beta-adrenergic blockade does not prevent hypoglycemia awareness in non-diabetic humans. Psychosom Med 51:165–172, 1989

Cameron OG, Buzan R, McCann DS: Symptoms of insulin-induced hypoglycemia in normal subjects. J Psychosom Res 32:41–49, 1988

Cameron OG, Gunsher S, Hariharan M: Venous plasma epinephrine levels and the symptoms of stress. Psychosom Med 52:411–424, 1990

Cechetto DF, Saper CB: Role of the cerebral cortex in autonomic function, in Central Regulation of Autonomic Functions. Edited by Loewy AD, Spyer KM. New York, Oxford University Press, 1990, pp 208–223

Chernigovskiy VN: Interoceptors. Washington, DC, American Psychological Association, 1967

Clausen-Sjobom N, Lins P, Adamson U, et al: Effects of metoprolol on the counterregulation and recognition of prolonged hypoglycemia in insulin-dependent diabetics. Acta Med Scand 222:57–63, 1987

Cole JO: Beta-adrenergic blockade and buspirone, in Psychiatry Update: The American Psychiatric Association Annual Review, Vol 3. Edited by Grinspoon L. Washington, DC, American Psychiatric Press, 1984, pp 490–503

Cook L, Davidson A, Davis DJ, et al: Epinephrine, norepinephrine, and acetylcholine as conditioned stimuli for avoidance behavior. Science 131:990–991, 1960

Cooper JR, Bloom FE, Roth RH: The Biochemical Basis of Neuropharmacology, 6th Edition. New York, Oxford University Press, 1991

Dales RE, Spitzer WO, Schechter MT, et al: The influence of psychological status on respiratory symptom reporting. Am Rev Respir Dis 139:1459–1463, 1989

Davis MR, Langer AW, Sutterer JR, et al: Relative discriminability of heartbeat-contingent stimuli under three procedures for assessing cardiac perception. Psychophysiology 23:76–81, 1986

Deacon SP, Karunanayake A, Barnett D: Acebutolol, atenolol, and propranolol and metabolic responses to acute hypoglycaemia in diabetics. Br Med J 2:1255–1257, 1977

Deanfield JE, Maseri A, Selwyn AP, et al: Myocardial ischemia during daily life in patients with stable angina: its relation to symptoms and heart rate changes. Lancet 2:753–758, 1983

Diamond EL, Massey KL, Covey D: Symptom awareness and blood glucose estimation in diabetic adults. Health Psychol 8:15–26, 1989

Donovan BT: Humors, Hormones, and the Mind. London, England, Stockton Press, 1988

Droste C, Meyer-Blankenburg M, Greenlee MW, et al: Effect of physical exercise on pain thresholds and plasma beta-endorphins in patients with silent and symptomatic myocardial ischemia. Eur Heart J 9 (suppl N):25–33, 1988

Esler M, Jennings G, Lambert G: Measurement of overall and cardiac norepinephrine release into plasma during cognitive challenge. Psychoneuroendocrinology 14:477–481, 1989

Essau CA, Jamieson JL: Heart rate perception in the type A personality. Health Psychol 6:43–54, 1987

Fehr FS, Stern JA: Peripheral physiological variables and emotion: the James-Lange theory revisited. Psychol Bull 74:411–424, 1970

Frankenhaeuser M: Behavior and circulatory catecholamines. Brain Res 31:241–262, 1971

French EB, Kilpatrick R: The role of adrenaline in hypoglycemia reactions in man. Clin Sci 14:639–651, 1955

Gonder-Frederick LA, Cox DJ, Bobbitt SA, et al: Mood changes associated with blood glucose fluctuations in insulin-dependent diabetes mellitus. Health Psychol 8:45–59, 1989

Greenstadt L, Shapiro D, Whitehead R: Blood pressure discrimination. Psychophysiology 23:500–509, 1986

Grigg L, Ashton R: Heart rate discrimination viewed as a perceptual process: a replication and extension. Psychophysiology 19:13–20, 1982

Heal DJ, Marsden CA: The Pharmacology of Noradrenaline in the Central Nervous System. Oxford, England, Oxford University Press, 1990

Heller SR, Herbert M, Macdonald IA, et al: Influence of sympathetic nervous system on hypoglycaemic warning symptoms. Lancet 2:359–385, 1987

Hoeldtke RD, Boden G, Shuman CR, et al: Reduced epinephrine secretion and hypoglycemia unawareness in diabetic autonomic neuropathy. Ann Intern Med 96:459–462, 1982

Hoffman BB, Lefkowitz RJ: Catecholamines and sympathomimetic drugs, in The Pharmacological Basis of Therapeutics, 8th Edition. Edited by Gilman AG, Rall TW, Nies AS, et al. New York, Pergamon, 1990, pp 187–220

Johnson DD, Dorr KE, Swenson WM, et al: Reactive hypoglycemia. JAMA 243:1151–1155, 1980

Johnston DW, Anastasiades P: The relationship between heart rate and mood in real life. J Psychosom Res 34:21–27, 1990

Jones GE, Hollandsworth JG: Heart rate discrimination before and after exercise-induced augmented cardiac activity. Psychophysiology 18:252–257, 1981

Jones GE, Jones KR, Cunningham RA, et al: Cardiac awareness in infarct patients and normals. Psychophysiology 22:480–487, 1985

Jones GE, Jones KR, Rouse CH, et al: The effect of body position on the perception of cardiac sensations: an experiment and theoretical implications. Psychophysiology 24:300–311, 1987

Katkin ES: Blood, sweat, and tears: individual differences in autonomic self-perception. Psychophysiology 22:125–137, 1985

Kellner R: Psychosomatic Syndromes and Somatic Symptoms. Washington, DC, American Psychiatric Association, 1991

Kielholz P: Beta-blockers and the Central Nervous System. Baltimore, MD, University Park Press, 1977

Krause M, Reinhardt D, Kruse K: Phaeochromocytoma without symptoms: desensitization of the α- and α-adrenoreceptors. Eur J Pediatr 147:121–122, 1988

Kroenke K: Symptoms in medical patients: an untended field. Am J Med 92 (suppl 1A):3S–6S, 1992

Lader MH, Tyrer PJ: Central and peripheral effects of propranolol and sotalol in normal human subjects. Br J Pharmacol 45:557–560, 1972

Lake CR, Ziegler MG: The Catecholamines in Psychiatric and Neurologic Disorders. Boston, MA, Butterworth Publishers, 1985

Lewis MJ, Jones DM, Dart AM, et al: The psychological side effects of acebutolol and atenolol. Br J Clin Pharmacol 17:364–366, 1984

Loewy AD, Spyer KM: Central Regulation of Autonomic Functions. New York, Oxford University Press, 1990

Ludwick-Rosenthal R, Neufeld RWJ: Heartbeat interoception: a study of individual differences. Int J Psychophysiol 3:57–65, 1985

Lyons W: Emotion. Cambridge, England, Cambridge University Press, 1980

Maas JW, Leckman JF: Relationships between central nervous system noradrenergic function and plasma and urinary MHPG and other norepinephrine metabolites, in MHPG: Basic Mechanisms and Psychopathology. Edited by Maas JW. New York, Academic Press, 1983, pp 33–43

MacKenzie ET, McCollock J, O'Keane M, et al: Cerebral circulation and norepinephrine: relevance of the blood-brain barrier. Am J Physiol 231:483–488, 1976

Marks V, Rose FC: Hypoglycemia. Oxford, England, Blackwell Scientific, 1981

Mason ST: Catecholamines and Behavior. Cambridge, England, Cambridge University Press, 1984

Mayes A: The physiology of fear and anxiety, in Fear in Animals and Man. Edited by Sluckin W. New York, Van Nostrand Reinhold, 1979, pp 24–55

Meltzer HY: Psychopharmacology: The Third Generation of Progress. New York, Raven, 1987

Montgomery WA, Jones GE: Laterality, emotionality, and heartbeat perception. Psychophysiology 21:459–465, 1984

Moses JL, Bradley C: Accuracy of subjective blood glucose estimation by patients with insulin-dependent diabetes. Biofeedback Self Regul 10:301–314, 1985

Neiss R: Reconceptualizing arousal: psychobiological states in motor performance. Psychol Bull 103:345–366, 1988

Noyes R Jr: Beta-adrenergic blocking drugs in anxiety and stress. Psychiatr Clin North Am 8:119–132, 1985

Oldendorff WH: Brain uptake of radiolabeled amino acids, amines, and hexoses after arterial injection. Am J Physiol 221:1629–1639, 1971

Oleson J, Hougard K, Hertz M: Isoproterenol and propranolol: ability to cross the blood-brain barrier and effects on cerebral circulation in man. Stroke 9:344–349, 1978

Palardy J, Havrankova J, LePage R, et al: Blood glucose measurements during symptomatic episodes in patients with suspected postprandial hypoglycemia. N Engl J Med 321:1421–1474, 1989

Pauli P, Hartl L, Marquardt C, et al: Heartbeat and arrhythmia perception in diabetic autonomic neuropathy. Psychol Med 21:413–421, 1991

Pennebaker JW, Cox DJ, Gonder-Frederick L, et al: Physical symptoms related to blood glucose in insulin-dependent diabetics. Psychosom Med 43:489–500, 1981

Pennebaker JW, Gonder-Frederick L, Stewart H, et al: Physical symptoms associated with blood pressure. Psychophysiology 19:201–210, 1982

Permutt MA: Postprandial hypoglycemia. Diabetes 25:719–733, 1976

Persson L, Sjobert L: Mood and somatic symptoms. J Psychosom Res 31:499–511, 1987

Razran G: The observable unconscious and the inferable conscious in current Soviet psychophysiology. Psychol Rev 68:81–147, 1961

Redmond DE: New and old evidence for the involvement of a brain norepinephrine system in anxiety, in Phenomenology and Treatment of Anxiety. Edited by Fann WE, Karacan I, Porkorny AD, et al. New York, SP Medical & Scientific Books, 1979, pp 153–203

Reed SD, Harver A, Katkin ES: Interoception, in Principles of Psychophysiology: Physical, Social, and Inferential Elements. Edited by Cacioppo JT, Tassinary LG. New York, Cambridge University Press, 1990, pp 253–291

Rose RM: Endocrine responses to stressful psychological events. Psychiatr Clin North Am 3:251–276, 1980

Roth WT, Tinklenberg JR, Doyle CM, et al: Mood states and 24-hour cardiac monitoring. J Psychosom Res 20:179–186, 1976

Rouse CH, Jones GE, Jones KR: The effect of body composition and gender on cardiac awareness. Psychophysiology 25:400–407, 1988

Schuster CR, Brady JV: The discriminative control of a food-reinforced operant by interoceptive stimulation. Pavlovian Journal of Higher Nervous Activity 14:448–458, 1964

Schuster CR, Johanson CE: Relationship between the discriminative stimulus properties and subjective effects of drugs, in Transduction Mechanisms of Drug Stimuli. Edited by Colpaert FC, Balster RL. Berlin, Springer-Verlag, 1988, pp 161–175

Schwartz NS, Clutter WE, Shah SD, et al: Glycemic thresholds for activation of glucose counterregulatory systems are higher than the threshold for symptoms. J Clin Invest 79:777–781, 1987

Service FJ: Hypoglycemic Disorders: Pathogenesis, Diagnosis, and Treatment. Boston, MA, GK Hall Medical Publishers, 1983

Service FJ, Dale AJD, Elveback LR, et al: Insulinoma: clinical and diagnostic features of 60 consecutive cases. Mayo Clin Proc 51:417–429, 1976

Skelton RW, Phillips AG, Miller JJ: Electrical stimulation of the perforant pathy input into the hippocampus as a discriminative stimulus, in Drug Discrimination: Applications in CNS Pharmacology. Edited by Colpaert FC, Slangen JL. Amsterdam, Elsevior Biomedical Press, 1982, pp 369–380

Sobotka PA, Mayer JH, Bauernfeind RA, et al: Arrhythmias documented by 24-hour continuous ambulatory electrocardiographic monitoring in young women without apparent heart disease. Am Heart J 101:753–758, 1981

Stark RP, McGinn AL, Wilson RF: Chest pain in cardiac-transplant recipients: evidence of sensory reinnervation after cardiac transplantation. N Engl J Med 324:1791–1794, 1991

Starkman MN, Zelnik T, Nesse RM, et al: Anxiety in patients with pheochromocytomas. Arch Intern Med 145:248–252, 1985

Starkman MN, Cameron OG, Nesse RM, et al: Peripheral catecholamine levels and the symptoms of anxiety: studies in patients with and without pheochromocytoma. Psychosom Med 52:129–142, 1990

Stolerman IP, Shine PJ, Samele C, et al: The drug discrimination database: 1951–1988. Rockville, MD, National Institute on Drug Abuse, 1989a

Stolerman IP, Rasul F, Shine PJ: Trends in drug discrimination research analyzed with a cross-indexed bibliography: 1984–1987. Psychopharmacology (Berl) 98:1–19, 1989b

Svensson TH: Peripheral, autonomic regulation of locus coeruleus noradrenergic neurons in brain: putative implications for psychiatry and psychopharmacology. Psychopharmacology (Berl) 92:1–7, 1987

Taylor LA, Rachman SJ: The effects of blood sugar level changes on cognitive function, affective state, and somatic symptoms. J Behav Med 11:279–291, 1988

Van Toller C: The Nervous Body: An Introduction to the Autonomic Nervous System and Behavior. New York, Wiley, 1979

Ziegler MG, Lake CR: Norepinephrine. Baltimore, MD, Williams & Wilkins, 1984

10

Hypertension and Adrenergic Functioning

RICHARD A. NELESEN, PH.D.
JOEL E. DIMSDALE, M.D.

The authors of the chapters in this volume discuss the psychobiology of various states such as depression or panic disorder that have clear-cut adrenergic dysfunction. There are other disorders as well where the interplay of psychobiology and sympathetic nervous system (SNS) activity is important. One of these disorders is hypertension, which is markedly affected by psychological stressors. Given that the SNS plays such a major role in response to stressors and that the SNS interdigitates with diverse hormonal regulatory systems, studying SNS physiology may improve comprehension of the mosaic of the pathogenesis of hypertension. The issues faced by hypertension researchers are thus highly relevant to in-

This work was supported in part by Grants HL36005, HL40102, HL07261, and RR00827 from the National Institutes of Health.

vestigators studying other aspects of the psychobiology of adrenergic functioning.

In the United States, essential hypertension affects 10%–20% of the adult white population and 20%–40% of the black population. Hypertension is an arbitrarily defined diagnosis that has diverse underlying pathophysiologies. Disturbances in adrenergic functioning underlie many cases of hypertension (Michel et al. 1990). Since this research area is extraordinarily vast, we do not attempt to review it comprehensively in this chapter. Instead, we discuss selected observations from human studies on resting and perturbed measures of SNS activity, as inferred from plasma, urinary, and platelet catecholamines as well as direct monitoring of sympathetic nerve fibers.

Research in adrenergic physiology has followed three phases. The initial phase examined end-organ response (such as blood pressure or heart rate) to various challenges. With improved assay capabilities, researchers then tracked the neurotransmitter released on stimulation of the sympathoadrenal-medullary system. More recently, researchers are concentrating on the end-organ receptors for such transmitters. Fortunately, peripheral blood lymphocytes are reasonable models for β-adrenergic receptors on the vascular wall, the heart, and the brain (Michel et al. 1990).

There are two diseases (experiments of nature) that highlight the impact of adrenergic functioning on blood pressure regulation. In pheochromocytoma, a chromaffin cell tumor present in about 0.1% of the population (Finton et al. 1984), excessive amounts of norepinephrine (NE), epinephrine, or both catecholamines are secreted episodically. Hypertension, either persistent or episodic, is an unvarying symptom of a pheochromocytoma, although pheochromocytomas account for less than 1% of hypertensive cases (Finton et al. 1984). High plasma and urinary levels of catecholamines are found, particularly during hypertensive episodes. It is interesting that although symptoms of anxiety are reported in 26%–50% of pheochromocytoma patients, the quality of the anxiety is different from the pathological anxiety present during panic attacks (Starkman et al. 1990).

The converse of pheochromocytoma is represented by patients with various forms of autonomic failure, such as that found in the Shy-Drager syndrome or Riley-Day syndrome (Bannister 1983; Maayan et al. 1987). Patients with these syndromes manifest severe orthostatic hypotension and deficient release of NE from sympathetic nerves tonically as well as in response to provocation (orthostatic challenge, stress) (Dancis 1983). Since the adrenergic receptors are upregulated as a response to the chronic deficit in adrenergic transmission, small doses of NE lead to large increases in blood pressure in these patients (Smith and Dancis 1964).

Despite the clear influence of the SNS on blood pressure regulation, it is a far more complicated task to establish how SNS alterations are responsible for maintaining essential hypertension. Not only are there many other physiological systems operating to influence blood pressure, but there are also major complexities involved even in measuring SNS activity (Dimsdale and Ziegler 1991).

In the rest of the chapter, we review some of the contradictory threads relating hypertension to SNS functioning, as inferred from plasma, urinary, and platelet catecholamine studies and from direct sympathetic nerve monitoring. From time to time, we discuss physiological observations under both basal and dynamic (or perturbed) conditions. The latter sorts of situations are diverse and include situations such as exercise, hypoxia, altered glucose metabolism, and exposure to behavioral stressors.

Catecholamines in Hypertension

Goldstein (1983) reviewed studies on plasma NE related to hypertension published in English between 1973 and 1981. Of 32 studies, 28 reported higher plasma NE in those who were hypertensive (HT) than in those who were normotensive (NT), but the difference in levels was significant in only 13 studies. In most of these studies, the level of NE in HTs was almost identical to the level of NE in NTs regardless of any statistical difference. In the studies not reporting significant differences, however, the control subjects had higher NE levels than the "control group" in studies that found significant differences. Scarce information was given in these articles concerning how control subjects were recruited, and Goldstein concluded that the choice of NT control subjects may be an important design feature that needed subsequent attention.

Despite these uneven findings, research into the relationship of NE and epinephrine levels in hypertension continued. Baseline as well as perturbed levels of catecholamines have been explored to elucidate the role of the sympathoadrenal-medullary system in the regulation of blood pressure.

We reviewed the literature from 1983 to 1990. The literature was gleaned primarily from a computer search and from studies cited in the references of papers found in the search. This search initially resulted in 825 papers that examined catecholamine levels in humans; 422 articles reported NE and epinephrine levels related to blood pressure. Of these 422 papers, 187 were relevant to hypertension; 93 articles examined secondary hypertension (i.e., pheochromocytoma, renal hypertension) and were excluded. Fifty papers reported results for HTs only (generally clinical trials involving a pharmacological intervention), leaving 31 articles that met our other review criteria. Our review criteria were that papers be published in English, that the studies use radioenzymatic or high-performance liquid chromatography techniques for analysis of catecholamine levels, that they include NT age-matched control subjects who were not drawn from laboratory personnel, that resting values were obtained after an appropriate recovery from venipuncture, and that the baseline values were reported when the subjects were challenged. Thirteen articles examined those who were borderline hypertensive (BHT); 18 studied those who were established hypertensive (EHT); and 4 papers included both groups. Catecholamine levels were examined at rest and during dynamic exercise, isometric exercise, or mental challenge. Figure 10–1 summarizes the results of this survey.

Resting Plasma Catecholamine Levels

Of the 13 studies that examined BHTs, 9 found this group to have higher NE levels than NTs at rest; 4 found that BHTs had lower NE than NTs. Of the 18 papers examining EHTs, 12 reported higher NE for these subjects; however, only 3 papers reported statistically significant differences. In the latter cases, BHTs and EHTs had lower NE levels than NTs (Feldt-Rasmussen et al. 1987; Kusaka et al. 1989; Vapaatalo et al. 1988). Thus, as reported previously, resting plasma NE levels appear to be slightly higher in most BHTs and equivocal in EHTs when compared with matched control subjects.

Epinephrine levels were reported in 18 of these papers. Eleven found higher levels in BHTs, and 6 found higher epinephrine levels in EHTs; however, no statistically significant differences were found. Seven papers showed lower epinephrine levels in EHTs than NTs (Feldt-Rasmussen 1987; Graafsma et al. 1989; Hamada et al. 1987; Jansen et al. 1989; Theorell et al. 1985; Vapaatalo et al. 1988; Wallin et al. 1987). In the four articles that examined both BHTs and EHTs, two reported higher epinephrine in BHTs (Eliasson et al. 1983; Vapaatalo et al. 1988), one reported higher levels in EHTs (Cerasola et al. 1987), and one reported no difference in

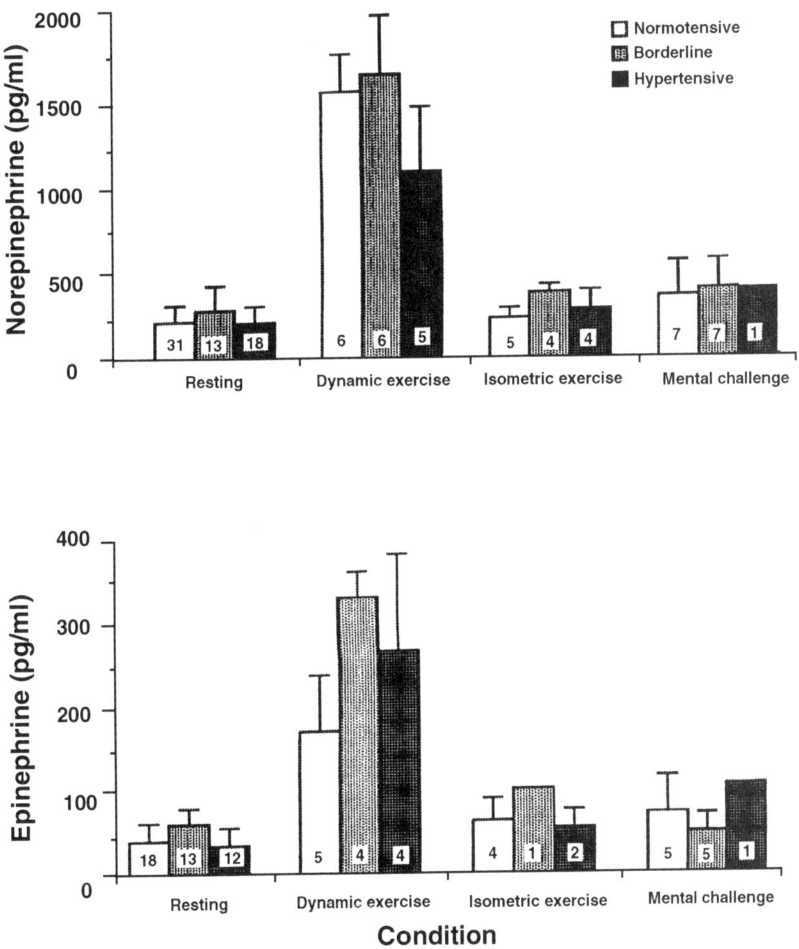

Figure 10–1. Summary of plasma catecholamine levels comparing normotensive, borderline hypertensive, and established hypertensive subjects. Data are presented as the mean and standard deviation of the studies reviewed. These data are the weighted mean and standard deviation derived from the data presented in the cited articles. The number in the bar represents the number of studies reporting findings for the condition.

epinephrine levels in both groups (Cerasola et al. 1988).

In summary, BHTs appear to have higher catecholamine levels at rest than either EHTs or NTs, whereas EHTs may actually have lower NE and epinephrine levels than NTs.

Norepinephrine and Epinephrine Responses to Exercise

Dynamic exercise elicits significant increases in NE and epinephrine. Only one paper found a significant difference between NTs and BHTs in response to dynamic exercise. The NE and epinephrine levels at peak exercise for BHTs were higher than for NTs (by 1.5 and 3 times, respectively) (Shultz et al. 1986). Isometric handgrip at 30% maximal effort for 3 minutes results in significant increases in the catecholamines for all groups. Ito et al. (1988) showed that BHTs had significantly larger increases in NE and epinephrine responses to the isometric handgrip than NTs. These results suggest that exercise, a potent sympathoadrenal-medullary system stimulus, evokes a catecholamine response that is larger in HTs.

Mental Challenge

Mental challenges, such as mental arithmetic, also elicit increases in NE and epinephrine, but 15 studies did not report significant differences between HTs and NTs. In many of the studies, HTs had a smaller catecholamine response than NTs (Bohlin et al. 1986; Dimsdale et al. 1987; Vapaatalo et al. 1988).

Race and Gender Effects

Dimsdale et al. (1990) examined catecholamine levels in male, black and white BHTs and NTs. They found that white NTs had significantly lower catecholamine levels than the other three groups, but there were no differences among the latter three groups. Other studies examining racial differences in catecholamine levels either examined NTs only, examined HTs only, or did not report actual NE and epinephrine levels (Bohlin et al. 1986; Spence et al. 1990; Tischenkel et al. 1989).

Gender differences in HTs and NTs have been reported very sparsely. Although women were included in many of the pharmacological intervention studies that examined HTs only, the catecholamine values for the women were reported only as part of the entire sample. The one study that reported a gender effect found women to have lower NE and epinephrine levels than men, but no differences were found between EHTs and NTs in either sex (Netter 1987). Thus, it appears that neither gender nor ethnicity has been studied widely as a factor affecting catecholamine levels in relation to hypertension (supporting the wisdom of National Institutes of Health regulations to mandate such studies).

Urinary Catecholamine Levels

Differences between HTs and NTs for urinary catecholamine levels were reported infrequently during the period of the review. The majority of these papers examined secondary hypertension. This relative lack of study can be attributed to the greater contemporary interest in plasma catecholamines and their reflection of acute sympathoadrenal-medullary system responses to stress. Many studies examined urinary epinephrine and vanillylmandelic acid to diagnose pheochromocytoma, but few studies examined urinary levels in BHTs or EHTs. As with the plasma catecholamine levels, urinary NE and epinephrine were found to be statistically lower in younger subjects compared with older subjects and in women relative to men (Lehmann and Keul 1986), but failed to show significant differences between HTs and NTs (Brown et al. 1985; Saito et al. 1990).

Platelet Norepinephrine

Platelets contain NE that can be measured by high-performance liquid chromatography or radioenzymatic techniques after concentrating and then homogenizing the platelet aliquot. Although platelets absorb ambient NE and epinephrine from the surrounding serum, the platelet NE level is not a marker of acute change. For example, exercising for 10 minutes, which increased plasma catecholamines significantly, was insufficient for increasing platelet catecholamine levels (Smith et al. 1985). On the other hand, there is some reason to believe that platelet NE levels may represent an integrated measure SNS activity. Platelet NE levels begin to reflect plasma levels after about 1 hour of increased levels (Zweifler and Fedorak 1985). Other studies show that platelet NE levels increase within 1 week after subjects have been placed on a low-salt diet, which elevates plasma NE levels (Kjeldsen et al. 1988b). Platelet levels, however, appear to be increased by large-scale or chronic perturbations in SNS activity. Pheochromocytomas are associated with increased levels of platelet NE and epinephrine (Berent et al. 1990; Zweifler and Julius 1982). Just as there is uncertainty about the apparent differences between HTs and NTs in plasma catecholamine levels, there is equivocal evidence for an association between blood pressure and platelet NE levels in patients with essential hypertension (Kjeldsen et al. 1988a; Zweifler and Julius 1982).

The promise of platelet catecholamine levels is presently limited because they are so difficult to measure. This difficulty translates into variability in assay measurement (roughly three times higher than for plasma catecholamine assays) (Schneider et al. 1987), which in turn limits the statistical power of observations unless there are large differences such as those found in comparing patients with pheochromocytomas with control subjects.

In conclusion, the findings in the literature about catecholamine levels have not changed greatly since Goldstein's review in 1983. About 70% of the literature finds that BHTs have higher resting NE levels (by about 62 pg/ml) and epinephrine levels (by about 20 pg/ml). EHTs, on the other hand, showed decidedly mixed results, with half of the literature reporting higher resting NE levels and half reporting lower or no difference. Similar results were reported for epinephrine. By reviewing only studies with matched-group controls, factors that can result in artificial group differences such as age and familiarity with the laboratory situation are controlled. These factors, which are independent predictors of catecholamine levels (Nelesen and Ziegler, Chapter 7, this volume), may explain the lack of positive findings as age and comfort with the laboratory have a stronger influence on catecholamine levels than essential hypertension. It appears, however, that BHTs are marked by increased catecholamine levels at rest, implying more adrenergic drive in this group relative to either NTs or EHTs.

Few groups have explicitly compared the catecholamine levels of NTs and HTs to challenges. Dynamic exercise produces a large increment in catecholamines and appears to lead to a bigger increase during the challenge for BHTs (Shultz et al. 1986). Isometric exercise elicits increases in NE and epinephrine that differentiate BHTs from NTs. Although mental challenges evoke changes in NE and epinephrine levels, differences across diagnostic groups are not consistently found.

The inability to find significant HT-NT differences in catecholamine levels under a variety of conditions may not be an accurate reflection of underlying differences in adrenergic function. The catecholamines are very labile and suffer from large intersubject variability for a variety of reasons, including assay error. Furthermore, the adrenergic system interacts with other systems in the maintenance of homeostasis. It may be that this interaction with other hormones, such as insulin, results in the conditions to develop or maintain hypertension.

Hyperinsulinemia and Adrenergic Function in Hypertension

Insulin resistance, hyperinsulinemia, and dyslipidemia are common findings in the presence of essential hypertension (Kaplan 1989). Since increased epinephrine levels antagonize the effects of insulin on peripheral glucose utilization, SNS activity may be involved in insulin resistance. Glucose metabolism during epinephrine infusion has been noted to fall by as much as 41% (Deibert and De Fronzo 1980; Ferrannini et al. 1990). Although catecholamine levels may be elevated among HTs, circulating catecholamines are not high enough to explain totally the insulin resistance seen in essential hypertension.

A second link with SNS activity stems from the fact that insulin stimulates plasma NE release in both lean and obese HTs (O'Hare 1988). Acute insulin infusion increases plasma NE and other indices of adrenergic activity such as heart rate, pulse pressure, and blood pressure (Rowe et al. 1981). Insulin also increases heart rate and systolic blood pressure directly, even after controlling for SNS activity (O'Hare 1988).

Hyperinsulinemia, besides stimulating NE release, increases renal sodium retention via increased reabsorption from the distal nephron (De Fronzo et al. 1979; Swislocki 1990). This continued expansion of the extracellular fluid volume as a result of the sodium retention may lead to an elevation in blood pressure ensuing from a rise in cardiac output (De Fronzo 1982). Thus, sodium retention and increased sympathetic activity may also account for the hypertensive action of insulin (Christlieb et al. 1987). Illustrating this point, HTs challenged with a 75-gram glucose load show marked increases in both NE and insulin relative to control subjects (Solt et al. 1990). These data suggest insulin resistance and hyperinsulinemia may mediate excess rates of NE release and be related to hypertension.

The fact that cortisol levels are elevated in HTs with hyperinsulinemia and insulin resistance as well suggests further consequences on adrenergic function, since increased cortisol sensitizes the adrenergic receptors to the catecholamines, resulting in increased blood pressure (Surwit and Feinglos 1984). Thus, insulin-resistant HTs may show signs of enhanced adrenergic activity despite low catecholamine levels.

It is clear that the interaction of insulin and adrenergic function needs further study, but the implication is strong that a significant interaction among these systems exists in the development or maintenance of essential hypertension.

Chromogranin

Because of all the complexities in interpreting plasma and urinary catecholamine levels, investigators have searched for other chemical markers of sympathetic nerve firing. Some years ago, there was considerable interest in dopamine β-hydroxylase as such a marker (Weinshilboum 1983). Dopamine β-hydroxylase is a crucial enzyme in catecholamine synthesis; it is readily measurable in plasma and is not susceptible to reuptake. Although its levels roughly track sympathetic transmission and may be altered in certain psychiatric illnesses, the major determinant of the dopamine β-hydroxylase level turns out to be a genetic characteristic or marker, rather than a marker of pathophysiological activity (Weinshilboum 1979).

There are, however, numerous other compounds released by chromaffin cells. In recent years, considerable data have been accumulated on the acidic protein chromogranin A. Chromogranin A is released by exocytosis, along with NE, from sympathetic nerves and from adrenomedullary cells (O'Connor and Bernstein 1984). Like dopamine β-hydroxylase, chromogranin A is relatively easily measured with a radioimmunoassay. Chromogranin A is present in the nanogram/milliliter range in plasma, is refractory to freeze-thaw degradation, and is not influenced by reuptake (O'Connor 1985). As a result, it offers considerable promise as a marker for sympathetic transmission. A number of studies have demonstrated that chromogranin A levels are appropriately high in the face of strong stimuli (insulin-induced hypoglycemia, vasodilation induced by amyl nitrate) (O'Connor et al. 1987). Some, but not all, studies have also found that chromogranin A levels are increased in hypertension (Dimsdale et al. 1989). We know of only one behavioral stressor study on chromogranin A levels. In this case, Dimsdale et al. (1989) observed that chromogranin A levels were not appreciably increased by mild behavioral challenges such as mental arithmetic tests. There are beginning studies on chromogranin A levels and major psychiatric illness; cerebrospinal fluid chromogranin A levels are apparently not correlated with behavioral ratings of psychopathology in schizophrenic patients (van Kammen et al. 1991).

Sympathetic Nervous System Fiber Activity

In the preceding sections, we concentrated heavily on various chemical measures of SNS activity. Since nerve transmission is both chemical and electrical, it comes as no surprise that investigators have attempted to sample the SNS electrical signal itself. Sympathetic fibers run in the relatively superficial peroneal nerve and thus can be monitored. The discomfort associated with this procedure is not different from electromyography, and it is relatively safe. According to Wallin and Fagius (1988), 5%–10% of the subjects report some paresthesias, which persist as long as 2 weeks following SNS recording. The signal from the nerve can be counted in terms of bursts per minute, bursts per 100 heartbeats, or bursts per minute times mean burst amplitude (Anderson et al. 1987).

There is a repeated unfortunate tendency to look for one unifying measure to quantify SNS activity. As Esler et al. (1988) noted, this has turned into a quest for the "Holy Grail" and has been equally successful. As discussed in previous chapters, plasma catecholamine levels are complex to interpret and reflect greatly on the local vascular bed from which the sample was drawn. The burgeoning research on direct SNS fiber monitoring has demonstrated similar nonuniformities of firing. Although there is heuristic value in regarding the SNS as a system that responds in uniform orchestrated fashion, it is more appropriate to acknowledge that within that orchestra the strings, woodwinds, brass, and percussion play different roles.

This is clearly demonstrated by differences in SNS fiber recording in skin and muscle. The two areas have somewhat different provoking triggers. For muscle, slight falls in diastolic pressure trigger increase firing. In contrast, skin sympathetic activity is particularly activated by behavioral stressors (Delius et al. 1972).

There are intriguing data suggesting that the nerve firing rate may be increased in some, but not all, HTs (Wallin 1989). In addition, HTs have a dramatically enhanced rate of sympathetic nerve firing to stimuli such as relative hypoxia or breath-holding (Somers et al. 1988) (see Figure 10–2). Despite all the artifacts associated with measuring catecholamine levels, it is comforting to find that SNS firing correlates well with plasma NE levels (Dimsdale and Ziegler 1991; Morlin et al. 1983).

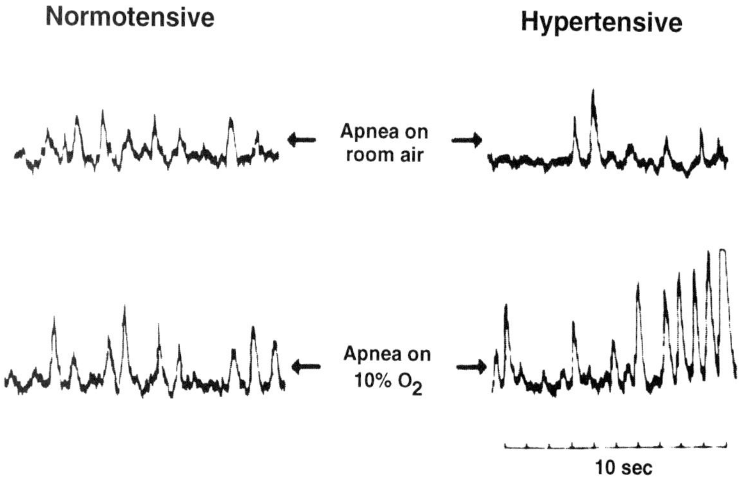

Figure 10–2. Sympathetic nerve activity during 10–15 seconds of breath-holding in a normotensive subject and a subject with borderline hypertension. Subjects were studied in room air and while breathing 10% oxygen. During the combination of hypoxia with breath-holding, the hypertensive subject demonstrated a profound increase in sympathetic nerve activity.

Source. Reprinted with permission from Somers V, Mark A, Abboud F: "Potentiation of Sympathetic Nerve Responses to Hypoxia in Borderline Hypertensive Subjects." *Hypertension* 11:608–612, 1988.

Conclusion

In this chapter, we have reviewed an admittedly selective assortment of topics related to adrenergic function and human hypertension. We have highlighted the continuing controversy about differences in plasma catecholamines in such patients. Acknowledging that the physiological meaning of such levels is hard to decode, the fact remains that plasma levels of epinephrine and NE are not appreciably higher in HTs at rest or in response to exercise or mental stressors. Even in studies of platelet catecholamines, where short-term fluctuations of levels are less of a problem, there are no consistent differences between HTs and NTs. It is ironic that as our tools for assessing SNS function get ever sharper, we are learning about more layers of complexity in this system. Now with the advent of direct intraneuronal monitoring of sympathetic fibers comes the bittersweet knowledge that the SNS cannot be described as a homogeneous orchestrated organ response system. Instead, different parts of this system fire under different provocations, each to its own drummer. The pervasive impact of all the drummers on blood pressure is a large one, as suggested by the fact that antihypertensive medications so frequently target this system. Documenting the nuances of adrenergic functioning on blood pressure regulation, however, will remain a continuing challenge.

References

Anderson E, Wallin B, Mark A: Dissociation of sympathetic nerve activity in arm and leg muscle during mental stress. Hypertension 9 (III):III.114–III.119, 1987

Bannister R (ed): Autonomic Failure: A Textbook of Clinical Disorders of the Autonomic Nervous System. Oxford, England, Oxford University Press, 1983

Berent H, Uchman B, Wocial B, et al: Platelet norepinephrine and epinephrine concentration in patients with pheochromocytoma. Am J Hypertens 3:618–621, 1990

Bohlin G, Eliasson K, Hjemdahl P, et al: Personal control over work pace-circulatory, neuroendocrine and subjective responses in borderline hypertension. J Hypertens 4:295–305, 1986

Brown MJ, Causon RC, Barnes VF, et al: Urinary catecholamines in essential hypertension: catecholamine analysis from patients in the medical research council trial for mild hypertension and from matched controls. Q J Med 57:637–651, 1985

Cerasola G, Cottone G, D'Ignoto L, et al: Role of epinephrine in the development of essential hypertension. Journal of Clinical Hypertension 3:670–680, 1987

Cerasola G, Cottone S, D'Ignoto G, et al: Sympathetic activity in borderline and established hypertension in the elderly. J Hypertens 6:S55–S58, 1988

Christlieb AR, Krolewski AS, Warram JH: Systemic hypertension, diabetes mellitus, and the kidney. Am J Cardiol 60:61I–65I, 1987

Dancis J: Familial dysautonomia (Riley-Day syndrome), in Autonomic Failure: A Textbook of Clinical Disorders of the Autonomic Nervous System. Edited by Bannister R. Oxford, England, Oxford University Press, 1983, pp 276–294

De Fronzo RA: Insulin secretion, insulin resistance, and obesity. Int J Obes 6 (suppl 1):73–82, 1982

De Fronzo RA, Diebert D, Hendler R, et al: Insulin sensitivity and insulin binding to monocytes in maturity-onset diabetes. J Clin Invest 63:939–946, 1979

Deibert DC, De Fronzo RA: Epinephrine induced insulin resistance in man. J Clin Invest 65:717–721, 1980

Delius W, Hagbarth K, Hongell A, et al: Manoeuvres affecting sympathetic outflow in human skin nerves. Acta Physiol Scand 84:177–186, 1972

Dimsdale J, Ziegler MG: What do plasma and urinary measures of catecholamines tell us about human response to stressors? Circulation 83 (suppl II):II36–II42, 1991

Dimsdale JE, Young D, Moore R, et al: Do plasma norepinephrine levels reflect behavioral stress? Psychosom Med 49:375–382, 1987

Dimsdale J, O'Connor D, Ziegler MG, et al: Does chromogranin-A respond to short-term mild physiological challenge? Neuropsychopharmacology 2:237–240, 1989

Dimsdale JE, Ziegler MG, Mills P, et al: Effects of salt, race, and hypertension on reactivity to stressors. Hypertension 16:573–580, 1990

Eliasson K, Hjemdahl P, Kahan T: Circulatory and sympathetic adrenal responses to stress in borderline and established hypertension. J Hypertens 1:131–139, 1983

Esler M, Jennings G, Korner P, et al: Assessment of human sympathetic nervous system activity from measurements of norepinephrine turnover. Hypertension 11:3–20, 1988

Feldt-Rasmussen B, Mathesen ER, Deckert T, et al: Central role for sodium in the pathogenesis of blood pressure changes independent of angiotensin, aldosterone and catecholamines in type 1 (insulin-dependent) diabetes mellitus. Diabetologia 30:610–617, 1987

Ferrannini E, Haffner SM, Stern MP: Essential hypertension: an insulin-resistant state. J Cardiovasc Pharmacol 15:S18–S25, 1990

Finton C, Chernow B, Keiser H: Pheochromocytoma: clinical considerations, in Norepinephrine: Frontiers of Clinical Neuroscience, Vol 2. Edited by Ziegler MG, Lake CR. Baltimore, MD, Williams & Wilkins, 1984, pp 27–37

Goldstein DS: Plasma catecholamines and essential hypertension: an analytical review. Hypertension 5:86–89, 1983

Graafsma SJ, Van Tits LJ, Van Herjst P, et al: Effects of isometric exercise on blood cell adrenoceptors in essential hypertension. J Cardiovasc Pharmacol 14:598–602, 1989

Hamada M, Kazatani Y, Shigematsu Y, et al: Enhanced blood pressure response to isometric handgrip exercise in patients with essential hypertension: effects of propranolol and prazosin. J Hypertens 5:305–309, 1987

Ito Y, Ando K, Noda H, et al: Evidence for increased sympatho-adrenomedullary activity in young subjects with borderline hypertension. Jpn Circ J 52:1326–1334, 1988

Jansen R, Lenders J, Thien T, et al: The influence of age and blood pressure on the hemodynamic and humoral response to head-up tilt. J Am Geriatr Soc 37:528–532, 1989

Kaplan NM: The deadly quartet. Arch Intern Med 149:1514–1520, 1989

Kjeldsen S, Gjesdal K, Leren P, et al: Decreased platelet-free dopamine and unchanged noradrenaline and adrenaline in essential hypertension. Thromb Haemost 60:251–254, 1988a

Kjeldsen S, Westheim A, Lande K, et al: Sodium depletion increases platelet and plasma catecholamines in hypertensive men. Hypertension 11:477–482, 1988b

Kusaka M, Atarashi K, Matsumoto K, et al: Effects of treatment with captopril on exercise tolerance and plasma catecholamines in elderly hypertensives. J Hypertens 7:559–561, 1989

Lehmann M, Keul J: Urinary excretion of free noradrenaline and adrenaline related to age, sex and hypertension in 265 individuals. Eur J Appl Physiol 55:14–18, 1986

Maayan C, Akselrod F, Akselrod S, et al: Evaluation of autonomic dysfunction in familial dysautonomia by power spectral analysis. J Auton Nerv Syst 21:51–58, 1987

Michel MC, Brodde OE, Insel PA: Peripheral adrenergic receptors in hypertension. Hypertension 16:107–120, 1990

Morlin C, Wallin G, Erickson BM: Muscle sympathetic activity and plasma noradrenaline in normotensive and hypertensive men. Acta Physiol Scand 119:117–121, 1983

Netter P: Psychological aspects of catecholaminc response patterns to pain and mental stress in essential hypertensive patients and controls. Journal of Clinical Hypertension 3:727–742, 1987

O'Connor D: Plasma chromogranin A: initial studies in human hypertension. Hypertension 7:I76–I79, 1985

O'Connor D, Bernstein K: Radioimmunoassay of chromogranin A in plasma as a measure of exocytotic sympathoadrenal activity in normal subjects and patients with pheochromocytoma. N Engl J Med 311:764–770, 1984

O'Connor D, Pandian M, Cervenka J, et al: What is the source and disposition of chromogranin A in human plasma? Clin Res 35:605A, 1987

O'Hare JA: The enigma of insulin resistance and hypertension. Insulin resistance, blood pressure, and the circulation. Am J Med 84 (3 Pt 1):505–510, 1988

Rowe JW, Young JB, Minaker KL, et al: Effect of insulin and glucose infusions on sympathetic nervous system activity in normal man. Diabetes 30:219–225, 1981

Saito I, Takeshita E, Hayashi S, et al: Comparison of clinic and home blood pressure levels and the role of the sympathetic nervous system in clinic-home differences. Am J Hypertens 3:219–224, 1990

Schneider R, Julius S, Moss G, et al: New markers for type A behavior: pupil size and platelet epinephrine. Psychosom Med 49:579–590, 1987

Shultz W, Hortnagl H, Magometschnigg D: Function of the autonomic nervous system in young untreated hypertensive patients. Int J Cardiol 10:133–140, 1986

Smith A, Dancis J: Exaggerated response to infused norepinephrine in familial dysautonomia. N Engl J Med 270:704–707, 1964

Smith C, Curtis L, Delamothe A, et al: The distribution of catecholamines between platelets and plasma in normal human subjects. Clin Sci 69:1–6, 1985

Solt V, Brown MR, Kennedy B, et al: Elevated insulin, norepinephrine and neuropeptide Y in hypertension. Am J Hypertens 8:823–828, 1990

Somers V, Mark A, Abboud F: Potentiation of sympathetic nerve responses to hypoxia in borderline hypertensive subjects. Hypertension 11:608–612, 1988

Spence JD, Manuck SB, Munoz C, et al: Hemodynamic and endocrine effects of mental stress in untreated borderline hypertension. Am J Hypertens 3:859–862, 1990

Starkman M, Cameron O, Neese R, et al: Peripheral catecholamine levels and the symptoms of anxiety. Psychosom Med 52:129–142, 1990

Surwit RS, Feinglos MN: Relaxation-induced improvement in glucose tolerance is associated with decreased plasma cortisol. Diabetes Care 7:203–204, 1984

Swislocki A: Insulin resistance and hypertension. Am J Med 300:104–115, 1990

Theorell T, Hjemdahl P, Ericsson F, et al: Psychological and physiological factors in relation to blood pressure at rest: a study of Swedish men in their upper twenties. J Hypertens 3:591–600, 1985

Tischenkel NJ, Saab PG, Schneiderman N, et al: Cardiovascular and neurohumoral response to behavioral challenge as a function of race and sex. Health Psychol 8:503–524, 1989

van Kammen D, Peters J, Yao J, et al: CSF chromogranin A-like immunoreactivity in schizophrenia: assessment of clinical and biochemical relationships. Schizophr Res 6:31–39, 1991

Vapaatalo H, Tuomisito M, Turyanma V, et al: Plasma catecholamines and production of thromboxane B2 by platelets after psychological and physical tests in normotensive and mild hypertensive subjects. Biomed Biochim Acta 47:S294–S298, 1988

Wallin B: Human sympathetic nerve activity and blood pressure regulation. Clin Exp Hypertens [A] 11:91–101, 1989

Wallin B, Fagius J: Peripheral sympathetic neural activity in conscious humans. Annu Rev Physiol 50:565–576, 1988

Wallin BG, Morlin C, Hjemdahl P: Muscle sympathetic activity and venous plasma noradrenaline concentrations during static exercise in normotensive and hypertensive subjects. Acta Physiol Scand 129:489–497, 1987

Weinshilboum R: Serum dopamine-beta-hydroxylase. Pharmacol Rev 30:133–166, 1979

Weinshilboum R: Biochemical genetics of catecholamines in humans. Mayo Clin Proc 58:319–330, 1983

Zweifler A, Fedorak I: Uptake and retention of catecholamines by human platelets in vitro. Clin Res 33:291A, 1985

Zweifler A, Julius S: Increased platelet catecholamine content in pheochromocytoma: a diagnostic test in patients with elevated plasma catecholamines. N Engl J Med 306:890–894, 1982

11

Adrenergic Mechanisms in Type A Behavior

ROBERT H. SCHNEIDER, M.D.

Every affection of the mind that is attended with either pain or pleasure, hope or fear, is the cause of an agitation whose influence extends to the heart.

—William Harvey, *Exercitatio Anatomica de Motu Cordis et Sanguinis in Animalibus*

I believe that the high pressure at which men live and the habit of working the machine with maximum capacity are responsible for arterial degeneration, rather than excesses in eating or drinking.

—Sir William Osler, *Lectures on Angina Pectoris and Allied States*

For the past 30 years, the type A behavior pattern has been proposed as a risk factor for coronary heart disease (CHD) (Cooper et al. 1981; Friedman and Rosenman 1959; Rosenman et al. 1964). In a review of the epidemiological literature on type A behavior and CHD, Matthews and Haynes (1986) concluded that type A behavior meets some, but not all of the epidemiologic criteria for causation. Type A confers a relative risk up to 3.0 for total coronary heart disease incidence. This was particularly true for middle-aged, white-collar men since there were few or no causation data for women, nonwhites, or young adults. Negative results from some relatively recent studies of type A behavior in high-risk or diseased populations have questioned the link between type A behavior and CHD (Dimsdale 1988; Krantz et al. 1989). However, a quantitative meta-analysis of prospective studies from both population and high-risk samples suggested that the association between type A and CHD remains significant in population samples when all available studies were considered.

Conceptually, the type A behavior pattern has been described as a multidimensional construct characterized by hard-driving, and achievement-oriented behavior, excessive job involvement, time urgency, aggressiveness, competitiveness, impatience, hostility, and vigorous paralinguistic and psychomotor activity (Rosenman and Friedman 1974). It has been suggested that components of type A behavior related to hostility and anger are perhaps more specific coronary-prone behaviors than global type A behavior. These suggestions are supported by data from clinical trials and coronary angiography studies (Costa et al. 1989; Dembroski and Czajkowski 1989; Matthews 1988). In this chapter, I review mainly studies on global type A behavior, since the majority of physiological studies to date on coronary-prone behavior have used this construct. Where applicable, relevant studies on the hostility component of type A behavior are covered.

Physiological Mechanisms

Ever since the first evidence for a relationship between type A behavior and CHD, there began a search for the possible physiological mechanisms underlying the association that would connect the mental and behavioral phenomenon with presumed physical health outcomes (Friedman and Rosenman 1964, 1975). There are several theories that attempt to explain the physiological mechanisms that mediate the relationship between type A behavior and CHD (Cooper et al. 1981). Given the theme of this volume and the preeminence of the sympathetic nervous system–related findings in type A behavior, I focus in this review on the investigations of adrenergic function and evidence for adrenergic dysfunction in type A behavior. Of course, abnormalities of other physiological systems may interact with the sympathetic nervous system. Some of these are briefly referred to below.

A widely held current hypothesis of coronary artery disease is that the initiation and progression of the atherosclerotic plaque are determined by various factors that contribute to injury of the arterial endothelium (Ross and Glomset 1973).

It has been frequently hypothesized that increased activity of the sympathetic nervous system could play a central role in producing the pathophysiology of CHD associated with type A behavior (Glass and Contrada 1984; Krantz and Manuck 1984; Williams et al. 1978). Excessive adrenergic activity could contribute to vascular injury and CHD through 1) shear forces produced by pressor responses, 2) accelerated lipid mobilization, 3) atherogenesis, 4) platelet aggregation and thrombosis, 5) myocardial toxicity, 6) ventricular hypertrophy, or 7) cardiac arrhythmia (Bassett 1982; Goldstein and Rafjer 1984). In clinical settings, the extent of increase in both norepinephrine (NE) and epinephrine activity has been associated with severity of myocardial infarction, cardiac arrhythmias, and congestive heart failure (Ganguly 1989). Animal models have confirmed many of these findings (Schneiderman 1983). Interestingly, a new hypothesis concerning the pathological mechanism of adrenergic-induced cardiovascular injury implicates oxidative stress from free radical byproducts of catecholamines (Ganguly 1989).

A substantial body of research has documented greater cardiovascular reactivity in type A subjects compared with type B subjects, especially when confronted with cognitively stressful or challenging stimuli (Houston 1988). These studies, on the average, have shown increases in both blood pressure and heart rate activity as a reaction to laboratory stressors. Harbin (1989) performed a meta-analysis of the published literature on type A behavior and cardiovascular reactivity. This quantitative review of 71 studies indicated that type A subjects responded to cognitive and psychomotor stimulus situations with significantly greater systolic and diastolic blood pressures and heart rate responses than did type B subjects. This relationship was not evident for females. The strength of the relationship depended on the instrument used to assess type A behavior. That is, the results were stronger when the Structured Interview was used compared with the

Jenkins Activity Survey. The age of the subject did not seem to be a major factor influencing type A versus type B differences in reactivity. As noted, it has been suggested that increased cardiovascular reactivity may be mediated at least in part by hyperactivity of the sympathetic nervous system. Such adrenergic hyperactivity could be monitored by changes in urinary catecholamines, plasma catecholamines, or other newer and novel markers of adrenergic activity. For these reasons, it appears that adrenergic activity may be significant for understanding the mechanisms whereby type A and related behaviors may lead to cardiovascular pathology. The evidence for each of the major areas of measurable sympathetic activity is reviewed in this chapter.

I do not comment on the validity and reliability of the collection and assay procedures used. The reader is referred elsewhere (Nelesen and Ziegler, Chapter 7, this volume) for technical discussion of these methods. Many of the studies reviewed here measured the association of type A behavior with physiological processes other than adrenergic indices, including blood pressure, heart rate, and other neurohormones. Since the focus here is on adrenergic mechanisms, these other measurements are not covered unless they highlight adrenergic mechanisms.

Measurement of Type A Behavior

It is appropriate to comment on the different methods of diagnosing type A behavior. Previous research has shown that structured interview assessments of type A behavior are superior in their prediction of physiological and clinical outcomes than the paper-and-pencil methods, the most popular of which is the Jenkins Activity Survey (Harbin 1989; Matthews 1988). Therefore, I primarily review studies that used the structured interview method for behavioral assessment. However, where they may highlight particular points, studies using the Jenkins survey will be mentioned and identified as such. Hostility and related component behaviors of type A behavior have been derived from a structured interview or from the Cooke-Medley scale of the Minnesota Multiphasic Personality Inventory. Both of these methods for assessment have been validated (Dembroski and Czajkowski 1989). Where relevant data are available, these are also mentioned.

Urinary Catecholamines

Among the first studies on type A behavior and sympathetic activity was that by Friedman et al. (1960). These early pioneers of the study of type A behavior measured urinary excretion of NE and epinephrine in a sample of white middle-class men showing either extreme type A or B behavior. Samples collected over 4 consecutive workdays demonstrated significantly higher levels of NE in the type A subjects. There were no differences in nocturnal levels. A subsequent analysis revealed that, during working hours, this group of type A subjects exhibited twice as much urinary vanillylmandelic acid as did the type B subjects (Byers et al. 1962). However, a subsequent study of Belgian white-collar workers (DeBacker et al. 1979) did not replicate the original Friedman et al. finding. Schlegel et al. (1980) studied vanillylmandelic acid excretion outside of the laboratory in type A and B postinfarction male outpatients and found no difference between A and B groups.

Frankenhaeuser and colleagues have conducted a series of experiments employing urinary samples of catecholamines under more controlled laboratory conditions. This group typically employed college students assessed for type A behavior with a Swedish questionnaire version of the Jenkins Activity Survey (Lundberg 1980). Therefore, the behavioral validity of these studies is open to question. These experimenters chose extreme (upper and lower quartiles) groups, and controlled for diurnal variation, smoking, caffeine, and other substances that might affect urinary catecholamine excretion. In the first study (Lundberg and Forsman 1979), there was no evidence of different levels of excretion of catecholamines in type A compared with type B subjects under conditions of "understimulation" or "overstimulation" with cognitive stress. A later study (Frankenhaeuser et al. 1980a) reported that type B subjects secreted more epinephrine during work time, whereas type A subjects excreted more epinephrine during the nonwork time. An interesting interpretation of the authors is that for type A subjects, inactivity is more stressful, whereas for type B subjects, work is more stressful. Frankenhaeuser et al. (1980b) again found no differences in epinephrine or NE excretion between type A and B subjects who performed on a laboratory stressor under varying control conditions.

In Sweden, Lundberg et al. (1989) published a report assessing 30 men and 30 women scored for type A behavior and components by the Videotaped Structured Interview on laboratory and field adrenergic responses to stress. Results showed that during laboratory stress male subjects with high hostility had greater reactivity of urinary epinephrine and NE than men with low hostility. In extreme type A and B men, there was a positive correlation between the type A component of competitiveness and urinary excretion of epinephrine and NE. However, the opposite effect tended to be seen in type A women.

Van Doornen (1980) studied 33 type A and B subjects (defined using the Jenkins Activity Survey) before and after their doctoral oral defenses. These authors reported higher urinary excretion of epinephrine in type A subjects compared with type B subjects. Lambert et al. (1987) used an 11-item self-report measure based on the Jenkins survey and measured urinary excretion of catecholamines. The results showed a lower level of urinary excretion of NE during the daytime in subjects scoring high on type A behavior. Again, because of use of the Jenkins survey, the validity of this finding remains questionable.

The most prominent feature of the studies reviewed in this section is that urinary excretion of catecholamines is not consistently associated with diagnoses of type A behavior. This holds true for both the field studies and laboratory experiments. Notable exceptions are the early results of Friedman and colleagues (Byers et al. 1962; Friedman et al. 1960) and to some extent Lundberg et al. (1989). Unpublished data from our group on urinary catecholamine excretion in type A behavior, corrected for creatinine clearance (renal function), confirm these mixed results (R. Schneider, R. Karunas, S. Julius, unpublished observations, 1984).

The conclusion appears to be that studies of urinary indices of adrenergic mechanisms do not elucidate or support the hypothesis of adrenergic mechanisms in type A behavior. However, it may be noted here that there are many studies that measured behaviors not identical to but related to type A behavior that do show association with catecholamines. For example, competitiveness and aggressiveness in sports has been shown to be associated with catecholamine excretion (Elmadjian 1963). Other examples are reviewed by Glass and Contrada (1984). Although these studies are not, of course, direct measures of type A behavior, their indirect support of the latter hypothesis may be cautiously considered. Urinary catecholamines and their metabolites have as their advantage a slower rate of change and therefore, at least theoretically, some ability to assess long-term changes in systemic sympathetic activity. Yet there may be other adrenergic indices that more accurately reflect chronic behavioral and adrenergic hyperreactivity in type A subjects. These are reviewed later.

Plasma Catecholamines

Plasma measurements of NE and epinephrine are most suited for determining short-term changes in sympathetic activity under laboratory conditions. Thus, most of the research on type A behavior and plasma catecholamines has assessed plasma catecholamine responsivity to acute stressors in the laboratory. An underlying assumption of these studies is that laboratory reactivity bears some relevance to real-life field patterns, an assumption that may or may not turn out to be true (Krantz and Manuck 1984; Schneider et al. 1987).

The first study was performed by Friedman et al. (1975). In this study, 15 pairs of middle-aged type A and type B men were brought into the laboratory and participated in a competitive and frustrating mental task with harassment for 15 minutes. Blood samples for NE and epinephrine were drawn by repeated venipuncture before, during, and after the task. For NE there was no significant difference between type A subjects and type B subjects at baseline; however, averaged levels of NE during and after the laboratory task were significantly higher for type A subjects than for type B subjects. Plasma epinephrine did not change significantly.

Another early study of plasma catecholamines and type A behavior was conducted by Simpson et al. (1974). Male college students were classified with the Jenkins Activity Survey, and blood samples were drawn before and after a physical exercise task. The results showed no differences between type A and type B subjects in baseline resting plasma NE. After exercise, however, type A subjects showed a trend toward higher NE levels. No differences were observed for plasma epinephrine. Platelet aggregation tests were also performed in this study, and type B subjects showed reduced aggregation response to NE after exercise. However, a clear relationship between platelet aggregation and other markers of adrenergic activity has not yet been established.

Glass and colleagues (Glass and Contrada 1984; Glass et al. 1980a, 1980b) reported a series of well-controlled experiments aimed at elucidating relationships among type A behavior, plasma catecholamines, and psychological stress conditions. In their first experiment, Glass et al. (1980a) reported that both type A and type B men demonstrated significant and similar increases in plasma NE and epinephrine while performing a competitive laboratory task. Addition of harassment or hostile conditions to the stressor potentiated plasma epinephrine responses of the type A subjects during the task. The authors attributed the differences in results of their study from the original Friedman et al. (1975) study to the differences in psychosocial environments between the two studies. In a second experiment, Glass et al. (1980a) reported again an increase in plasma epinephrine concentration in type A subjects performing on a video game in the laboratory irrespective of the presence of an incentive or no incentive. In a third experiment, Glass et al. (1980b) studied 30 type A and type B male volunteers who were university students or transit authority workers. The subjects participated in a task-overload laboratory situation performing two demanding cognitive tasks simultaneously. Contrary to previous studies, type A subjects showed reliably higher baseline epinephrine concentrations than type B subjects. Type A subjects also showed greater elevations of plasma epinephrine consistently across four trials of the task. NE patterns were similar although only marginally significant. Contrada et al. (1982) reported a well-controlled experiment using complex stimulus-and-demand conditions that involved contingent or noncontingent reinforcement combined with high- or low-frequency aversive stimulation. Although type A subjects generally reacted more than type B subjects on plasma catecholamines, the combination of demand conditions yielded inconsistent results regarding situational specificity. From these experiments, Glass and Contrada (1984) generally concluded that differing challenges and stressors in the laboratory can evoke different plasma catecholamine responses in type A and type B subjects. Differences in catecholaminergic reactivity may emerge only under particularly relevant circumstances, those that evoke or involve type A characteristics (such as a high degree of challenge, competition, or hostility). Similarly, situational specificity of blood pressure and heart rate hyperreactivity in type A behavior has been suggested (Harbin 1989; Krantz and Manuck 1984; Schneider et al. 1989).

Williams et al. (1982) studied 21 type A and type B male undergraduate students with continuous blood sampling for catecholamines during a series of tasks in the laboratory. One task emphasized mental work (arithmetic) and another emphasized sensory intake (reaction time). This experiment attempted to develop further the situational specificity hypothesis of type A reactivity. The results showed that type A subjects exhibited greater increases than did the type B subjects in plasma NE and epinephrine on the mental work task but not on the sensory intake task. (However, type A subjects showed greater plasma testosterone responses during the reaction time task.)

There are some experiments that have reported greater reactivity of type B subjects on indices of adrenergic activity. For example, Hilton et al. (1983) reported that type B subjects demonstrated greater elevations of plasma NE under conditions that entailed thinking about and expressing experiences of strong affect. However, there did not appear to be any type A demand or hostile conditions involved. It has been reported by Zahn et al. (1982) that the minimally challenging task of standing upright for 5 minutes produced greater increments in plasma NE in type B subjects compared with type A subjects.

The studies reviewed in this section generally indicate a lack of differences at baseline or rest between type A and B subjects in plasma epinephrine or NE. Yet A/B differences in reactivity of plasma catecholamines to specific types of laboratory challenge stress have been consistently, although not unanimously, observed. The type of tasks most effective in eliciting these responses involved cognitive challenge, competition, harassment, and/or hostility. Harbin (1989) conducted a quantitative review (meta-analysis) of the relationship between type A behavior and physiological reactivity. This meta-analysis included 14 different experiments on NE reactivity and 13 experiments on epinephrine reactivity. Combining the results of these studies showed a significant difference in plasma NE reactivity. When all of the published literature on catecholamine reactivity in type A subjects was analyzed, there was a significant increase in plasma NE ($P = .03$) across studies, whereas there was a nonsignificant difference in epinephrine reactivity.

It has been suggested that NE may be more sensitive to physical effort and physical stresses, whereas epinephrine may have greater sensitivity to psychological reactivity (Frankenhaeuser 1975; Ward et al. 1983). An older hypothesis associated epinephrine with fear and anxiety and NE elevations with anger and aggressiveness (Ax 1953; Elmadjian 1963; Funkenstein 1956; Silverman and Cohen 1960). However, the data reviewed herein have not been able to support or consistently to reproduce this hypothesis. As discussed, NE reactivity to cognitive stressors was observed more consistently than epinephrine in type A behavior reactivity (Harbin 1989).

Newer and Novel Markers of Adrenergic Activity

Forearm Muscle Blood Flow

Williams and colleagues (Muranaka et al. 1988a; Williams et al. 1984) used the techniques of forearm blood flow and vascular resistance to indicate adrenergic activity. In these studies, type A subjects exhibited larger increases in forearm skeletal muscle blood flow than did type B subjects during mental arithmetic but not reaction time. This was interpreted as greater β-adrenergic activation during cognitive stress in type A behavior. However, in response to a pharmacological stress (i.e., isoproterenol infusion), type A men did not differ from type B men on forearm blood or plasma cyclic adenosine monophosphate levels or heart rate (Muranaka et al. 1988a). This was taken to indicate a lack of A/B difference in receptor sensitivity.

Electrocardiographic Changes

Van Egeren et al. (1983) selected subjects for type A and B behavior and tested them on an interactive computer game task with electrocardiographic (ECG) monitoring. Social stimulation elicited greater ST-segment depression and reduction of T-wave and R-wave amplitudes in type A compared with type B subjects. These changes may be indicative of greater cardiac sympathetic activity under stress in type A behavior. In older healthy type A subjects, Scher et al. (1986) also found greater reduction in T-wave amplitude during laboratory stress. These authors suggested that type A subjects experience excessive neurally mediated sympathetic stimulation of the myocardium during mental work. Contrada et al. (1989) reported pharmacological and behavioral data that would support the hypothesis that the T-wave of the ECG is significantly affected by β-sympathetic activity on the heart. These data support the above hypothesis of type A behavior sympathetic activation.

Williams and colleagues (Muranaka et al. 1988a, 1988b) performed a series of experiments on type A behavior and the autonomic nervous system that attempted to define a set of noncognitively mediated, basic physiological differences between type A and B subjects. Rather than expose subjects to cognitive challenges, these studies employed pharmacological and behavioral stimuli to assess autonomic-mediated responses. One line of evidence that suggested noncognitive mediation of reactivity in type A behavior was the finding that during general anesthesia, type A subjects exhibited greater cardiovascular arousal than type B subjects (Krantz and Durel 1983). In the first study of this series, Muranaka et al. (1988b) evaluated cardiovascular and ECG responses of young type A and B male subjects to intravenous infusions of isoproterenol, a β-adrenergic agonist. The major finding was a more prolonged attenuation in ECG T-wave amplitude among the type A men in response to the drug infusion. The addition of hostility scores to type A behavior resulted in improved discrimination between subjects with high and low T-wave amplitude recovery. As noted earlier, other β-adrenergically mediated responses (i.e., heart rate, forearm blood flow, and plasma cyclic adenosine monophosphate levels) did not differ between the type A and B subjects. A possible explanation offered for the attenuated recovery of T-wave amplitude in type A subjects was diminished parasympathetic antagonism of sympathetically mediated electrophysiological effects on the myocardium (Williams 1989).

Muranaka et al. (1988a) further evaluated these phenomena. Type A and B groups were exposed to a β-adrenergic stimulus (mental arithmetic) and a combined α-adrenergic/vagal stimulus (cold face stimulus). The two groups did not differ in their response to the β-adrenergic stimulus on heart rate, blood pressure, forearm blood flow, or vascular resistance responses. However, with the cold face stimulus, the type B subjects demonstrated more prolonged heart rate and blood flow responses to the α-adrenergic/vagal stimulus. This further suggested that type A and type B subjects may be differentiated in their autonomic responses to laboratory stress.

Platelet Catecholamines

One problem with employing plasma catecholamines or other acute markers of adrenergic activity is that they reflect sympathetic nervous system activity during a relatively short period of time, from seconds to minutes, thus giving a "snapshot" view of the sympathetic nervous system (Ziegler 1989). Although this problem has been addressed with the use of urinary catecholamines, there are also limitations of this measure that have made it an imperfect long-term marker of sympathetic activity. For example, most urinary catecholamines are not derived from plasma (Ziegler 1989). Another technique has been proposed and utilized by our group in the study of type A behavior—that is, platelet catecholamine content (Schneider et al. 1987). The rationale for this measure is that blood platelets take up, concentrate, and store NE in vivo and in vitro. Platelet catecholamine content may reflect average circulating plasma catecholamine levels from several days' time. Blood platelets continually accumulate catecholamines from plasma. Whereas plasma catecholamine levels fluctuate rapidly, platelet catecholamines appear to demonstrate a slower turnover, making them potentially useful as indices of long-term sympathetic activity (Chamberlain et al. 1990; Schneider et al. 1987).

We reported that platelet epinephrine content was greater in type A subjects than in type B subjects (Schneider et al. 1987). There was a nonsignificant tendency for platelet NE to be elevated in type A subjects. At the same time, there were no differences in resting plasma catecholamine levels. None of the interview-derived components of type A behavior correlated with platelet catecholamines. These results suggested that the type A subjects may have been subject to a repeated elevation of plasma catecholamines outside of the laboratory in their daily lives. This finding was reflected in their elevated platelet catecholamine contents measured in the laboratory.

Pupillary Responses

Pupil dilation and constriction are controlled by the sympathetic and parasympathetic nervous systems, respectively. Pupillometry provides a measure of α-adrenergic and cholinergic tone outside of the vascular system (Janisse 1977; Lowenstein and Lowenfeld 1962; Rubin 1974). Pupillometry has the additional advantage of being noninvasive and readily assessed. Thus, for the purpose of utilizing greater sensitivity and specificity, for ease of use, and for confirming previous adrenergic findings with a new and independent marker of sympathetic activation, our group (Schneider et al. 1987) applied the technique of pupillometry to a group of middle-aged, working type A and B subjects. We measured both resting pupil size and mydriatic responses to an α-adrenergic agonist, phenylephrine, in 33 subjects assessed by the Structured Interview. The results showed that type A subjects consistently showed larger pupil diameters than type B subjects without stimulation; there was no difference in pupillary dilation to local phenylephrine. Simultaneous with significant differences in pupil size at rest, there were no differences in plasma catecholamine also at rest. These results provided a nonvascular indication of increased resting sympathetic nervous system tone in type A subjects, yet the possibility of decreased cholinergic tone could not be ruled out. The lack of difference in responses to phenylephrine suggested no difference in α_1-adrenergic receptor sensitivity between the two groups.

In another study of pupillary responses, using the Jenkins Activity Survey for behavioral typing, Malcolm et al. (1984) reported that type A subjects showed an altered pupillary constriction pattern to light stimulation under laboratory stress conditions compared with type B subjects. The authors interpreted the results as indicating that type A subjects had stronger sympathetic responses than did type B subjects. These findings tend to support our finding (Schneider et al. 1987) of increased sympathetic activation. Although our study (Schneider et al. 1987) did not observe a difference between behavioral groups in response to phenylephrine eye drops, this procedure was not performed during cognitive stress in the laboratory as was done by Malcolm et al.

Adrenergic Receptors

Kahn et al. (1987) studied lymphocyte β_2-receptor sensitivity and platelet α_2-receptor sensitivity in 17 young men with a positive family history of CHD rated for type A behavior. The results showed that higher interview scores of type A behavior correlated with greater β-receptor density and inversely with the ratio of α_2-receptor:β_2-receptor density. The finding of β_2-receptor densities suggested that type A behavior was associated with diminished chronic β_2-adrenergic stimulation. The results of this work also suggested that type A behavior may be associated with increased α_2-adrenergic activity resulting in the down-regulation of the platelet α_2-receptors. β_2-Receptors cause vasodilation, and α_2-receptors play a role in vasoconstriction (Mills and Dimsdale 1988). The findings of Kahn et al., if confirmed, would indicate enhanced α-mediated vasoconstriction in type A subjects, which could relate to coronary artery vasoconstriction and spasm. The latter are important mechanisms in acute coronary events. However, these findings are not consistent with our adrenergic receptor–related findings (Schneider et al. 1987) and those of Muranaka et al. (1988b). In our laboratory, Mills et al. (1987, 1989) measured lymphocyte β-adrenergic receptors in subjects also rated for type A behavior and found no differences between type A and type B subjects. Although there were differences between subjects practicing stress management (the Transcendental Meditation Program) and those who were not.

Summary of Newer and Novel Markers

The results of a wide variety of novel adrenergic markers fairly consistently indicate alterations of different arms of the sympathetic nervous system in type A subjects. Abnormalities have ranged from pupil size to myocardial electrophysiology to platelet catecholamines and possibly to receptors. These findings elaborate and extend earlier studies with measures of urinary and plasma catecholamines. More research is warranted in these areas.

Antiadrenergic Treatments of Type A Behavior

Because the type A behavior pattern has been considered to be a risk factor for CHD, there have been numerous attempts to modify the type A behavior pattern psychologically, behaviorally, and pharmacologically to prevent CHD either primarily or secondarily (Levenkron and Moore 1988; Matthews 1988; Nunes et al. 1987). Only a few of the nonpharmacological studies have measured physiological outcomes directly relatable to the sympathetic nervous system (Levenkron and Moore 1988). Those studies that specifically evaluated adrenergic indices are reviewed individually in more detail as well as the antiadrenergic pharmacological studies.

Psychological and Behavioral Interventions

Levenkron et al. (1983) evaluated 38 subjects assessed by self-report for type A behavior over an 8-week period. The results showed no change in reactivity in blood pressure or heart rate. Roskies et al. (1986) used a cognitive-behavioral program in 107 type A and B subjects for 10 weeks. Their results showed no changes in blood pressure, heart rate, or plasma catecholamine reactivity. Another report from Roskies' group indicated a failure to alter cardiovascular reactivity with behavioral stress management or exercise programs (Seraganian et al. 1987). Blumenthal et al. (1990) found that aerobic exercise training was associated with diminished cardiovascular responses to mental stress in the laboratory in type A men. This may have been mediated by increased parasympathetic tone, which may have counterbalanced sympathetic activation to the heart.

Nunes et al. (1987) performed a meta-analysis of the literature of controlled studies using psychological and behavioral treatments for type A behavior. Their review of 18 studies concluded that there was a mean reduction in type A scores by one-half standard deviation. When clinical CHD morbidity and mortality outcome was assessed (by only two studies), there was a lack of significant reduction in morbidity or mortality after 1–2 years; after 3 years, however, there was an approximately 50% reduction in coronary events. The physiological mechanisms of these treatments are as yet not well elucidated. However, it is possible that these various psychological treatments aimed at stress management and behavior modification may have modified some elements of sympathetic nervous system reactivity observed in type A subjects and discussed above.

Along these lines, we have found in our laboratory that type A subjects who practiced the Transcendental Meditation Program showed reduced cardiovascular reactivity to mental stress compared with type A subjects who did not practice this program. Reductions in sympathetic tone and adrenergic receptor sensitivity (Mills 1987; Mills et al. 1989) may contribute to the physiology of this effect. Although these studies on practitioners of Transcendental Meditation did not include clinical outcome data, another epidemiological study on Transcendental Meditation practitioners suggests an 80% reduction in the incidence of heart disease in this population (Orme-Johnson 1987).

Pharmacological Interventions

The first systematic investigation of pharmacological treatment of type A behavior comes from Krantz et al. (1982), who performed a cross-sectional study that compared behavioral and psychophysiological characteristics of coronary patients medicated with and without propranolol, a nonspecific β-adrenergic blocker. Their results indicated that patients taking propranolol demonstrated significantly lower intensity of type A behavior than patients not taking propranolol. These effects were not seen in subjects chronically medicated with diuretics, nitrates, or other central nervous system active drugs. Analysis of components of type A behavior indicated that the propranolol patients showed lower hostility and other speech stylistics of type A behavior. In addition, the propranolol patients showed lower heart rate and rate-pressure product responses to a laboratory stressor.

Schmeider et al. (1983) randomly assigned 19 white male hypertensive patients to treatment with either atenolol (β_1-specific blocker) or a diuretic (hydrochlorothiazide/amiloride). After 4 weeks or more of therapy, patients treated with the β-blocker showed a reduction in type A behavior and attenuated cardiovascular reactivity. Rudel et al. (1986) treated hypertensive patients randomly with either oxprenolol, another β-adrenergic blocker, or the calcium-channel blocker, nitrendipine, for 6 months. Patients on β-blockers showed lower hostility than patients on the calcium-channel blocker. There was no difference in global type A behavior.

Krantz et al. (1988) randomly assigned 46 male hypertensive subjects to receive either propranolol, atenolol, hydrochlorothiazide, or a placebo for 6 weeks. Propranolol is a nonselective β-adrenergic blocker that penetrates into the central nervous system more completely than atenolol. The latter drug is also β_1-selective. Results indicated that the diuretic- and placebo-treated subjects did not differ on cardiovascular reactivity to cognitive stress in the laboratory. However, subjects taking either β-blocker had reduced heart rate but not blood pressure reactivity to mental stress. On type A behavior, there was only a reduction in one of the components, explosive speech, with β-blockers. Of the subjects receiving propranolol, 58% showed reductions in global type A behavior, whereas 25% of the subjects receiving atenolol and 18% of control subjects showed reductions. These results did not replicate the Schmeider et al. (1983) study with atenolol. The fact that Krantz et al. (1988) found that the lipophilic and nonselective β-blocker propranolol had a greater effect in reducing type A behavior than the hydrophilic and cardioselective β-blocker atenolol suggests that central nervous system adrenergic mechanisms are more important in type A behavior than peripheral adrenergic mechanisms. However, although atenolol is considered to be hydrophilic, it does penetrate the central nervous system to some extent; therefore these effects cannot be ruled out. Chronic treatment with both β-blockers reduced heart rate reactivity, but not blood pressure reactivity, to mental stress. This suggests peripheral mechanisms for excessive cardiovascular reactivity.

We (Schneider et al. 1985) followed up our findings of adrenergic abnormalities in type A behavior with a randomized, double-blind, placebo-controlled trial of low-dose β-adrenergic blocker treatment with metoprolol in a nonclinical sample of healthy type A and B subjects. In all subjects, metoprolol treatment was associated with lower levels of platelet epinephrine content. This could have been due to reduced sympathetic activity in type A subjects, who prior to treatment had higher levels of platelet epinephrine than type B subjects. Secondly, when component behaviors were analyzed, our work found a trend toward reduction of hostility in the metoprolol group. There was a pattern of reduced NE reactivity to mental stress associated with the reduction of hostility.

Krantz et al. (1987) studied the acute effects of drugs that stimulate or block adrenergic activity in type A behavior and cardiovascular reactivity to mental stress. In this study, either propranolol, isoproterenol, or a placebo was infused intravenously into 12 healthy males. The results indicated that the β-agonist, isoproterenol, increased cardiovascular reactivity to the stressors and was associated with increased hostility. However, unexpectedly, global type A behavior scores and several components were lowered by isoproterenol and unaffected by propranolol. These findings differ from studies of chronic treatment with β-blockers reviewed earlier. The authors speculated that isoproterenol may have caused its behavioral effects through the potentiation of emotion and/or adrenergic and organ reactivity.

Summary of Antiadrenergic Treatments

The rationale for antiadrenergic treatment of type A behavior and related components appears to be somewhat merited. The results of the pharmacological studies suggest that certain aspects of the type A behavior pattern and adrenergic reactivity may be modified with chronic β-blocker treatment. However, given that the majority of type A persons in the population are probably nonclinical (no apparent disease), the likelihood of widespread popular acceptance of chronic drug treatment seems low. Thus, nonpharmacological behavioral means (Levenkron and Moore 1988; Nunes et al. 1987) appear to hold promise in this respect. However, their physiological effectiveness generally has not been strongly demonstrated. Notable exceptions are the reductions in clinical heart disease found with intensive behavioral counseling (Friedman et al 1986) and the Transcendental Meditation Program (Orme-Johnson 1987; Orme-Johnson and Schneider 1987).

References

Ax AF: The physiological differentiation between fear and anger in humans. Psychosom Med 15:433–442, 1953

Bassett JR: Psychological stress and the coronary artery in ischemic heart disease, in The Coronary Artery. Edited by Kalsner S. New York, Oxford University Press, 1982, pp 474–501

Blumenthal JA, Fredrikson M, Kuhn CM, et al: Aerobic exercise reduces levels of cardiovascular and sympathoadrenal responses to mental stress in subjects without prior evidence of myocardial ischemia. Am J Cardiol 65:93–98, 1990

Byers SO, Friedman M, Rosenman RH, et al: Excretion of 3-methoxy-4-hydroxymandelic acid in men with behavior pattern associated with high incidence of coronary artery disease. Federation Proceedings 21:99–101, 1962

Chamberlain KG, Pestell RG, Best JD: Platelet catecholamine contents are cumulative indexes of sympathoadrenal activity. American Journal of Physiology 259:E141–E147, 1990

Contrada RJ, Glass DC, Krakoff LR, et al: Effects of control over aversive stimulation and Type A behavior on cardiovascular and plasma catecholamine responses. Psychophysiology 19:408–419, 1982

Contrada RJ, Krantz DS, Durel LA, et al: Effects of beta adrenergic activity on T-wave amplitude. Psychophysiology 26:488–492, 1989

Cooper T, Detre T, Weiss SM: Coronary prone behavior and coronary heart disease: a critical review. Circulation 63:1199–1225, 1981

Costa PT, McCrae RR, Dembroski TM: Agreeableness versus antagonism: explication of a potential risk factor for CHD, in In Search of Coronary-Prone Behavior: Beyond Type A. Edited by Siegman AW, Dembroski TM. Hillsdale, NJ, Lawrence Erlbaum, 1989, pp 41–60

DeBacker G, Kornitzer M, Kittel F, et al: Relation between coronary-prone behavior pattern, excretion of urinary catecholamines, heart rate, and heart rhythm. Prev Med 8:14–22, 1979

Dembroski TM, Czajkowski SM: Historical and current developments in coronary-prone behavior, in In Search of Coronary-Prone Behavior: Beyond Type A. Edited by Siegman AW, Dembroski TM. Hillsdale, NJ, Lawrence Erlbaum, 1989, pp 21–35

Dimsdale JE: A perspective on Type A behavior and coronary disease. N Engl J Med 318:110–112, 1988

Elmadjian F: Excretion and Metabolism of Epinephrine and Norepinephrine in Various Emotional States. Proceedings of the 5th Pan American Congress on Endocrinology. Lima, Peru, 1963, pp 341–370

Frankenhaeuser M: Experimental approaches to the study of catecholamines and emotion, in Emotions: Their Parameters and Measurement. Edited by Levi L. New York, Raven, 1975, pp 209–234

Frankenhaeuser M, Lundberg U, Forsman L: Dissociation between sympathetic-adrenal and pituitary-adrenal responses to an achievement situation characterized by high controllability: comparison between Type A and Type B males and females. Biol Psychol 10:79–91, 1980a

Frankenhaeuser M, Lundberg U, Forsman L: Note on arousing Type A persons by depriving them of work. J Psychosom Res 24:45–47, 1980b

Friedman M, Rosenman RH: Association of specific overt behavior pattern with blood and cardiovascular findings. JAMA 169:1286–1296, 1959

Friedman M, St George S, Byers SO, et al: Excretion of catecholamines, 17-ketosteroids, 17-hydroxy-corticoids and 5-hydroxindole in men exhibiting a particular behavior pattern (A) associated with high incidence of clinical coronary artery disease. J Clin Invest 39:758–764, 1960

Friedman M, Byers SO, Diamant J, et al: Plasma catecholamine response of coronary-prone subjects (Type A) to a specific challenge. Metabolism 24:205–210, 1975

Friedman M, Thorensen CE, Gill JJ, et al: Alteration of Type A behavior and its effect on cardiac recurrences in post-myocardial infarction patients: summary results of the Recurrent Coronary Prevention Project. Am Heart J 112:653–655, 1986

Funkenstein DH: Norepinephrine-like and epinephrine-like substances in relation to human behavior. J Nerv Ment Dis 124:58–65, 1956

Ganguly PK: Catecholamines and cardiovascular disorders: pathophysiologic considerations. Am Heart J 118:868–872, 1989

Glass DC, Contrada RJ: Type A behavior and catecholamines: a critical review, in Norepinephrine Frontiers of Clinical Neuroscience, Vol 2. Edited by Ziegler MG, Lake CR. Baltimore, MD, Williams & Wilkins, 1984, pp 346–367

Glass DC, Krakoff LR, Contrada R, et al: Effect of harassment and competition upon cardiovascular and catecholamine response in Type A and B individuals. Psychophysiology 17:453–463, 1980a

Glass DC, Krakoff LR, Finkelman J, et al: Effect of task overload upon cardiovascular and plasma catecholamine responses in Type A and B individuals. Basic and Applied Social Psychology 1:199–218, 1980b

Goldstein RE, Rafjer SI: Effects of catecholamines in clinical heart disease, in Norepinephrine Frontiers of Clinical Neuroscience, Vol 2. Edited by Ziegler MG, Lake CR. Baltimore, MD, Williams & Wilkins, 1984, pp 327–345

Harbin TJ: The relationship between the Type A behavior pattern and physiological responsivity: a quantitative review. Psychophysiology 26:110–119, 1989

Harvey W: Exercitatio Anatomica de Motu Cordis et Sanguinis in Animalibus. Francofurti, Sumptibus Guilielmi Fitzeri, 1628

Houston BK: Cardiovascular and neuroendocrine reactivity in global type A behavior, in Type A Behavior Pattern: Research, Theory and Intervention. Edited by Houston BK, Snyder CR. New York, Wiley, 1988, pp 212–253

Janisse MP: Pupillometry: The Psychology of the Pupillary Response. Washington, DC, Hemisphere, 1977

Kahn JP, Gully RJ, Cooper TB, et al: Correlation of Type A behavior with adrenergic receptor density: implications for coronary artery disease pathogenesis. Lancet 2:937–939, 1987

Krantz DS, Durel LA: Psychobiological substrates of the Type A behavior pattern. Health Psychol 2:393–411, 1983

Krantz DS, Manuck SB: Acute psychophysiologic reactivity and risk of cardiovascular disease: a review and methodologic critique. Psychol Bull 96:435–444, 1984

Krantz DS, Durel LA, Davia JE, et al: Propranolol medication among coronary patients: relationship to Type A behavior and cardiovascular stress. Journal of Human Stress 8:4–12, 1982

Krantz DS, Contrada RJ, LaRiccia PJ, et al: Effects of beta-adrenergic stimulation and blockade on cardiovascular reactivity, affect, and Type A behavior. Psychosom Med 49:146–158, 1987

Krantz DS, Contrada RJ, Durel LA, et al: Comparative effects of two beta blockers on cardiovascular reactivity in Type A behavior in hypertensives. Psychosom Med 50:615–626, 1988

Krantz DS, Schneiderman N, Chesney MA, et al: Biobehavioral research on cardiovascular disorders: task force reports. Health Psychol 8:737–746, 1989

Lambert WW, Macevoy B, Klackenberg-Larsson I, et al: The relation of stress hormone excretion to Type A behavior and to health. Journal of Human Stress 13:128–135, 1987

Levenkron JC, Moore LG: The Type-A behavior pattern: issues for intervention research. Annals of Behavioral Medicine 10:78–83, 1988

Levenkron JC, Cohen JH, Mueller HS, et al: Modifying the Type A coronary-prone behavior pattern. J Consult Clin Psychol 51:192–204, 1983

Lowenstein O, Lowenfeld IE: The pupil, in The Eye, Vol 3. Edited by Davson H. New York, Academic Press, 1962, pp 8–20

Lundberg U: Type A behavior and its relation to personality variables in Swedish male and female university students. Scand J Psychol 21:133–138, 1980

Lundberg U, Forsman L: Adrenal-medullary and adrenal-cortical responses to understimulation and overstimulation: comparison between Type A and Type B persons. Biol Psychol 9:79–89, 1979

Lundberg U, Hedman M, Melin B, et al: Type A behavior in healthy males and females as related to physiological reactivity in blood lipids. Psychosom Med 51:113–122, 1989

Malcolm AT, Janisse MP, Dyck DG: Type A behavior, heart rate and pupillary response: effects of cold pressor and ego threat. J Psychosom Res 28:27–34, 1984

Matthews KA: Psychological perspectives on the Type A behavior pattern. Psychol Bull 91:293–323, 1982

Matthews KA: Coronary heart disease and Type A behaviors: update on an alternative to the Boothe-Kewley and Friedman (1987) Quantitative Review. Psychol Bull 3:373–380, 1988

Matthews KA, Haynes SG: Type A behavior pattern and coronary disease risk: update and critical evaluation. Am J Epidemiol 123:923–959, 1986

Mills PJ: Cardiovascular and adrenergic reactivity and beta-adrenergic receptor sensitivity in practitioners of the Transcendental Meditation program and Type A behavior. Unpublished doctoral dissertation, Maharishi International University, Fairfield, Iowa, April 1987

Mills PJ, Dimsdale JE: The promise of receptor studies in psychophysiologic research. Psychosom Med 50:555–566, 1988

Mills PJ, Schneider RH, Hill D, et al: Lymphocyte beta-adrenergic receptors and cardiovascular responsivity in TM participants and Type A behavior. Psychosom Med 49:211, 1987

Mills PJ, Schneider RH, Hill D, et al: Beta-adrenergic receptor sensitivity in subjects practicing Transcendental Meditation. J Psychosom Res 34:29–33, 1989

Muranaka M, Monou H, Suzuki J, et al: Physiological responses to catecholamine infusions in Type A and Type B men. Health Psychol 7:145–163, 1988a

Muranaka M, Lane JD, Suarez C, et al: Stimulus-specific patterns of cardiovascular reactivity in Type A and B subjects: evidence for enhanced vagal reactivity in Type B. Psychophysiology 25:330–338, 1988b

Nunes EV, Frank KA, Kornfeld DS: Psychologic treatment for the Type A behavior pattern and for coronary heart disease: a meta-analysis of the literature. Psychosom Med 48:159–173, 1987

Orme-Johnson DW: Medical care utilization and the Transcendental Meditation program. Psychosom Med 49:493–507, 1987

Orme-Johnson DW, Schneider RH: Reduced health care utilization in Transcendental Meditation practitioners: Maharishi Ayurveda as preventive medicine. Paper presented at the Society for Behavioral Medicine Annual Meeting, Washington, DC, March 1987

Osler W: Lectures on Angina Pectoris and Allied States. New York, Appleton and Company, 1901

Rosenman RH, Friedman M: Neurogenic factors in the pathogenesis of coronary heart disease. Med Clin North Am 58:269–279, 1974

Rosenman RH, Friedman M, Straus R, et al: A predictive study of coronary heart disease: the Western Collaborative Group Study. JAMA 189:15–22, 1964

Rosenman RH, Brand RJ, Jenkins CD, et al: Coronary heart disease in the Western Collaborative Group Study: final follow-up experience of 8-1/2 years. JAMA 233:872–877, 1975

Roskies E, Seraganian P, Oseasohn R, et al: The Montreal Type A intervention project: major findings. Health Psychol 5:45–69, 1986

Ross R, Glomset J: Atherosclerosis and the arterial smooth muscle cell. Science 180:1332–1339, 1973

Rubin LS: The utilization of pupillometry in the differential diagnosis and treatment of psychotic and behavioral disorders, in Pupillary Dynamics and Behavior. Edited by Janisse MP. New York, Plenum, 1974, pp 75–134

Rudel H, Schmeider R, Langewitz W, et al: Effective anti-hypertensive therapy with beta-blockers on Type A behavior in hostility (abstract). Psychosom Med 48:295, 1986

Scher H, Hartman LM, Furedy JJ, et al: Electrocardiographic T-wave changes are more pronounced in Type A than in Type B men during mental work. Psychosom Med 48:159–166, 1986

Schlegel RP, Wellwood JK, Copps BE, et al: The relationship between perceived challenge and daily symptom reporting in Type A vs. Type B postinfarct subjects. J Behav Med 3:191–204, 1980

Schmeider RS, Friedrich G, Neus H, et al: The influence of beta-blockers on cardiovascular reactivity and Type A behavior pattern in hypertensives. Psychosom Med 45:417–423, 1983

Schneider RH, Julius S, Moss G: Physiological correlates of the Type A coronary prone behavior pattern and the influence of metoprolol (Lopressor): a preliminary report, in Towards Preventive Treatment of Coronary Prone Behavior. Edited by Zanchetti A, Turner P. Bern, Switzerland, Hans Huber Publishers, 1985, pp 42–48

Schneider RH, Julius S, Moss GE, et al: New markers for Type A behavior: pupil size and platelet epinephrine. Psychosom Med 49:570–590, 1987

Schneider RH, Julius S, Karunas R: Ambulatory blood pressure monitoring and laboratory reactivity in Type A behavior and components. Psychosom Med 51:290–305, 1989

Schneiderman N: Animal behavior models of coronary heart disease, in Handbook of Psychology and Health. Edited by Krantz DS, Baum A, Singer JE. Hillsdale, NJ, Lawrence Erlbaum Associates, 1983, pp 19–48

Seraganian P, Roskies E, Hanley JA, et al: Failure to alter psychophysiological reactivity in Type A men with physical exercise or stress management programs. Psychology and Health 1:195–213, 1987

Silverman AJ, Cohen SI: Affect and vascular correlates of catecholamines. Psychiatric Research Reports 132:6–30, 1969

Simpson MT, Olewine DA, Jenkins CS, et al: Exercise-induced catecholamines and platelet aggregation in the coronary-prone behavior patterns. Psychosom Med 36:476–487, 1974

Van Doornen LJP: The role of psychosocial factors in coronary heart disease. Psychother Psychosom 34:204–215, 1980

Van Egeren LF, Fabrega H, Thornton DW: Electrocardiographic effects of social stress on coronary-prone (Type A) individuals. Psychosom Med 45:195–203, 1983

Ward MM, Mefford IN, Parker SD, et al: Epinephrine and norepinephrine responses in continuously collected human plasma to a series of stressors. Psychosom Med 45:471–486, 1983

Williams RB: Biological mechanisms mediating the relationship between behavior and coronary heart disease, in In Search of Coronary-Prone Behavior: Beyond Type A. Edited by Siegman AW, Dembroski TM. Hillsdale, NJ, Lawrence Erlbaum, 1989, pp 195–204

Williams RB, Friedman G, Glass DC, et al: Summary statement: mechanisms linking behavioral and pathophysiological process, in Coronary Prone Behavior. Edited by Dembroski TM, Weiss SM, Shields JL, et al. New York, Springer-Verlag, 1978, pp 119–128

Williams RB, Lane JD, Kuhn CM, et al: Type A behavior and elevated physiological and neuroendocrine responses to cognitive tasks. Science 218:483–485, 1982

Zahn TP, Haynes SG, Mikkelsen ES, et al: Effects of postural and psychological stress on plasma norepinephrine and autonomic variables in Type A and Type B men. Bethesda, MD, Laboratory of Psychology and Psychopathology, National Institute of Mental Health, 1982

Ziegler MG: Catecholamine measurement in behavioral research, in Handbook of Research Methods in Cardiovascular Behavioral Medicine. Edited by Schneiderman N, Weiss SM, Kaufmann PG. New York, Plenum, 1989, pp 167–183

12

Adrenergic Mechanisms in Behavioral Interventions

ROBERT R. FREEDMAN, PH.D.

Although adrenergic mechanisms have been extensively studied in many pathological states, relatively little attention has been given to the role of these mechanisms in behavioral interventions that are used to treat various pathologies. Some studies have been done in which measures of adrenergic function, such as plasma catecholamines or adrenoceptors on blood cells, have been obtained during progressive relaxation, meditation, or biofeedback. These studies form the basis of the present chapter. Work in my laboratory has focused on adrenergic mechanisms in relaxation treatment for menopausal hot flashes and biofeedback treatment for Raynaud's disease; these areas are reviewed in detail. Since the rationale for some behavioral treatments is the reduction of physiological reactivity to stress, I begin the chapter with a brief discussion of adrenergic receptor regulation during mental stress.

Regulation of Adrenergic Receptors During Stress

Catecholamines produce a wide variety of physiological effects by binding to α- and β-adrenergic receptors. The responsiveness of these receptors is regulated by exposure to agonists, among other factors. As reviewed elsewhere (Goldstein, Chapter 8, this volume), adrenergic receptor regulation has been demonstrated during periods of acute stress in animals, and in chronic psychological states such as panic disorder and type A personality. Only recently has adrenoceptor regulation been studied during laboratory and naturalistic stressors in humans (Mills and Dimsdale 1988).

Graafsma et al. (1989) subjected normotensive and hypertensive subjects to mental arithmetic while blood was drawn for subsequent measurement of plasma catecholamines, platelet α_2-adrenoceptors, and lymphocyte β_2-adrenoceptors. Significant stress-induced elevations in plasma epinephrine (74%), norepinephrine (NE) (25%), and β_2-receptor density (23%) were found, with no significant changes in platelet α_2-receptor measurements. An additional study from the same laboratory (Graafsma et al. 1988) replicated the above findings and also showed that the elevation in β_2-receptor density did not occur in adrenalectomized subjects, who had no measurable levels of plasma epinephrine. These findings suggest that the up-regulation of β_2-receptors during mental arithmetic is due to increased plasma epinephrine levels.

Two other studies found similar results using different methods. Pacák et al. (1989) divided a group of male subjects into those who were high and those who were low in pressor responsiveness to the Stroop test and a numerical task. Those in the high-responsive group had significantly higher lymphocyte β_2-receptor density than those in the low-responsive group, although the number of subjects was very small. Mills et al. (1990b) found that the heart rate response to mental arithmetic was significantly and positively correlated with basal levels of lymphocyte β_2-receptor density and isoproterenol-stimulated cyclic adenosine monophosphate accumulation.

In our laboratory, we studied 30 first-year medical students during a stressful period (final exams) and 1 month earlier. During final exam week, platelet α_2-receptor binding affinity was significantly reduced (i.e., the dissociation constant increased), whereas levels of plasma catecholamines and reported anxiety were significantly increased, compared with the earlier period. Students showing the greatest increases in plasma NE and in reported anxiety also demonstrated the greatest reductions in α_2-receptor binding affinity. There were no significant changes in leukocyte β_2-receptor measurements.

The previous studies suggest that lymphocyte β_2-receptor density increases during certain mental stressors. Although the underlying physiological mechanisms are not known with certainty, the studies of Graafsma et al. (1988, 1989) suggest that elevated levels of plasma epinephrine are involved. Since Brodde et al. (1986) have shown that β_2-adrenoceptors from lymphocytes and from right atrial tissue are regulated in concert by nonspecific agonists such as epinephrine and isoproterenol, it is possible that increased heart rate responsiveness during mental stress is mediated, in part, by β_2-receptor up-regulation. This reasoning is supported by the data of Mills et al. (1990b) and by an additional study in which the heart rate response to mental stress was significantly correlated with plasma epinephrine levels and with isoproterenol sensitivity (Eisenhofer et al. 1985).

The relationships between stress and receptor regulation are no doubt influenced by the nature and time course of the stressor(s) involved. Although our study of examination stress did not find changes in β_2-receptor measurements, the duration of the stressor was on the order of several days (i.e., exam week). Studies finding increased β_2-receptor density employed stressors lasting several minutes, such as mental arithmetic. Conversely, our investigation found desensitization of platelet α_2-receptors during exam week, whereas Graafsma et al. (1988, 1989) found no such relationship during 5 minutes of mental arithmetic. In vivo studies reviewed elsewhere (Raymond et al. 1990) show that different mechanisms of adrenoceptor regulation operate over time periods ranging from minutes to days. Adrenergic receptors are clearly an important link between humoral signals and end-organ responsiveness and warrant further investigation in studies of stress and relaxation.

Meditation and the "Relaxation Response"

Transcendental meditation (TM) is a stress-reduction technique characterized, in some studies, by reductions in oxygen consumption, heart rate, and electrodermal activity (Wallace 1970). The relaxation response is an abbreviated version of TM in which the subject sits in a relaxed position in a quiet environment, closes the eyes, and silently repeats a simple phrase. Hoffman et al. (1982) measured plasma NE, blood pressure, and heart rate during orthostatic stress before and after elicitation of the relaxation response. Subjects who practiced the relaxation response had significantly higher levels of plasma NE than those who did not; the heart rates and blood pressures of the two groups were similar. The authors interpreted their findings to imply reduced end-organ responsivity to NE resulting from the relaxation response. These results were partially replicated by Morrell and Hollandswirth (1986), who studied meditators and nonmeditators during venipuncture and isometric handgrip. During venipuncture, the meditators had significantly higher plasma NE levels and lower heart rates and diastolic blood pressures compared with the nonmeditators. However, there were no significant group differences during handgrip. Similar results were reported by Lang et al. (1979), who compared experienced with less-experienced meditators during a variety of behavioral tasks. The experienced meditators had significantly higher plasma NE levels and lower heart rates than the less-experienced meditators during most conditions.

The above studies suggest that various forms of TM produce increased plasma NE levels and reduced cardiovascular responsivity, but shed no light on the intervening mechanisms. Therefore, Mills et al. (1990a) measured lymphocyte β_2-receptors and plasma catecholamines in experienced TM practitioners and naive control subjects. The TM subjects had a significantly lower percentage of β_2-receptors in the high affinity state compared with the control subjects, with no group differences in total receptor number, dissociation constant, or catecholamine levels. Although this study is limited by its cross-sectional design, it does suggest that down-regulation of β_2-receptors in TM might partially explain the reduced cardiovascular responsiveness despite higher NE levels. Further research in this area would be of great interest.

Progressive Muscle Relaxation

Progressive muscle relaxation (PMR) is a procedure developed by Jacobson (1938) in which sympathetic outflow is presumably reduced through a systematic tensing and relaxing of various gross muscle groups. The method was later modified by Bernstein and Borkovec (1973) to include a series of slow, deep-breathing exercises. Hjemdahl et al. (1989) measured muscle sympathetic nerve activity, arterial and venous catecholamines, cardiac output, and pulmonary arterial and venous pressures in 10 subjects during brief PMR. Significant declines in muscle sympathetic nerve activity and femoral venous NE were found during PMR, with no significant changes in the other measures.

Cottier et al. (1984) studied borderline and mild hypertensive subjects given PMR or a placebo control over 22 weeks. The PMR group showed significant reductions in mean home blood pressure, plasma epinephrine, and oxygen consumption, whereas the control group did not. There were no significant group changes in heart rate or plasma NE, although subjects with the highest initial NE levels showed the best reductions in blood pressure as well as significant declines in NE.

Mathew et al. (1981) measured platelet activity in 20 patients with generalized anxiety disorder and in 20 control subjects before and after 4 weeks of home relaxation exercises and electromyogram (EMG) biofeedback. The treatment procedures were not described in detail. Pretreatment trait anxiety scores and platelet monoamine oxidase activity were significantly higher in the patients than in the control subjects and declined significantly after treatment. Control subjects showed no significant changes.

The above studies thus far represent the accumulated scientific literature in which adrenergic mechanisms have been specifically studied during relaxation procedures. Using elegant methods, Hjemdahl et al.'s (1989) study provided clear evidence that muscle nerve sympathetic activity and plasma NE decreased during brief relaxation. However, since no control group was included, it is not possible to determine whether these changes were due to the relaxation procedure itself or to nonspecific factors such as adaptation to the laboratory situation. The study of Cottier et al. (1984) did employ a control group and provides reasonably convincing evidence that reduced sympathetic activation can be achieved in those hypertensive patients having the highest initial levels of sympathetic tone. The study of Mathew et al. (1981) is difficult to evaluate because the treatment was insufficiently described and because the etiologic significance of platelet monoamine oxidase in anxiety disorders is not known. Further research to verify the presence or absence of adrenergic changes during PMR and related procedures is clearly needed.

Menopausal Hot Flashes as a Hyperadrenergic State

Hot flashes, or flushes, are the most common symptom of the climacteric and are reported by the vast majority of menopausal women. The hot flash is subjectively experienced as a sensation of internal heat, spreading upward from the chest to affect the neck and face. Previous studies (Casper and Yen 1985; Kronenberg et al. 1984; Tataryn et al. 1980) have shown that hot flashes are accompanied by peripheral vasodilation, tachycardia, sweating, and increased skin conductance.

Although hot flashes accompany the estrogen withdrawal that occurs at menopause and may be related to altered hypothalamic function, their pathophysiology is not understood (Casper and Yen 1985). Hot flashes begin rapidly after bilateral oophorectomy, can be induced by abrupt estrogen withdrawal, and can be ameliorated with estrogen replacement therapy (Ravnikar 1990). However, estrogens per se do not appear responsible for hot flashes since there is a lack of correlation between menopausal symptoms and plasma (Askel et al. 1977), urinary (Asch and Greenblatt 1977), and vaginal (Stone et al. 1975) estrogen values.

Although it was thought that luteinizing hormone and luteinizing hormone–releasing hormone (LH-RH) pulses were involved in triggering hot flashes, (Casper et al. 1979), subsequent studies disproved these hypotheses (Gambone et al. 1984). It was then suggested that hot flashes are not triggered by altered LH-RH secretion per se, but by altered afferent input to LH-RH neurons, and NE emerged as a likely candidate for this mechanism.

The central noradrenergic system is accepted as the major activator of luteinizing hormone secretion (McCann and Moss 1975). Locally injected NE stimulates luteinizing hormone secretion (Vijayan and McCann 1978) and releases LH-RH from the hypothalamus in vitro (Negro-Vilar and Ojeda 1978). Increased NE turnover in the hypothalamus is one mechanism believed responsible for the activation of central thermoregulatory mechanisms (Cox and Lomax 1977) and may be responsible for the peripheral events of the hot flash that are characteristic of a heat dissipation response (Tataryn et al. 1980).

Animal studies have shown that yohimbine increases NE release by blocking inhibitory, presynaptic α_2-adrenergic receptors (Goldberg and Robertson 1983; Starke et al. 1989). These autoreceptors mediate the turnover of NE through a feedback mechanism, and a reduction in their number or sensitivity would result in increased NE release. Some clinical studies have shown that clonidine ameliorates hot flashes in menopausal women (Nagamani et al. 1987). We therefore conducted an acute laboratory investigation in which we administered intravenous clonidine and yohimbine to menopausal women with and without hot flashes (Freedman et al. 1990).

We have previously shown that hot flashes can be reliably provoked in the laboratory by peripheral heating and detected by large magnitude (≥ 2 μmho/30 sec) sternal skin conductance increases. Here we demonstrated that yohimbine, an α_2-antagonist, provokes hot flashes and that clonidine, an α_2-agonist, inhibits heat-induced hot flashes. Subjects were nine symptomatic menopausal women with frequent hot flashes documented by 24-hour ambulatory monitoring of sternal skin conductance. Six asymptomatic menopausal women served as a comparison group. They were prescreened by ambulatory monitoring to have no skin conductance responses and did not differ in age, weight, time since last period, or maturation index from the symptomatic group. In two blind laboratory sessions, subjects received either 1 μg/kg clonidine hydrochloride intravenous or placebo, followed by a 60-minute waiting period and then by 45-minute peripheral heating. In two additional blind sessions, subjects received intravenous yohimbine (.032–.128 μg/kg) or placebo. Hot flashes were objectively detected by computer, using the sternal skin conductance response. Clonidine significantly ($P = .01$) increased the amount of heating time needed to provoke a hot flash compared with placebo (40.6 ± 3.0 minutes versus 33.6 ± 3.6) (Figure 12–1) and reduced the number of hot flashes that did occur (two versus eight). A significantly greater number of hot flashes occurred during yohimbine sessions than in the corresponding placebo sessions (six versus zero, $P = .015$). These data support the hypothesis that α_2-adrenoceptors within the central noradrenergic system are involved in triggering menopausal hot flashes.

Further human studies have shown that yohimbine elevates (Charney et al. 1982) and clonidine reduces (Leckman et al. 1980) plasma levels of 3-methoxy-4-hydroxyphenylglycol (MHPG). MHPG is the major metabolite of brain NE and has been used as an index of central NE turnover. We therefore measured plasma MHPG before, during, and after hot flashes in eight postmenopausal symptomatic women and at comparable time points in seven postmenopausal asymptomatic women (Freedman and Woodward 1992a).

Subjects were supine in a 23°C laboratory while a 19-gauge butterfly needle was inserted in an antecubital vein and maintained patent with a slow drip of sterile saline solution. After 20 minutes, 10 ml of blood was drawn for basal measurements. Subjects then had two 42°C heating pads applied to the torso for 45 minutes to elicit hot flashes. Hot flashes were defined by an increase in sternal skin conductance ≥ 2 µmho/30 seconds, recorded on a polygraph and scored by computer. When a hot flash occurred, 10 ml of blood was again drawn; a final 10-ml sample was drawn at the end of the 45-minute heating period. Blood samples were then centrifuged and stored at −70°C in vials containing trideuterated MHPG as an internal standard. Subsequent MHPG determinations were performed in duplicate by gas chromatography-mass spectrometry.

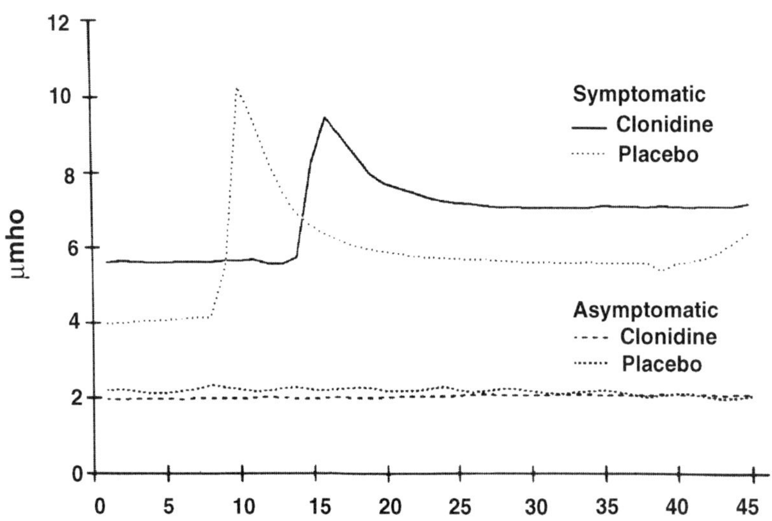

Figure 12–1. The occurrence of a hot flush during body heating was delayed after 1 µg/kg clonidine, compared with placebo. No hot flushes occurred in an asymptomatic woman. The heating pads were applied at time zero; clonidine or placebo was injected 60 minutes before this.
Source. Reprinted with permission from Freedman RR, Woodward S, Sabharwal SC: "α_2-Adrenergic Mechanism in Menopausal Hot Flushes." *Obstetrics and Gynecology* 76:573–578, 1990. Copyright 1990 by The American College of Obstetricians and Gynecologists.

In the symptomatic women, six hot flashes occurred in the resting period and seven during the heating period. No hot flashes occurred in the asymptomatic women during either period. MHPG levels were significantly higher in symptomatic than asymptomatic women at all time points ($P < .0001$). In the symptomatic women, MHPG increased significantly ($P < .001$) during resting and heat-induced hot flashes. There were no significant changes in the asymptomatic women.

Thus, we have shown that hot flashes can be provoked with an α_2-adrenergic antagonist and reduced with an α_2-adrenergic agonist. These findings suggest that central α_2-adrenergic receptors are involved in the initiation of menopausal hot flashes. Consistent with this mechanism, we then demonstrated that levels of plasma MHPG are significantly higher in symptomatic than asymptomatic women and that these levels increase further during hot flashes. Our findings thus support the hypothesis that central NE activation is increased in women with menopausal hot flashes due to a reduction in inhibitory α_2-adrenoceptors and suggest that methods that reduce this activation should ameliorate hot flashes.

Relaxation Treatment of Menopausal Hot Flashes

Though estrogen replacement therapy is an effective treatment for hot flashes, it is contraindicated for women who are at increased risk for cancer. We therefore evaluated PMR, using the method of Bernstein and Borkovec (1973), as a treatment for these women (Germaine and Freedman 1984).

Fourteen women reporting daily hot flashes were randomly assigned to receive six weekly training sessions of PMR or of alpha-electroencephalogram biofeedback (placebo-control procedure). Treatment effectiveness was evaluated by pre- and posttreatment heating tests and by self-report. Following treatment, the latency to onset of hot flashes during heat stress was significantly increased in relaxation subjects but not in control subjects. Reported hot flash frequency was significantly reduced in the relaxation subjects (6.6/day pretreatment to 3.0 posttreatment, $P < .01$), but not in the control subjects (5.9 pretreatment to 7.0 posttreatment). Respiration rate was measured during the heating tests. During the pretreatment test, neither group showed significant changes. Following treatment, the relaxation group maintained significantly slower breathing rates during heating (14.1 breaths/minute, baseline; 8.5 breaths/minute, heat, $P < .05$), whereas the control group showed no significant changes.

These findings suggested that slow, deep breathing might be the active component of PMR treatment for hot flashes and prompted a second study, the results of which are given here (Freedman and Woodward 1992b). Patients were randomly assigned, in equal numbers, to receive eight sessions of training in PMR, paced respiration, or alpha-electroencephalogram biofeedback (placebo control). Hot flashes were counted by computer from 24-hour ambulatory skin conductance level (SCL) recordings made before and after treatment. Patients were asked to relax voluntarily in the laboratory before and after treatment while their respiration was monitored with a metabolic cart. There were no group differences in frequency of pretreatment hot flashes; only the paced respiration group showed a significant change from pre- to posttreatment (mean = 15.8 ± 4 hot flashes/day pretreatment versus 8.0 ± 2 hot flashes/day posttreatment, $P < .05$). During the pretreatment voluntary relaxation test, no group showed significant changes. During the posttreatment test, only the paced respiration group showed significant changes: a period effect ($P < .05$) for respiration rate (11.9 ± 1.4 breaths/minute baseline, 8.5 ± .9 relaxation, 10.6 ± 1.5 recovery). The paced respiration group also had significant overall session effects for respiration rate (13.9 ± .5 pretreatment versus 10.3 ± .8 posttreatment) and tidal volume (.44 ± .01 L/breath pretreatment versus .57 ± .06 posttreatment). Thus, the paced respiration group was the only group to show both changes in respiration and in hot flash frequency.

From these results, we conclude that slow, deep breathing is the active component or one of the active components of behavioral treatments for hot flashes and that this technique in some way modulates the elevated central NE activation in women with hot flashes. However, the connection between slow, deep breathing and reduced noradrenergic activity remains to be made. In cross-sectional studies, Schaefer (1958) compared slow, deep breathing subjects having high alveolar CO_2 levels with fast, shallow breathing subjects having low CO_2 levels. The former group had slower resting heart rates, reduced adrenal cortical activity, and decreased autonomic responsiveness compared with the former group, but catecholamines and adrenergic receptors were not measured. These parameters are being investigated in our current studies.

Raynaud's Disease: A Peripheral Adrenergic Disorder

Physiological Control of Finger Blood Flow

The digital vasculature is almost entirely cutaneous and plays a fundamental role in the regulation of body temperature. The palmar surface and tip of the finger are rich in arteriovenous anastomoses, which function in parallel with the capillary bed. These shunts have the capacity to vary rapidly their lumen size and rate of blood flow in response to changes in external temperature. This is accomplished mainly through sympathetic adrenergic vasoconstrictor nerves (Figure 12–2). Body cooling causes reflex finger vasoconstriction through increased neural activity, and, conversely, body heating produces vasodilation through withdrawal of this activity. Finger capillary blood flow is somewhat affected by sympathetic nervous system activity, although to a lesser extent than arteriovenous shunt flow (Coffman 1972). There are no known vasodilating nerves in the human finger, although such nerves do exist in the skin of the forearm (Shepherd 1963).

Finger blood flow is also controlled through the interactions of circulating vasoactive substances with α- and β-adrenergic receptors to produce vasoconstriction and vasodilation, respectively. Circulating catecholamines, released from the adrenal medulla and from other nerve endings "upstream," act at α-adrenergic receptors to produce vasoconstriction. These α-receptors are probably closer to the lumen than those which respond to NE released from sympathetic nerve endings. A β-adrenergic vasodilating mechanism has recently been discovered in the finger, but an endogenous ligand that acts at these β-receptors has not yet been found.

The sensitivities of vascular adrenergic receptors change according to temperature and represent one means of local control of blood flow. Other local influences include changes in blood gases and metabolism, myogenic tone, and the axon reflex (Vanhoutte 1980).

Pathophysiology of Raynaud's Disease

Raynaud's disease is a disorder of the peripheral circulation in which episodic vasospasms occur in the fingers, toes, and, occasionally, the nose and ears. The symptoms are triggered by cold, emotional stress, or

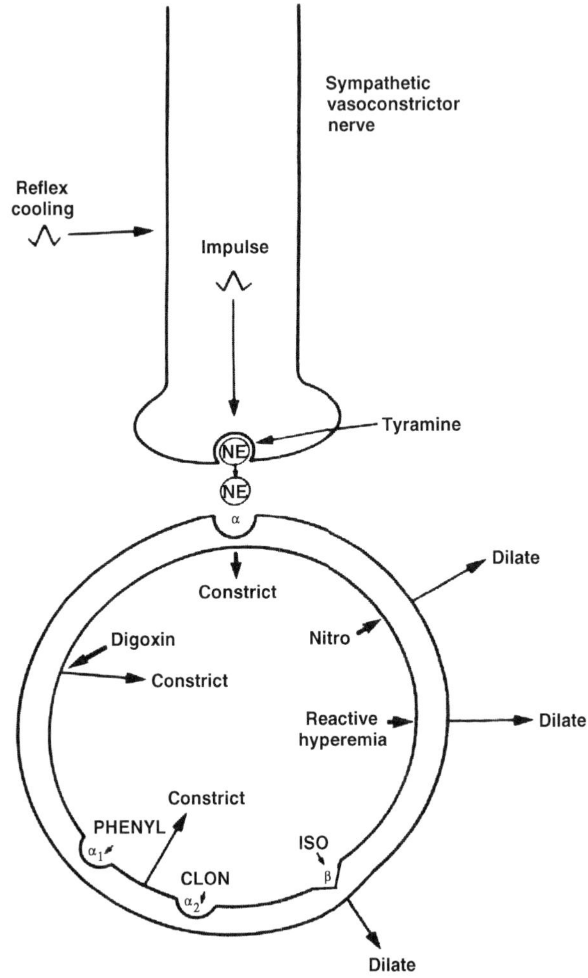

Figure 12–2. Control of digital blood flow. Reflexive cooling causes digital vasoconstriction through the liberation of norepinephrine (NE) from sympathetic nerve endings, which can be reduced by indirect heating. Injection of tyramine causes vasoconstriction by displacing NE from the nerves. Vasoconstriction can also be caused by the interaction of circulating NE or synthetic agonists such as phenylephrine (PHENYL) or clonidine (CLON) with α_1- and α_2-adrenergic receptors. Synthetic β-adrenergic agonists such as isoproterenol (ISO) cause vasodilation through interaction with β-receptors. Reactive hyperemia produces vasodilation through the accumulation of unknown compounds during ischemia.
Source. Reprinted with permission from Freedman RR: "Raynaud's Disease," in *Handbook of Clinical Psychophysiology.* Edited by Turpin G. London, Wiley, 1989, pp 469–495. Copyright 1989 by John Wiley & Sons Ltd.

both (Freedman and Ianni 1983) and are characterized by blanching followed by cyanosis and rubor.

Although the etiology of Raynaud's disease is unknown, two major theories have attempted to explain it. Raynaud (1888) felt that sympathetic nervous system hyperactivity caused an increased vasoconstrictor response to cold. Lewis (1929) hypothesized a "local fault" in which small peripheral blood vessels were oversensitive to local cooling. The present evidence tends to favor the theory of Lewis over that of Raynaud. Studies of plasma catecholamines in idiopathic Raynaud's disease have not generally supported Raynaud's theory since levels in patients were found to be lower (Surwit et al. 1983), higher (Peacock 1959), or not different from those in control subjects (Kontos and Wasserman 1969). Moreover, Fagius and Blumberg (1985) employed microelectrode recordings of skin nerve sympathetic activity and found no differences between Raynaud's patients and control subjects during reflex cooling.

We recently compared finger blood flow responses of Raynaud's disease patients and control subjects to dose-ranging brachial artery infusions of phenylephrine, an α_1-agonist; clonidine, an α_2-agonist; isoproterenol, a β-agonist; and tyramine, which causes vasoconstriction by releasing NE from sympathetic nerve endings (Freedman et al. 1989b). Twenty-eight patients with idiopathic Raynaud's disease—24 women and 4 men 22–69 years old—were matched by age and baseline finger blood flow levels with 24 control women and 5 control men, 22–66 years old. Subjects were supine in a 23°C room while finger blood flow was measured with venous occlusion plethysmography, and the above four compounds were given by pump, in successively doubling doses, through a brachial artery catheter. The doses employed did not change blood pressure or heart rate. Analyses of covariance showed that phenylephrine caused dose-related vasoconstriction ($P < .005$) that was greater in the patients than in the control subjects ($P < .002$). Similarly, dose-related vasoconstriction produced by clonidine ($P < .02$) was greater in patients than in control subjects ($P < .05$). There were no significant differences in responses to isoproterenol or tyramine. Thus, equal levels of circulating catecholamines would cause a greater degree of peripheral vasoconstriction in Raynaud's disease patients than in matched control subjects.

Lewis' (1929) evidence was based on the induction of vasospastic attacks in locally cooled fingers that had been anesthetized or sympathetically denervated. However, it is likely that most of his patients had advanced forms of the disease and that their vascular changes were therefore of a secondary nature. We tested Lewis' theory in a controlled laboratory investigation (Freedman et al. 1989a). Using a combination of environmental and local cooling, we induced vasospastic attacks of Raynaud's phenomenon in 9 of 11 patients with idiopathic Raynaud's disease and in 8 of 10 patients with scleroderma. Two fingers on one hand were anesthetized by local injection of lidocaine hydrochloride (Xylocaine), and the effectiveness of nerve blocks was verified by plethysmography. The frequency of vasospastic attacks in nerve-blocked fingers was not significantly different from that in the corresponding intact fingers on the contralateral hand. These findings show that the vasospastic attacks of Raynaud's disease and Raynaud's phenomenon can occur without the involvement of efferent digital nerves and argue against the etiologic role of sympathetic hyperactivity.

Thus, idiopathic Raynaud's disease appears to be characterized by increased responsiveness of peripheral vascular α-adrenergic receptors in the absence of abnormal elevations in generalized sympathetic activity. Animal studies (Flavahan et al. 1985) and human studies (Freedman and Sabharwal 1989) have shown that local cooling increases the responsiveness of peripheral vascular α_2-adrenoceptors, which could explain the triggering mechanism of Raynaud's disease vasospasms. This hypothesis is currently under investigation.

Behavioral Treatment of Raynaud's Disease

Despite the evidence that peripheral vascular α-adrenergic responsiveness is elevated in Raynaud's disease, it has not yet been possible to design pharmacological agents that will specifically target the digital blood vessels. Thus, side effects such as hypotension, headaches, and flushing have limited the usefulness of some drugs. In contrast, behavioral treatments for Raynaud's disease have been very promising and appear to have no adverse effects.

Behavioral treatments for Raynaud's disease have endeavored to increase finger temperature, or blood flow, or both and ameliorate symptoms through relaxation procedures such as autogenic training (Surwit et al. 1978) or through finger temperature biofeedback (Freedman et al. 1983,

1988). In the first controlled study (Surwit et al. 1978), patients were randomly assigned to receive either autogenic training (self-suggestion of warm imagery) alone or in combination with temperature feedback. In addition, for a 1-month period, half the subjects served as a waiting list control group for the other half and then received treatment. Subjects as a whole showed improvement on a cold stress test and reported decreased attack frequencies (10%–32%), but there were no group differences on these measures. Similar results were found in a subsequent study (Keefe et al. 1980) comparing progressive relaxation, autogenic training, or a combination of both.

Since the prior investigations utilized various combinations of behavioral procedures, we conducted a study (Freedman et al. 1983) in which the effects of temperature feedback alone were compared with those of autogenic training or frontalis EMG feedback. A fourth procedure was studied, in which temperature feedback was administered during mild cold stress to the finger. Patients who received temperature feedback alone or under cold stress showed significant elevations in finger temperature and significant declines in reported symptom frequency the following winter (66.8% and 92.5%, respectively). Those who received EMG feedback or autogenic training showed neither significant vasodilation nor symptomatic improvement, but did demonstrate declines in muscle tension, heart rate, and reported stress levels. Patients in the temperature feedback groups did not show declines in the latter three measures.

Thus, the effects of finger temperature biofeedback are physiologically different from those of autogenic training, frontalis EMG feedback, or simple instructions to increase finger temperature. Temperature feedback produces digital vasodilation without bradycardia or decreased muscle tension, whereas the other techniques do produce bradycardia and lower EMG levels but not increased finger temperature. Research has uncovered a β-adrenergic mechanism that may explain increased digital blood flow in the absence of decreased generalized physiological arousal (Cohen and Coffman 1981). We therefore tested the involvement of this mechanism during temperature feedback in control subjects and Raynaud's disease patients by local β-blockade of the vasodilation with intra-arterial infusions of propranolol (Freedman et al. 1988).

We first randomly assigned 18 patients with idiopathic Raynaud's disease and 16 control subjects to receive 10 sessions of temperature feedback or autogenic training. Following training, in a separate session, a catheter was placed in the right brachial artery using a local anesthetic and connected to two infusion pumps, one containing 0.9% saline solution and the other containing propranolol. The pumps were housed in a sound-proofed box and controlled remotely by the experimenter from a polygraph in a separate room. Finger blood flow was measured in both hands with venous occlusion plethysmography. After baseline recordings, the temperature-feedback signals or autogenic tape were activated. After 6 minutes, the infusion was switched from saline to propranolol (0.5 mg/min) for 2 minutes, then saline for 4 minutes, propranolol (same dose) for 2 minutes, and saline for 4 minutes. These changes occurred without the subject's knowledge. Significant bilateral vasodilation occurred in the temperature-feedback subjects and was significantly reduced by propranolol in the infused but not the control hand. The magnitudes of these effects were not significantly different in the Raynaud's disease and control subjects. There were no significant blood flow changes at all in the patients or control subjects who received autogenic training. There were no significant heart rate or blood pressure changes in the temperature-feedback group; the autogenic group showed a significant decline in heart rate during the autogenic instructions. Thus, a β-adrenergic mechanism is involved in feedback-induced vasodilation.

The only known efferent vasomotor nerves in human fingers are adrenergic (Figure 12–2); neurogenic vasoconstriction is caused by the interaction of released NE with postjunctional α-adrenergic receptors. Our finding of a β-adrenergic mechanism in temperature biofeedback thus raised the question of whether feedback-induced vasodilation is neurally mediated. Since it is possible to block the digital nerves by local injection of an anesthetic, a method was available to test this hypothesis. Since digital nerve blockade raises finger blood flow to near ceiling levels, we reduced it to mid-range by infusing NE (0.25 μg/min) in the right brachial artery. To control for all of the manipulations, we measured blood flow in three fingers: right, nerve-blocked plus NE; right, no block plus NE; and left, no block, no NE. We then repeated the propranolol infusion of the previous study (Figure 12–3). In two separate studies of control subjects ($N = 8$ and $N = 9$) and a subsequent study with Raynaud's disease patients ($N = 10$), we found that vasodilation produced by temperature feedback was not attenuated by nerve blockade or NE, but was reduced by propranolol. Thus, the β-adrenergic vasodilating mechanism of temperature feedback does not appear to be mediated through the digital nerves.

To examine possible changes in sympathetic nervous system activity during temperature feedback and autogenic training further, we recently measured plasma levels of epinephrine and NE during these procedures (Freedman et al. 1991). An intravenous needle was inserted into a vein on the back of the hand and connected to a Cormed blood withdrawal pump through nonthrombogenic tubing. The pump was located in an adjacent room, so that blood could be collected without observation by the subject.

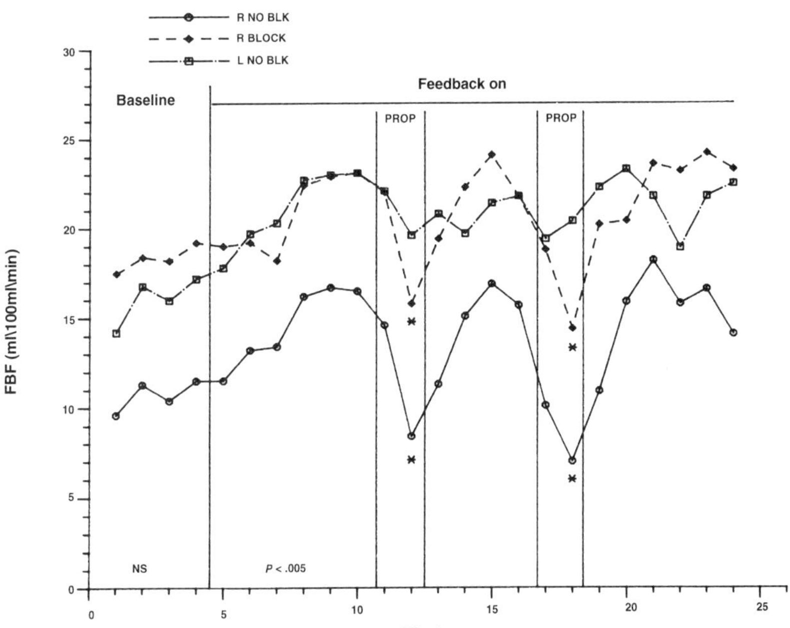

Figure 12–3. Blood flow (FBF) in three fingers during last 4 minutes of baseline period and subsequent temperature feedback period. Nerves in right second finger (R Block) were anesthetized. During baseline period FBF did not significantly change (NS). When feedback was activated, FBF in all three fingers increased significantly ($P < .005$). Propranolol (PROP) infused in right brachial artery (.5 mg for 2 minutes) caused significant ($P < .01$) reduction in FBF in infused hand (*) but not left hand.
Source. Reprinted with permission from Freedman RR, Sabharwal SC, Ianni P, et al: "Nonneural Beta-Adrenergic Vasodilating Mechanism in Temperature Biofeedback." *Psychosomatic Medicine* 50:394–401, 1988. Copyright 1988 by Williams & Wilkins.

Plasma levels of epinephrine and NE were subsequently analyzed by the high-performance liquid chromatography method with electrochemical detection. Thirty-one patients with idiopathic Raynaud's disease were randomly assigned to receive eight 32-minute sessions of finger temperature feedback ($N = 16$) or autogenic training ($N = 15$) over 28 days. During sessions 1 and 8, blood was continuously drawn. During training, significant temperature and blood flow elevations were shown by feedback patients ($P < .001$) but not by autogenic patients. There were no significant effects for NE and epinephrine for either group (Figure 12–4).

This investigation provides no evidence that sympathetic nervous system activation, as reflected by measures of plasma catecholamines, is reduced during behavioral treatments for primary Raynaud's disease. These results are consistent with previous findings that catecholamine levels are not consistently elevated in these patients (Kontos and Wasserman 1969; Surwit and Allen 1983), that vasospastic attacks can be induced during sympathetic nerve blockade (Freedman et al. 1989a), and that a sensitization of peripheral vascular α-adrenergic receptors is involved (Freedman et al. 1989b). Furthermore, our findings are consistent with evidence that feedback-induced vasodilation is mediated, in part, through a nonneural, β-adrenergic mechanism (Freedman et al. 1988). Future research should attempt to identify the ligand responsible for this vasodilation and its precise mechanism of action.

Conclusion

Research on the role of adrenergic mechanisms in behavioral treatments is still in its infancy. Some evidence exists suggesting that TM practice might be accompanied by elevated plasma NE levels, down-regulation of β_2-receptors, and reduced cardiac responsivity, but further work is clearly needed to substantiate this chain of events. In contrast, there is some evidence to show that PMR is accompanied by reduced NE and muscle sympathetic nerve activity levels, but adrenoceptors have not yet been studied in this technique. Evidence from our laboratory has implicated increased central NE activation as a trigger for menopausal hot flashes. We currently hypothesize that slow, deep breathing reduces this activation through changes in α_2-adrenoceptors, but this mechanism remains to be demonstrated. Finally, we have shown that Raynaud's disease is characterized by increased responsiveness of peripheral vascular α_1- and α_2-adrenoceptors, without elevations in plasma catecholamines or skin nerve sympathetic activity. Patients benefit from feedback-induced vasodilation, which operates, in part, through a β-adrenergic mechanism and not through reduced sympathetic activity. Further work on this mechanism is currently in progress.

Although it was formerly believed that most behavioral treatments were simply different forms of "relaxation," the work reviewed here suggests that that is not the case. The recent use of pharmacological probes to examine the physiological mechanisms of behavioral treatments has begun

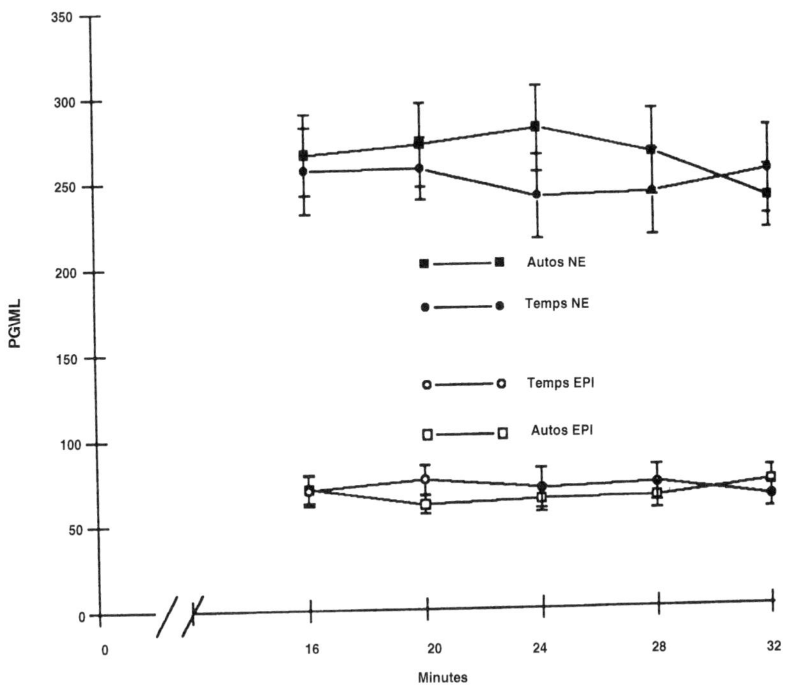

Figure 12–4. Plasma levels of norepinephrine (NE) and epinephrine (EPI) for both treatment groups, averaged across sessions 1 and 8. Means ± standard error.
Source. Reprinted with permission from Freedman RR, Keegan D, Migály P, et al: "Plasma Catecholamines During Behavioral Treatments for Raynaud's Disease." *Psychosomatic Medicine* 53:433–439, 1991. Copyright by Williams & Wilkins.

to implicate changes in adrenergic receptors that were previously unsuspected. The combination of biochemical and pharmacological methods with those of behavioral psychology holds great promise for future research.

References

Asch RH, Greenblatt RB: Steroidogenesis in the postmenopausal ovary. Clin Obstet Gynecol 4:85–106, 1977

Askel S, Schomberg DW, Tyrey L, et al: Vasomotor symptoms, serum estrogens, and gonadotropin levels in surgical menopause. Am J Obstet Gynecol 126:165–169, 1977

Bernstein DA, Borkovec TD: Progressive Relaxation Training: A Manual for the Helping Professions. Champaign, IL, Research Press, 1973

Brodde OE, Kretsch R, Ikezono K, et al: Human beta-adrenoreceptors: relation of myocardiol and lymphocyte beta-adrenoreceptor density. Science 231:1584–1585, 1986

Casper RF, Yen SSC: Neuroendocrinology of menopausal flushes: an hypothesis of flush mechanism. Clin Endocrinol (Oxf) 22:293–312, 1985

Casper RF, Yen SSC, Wilkes MM: Menopausal flushes: a neuroendocrine link with pulsatile luteinizing hormone secretion. Science 205:823–825, 1979

Charney DS, Heninger GR, Sternberg DE: Assessment of α_2 adrenergic autoreceptor function in humans: effects of oral yohimbine. Life Sci 30:2033–2041, 1982

Coffman JD: Total and nutritional blood flow in the finger. Clin Sci 42:243–250, 1972

Cohen R, Coffman J: Beta-adrenergic vasodilator mechanism in the finger. Circ Res 49:1196–1201, 1981

Cottier C, Shapiro K, Julius S: Treatment of mild hypertension with progressive muscle relaxation. Arch Intern Med 144:1954–1958, 1984

Cox B, Lomax P: Pharmacological control of temperature regulation. Annu Rev Pharmacol Toxicol 17:341–353, 1977

Eisenhofer G, Lambie DG, Johnson RH: β-Adrenoceptor responsiveness and plasma catecholamines as determinants of cardiovascular reactivity to mental stress. Clin Sci 69:483–492, 1985

Fagius J, Blumberg H: Sympathetic outflow to the hand in patients with Raynaud's phenomenon. Cardiovasc Res 19:246–253, 1985

Flavahan NA, Lindblad TJ, Verbeuren JT, et al: Cooling and α_1- and α_2-adrenergic responses in cutaneous veins: role of receptor reserve. Am J Physiol 249:H950–H955, 1985

Freedman RR, Ianni P: Role of cold and emotional stress in Raynaud's disease and scleroderma. BMJ 287:1499–1502, 1983

Freedman RR, Sabharwal SC: Cooling augments α_2 and depresses α_1-mediated vasoconstriction in man. Journal of Vascular Medicine and Biology 1:171, 1989

Freedman RR, Woodward S: Elevated α_2-adrenergic responsiveness in menopausal hot flashes, in Thermoregulation: The Pathophysiological Bases of Clinical Disorders. Edited by Lomax P, Schonbaum E. Basel, Karger, 1992a, pp 6–9

Freedman RR, Woodward S: Behavioral treatment of menopausal hot flashes: evaluation by ambulatory monitoring. Am J Obstet Gynecol 167:436–439, 1992b

Freedman RR, Ianni P, Wenig P: Behavioral treatment of Raynaud's disease. J Consult Clin Psychol 151:539–549, 1983

Freedman RR, Sabharwal SC, Ianni P, et al: Nonneural beta-adrenergic vasodilating mechanism in temperature biofeedback. Psychosom Med 50:394–401, 1988

Freedman RR, Sabharwal SC, Desai N, et al: Increased α-adrenergic responsiveness in idiopathic Raynaud's disease. Arthritis Rheum 33:61–65, 1989a

Freedman RR, Mayes MD, Sabharwal SC: Induction of vasospastic attacks despite digital nerve block in Raynaud's disease and phenomenon. Circulation 80:859–862, 1989b

Freedman RR, Woodward S, Sabharwal SC: α_2-Adrenergic mechanism in menopausal hot flushes. Obstet Gynecol 76:573–578, 1990

Freedman RR, Keegan D, Migály P, et al: Plasma catecholamines during behavioral treatments for Raynaud's disease. Psychosom Med 53:433–439, 1991

Gambone J, Meldrum DR, Laufer L, et al: Further delineation of hypothalamic dysfunction responsible for menopausal hot flashes. J Clin Endocrinol Metab 56:1097–1102, 1984

Germaine LM, Freedman RR: Behavioral treatment of menopausal hot flashes: evaluation by objective methods. J Consult Clin Psychol 52:1072–1079, 1984

Goldberg M, Robertson D: Yohimbine: a pharmacological probe for study of the α_2-adrenoceptor. Pharmacol Rev 35:143–180, 1983

Graafsma SJ, Lenders JW, Peters JH, et al: Role of adrenaline in the short-term upregulation of beta-adrenoceptors in essential hypertensive and adrenalectomized females. J Hypertens 6:S578–S580, 1988

Graafsma SJ, Lambertus JVT, van Heijst P, et al: Adrenoceptors on blood cells in patients with essential hypertension before and after mental stress. J Hypertens 7:519–524, 1989

Hjemdahl P, Fagius J, Freyschuss, et al: Muscle sympathetic activity and norepinephrine release during mental challenge in humans. Am J Physiol 257:E654–E664, 1989

Hoffman JW, Benson H, Arns PA, et al: Reduced sympathetic nervous system responsivity associated with the relaxation response. Science 215:190–192, 1982

Jacobson E: Progressive Relaxation. Chicago, IL, University of Chicago Press, 1938

Keefe F, Surwit R, Pilon R: Biofeedback, autogenic training, and progressive relaxation in the treatment of Raynaud's disease: a comparative study. Journal of Applied Behavioral Science 13:3–11, 1980

Kontos HA, Wasserman AJ: Effect of reserpine in Raynaud's phenomenon. Circulation 39:256–266, 1969

Kronenberg F, Cote LJ, Linkie DM, et al: Menopausal hot flashes: thermoregulatory, cardiovascular, and circulating catecholamine and LH changes. Maturitas 6:31–43, 1984

Lang R, Dehof K, Meurer KA, et al: Sympathetic activity and transcendental meditation. J Neural Transm 44:117–135, 1979

Leckman JF, Maas JW, Redmond E Jr, et al: Effects of oral clonidine on plasma 3-methoxy-4-hydroxyphenethyleneglycol (MHPG) in man: preliminary report. Life Sci 26:2179–2185, 1980

Lewis T: Experiments relating to the peripheral mechanism involved in spasmodic arrest of circulation in the fingers, a variety of Raynaud's disease. Heart 15:7–101, 1929

Mathew RJ, Ho BT, Kralik P, et al: Anxiety and platelet MAO levels after relaxation training. Am J Psychiatry 138:371–373, 1981

McCann SM, Moss RL: Putative neurotransmitters involved in discharging gonadotropin-releasing neurohormones and the action of LH-releasing hormone on the CNS. Life Sci 16:833–852, 1975

Mills PJ, Dimsdale JE: The promise of receptor studies in psychophysiologic research. Psychosom Med 50:555–566, 1988

Mills PJ, Schneider RH, Hill D, et al: Beta-adrenergic receptor sensitivity in subjects practicing transcendental meditation. J Psychosom Res 34:29–33, 1990a

Mills PJ, Dimsdale JE, Ziegler MG, et al: Beta-adrenergic receptors predict heart rate reactivity to a psychosocial stressor. Psychosom Med 52:621–623, 1990b

Morrell EM, Hollandswirth JG Jr: Norepinephrine alterations under stress conditions following the regular practice of meditation. Psychosom Med 48:270–277, 1986

Nagamani M, Kelver ME, Smith ER: Treatment of menopausal hot flashes with transdermal administration of clonidine. Am J Obstet Gynecol 156:561–565, 1987

Negro-Vilar A, Ojeda SR: In vitro release of somatostatin and LRH by MBH and ME fragments: effect of catecholamines. Federation Proceedings 37:445, 1978

Pacák K, Nedvídková J, Horváth M, et al: High cardiovascular reactivity to psychological challenge associated alternatively with high density of beta-adrenergic receptors or with high resting level of catecholamines. Act Nerv Super (Praha) 31:96–99, 1989

Peacock J: A comparative study of the digital cutaneous temperatures and hand blood flows in the normal hand, primary Raynaud's disease, and primary acrocyanosis. Clin Sci 8:25–33, 1959

Ravnikar V: Physiology and treatment of hot flushes. Obstet Gynecol 75:3S–17S, 1990
Raymond JR, Hnatowich M, Lefkowitz RJ, et al: Adrenergic receptors—models for regulation of signal transduction processes. Hypertension 15:119–131, 1990
Raynaud M: New Research on the Nature and Treatment of Local Asphyxia of the Extremities. Translated by Barlow T. London, The New Sydenham Society, 1888
Schaefer KE: Respiratory pattern and respiratory response to CO_2. J Appl Physiol 13:1–14, 1958
Shepherd JT: Physiology of the Circulation in Human Limbs in Health and Disease. Philadelphia, PA, WB Saunders, 1963
Starke K, Gothert M, Kilbringer H: Modulation of neurotransmitter release by presynaptic autoreceptors. Physiol Rev 69:864–989, 1989
Stone SC, Mickal A, Rye PH: Postmenopausal symptomatology, maturation index, and plasma estrogen levels. Obstet Gynecol 45:625–627, 1975
Surwit RS, Allen LM: Neuroendocrine response to cold in Raynaud's syndrome. Life Sci 32:995–1000, 1983
Surwit R, Pilon R, Fenton C: Behavioral treatment of Raynaud's disease. J Behav Med 1:323–325, 1978
Surwit R, Allen L, Gilgor R, et al: Neuroendocrine response to cold in Raynaud's syndrome. Life Sci 32:995–1000, 1983
Tataryn IV, Lomax P, Bajorek JG, et al: Postmenopausal hot flushes: a disorder of thermoregulation. Maturitas 2:101–107, 1980
Vanhoutte PM: Physical factors of regulation, in Handbook of Physiology, Vol 2. Edited by Bohr D, Somylo A, Sparks H. Baltimore, MD, American Physiological Society, 1980, pp 443–474
Vijayan E, McCann SM: Reevaluation of the role of catecholamines in control of gonadotropin and prolactin release. Neuroendocrinology 25:150–165, 1978
Wallace RK: Physiological effects of transcendental meditation. Science 167: 1751–1754, 1970

Section III

Psychiatric Disorders

13

Central Nervous System Adrenergic Function in Aging and Alzheimer's Disease

ALAN M. MELLOW, M.D., PH.D.
STEPHEN M. ARONSON, M.D.

The aging process has considerable impact on all brain neurotransmitter systems (Gottfries 1990; Simpkins and Millard 1987); adrenergic systems are no exception. An understanding of the effects of aging on adrenergic structure and function in the central nervous system (CNS) is particularly important since these systems have been implicated in functions such as mood, arousal, and memory, all of which may have alterations in expression in elderly individuals. In particular, there is a large body of data indicating alterations in adrenergic markers in Alzheimer's disease and related dementias, which constitute the most common neuropsychiatric syndromes in the elderly (Katzman and Jackson 1991; Katzman and Saitoh 1991).

In this chapter, we review 1) reports from both human pathological investigations, as well as biochemical, electrophysiologic, and behavioral studies from the animal literature, examining age-related adrenergic changes in the CNS; 2) histopathological and biochemical data describing changes in adrenergic systems in Alzheimer's disease; and 3) the possible role of adrenergic systems in human memory dysfunction in old age, and the clinical implications for the treatment of Alzheimer's disease.

Age-Related CNS Adrenergic Changes

Pathology Studies

The most notable finding of pathology in aging brains, which is thought to be the most important substrate for altered adrenergic function, is the loss of neurons in the locus coeruleus. The locus coeruleus is a collection of norepinephrine-containing neurons in the upper pons that projects widely to cortical and subcortical structures (Amaral and Sinnamon 1977). Several quantitative and topographic studies (Marcyniuk et al. 1989; Tomonaga 1983; Vijayashankar and Brody 1979) have demonstrated significant age-related decreases in locus coeruleus neurons, in the absence of neurological disease.

Neurochemical Studies

Biochemical indices of adrenergic function, including brain norepinephrine levels (Finch 1973; Goldman-Rakic and Brown 1981) and noradrenergic synthetic enzymes (McGeer et al. 1971; Reis et al. 1977), have been noted to be significantly decreased with age in a variety of brain regions in several animal species and in humans. The data concerning changes in adrenergic receptors in aging brains are, however, somewhat controversial. The availability of structurally specific radioligands has made possible the examination of the binding characteristics of the several known subtypes of adrenergic receptors. Reports of changes in α-adrenergic receptors have been conflicting. Early studies showed decreases in both α_1- and α_2-adrenergic brain receptor binding in aged animals (Hamilton et al. 1984; Misra et al. 1980), but several investigators failed to find such changes (De Blasi et al. 1982; Maggi et al. 1979). In a study using the highly specific agonist ligand ^{3}H-labeled *p*-aminoclonidine, Kalaria and Andorn (1991) demonstrated a significant decrease in α_2-receptor binding in prefrontal cortex in aged human brain. Such a finding is consistent with the loss of presynaptic receptors located on terminals from neurons originating in the locus coeruleus. The same group found no age-related change in α_1-receptor binding in rat (Harik et al. 1991) and human (Kalaria 1989) cerebral cortex. In a quantitative autoradiographic regional analysis, and using a different radioligand, Burnett and Zahniser (1989) found significant age-correlated decreases in α_1-receptor density in the thalamus and olfactory tubercle of the rat. They also found decreases in cerebral cortex, but only in the oldest age group.

With respect to β-adrenergic receptor binding, the data are also mixed. Some investigators (De Blasi et al. 1982; Misra et al. 1980) have shown age-related decreases in β-receptor binding in rat brain; others have not (Harik et al. 1991; Maggi et al. 1979). Pittman et al. (1980) found no changes in cerebral cortex, but in cerebellum found decreases in β_2-receptor binding and increases in β_1-receptor binding. Kalaria et al. (1989) found a weak negative correlation with age for β_1-receptor binding in human prefrontal cortex.

Clearly, more work is needed to clarify the nature of the changes in adrenergic binding sites in the aging brain and to gain more insight into species differences. It has been suggested that some of the discrepancies found may be explained by the presence of both agonist and antagonist configurations, at least for some α-receptors (Gelbmann and Muller 1990). It is also important to note that measurements of specific binding sites may yield incomplete data on the functional consequences of such changes, since there may exist age-related changes in postreceptor signal transduction mechanisms. This is a less well studied area of research; age-related changes have been reported, however, for adrenergic receptor-mediated phosphoinositide turnover (Burnett et al. 1990) and modulation of adenylate cyclase activity (Nomura et al. 1986; Scarpace and Abrass 1988).

Since aging is often associated with diminished adaptation, it is reasonable to ask, in the context of receptor binding studies, whether the capacity for regulation of adrenergic receptors is altered with the aging process. Greenberg and Weiss (1978) found a decreased capacity of rat pineal β-receptors to upregulate in response to light exposure. In a separate study (Greenberg and Weiss 1979), the same authors found aged rats to have impaired ability to increase β-receptor density in pineal, cerebral cortex, and cerebellum, in response to presynaptic catecholamine depletion with reserpine. In a study measuring isoproterenol-induced cyclic adenosine monophosphate (cAMP) generation in brain slices, Harik and Sylvia (1985) found aged rats to have diminished ability to increase the receptor-mediated response after denervation. In contrast, Harik et al. (1991) found that aged rats had preserved ability to upregulate both α-receptors *and* β-receptors in cerebral cortex in response to denervation by locus coeruleus lesioning. Thus, CNS adrenergic receptor adaptation is impaired with aging in some, but not all, cases.

To summarize the known neurochemical data, although there is some variation depending on species and experimental conditions, the overall evidence is for a significant decrease in the biochemical substrates of central adrenergic neurotransmission with increasing age.

Electrophysiological Studies

As might be predicted from the previous data, most studies examining the functional status of adrenergic neurons indicate a decrease in activity with age. Jones and Olpe (1983) found rostral cerebral cortical neurons in aged rats to have a diminished electrophysiological response to iontophoretically applied norepinephrine. Bickford-Wimer et al. (1988) found a progressive age-related decline in postsynaptic responsiveness to local application of norepinephrine to hippocampal pyramidal cells in the rat. In a pharmacological study, Parfitt and Bickford-Wimer (1990) found an age-related decline in the functional sensitivity of cerebellar Purkinje neurons to a β_2-adrenergic agonist, with preservation of the response to a β_1-agonist. Using the model of the Purkinje neuron, the same investigators examined electrophysiological correlates of receptor adaptation in young and aged rats (Bickford-Wimer et al. 1987). In young rats, chronic desipramine administration enhanced noradrenergic function, as measured by the electrophysiological response to norepinephrine. Aged rats showed no enhancement in response after desipramine treatment, thus providing more evidence of age-related impairment in adrenergic adaptation.

Behavioral Studies

Although the brain cholinergic system is thought to be the most prominent in mechanisms of learning and memory, there is some evidence for involvement of adrenergic mechanisms (Leslie et al. 1985; Pontecorvo et al. 1988). In addition, there are data suggesting that adrenergic interventions may affect memory function in aged animals. Arnsten and Goldman-Rakic (1985) demonstrated a partial reversal of cognitive deficit in aged rhesus monkeys after treatment with the α_2-adrenergic agonist clonidine. In a later study (Arnsten et al. 1988), similar results were obtained with a related agent, guanfacine, although in a study in rats (Sirvio et al. 1991), memory improvement was not noted. When taken in combination with data regarding adrenergic deficits in Alzheimer's disease (see later), these results may have implications for adrenergic interventions in human memory disorders (McEntee and Crook 1990).

CNS Adrenergic Changes in Alzheimer's Disease

Pathology and Neurochemical Studies

As with studies of normal memory processes, the cholinergic system has dominated work on Alzheimer's disease for the past decade (Davidson et al. 1991; Giacobini 1990). There is, however, considerable evidence for changes in adrenergic structure and function in this disorder. In a number of quantitative pathological studies (Bondareff et al. 1982; Mann et al. 1981; Tomlinson et al. 1981), locus coeruleus neuron loss was noted to be significantly more than accounted for by the aging process alone, and this was specific, not being observed in multi-infarct dementia (Mann et al. 1981). Neurochemical studies have repeatedly demonstrated significant depletion of norepinephrine from a variety of brain areas receiving locus coeruleus innervation (Arai et al. 1984; Baker and Reynolds 1989). As with the cholinergic system, norepinephrine depletion has been reported to correlate with the degree of intellectual deterioration of Alzheimer's disease patients (Adolfsson et al. 1979), although this may not be as robust a finding as with the former, since loss of dopamine β-hydroxylase (a noradrenergic synthetic enzyme) activity was found not to correlate with clinical severity (Perry et al. 1981).

Adrenergic receptor binding in Alzheimer's disease has been examined in detail. Shimohama et al. (1986) found reductions in α_1-receptors in hippocampus and cerebellum and reductions in α_2-receptors in the nucleus basalis of Meynert. In contrast, Kalaria (1989) found significant loss of α_1-receptor binding in prefrontal cortex, but not in hippocampus or cerebellum. In a separate study, Kalaria and Andorn (1991) found α_2-receptor binding diminished selectively in prefrontal cortex. As in normal aging, this was postulated as being due to loss of presynaptic receptors from locus coeruleus inputs, but the authors could not rule out selective postsynaptic deficits. In studies of β-adrenergic receptors, Shimohama et al. (1987) found significant decreases in Alzheimer's disease patients in β_1-receptors in hippocampus and increases in the nucleus basalis of Meynert and cerebellum. They also found reduced β_2-receptor binding in thalamus, the nucleus basalis of Meynert, and cerebellum, and increased binding in hippocampus and putamen. Similarly, Kalaria et al. (1989) found increases in β_2-receptor binding in hippocampus and prefrontal cortex. It is postulated that the increases in β_2-receptors found in hippocampus represent an up-regulation, secondary to deafferentation from locus coeruleus inputs; it is further suggested that this finding may be specific for Alzheimer's disease, rather than on a continuum of the aging process.

In summary, there is a large body of histopathological and biochemical data suggesting significant deficits in CNS adrenergic markers in Alzheimer's disease. Changes in adrenergic receptor binding are consistent, in part, with the loss of noradrenergic inputs from locus coeruleus neurons.

Clinical Implications

Although strategies for treatment in Alzheimer's disease have traditionally emphasized enhancement of cholinergic function (Davidson et al. 1991), there is now considerable interest in alternative interventions (Davidson and Stern 1991). The adrenergic deficits in Alzheimer's disease cited earlier, as well as the possible role for adrenergic systems in age-related memory loss (see earlier), has prompted investigations into the clinical efficacy of adrenergic agents in this disorder (Schlegel et al. 1989). In an intriguing line of work, Haroutunian et al. (1990) and Santucci et al. (1991) have examined memory deficits in animals with combined noradrenergic and cholinergic CNS lesions. These deficits were optimally restored after combination treatment with both cholinergic and adrenergic agents. Since dual lesions in both transmitter systems may better approximate the state of the brain in Alzheimer's disease, this combined pharmacotherapy strategy is now being pursued by several groups (Davidson et al. 1989; Sunderland et al. 1992).

Conclusion

Adrenergic function in the CNS undergoes significant change during the aging process, primarily with an overall decrease in activity, as well as impairment in adaptive responses. In addition, reduction in CNS adrenergic function may contribute to age-related memory impairment. Finally, there are significant deficits in markers of adrenergic function in the brains of patients with Alzheimer's disease. A more detailed understanding of the nature of these phenomena may lead to more effective treatments of psychiatric syndromes in the elderly, and particularly may serve as a basis for novel pharmacotherapeutic strategies in Alzheimer's disease.

References

Adolfsson R, Gottfries CG, Roos BE, et al: Changes in the brain catecholamines in patients with dementia of Alzheimer type. Br J Psychiatry 135:216–223, 1979

Amaral DG, Sinnamon HM: The locus coeruleus: neurobiology of a central noradrenergic nucleus. Prog Neurobiol 9:147–196, 1977

Arai H, Kosaka K, Iizuka R: Changes of biogenic amines and their metabolites in postmortem brains from patients with Alzheimer-type dementia. J Neurochem 43:388–393, 1984

Arnsten AF, Goldman-Rakic P: Alpha 2-adrenergic mechanisms in prefrontal cortex associated with cognitive decline in aged nonhuman primates. Science 230:1273–1276, 1985

Arnsten AF, Cai JX, Goldman-Rakic PS: The alpha-2 adrenergic agonist guanfacine improves memory in aged monkeys without sedative or hypotensive side-effects: evidence for alpha-2 receptor subtypes. J Neurosci 8:4287–4298, 1988

Baker GB, Reynolds GP: Biogenic amines and their metabolites in Alzheimer's disease: noradrenaline, 5-hydroxytryptamine and 5-hydroxyindole-3-acetic acid depleted in hippocampus but not in substantia innominata. Neurosci Lett 100:335–339, 1989

Bickford-Wimer PC, Parfitt K, Hoffer BJ, et al: Desipramine and noradrenergic neurotransmission in aging: failure to respond in aged laboratory animals. Neuropharmacology 26:597–605, 1987

Bickford-Wimer PC, Miller JA, Freedman R, et al: Age-related reduction in responsiveness of rat hippocampal neurons to locally applied monoamines. Neurobiol Aging 9:173–199, 1988

Bondareff W, Mountjoy CQ, Roth M: Loss of neurons of origin of the adrenergic projection to cerebral cortex (nucleus locus ceruleus) in senile dementia. Neurology 32:164–168, 1982

Burnett DM, Zahniser NR: Region-specific loss of alpha 1-adrenergic receptors in rat brain with aging: a quantitative autoradiographic study. Synapse 4:143–155, 1989

Burnett DM, Bowyer JF, Masserano JM, et al: Effect of aging on alpha-1 adrenergic stimulation of phosphoinositide hydrolysis in various regions of rat brain. J Pharmacol Exp Ther 255:1265–1270, 1990

Davidson M, Stern RG: The treatment of cognitive impairment in Alzheimer's disease: beyond the cholinergic approach. Psychiatr Clin North Am 14:461–482, 1991

Davidson M, Bierer LM, Kaminsky R, et al: Combined administration of physostigmine and clonidine to patients with dementia of the Alzheimer type: a pilot safety study. Alzheimer Dis Assoc Disord 3:224–227, 1989

Davidson M, Stern RG, Bierer LM, et al: Cholinergic strategies in the treatment of Alzheimer's disease. Acta Psychiatr Scand Suppl 366:47–51, 1991

De Blasi A, Cotecchia S, Mennini T: Selective changes of receptor binding in brain regions of aged rats. Life Sci 31:335–340, 1982

Finch CE: Catecholamine metabolism in the brains of ageing male mice. Brain Res 52:261–276, 1973

Gelbmann CM, Muller WE: Specific decrease of high-affinity agonist states of alpha 2-adrenoceptors in the aging mouse brain. Journal of Neural Transmission: General Section 79:131–136, 1990

Giacobini E: The cholinergic system in Alzheimer disease. Prog Brain Res 84:321–332, 1990

Goldman-Rakic P, Brown RM: Regional changes of monoamines in cerebral cortex and subcortical structures of aging rhesus monkeys. Neuroscience 6:177–187, 1981

Gottfries CG: Neurochemical aspects on aging and diseases with cognitive impairment. J Neurosci Res 27:541–547, 1990

Greenberg LH, Weiss B: Beta-adrenergic receptors in aged rat brain: reduced number and capacity of pineal gland to develop supersensitivity. Science 201:61–63, 1978

Greenberg LH, Weiss B: Ability of aged rats to alter beta adrenergic receptors of brain in response to repeated administration of reserpine and desmethylimipramine. J Pharmacol Exp Ther 211:309–316, 1979

Hamilton CA, Howe CA, Reid JL: Changes in brain alpha-adrenoceptors with increasing age in rabbits. Brain Res 322:177–179, 1984

Harik SI, Sylvia AL: Effect of aging on the response of the cerebral cortex to noradrenergic denervation. Gerontology 31:285–292, 1985

Harik SI, Sromek SM, Kalaria RN: Alpha- and beta-adrenergic receptors of the rat cerebral cortex and cerebral microvessels in aging, and their response to denervation. Neurobiol Aging 12:567–573, 1991

Haroutunian V, Kanof PD, Tsuboyama G, et al: Restoration of cholinomimetic activity by clonidine in cholinergic plus noradrenergic lesioned rats. Brain Res 507:261–266, 1990

Jones RS, Olpe HR: Altered sensitivity of forebrain neurones to iontophoretically applied noradrenaline in aging rats. Neurobiol Aging 4:97–99, 1983

Kalaria RN: Characterization of [^{125}I]HEAT binding to alpha 1-receptors in human brain: assessment in aging and Alzheimer's disease. Brain Res 501:287–294, 1989

Kalaria RN, Andorn AC: Adrenergic receptors in aging and Alzheimer's disease: decreased alpha 2-receptors demonstrated by [^{3}H]*p*-aminoclonidine binding in prefrontal cortex. Neurobiol Aging 12:131–136, 1991

Kalaria RN, Andorn AC, Tabaton M, et al: Adrenergic receptors in aging and Alzheimer's disease: increased beta 2-receptors in prefrontal cortex and hippocampus. J Neurochem 53:1772–1781, 1989

Katzman R, Jackson JE: Alzheimer disease: basic and clinical advances. J Am Geriatr Soc 39:516–525, 1991

Katzman R, Saitoh T: Advances in Alzheimer's disease. FASEB J 5:278–286, 1991

Leslie FM, Loughlin SE, Sternberg DB, et al: Noradrenergic changes and memory loss in aged mice. Brain Res 359:292–299, 1985

Maggi A, Schmidt MJ, Ghetti B, et al: Effect of aging on neurotransmitter receptor binding in rat and human brain. Life Sci 24:367–373, 1979

Mann DMA, Yates PO, Hawkes J: The noradrenergic system in Alzheimer and multi-infarct dementias. J Neurol Neurosurg Psychiatry 45:113–119, 1981

Marcyniuk B, Mann DMA, Yates PO: The topography of nerve cell loss from the locus coeruleus in elderly persons. Neurobiol Aging 10:5–9, 1989

McEntee WJ, Crook TH: Age-associated memory impairment: a role for catecholamines. Neurology 40:526–530, 1990

McGeer EG, Fibiger HC, McGeer PL, et al: Aging and brain enzymes. Exp Gerontol 6:391–396, 1971

Misra CH, Shelat HS, Smith RC: Effect of age on adrenergic and dopaminergic receptor binding in rat brain. Life Sci 27:521–526, 1980

Nomura Y, Kitamura Y, Kawai M, et al: Alpha 2-adrenoceptor-GTP binding regulatory protein-adenylate cyclase system in cerebral cortical membranes of adult and senescent rats. Brain Res 379:118–124, 1986

Parfitt KD, Bickford-Wimer P: Age-related subsensitivity of cerebellar Purkinje neurons to locally applied beta 1-selective adrenergic agonist. Neurobiol Aging 11:591–596, 1990

Perry EK, Tomlinson BE, Blessed G, et al: Neuropathological and biochemical observations on the noradrenergic system in Alzheimer's disease. J Neurol Sci 51:279–287, 1981

Pittman RN, Minneman KP, Molinoff PB: Alterations in beta 1- and beta 2-adrenergic receptor density in the cerebellum of aging rats. J Neurochem 35:273–275, 1980

Pontecorvo MJ, Clissold DB, Conti LH: Age-related cognitive impairments as assessed with an automated repeated measures memory task: implications for the possible role of acetylcholine and norepinephrine in memory dysfunction. Neurobiol Aging 9:617–625, 1988

Reis DJ, Ross RA, Joh TH: Changes in the activity and amounts of enzymes synthesizing catecholamines and acetylcholine in brain, adrenal medulla, and sympathetic ganglia of aged rat and mouse. Brain Res 136:465–474, 1977

Santucci AC, Haroutunian V, Davis KL: Pharmacological alleviation of combined cholinergic/noradrenergic lesion-induced memory deficits in rats. Clin Neuropharmacol 14 (suppl 1):S1–S8, 1991

Scarpace PJ, Abrass IB: Alpha- and beta-adrenergic receptor function in the brain during senescence. Neurobiol Aging 9:53–58, 1988

Schlegel J, Mohr E, Williams J, et al: Guanfacine treatment of Alzheimer's disease. Clin Neuropharmacol 12:124–128, 1989

Shimohama S, Taniguchi T, Fujiwara M, et al: Biochemical characterization of alpha-adrenergic receptors in human brain and changes in Alzheimer-type dementia. J Neurochem 47:1295–1301, 1986

Shimohama S, Taniguchi T, Fujiwara M, et al: Changes in beta-adrenergic receptor subtypes in Alzheimer-type dementia. J Neurochem 48:1215–1221, 1987

Simpkins JW, Millard WJ: Influence of age on neurotransmitter function. Endocrinol Metab Clin North Am 16:893–917, 1987

Sirvio J, Riekkinen PJ, Vajanto I, et al: The effects of guanfacine, alpha-2 agonist, on the performance of young and aged rats in spatial navigation task. Behav Neural Biol 56:101–107, 1991

Sunderland T, Molchan SE, Lawlor BA, et al: A strategy of "combination chemotherapy" in Alzheimer's disease: rationale and preliminary results with physostigmine and deprenyl. International Psychogeriatrics 4 (suppl 2):291–309, 1992

Tomlinson BE, Irving D, Blessed G: Cell loss in the locus coeruleus in senile dementia of Alzheimer type. J Neurol Sci 49:419–428, 1981

Tomonaga M: Neuropathology of the locus ceruleus: a semi-quantitative study. J Neurol 230:231–240, 1983

Vijayashankar N, Brody H: A quantitative study of the pigmented neurons in the nuclei locus coeruleus and subcoeruleus in man as related to aging. J Neuropathol Exp Neurol 38:490–497, 1979

14

Norepinephrine in Schizophrenia

JOHN A. GURKLIS, JR., M.D.
DANIEL P. VAN KAMMEN, M.D., PH.D.

Since the discovery that they affect behavior, monoamines have been studied in the major psychiatric disorders. Early on, norepinephrine (NE) was hypothesized to be related to arousal in schizophrenia. However, because clinical response and potency of antipsychotic agents were correlated with dopamine receptor binding rather than with NE α-receptor binding, dopamine was hypothesized to be the main neurotransmitter involved in schizophrenia. This hypothesis has been useful in developing new antipsychotic agents and for structuring the study of schizophrenia. So far, dopamine-related studies have raised more questions than answers. Other neurotransmitter systems such as those involving NE (Hornykiewicz 1982), serotonin (van Praag et al. 1987), glutamate (Javitt and Zukin 1991), gamma-aminobutyric acid (Roberts 1973; van Kammen and Gelernter 1987), and the neuropeptides have been investigated to determine their role in schizophrenia. In this chapter, we examine the roles of NE in schizophrenia.

Although some investigators have suggested that a primary NE disturbance could explain schizophrenia, most clinical researchers have tested the involvement of NE to extend the dopamine hypothesis. Further, some have suggested a more integrated view of brain function where the various neurotransmitter systems are interconnected. Hence it is possible that a complicated illness such as schizophrenia may come about because of changes in the dynamic interactions of these neurotransmitters.

In this context, we explore the data on disturbances of the NE system that have been reported in schizophrenia. We review data from autopsy, cerebrospinal fluid (CSF), plasma, and urine studies as well as studies that look at the effects of drugs on the NE system. We selectively review some of the animal literature that highlight interactions between NE and other neurotransmitter systems, particularly dopamine. Finally, we attempt to formulate a current hypothesis on the role of NE in schizophrenia.

Autopsy Studies

Autopsy studies have provided evidence of NE abnormalities in schizophrenia. Farley et al. (1978) were the first to examine NE in the brains of patients with schizophrenia. They found that NE levels were elevated in the ventral septum, the bed of the stria terminalis, the nucleus accumbens, and the mammillary bodies in 4 patients with chronic paranoid schizophrenia compared with 12 control subjects. They did not find changes in the hypothalamus or the medial olfactory area. These results suggested a possible malfunction of limbic noradrenergic mechanisms in paranoid schizophrenia. Bird et al. (1979) examined the brains of 46 schizophrenic patients and 31 control subjects and found that there were significantly increased levels of dopamine and NE in the nucleus accumbens and anterior perforated substance in the schizophrenic patients; they did not find differences in the caudate, septal nuclei, and amygdala central nucleus. The increases in NE were significant only if parametric statistics were used; the dopamine findings were significant with both parametric and nonparametric tests. Levels of dopamine and NE in the anterior perforated substance were found to correlate positively. Kleinman et al. (1982) examined the brains of 10 patients with chronic paranoid schizophrenia, 8 with chronic undifferentiated schizophrenia, 14 control subjects, and 8 patients with other psychiatric disorders, measuring catecholamines and their metabolites. They did not observe differences in the dopamine or dopamine metabolite levels. NE and 3-methoxy-4-hydroxyphenylglycol (MHPG) concentrations were increased in the nucleus accumbens of the chronic paranoid schizophrenic patients compared with the other groups. NE and MHPG were also increased in the hypothalamus of the chronic paranoid schizophrenic and chronic undifferentiated patients compared with control subjects. The finding that the NE increases were stronger in the chronic paranoid schizophrenic patients than in the chronic undifferentiated schizophrenic patients, while both groups had similar medication exposure, argued against the idea that these findings were caused by neuroleptic treatment alone. Bridge et al. (1985) examined brains from 13 elderly schizophrenic and 6 elderly control patients and found that hypothalamic NE was increased in the schizophrenic patients compared with control subjects. Demented schizophrenic patients had lower hypothalamic NE than nondemented schizophrenic patients; however, even the demented schizophrenic patients had higher NE levels than control subjects.

Lohr and Jeste (1988) examined the morphology of the locus coeruleus (LC) in 13 control subjects, 15 leukotomized schizophrenic patients, and 11 leukotomized nonschizophrenic patients. The LC is a pigmented brainstem nucleus that is one of the major NE-containing structures in the brain. The authors found an age-related neuronal loss in the LC and no significant differences in the different groups. However, they found a trend for decrease in the LC volume in the schizophrenic group compared with the other groups, and this trend did reach statistical significance when only those schizophrenic patients who met DSM-III (American Psychiatric Association 1980) criteria for schizophrenia were considered.

Taken as a group, these studies point to a potential altered NE status in the brains of patients with schizophrenia.

Cerebrospinal Fluid Studies

CSF studies of schizophrenic patients and control subjects have shown abnormalities in the concentration of NE in schizophrenic patients. Lake et al. (1980) compared CSF NE concentrations in 29 control subjects and

35 schizophrenic patients who had been drug free at least 2 weeks and found that schizophrenic patients had significantly higher CSF NE than control subjects and that this increase was especially pronounced in paranoid schizophrenic patients. The authors speculated that overflow from the periventricular regions may reflect abnormally high release or diminished enzymatic destruction of NE in schizophrenic patients. Kemali et al. (1982) compared CSF NE, dopamine, and epinephrine in 10 control subjects and 8 schizophrenic patients who had been drug free for at least 15 days. They found that NE was significantly increased in the schizophrenic patients, especially the paranoid patients, whereas dopamine and epinephrine were not significantly different. The authors speculated that the increased CSF NE levels reflected a rise of central NE activity but were uncertain about the pathophysiological significance. Gomes et al. (1980) examined CSF concentrations of NE, 5-hydroxyindoleacetic acid (5-HIAA), homovanillic acid (HVA), and cyclic adenosine monophosphate (cAMP) in 10 patients with acute schizophrenia, 11 patients with chronic schizophrenia, 19 patients with psycho-organic syndromes, 20 patients with personality disorders, and 18 control subjects. They found a significant increase in NE and cAMP in the group of chronic schizophrenic patients compared with all other groups. The authors concluded that medications were not the cause of the increased NE as about half of the psychiatric patients were on medications and half were not at the time of the lumbar puncture. However, all 11 chronic patients were on antipsychotic agents. The authors interpreted the results as evidence of noradrenergic overactivity, which may possibly reflect a primary abnormality in schizophrenia. However, they also acknowledged that because there were complex interactions among other neurotransmitters, their findings could represent secondary phenomena.

There are also some studies that do not show differences in CSF NE or attribute such differences to neuroleptic medications. Gattaz et al. (1983) studied the CSF of 28 schizophrenic patients—13 who were unmedicated for at least 4 weeks and 15 who were medicated with neuroleptics at least 3 weeks—as well as 15 control subjects. There was a trend ($P < .10$) for NE to be elevated in the unmedicated schizophrenic patients. However, in the medicated schizophrenic patients, NE was significantly elevated compared with the control subjects. CSF dopamine was not different between the control subjects and unmedicated schizophrenic patients but was increased in the medicated patients. The authors concluded that the increased dopamine and NE could be due in part to neuroleptic medication, ignoring the trend of higher NE in the unmedicated patients compared with the control subjects. They also noted a positive correlation between CSF levels of dopamine and NE, supporting a relationship between these two neurotransmitter systems.

Linnoila et al. (1983) studied the reliability of CSF measures of NE and other metabolites in 15 schizophrenic patients using repeated drug-free lumbar punctures. Their main finding was that there was considerable intraindividual variation in the concentrations of NE and the major monoamine metabolites from one drug-free evaluation to another. They were not able to find reasons for the variability at the time. More about this variability is discussed later.

State-dependent changes in NE could be one possible explanation for this variability. van Kammen et al. (1989a) studied 32 schizophrenic patients both stabilized on haloperidol and after the patients were drug free for up to 6 weeks. In the drug-free phase, patients were divided into those that relapsed in that time and those that did not. Relapsing patients higher had CSF NE and MHPG than nonrelapsing patients; there were no differences between the two groups in CSF levels of HVA or 5-HIAA. Of interest is that in the initial haloperidol phase prior to the drug-free phase, patients who would go on to relapse already had higher CSF NE than those who would not relapse. There were no significant differences in MHPG, HVA, or 5-HIAA between these patients on haloperidol. The authors concluded that elevated CSF NE levels during neuroleptic treatment may predict behavioral decompensation after discontinuation of medication. In an extension of this work, van Kammen et al. (1990) found that in relapsing patients, increased CSF NE and MHPG were significantly positively correlated with increases in negative symptoms and psychosis ratings. In relapsing patients, NE was elevated prior to and after relapse; in the nonrelapsing patients, NE was not elevated prior to haloperidol withdrawal and actually declined following withdrawal.

The interpretations of the results of these studies have ranged from showing evidence for noradrenergic overactivity as a possible primary abnormality in chronic schizophrenia (Gomes et al. 1980), to an increase in central noradrenergic activity as a finding of uncertain significance (Kemali et al. 1982; Lake et al. 1980), to findings suggesting that the increases were due to lingering or active neuroleptic effects (Gattaz et al. 1983). There are also questions about the stability of CSF measurements over time, and one possible explanation may be state-dependent changes in neurotransmitter levels.

Plasma Norepinephrine Studies

Plasma studies of NE show that plasma NE is often elevated in patients with schizophrenia. Although plasma is more accessible than CSF, plasma NE has a short half-life (2–3 minutes), fluctuates rapidly, and does not represent brain NE but mainly NE from systemic vascular and nonvascular sources. Actually arterial NE may better reflect autonomic nervous activity. Kemali et al. (1982) looked at plasma NE and dopamine in 46 schizophrenic patients who had been drug free for at least 15 days, 45 healthy control subjects, and 10 neurological control subjects. Plasma NE levels were found to be significantly higher in the schizophrenic patients, especially those with paranoid schizophrenia. They ascribed the increased plasma NE to the higher nonspecific arousal level seen in schizophrenic patients, some of whom had increased pulse rates. Other possibilities for the increase were medication effects or the stress of the blood drawing procedure, especially in the paranoid patients. In further work, Kemali et al. (1985) examined the 24-hour patterns of plasma NE, epinephrine, dopamine, prolactin, cortisol, and growth hormone in 23 schizophrenic patients (2 weeks drug free) and in 10 psychiatric control subjects. They found a significant increase in plasma NE but no increases in epinephrine, dopamine, or prolactin. The increase in plasma NE was seen only during the waking hours, and this was interpreted as an indication that the state of increased autonomic arousal may be tied to the stress of daytime social interactions.

Dajas et al. (1983) studied plasma NE in 32 acute schizophrenic patients and control subjects. Five of the patients were drug free. Plasma NE was significantly higher in the patients, on or off drug, compared with the control subjects. Plasma NE correlated significantly with global psychopathology, paranoia, and positive symptoms, but not with negative symptoms or hallucinatory symptomatology. The five patients who had been drug free had a second blood draw; plasma NE had decreased significantly but was still above normal. The authors concluded that high levels of sympathetic activity can be found in schizophrenia in association with acute symptomatology, especially paranoid symptoms. Albus et al. (1982) studied 12 chronic schizophrenic patients who were drug free at least 4 weeks and 63 control subjects on measures of autonomic functions and situational reactivity. They found that during resting conditions the schizophrenic patients had significantly elevated plasma NE and cortisol levels. These patients had a higher baseline autonomic arousal level as indicated by increased heart rate, electromyographic activity, and skin conductance levels. During activating tasks such as mental arithmetic, control subjects had increases in plasma NE and increases in heart rate. Schizophrenic patients did not show any rise in plasma NE or in heart rate. The authors concluded that the schizophrenic patients showed higher nonspecific activation at baseline testing and that they had a reduced ability to react to external stress. The question of medication effects on plasma NE was studied by Naber et al. (1985). They studied 36 patients with chronic paranoid schizophrenia during maintenance treatment and after 12 days of medication withdrawal. They noted that during the medication phase, plasma NE was elevated, which came down after 12 days of medication withdrawal but was still elevated above the levels of the control subjects. Of note there was much variation in the NE levels and the clinical status of the patients off medication. The withdrawal condition showed some patients improving and others worsening.

In all, plasma studies show data for increased plasma NE in schizophrenia. The increased plasma NE is often ascribed to general overarousal, but sometimes is correlated with psychopathology ratings. Although this is an interesting measure, how well it reflects central activity remains unclear.

MHPG Studies

MHPG is the major NE metabolite that can be measured in the brain, CSF, plasma, and urine. There have been a number of studies looking at MHPG in schizophrenic patients on and off medications. Sedvall et al. (1976) measured CSF MHPG and HVA in schizophrenic patients before treatment and at 2 and 4 weeks after treatment with chlorpromazine or thiothixene. With chlorpromazine they observed a significant increase in CSF HVA and a significant decrease in CSF MHPG. These changes were significantly correlated with clinical improvement. With thiothixene treatment they observed an increase in CSF HVA but no significant change in CSF MHPG. When they examined the change in the HVA/MHPG ratio in both drug-treatment groups, they found that the greater the change in the HVA/MHPG ratio, the greater the clinical improvement. The authors concluded that the antipsychotic action of neuroleptic drugs was related to their interaction with central dopaminergic as well as noradrenergic mechanisms. Kirch et al. (1988) measured plasma catecholamine metabolites from 22 patients with chronic schizophrenia while on a stable dose of neuroleptic medication and after a 6-week drug-free period. They found a significant increase in MHPG and a trend toward an increase in HVA off drug. They did not find significant correlations between plasma MHPG or HVA concentrations and the corresponding psychopathology ratings. Ko et al. (1988) examined plasma MHPG in 14 schizophrenic patients during medication and serially off medication for at least 6 weeks. They found that MHPG plasma levels changed with clinical state; that is, they noted higher MHPG levels when patients were in a more psychotic state and lower levels when patients were less psychotic. Compared with the 22 control subjects, paranoid schizophrenic patients had higher plasma MHPG. There was a trend for higher plasma MHPG values in paranoid schizophrenic patients as compared with those with undifferentiated schizophrenia. Their findings suggested that plasma MHPG may reflect psychosis-related changes in NE function in schizophrenia. This idea that state dependency may influence NE activity is explored later.

Chang et al. (1990) measured plasma HVA and MHPG in 33 schizophrenic patients before and 6 weeks after haloperidol treatment. They divided patients into responding patients and nonresponding patients and found that patients who responded to haloperidol had higher baseline HVA and that this went down after treatment. The authors observed a similar but nonsignificant trend for plasma MHPG. They suggested possible subtypes of schizophrenia, one with increased dopamine activity and one without. They also noted a significant correlation between plasma HVA and MHPG. They speculated that because of the trend of MHPG to parallel changes in HVA that the NE system may also be abnormal in some schizophrenic patients and that the dopamine and NE system were interrelated. Bowers et al. (1987) looked at plasma catecholamine metabolite levels in 47 psychotic inpatients who required antipsychotic treatment. This group included patients with schizophrenia as well as atypical affective disorders with psychotic features. They found that positive responding patients had significantly higher pretreatment HVA levels and a trend toward higher MHPG values as compared with the nonresponding patients. Pickar et al. (1990) examined the CSF and plasma of 22 drug-free, highly symptomatic schizophrenic patients and 33 control subjects. They observed that although CSF and plasma HVA levels have significant relationships to psychosis, there were no differences between the groups in CSF levels of HVA, NE, or 5-HIAA. The only difference between the two groups was significantly increased CSF MHPG in the schizophrenic patient group. They also observed that plasma MHPG was positively correlated with negative symptoms in the patient group.

Urine Studies

Urine represents another accessible material for studying the neurotransmitter metabolites in schizophrenia. Karoum et al. (1987) examined the urinary metabolites of dopamine and NE in 20 chronic schizophrenic patients who had been drug free at least 2 weeks and again when they were on medication. They compared these to 16 control subjects. They collected multiple sets of 24-hour urines and measured the sum dopamine (dopamine + 3,4-dihydroxyphenylacetic acid [DOPAC] + HVA) and sum NE (NE + MHPG + normetanephrine + vanillylmandelic acid) and expressed their results in a ratio of sum dopamine to sum NE. Sum dopamine was significantly reduced in drug-free patients, whereas sum NE was not different from control subjects. The drug-free ratio was lower than in control subjects. The sum dopamine/sum NE ratio normalized when the patients were treated with antipsychotic agents. The authors raised the possibility that schizophrenia may be associated with an imbalance between noradrenergic and dopaminergic activity and that neuroleptic medication preferentially stimulated dopamine excretion, thereby balancing the activity between the two amines. Mueser et al. (1986) measured MHPG/creatinine ratios in 58 psychiatric inpatients (21 of whom were schizophrenic patients) who were acutely admitted and could be maintained drug free for 1–3 weeks. In the schizophrenic group, paranoid schizophrenic patients had a significantly higher MHPG/creatinine ratio than nonparanoid schizophrenic patients. The authors also found that the MHPG/creatinine ratio was positively correlated with levels of idiosyncratic behavior in all of the patient groups. They speculated that behaviors such as repetitive movements, self-verbalization, or bizarre behavior present in some schizophrenic, schizoaffective, or affective patients may be mediated by the noradrenergic system. Kemali et al. (1982) examined urinary catecholamines in 16 schizophrenic patients and 20 control subjects and found no differences in the levels of urinary NE, dopamine, or epinephrine. In addition, they found no correlations between urinary and plasma NE.

Platelet α-Adrenergic Receptors

α-Adrenergic receptors are found both presynaptically and postsynaptically in the brain on neurons; they are also found on platelets. Platelets represent a peripheral way to study α-receptor function. On platelets, there is a complex that consists of an α_2-adrenergic receptor coupled with a prostaglandin E1 (PGE1) receptor, and these are connected to an adenylate-cyclase complex that converts adenosine triphosphate to cAMP. Stimulation of the α-receptor leads to inhibition of cAMP production; PGE1 stimulation leads to increased cAMP production (Kafka and Paul 1986). Kafka et al. (1980) studied platelet α-adrenergic receptor number and function in patients with schizophrenia, affective illness, hypertension, and other diseases and in control subjects. They found that compared with control subjects, patients with schizophrenia, affective illness, and hypertension had increased numbers of platelet α-receptors. Platelet α-receptor function was studied by measuring PGE1-stimulated cAMP production. Decreased cAMP production was seen in the platelets of the schizophrenic and hypertensive patients, but not those of the affective illness patients or the control subjects. The authors speculated that if these changes observed in platelets occur in neurons, adrenergic transmission in the brain might change and thereby contribute to the symptomatology of these illnesses. Kafka and van Kammen (1983) in further work found a trend for increased α-receptors in patients with schizophrenia compared with control subjects. They also found evidence of decreased PGE1-stimulated cAMP production and speculated that if production of cAMP in neurons is similar to that in platelets, diminished cAMP production may be associated with a vulnerability to schizophrenia.

Rotrosen et al. (1980) found that PGE1-stimulated cAMP accumulation in platelets was decreased in schizophrenic patients compared with control subjects. They speculated that because E-type prostaglandins appear to antagonize central nervous system catecholamine transmission, a prostaglandin deficiency could provide a basis for the hyperdopaminergic activity postulated to underlie schizophrenic symptoms. Kanof et al. (1987) studied 35 actively psychotic male schizophrenic patients and found that platelet cAMP response to PGE1 stimulation was negatively correlated with global symptom severity. They speculated that if the subsensitivity of platelet prostaglandin E receptors extended to the brain, schizophrenic patients could have an impairment in the ability of endogenous prostaglandin E to inhibit dopaminergic transmission and that such an impairment could be involved in psychotic symptoms.

Studies with platelets are useful if they reflect basic mechanisms similar to what is going on in the brain; however, platelet α_2-receptors tend to respond to excess NE with the development of tolerance and may be different from brain α_2 receptors. The exact nature of the deficit in α-receptors, prostaglandin receptors, or the adenylate-cyclase complex remains uncertain.

Noradrenergic Drugs in Schizophrenia

The search for drugs other than typical antipsychotic agents in the treatment of schizophrenia has lead to trials of drugs that work primarily on the NE system. Trials of clonidine (Elizur et al. 1980; Freedman et al. 1982; Jimerson et al. 1980; Simpson et al. 1967; van Kammen et al. 1989b; Zubenko et al. 1984), propranolol (Gardos et al. 1973; King et al. 1983; Peet et al. 1981; Pugh et al. 1983; Yorkston et al. 1977), and prazosin (Hommer et al. 1984) and experiments with yohimbine (Glazer et al. 1987; Holmberg and Gershon 1961) have been done. So far none of these drugs has been shown to have a consistent antipsychotic action; however, they were effective in some patients. Continued investigations may be fruitful if it is possible to identify biochemically which patients are likely to respond to NE drugs.

Clonidine is an α_2-receptor agonist that decreases the release of NE by stimulating postsynaptic and presynaptic α_2-receptors. Simpson et al. (1967) found no antipsychotic activity in a clonidine trial of nine chronic schizophrenic patients after a 1-month medication-free period followed by a 12-week clonidine phase. Jimerson et al. (1980) studied two schizophrenic patients on clonidine and found that they became more agitated and aggressive without improvement in psychosis. Elizur et al. (1980) conducted a clonidine trial in two schizophrenic patients and found that one patient got worse and the other showed some improvement and then subsequently got worse. Zubenko et al. (1984) found that three psychotic bipolar patients remitted on clonidine.

Some of the above studies looked at treatment-resistant patients, a group in which improvement may not be expected. Freedman et al. (1982) studied eight schizophrenic patients with tardive dyskinesia in a double-blind, cross-over study comparing placebo, clonidine, and standard antipsychotics. The patients were not treatment resistant. The authors found that the patients did as well on clonidine as they did on standard antipsychotic medication and that the tardive dyskinesia improved with clonidine. van Kammen et al. (1989b) found that 4 of 13 drug-free relapsed schizophrenic patients improved with a 5-week, double-blind clonidine trial. CSF NE and MHPG decreased and CSF HVA increased with clonidine treatment, but no effect on 5-HIAA was noted. All the responding patients were paranoid schizophrenic patients. Examination of pretreatment biological variables indicated that patients likely to respond to clonidine were likely to have "normal" CSF NE and normal or high α_2-receptor activity. α_2-Receptor activity was assessed using a clonidine challenge test where greater functional α_2-activity produces higher peak growth-hormone levels in response to the challenge dose of clonidine. The authors also noted that those who responded were more likely to be in an early or prodromal phase of psychotic relapse as indicated by higher emotional withdrawal and depression ratings. The authors hypothesized that clonidine treatment may decrease symptoms by decreasing NE release and normalizing NE transmission. This decrease may be important only in a small subgroup of patients with normal or elevated α_2-receptor function, whereas patients with subsensitive α_2-receptor function may not have normalization of NE transmission with clonidine. Studies in patients following antipsychotic withdrawal using clonidine as a potential protective agent against relapse are needed.

Propranolol is a β-adrenoreceptor blocking agent that has been used in several trials as an antipsychotic agent or as an augmenting medication in addition to neuroleptic treatment. In a study of 12 treatment-resistant schizophrenic patients kept drug free before a 1-month propranolol trial, King et al. (1983) found no antipsychotic effect of propranolol with doses up to 960 mg/day level. CSF HVA was increased, and there was a similar trend in CSF NE. Gardos et al. (1973) studied eight treatment-resistant chronic schizophrenic patients after a 4-week drug-free period on 120–720 mg propranolol per day for 6–10 weeks. They did not find an antipsychotic effect; they did, however, see an anxiolytic effect. In a double-blind trial involving 53 hospitalized chronic schizophrenic patients randomized to propranolol, chlorpromazine, or placebo for up to 3 months, Peet et al. (1981) found that propranolol had no important advantages over placebo, whereas chlorpromazine had a beneficial effect. They hypothesized that increased antipsychotic levels, caused by a pharmacokinetic interaction between propranolol and the antipsychotic, brought about more improvement. D. van Kammen et al. (D. van Kammen, C. Schultz, D. Hommer, W. Bunney, unpublished data, January 1982) found that lower pretreatment cAMP response to the β-receptor agonist isoproterenol in lymphocytes was associated with stronger antipsychotic response to propranolol.

The above studies indicate that propranolol alone is not likely to be an effective antipsychotic agent; however, studies in which propranolol is added to ongoing antipsychotic treatment showed some benefit. Yorkston et al. (1977) conducted a study where 14 chronic schizophrenic patients were assigned to either propranolol or placebo in addition to their antipsychotic regimen for 12 weeks. The authors found that although both groups improved, the propranolol group had improved significantly more. Pugh et al. (1983) studied 41 chronic schizophrenic patients maintained on neuroleptics during a 12-week double-blind phase where patients received either propranolol or placebo. The authors found no Brief Psychiatric Rating Scale score differences between the two groups, but there was decreased irritability and some global improvement in the propranolol group. The authors speculated that this improvement may have occurred because of the pharmacokinetic interaction between propranolol and the neuroleptic, where propranolol can increase serum levels of the antipsychotic medication. However, there is no evidence that raising antipsychotic drug levels will lead to an increased antipsychotic effect.

Prazosin is a specific α_1-noradrenergic receptor antagonist used in the treatment of hypertension. Hommer et al. (1984) studied seven schizophrenic patients in a double-blind, placebo-controlled trial of prazosin. No therapeutic effects were noted, but psychophysiological evaluation showed that prazosin produced an increase in autonomic arousal. α_1-Noradrenergic receptors may be involved in the effect of clozapine. Chiodo and Bunney (1985) demonstrated in rats that chronic administration of haloperidol caused depolarization blockade in the A9 dopamine neurons (innervating the striatum) and the A10 dopamine neurons (innervating the nucleus accumbens) and that chronic administration of clozapine caused depolarization blockade only in the A10 dopamine neurons. The authors then looked at the effect of the combination of prazosin and haloperidol and found that chronic treatment resulted in depolarization block in the A10 but not the A9 neurons, similar to the effect seen with clozapine. In chronically treated rats, Lane et al. (1988) found reductions in the basal levels of dopamine release in the striatum and nucleus accumbens with haloperidol, whereas clozapine or the combination of haloperidol and prazosin resulted in reductions of basal dopamine release in the nucleus accumbens but not the striatum. The results of these experiments suggest that α_1-noradrenergic receptor-blocking properties of clozapine may in part mediate its differential actions on the nigrostriatal and mesolimbic dopamine release, and this action may contribute to the lack of extrapyramidal side effects.

Yohimbine is an α_2-receptor antagonist that increases noradrenergic function. Holmberg and Gershon (1961) tested the autonomic and psychic effects of yohimbine on control subjects and on psychiatric patients. They gave five schizophrenic patients injections of 20–40 mg of yohimbine im three times per week. Two of these patients had a minimal overall response to these injections, and the other three patients had an activation of their psychosis with increased delusions, hallucinations, and disturbed behavior. Glazer et al. (1987) administered a 20-mg oral yohimbine challenge test to 16 control subjects and 18 schizophrenic patients who had been drug free at least 1 month. They observed that the schizophrenic patients showed a statistical trend toward yohimbine-induced increases in plasma MHPG levels and systolic blood pressure. Six of the schizophrenic patients but no control subjects had a notable dysphoric arousal reaction with fearfulness, restlessness, anxiety, and sadness. The authors concluded that these results pointed toward NE involvement in autonomic arousal rather than in core psychotic process. Conceivably, these patients were in a stable clinical state, which would have protected them from becoming more psychotic, comparable to the differential amphetamine effect (van Kammen et al. 1982).

Taken together, these studies of noradrenergic drugs show no consistent effective antipsychotic response; however, the trials were often small, and the subjects were often stable treatment-resistant patients, which may have made beneficial effects unlikely. The fact that there was some improvement or worsening indicates that further exploration of the NE system in schizophrenia is warranted and that perhaps the right NE drugs have not yet been found. These studies are also made difficult by the fact that tolerance develops to α-receptor stimulation and by our inability to identify whether those who are responding are true responses or responses to placebo. Further biological identification of such responders may help future studies. A more spurious interpretation is that the therapeutic effects were very limited.

Animal Studies

Animal studies on the catecholamine system may give some hints about the meaning of NE changes seen in schizophrenia. Such studies provide an opportunity to investigate the neurotransmitter systems with lesion experiments and antipsychotic drugs. Antipsychotics affect not only the dopamine system but also the NE system as well as many other neurotransmitters. They provide evidence that brain neurotransmitter systems are interconnected. Here we review some of these studies and discuss their implications for schizophrenia.

Taghzouti et al. (1988) demonstrated that the NE system interacts with the ascending dopamine system in cortical and subcortical structures in the rat. The ventral mesencephalic tegmentum is the major dopamine innervation to the limbic system (nucleus accumbens) and prefrontal cortex. A lesion to the ventral mesencephalic tegmentum resulted in a syndrome that consisted of locomotor hyperactivity and an inability to make spontaneous changes in behavior. They showed that this syndrome, caused by lesions to the dopamine system, could be reversed by lesions to the pedunculus cerebellaris superior, which interrupts the dorsal noradrenergic ascending pathway. The authors concluded that noradrenergic innervation has a permissive role in the behavioral expression of the dopamine system. They speculated that behavioral disturbances could result from an imbalance between the activities of the interrelated systems. The functional recovery of the ventral mesencephalic tegmentum syndrome by a lesion to the NE system provided insights into the antagonistic properties of the ascending dopamine and NE neurons. Glowinski et al. (1988) reported that NE neurons play a role in regulating cortical type 1 dopamine (D_1) receptors, and that cortical D_1 receptors are also under regulation of serotonergic and glutamatergic neurons. They also demonstrated that an intact NE system is required for the development of the supersensitivity of D_1 receptors.

The role of NE-containing neurons located in the pontine nucleus coeruleus in behavior has been explored by Aston-Jones and Bloom (1981) in unanesthetized rats. They proposed that the LC-NE system may function more in phasic processes than in modulation of the tonic arousal level. LC activity increased with auditory, visual, and somatosensory stimuli. They proposed that this LC-NE discharge may enhance activity within target cell systems responding primarily to external stimuli and suppress central nervous system activity related to tonic, vegetative functions. The LC-NE system may bias global behavioral orientation between stimuli in the external versus internal environments. Svensson (1987) noted that in schizophrenia there are problems in the attribution of significance and discrimination of sensory environmental stimuli. He reviewed the literature that showed that the LC reacted to internal environment changes such as blood pressure and bowel contractions. Svensson suggested that a disorder of the LC could be responsible for the intensified alarm reactions, strange aggressive reactions to touch, and unique physical complaints that are seen in schizophrenia. Patients often misinterpret their environment and are hypervigilant, suggesting that something is amiss in the systems that are involved in processing information from the internal and external environments.

Yamamoto et al. (1991) studied the skin conductance response in rats that were treated with intraventricular injections of 6-hydroxydopamine, a neurotoxin that selectively destroys noradrenaline nerve terminals leading to denervation supersensitivity at the synapse. Normally skin conductance response increases when there is a new stimulus and, as the stimulus is repeatedly presented, there is habituation. Compared with sham-injected rats, the rats injected with 6-hydroxydopamine had an initial disappearance of the skin conductance response lasting a few days; when it later reappeared, there was decreased habituation. In patients with schizophrenia, nonresponsiveness and nonhabituation are seen in skin conductance response studies, and the authors speculated that some dysfunction of the NE system may explain the abnormalities seen in schizophrenia. This nonresponsiveness may be a model for a certain state in the schizophrenic process.

Animal experiments have also explored the effects of neuroleptics and other drugs that interact with the NE system. Freedman (1977) showed that when stimulated, LC-NE projections to cerebellar Purkinje cells caused an inhibition of the Purkinje cell firing rates. Fluphenazine and other antipsychotic agents reversed this inhibition relative to their clinical potencies. This indicates that NE activity could be involved in the antipsychotic drug effect. Freedman also postulated a role for the LC in modulating responsiveness of other systems and speculated that hyperactivity in the NE projections to the brain stem, cerebral cortex, and cerebellum may be responsible for these neuronal structures becoming overly sensitive to afferent inputs, leading to a clinical syndrome of the patient being flooded with stimuli. Neuroleptics that act on both dopamine and NE may cause apparent indifference, as well as a slowing of responses to external stimuli. Further evidence for the interaction of the NE and dopamine systems comes from Grenhoff and Svensson (1988), who found that clonidine, a noradrenergic α_2-agonist, regularized the rate of dopaminergic substantia nigra cell firing in the anesthetized rat and that this effect was reversed by yohimbine, an α_2-antagonist. Dickinson et al. (1988), in an interesting set of experiments, examined how α_1- and α_2-adrenoreceptor antagonists differentially influence amphetamine- and apomorphine- (dopamine receptor agonists) induced behaviors in rats. They showed that dopamine-induced locomotor hyperactivity and stereotypy were differentially influenced by both α_1- and α_2-adrenergic antagonists. This suggested that NE modulates the expression and character of behavior by influencing dopamine function in certain brain areas. This is interesting in that it shows that the dopamine system is sensitive to NE manipulation, providing evidence for the interaction between NE and dopamine. These findings gave further support for the idea that NE systems and dopamine systems interact and that this interaction may be the reason that clonidine is sometimes effective in some neuropsychiatric disorders such as mania (Zubenko et al.

1984), schizophrenia (Freedman et al. 1982; van Kammen et al. 1989b), and tardive dyskinesia (Freedman et al. 1982).

Dinan and Aston-Jones (1985) showed that chronic (i.e., 21-day) administration of haloperidol 0.5 mg/kg to rats caused suppression of spontaneous activity in 20% of LC neurons. Most likely this was not through dopamine activity as there is little dopamine innervation to the LC, but probably through antagonism of α_2-receptors. Haloperidol has profound effects on LC neurons, indicating a possible role of the brain NE system in clinical effects or side effects of this neuroleptic.

These animal studies show that neuroleptic medication has effects not only on the dopamine system but also the NE system relative to their antipsychotic potential. The animal research also provides direct evidence for the role of NE as a regulator of the dopamine system. There seems to be a hierarchy of how dopamine and NE interact. When examining the ventral mesencephalic tegmentum syndrome or the effect of amphetamines, lesions of the NE system alone do not cause deficits but rather NE effects are seen only when the dopamine system is altered (Taghzouti et al. 1988). Dopamine supersensitivity effects are seen only when the NE system is at least partially intact (Glowinski et al. 1988). These experiments indicate that NE may serve as a modulator for the dopamine-dependent behaviors.

Norepinephrine Hypotheses in Schizophrenia

Hypotheses about the involvement of NE in schizophrenia have been around for nearly 20 years. Stein and Wise (1971) proposed that endogenous production of the neurotoxin 6-hydroxydopamine could lead to destruction of the noradrenergic reward system and that this was related to the symptoms of schizophrenia and the slow downhill deterioration seen in the illness. Alternatively, Wise and Stein (1973) speculated that a loss of dopamine β-hydroxylase, the enzyme that converts dopamine and NE, could lead to a similar weakness of the noradrenergic system. Later studies (Cross et al. 1978; Crow et al. 1979; Wyatt et al. 1975) were unable to replicate these findings of decreased dopamine β-hydroxylase activity in similar small-number studies.

Hornykiewicz (1982) reviewed data supporting an interaction between NE and dopamine in schizophrenia. He pointed out that the main NE-containing cell body of the LC heavily innervates the limbic system, even more so than dopamine. NE is thought to play a role in multiple brain functions, including stress, rage, aggressiveness, locomotor behavior, arousal, reward, memory, sleep-wakefulness cycles, neuroendocrine control, and autonomic control of temperature and the cardiovascular system. From these observations, he proposed that NE plays a role in regulating the function of other neurotransmitter systems by setting gain or adjusting bias. Thus NE can modulate both inhibitory and excitatory neurotransmitter systems.

Earlier, Antelman and Caggiula (1977) had reviewed the literature on the role of NE-dopamine interactions in behavior. They concluded that NE facilitates dopamine-dependent behaviors and that these neurotransmitters respond in an adaptive fashion to stimuli that perturb the homeostasis in the brain. Under conditions of high stress or impaired noradrenergic function, NE is unable to sustain its modulatory influence on dopamine, and the dopamine system becomes deranged. The authors also pointed out that the dopamine versus NE debate may have distracted investigators and caused them not to pursue an interactive hypothesis.

Svensson (1987) noted that the LC responds with increased activity when there is novel exterior stimulation, but also reacts to internal body changes such as changes in blood pressure or bowel function. As discussed earlier, there are problems in discriminating the significance of stimuli from the internal and external environments in schizophrenia. Svensson speculated that in schizophrenia internal stimuli may be misperceived as coming from the outside and may therefore cause alarm and perhaps aggressive behavior in a paranoid patient without any obvious external triggering stimulus. A dysfunction of the noradrenergic system that modulates that response to stimuli may potentially explain some of the behaviors seen in schizophrenic patients.

State-Dependent Changes in Norepinephrine

With evidence of increases in NE and its metabolites in schizophrenia, in addition to the evidence that NE is involved in the regulation of dopamine-expressed behaviors from the animal literature, one can speculate that noradrenergic function may be altered in schizophrenia. Linnoila et al. (1983) noted that levels of NE in the CSF of drug-free schizophrenic patients had more variability than did measures of CSF HVA, MHPG, and 5-HIAA. These variations in the NE system may be important in the expression of the illness. This variability suggests that rather than looking at neurotransmitter levels as stable entities that are going to be elevated or decreased in a given disease, it may be more useful to propose that the neurotransmitters may fluctuate as the clinical state changes, as these neurotransmitters are what allow expression of normal and abnormal behaviors. It is unknown, however, how much CSF NE reflects NE availability at the synapse, but it is a reasonable assumption for investigation that there is some relationship. It is also possible that CSF NE relates to activity of the patient just before the lumbar puncture or has fluctuations like those observed with plasma NE.

The variable response to *d*-amphetamine infusions in drug-free schizophrenic patients supports the notion of the state-dependent influence of biochemical (i.e., NE) activity. van Kammen et al. (1982) reviewed the literature on the response to *d*- or *dl*-amphetamine in drug-free schizophrenic patients and reviewed the numbers of those that worsened ($n = 72$), improved ($n = 81$), and did not change ($n = 132$). They went on to do a placebo-paired, double-blind *d*-amphetamine (20-mg) study on 45 schizophrenic patients that had been drug free at least 3 weeks (mean, 34 days). They also collected CSF the week before the infusion and measured DOPAC, HVA, and MHPG. They found that 13 patients improved, 18 worsened, and 14 showed no change. Although there were no differences in the dopamine metabolites in the CSF of the patients that worsened and those that improved, CSF MHPG tended to be higher in patients that worsened. The authors concluded that NE activity, as expressed by MHPG, may be a critical factor during clinical states and may determine whether dopamine stimulation will lead to a change in psychosis. Higher CSF MHPG in patients that worsened after the infusion could indicate a derangement in the NE system that allowed dopamine stimulation to be expressed as worsening.

There have been several studies of CSF NE in schizophrenic patients on and off neuroleptic medication, and the results have been conflicting. Sternberg et al. (1981) examined the effect of pimozide treatment on CSF NE; 33 drug-free (at least 3 weeks) psychotic schizophrenic patients had significantly higher baseline CSF NE values than did control subjects. They were then treated with pimozide for at least 4 weeks, and there was a significant decrease in CSF NE values. There were significant correlations between the decrease in CSF NE and the decrease in global psychosis ratings. Six patients after subsequent withdrawal from pimozide received repeat lumbar punctures. The authors found that the longer the drug-free interval, the higher the NE levels. The authors concluded that the pimozide-induced decrease in CSF NE and the decrease in psychosis suggested a role for the noradrenergic system in the antipsychotic actions of neuroleptic medication. Zander et al. (1981) studied 17 chronic schizophrenic patients under chronic neuroleptic treatment for 13 years and after they had been withdrawn from medication for 30 days. Thirteen patients completed the paired study conditions, having repeated lumbar punctures and blood draws. Of these 13 patients, 8 improved slightly and 5 worsened slightly. Four patients dropped out because of relapse, and these patients were not able to be retested drug free, so no comment could be made on the biochemical state in the relapsing patients. HVA and MHPG in the CSF did not change significantly, but plasma NE was elevated during chronic treatment and significantly decreased after neuroleptic withdrawal. The four patients that relapsed and were therefore unable to be tested possibly represented the most interesting potential information. Bagdy and Perenyi (1984) obtained the CSF of 15 chronic schizophrenic patients who had been treated for at least 3 years continuously with neuroleptics and then after 2 weeks off medication. They observed that dopamine, DOPAC, HVA, and NE were decreased after discontinuation. They noted that the patients who did not show a decrease in catecholamine levels were the treatment-nonresponding patients who had required continuous hospitalization. The authors concluded that these persistent elevated levels of CSF monoamine metabolites after withdrawal of neuroleptics may predict poor treatment response.

van Kammen et al. (1989a) measured CSF NE, MHPG, HVA, and 5-HIAA in 32 schizophrenic patients during haloperidol treatment and again after up to 6 weeks drug free. To examine state-dependent issues, they divided patients into two groups: patients that relapsed within the 6 weeks drug free and those who did not relapse. They found that the patients who relapsed had significantly higher CSF NE during haloperidol treatment. When they relapsed after haloperidol withdrawal, they had significantly higher CSF NE and MHPG, without such changes in CSF HVA and 5-HIAA. The patients who went on to relapse were not able to be differentiated by psychosis ratings from those that did not relapse prior to drug withdrawal. In patients who did not relapse, there was a decrease in CSF NE together with a small significant decrease in psychosis ratings. These data indicated that elevated CSF NE levels during neuroleptic treatment may predict behavioral decompensation after discontinuing medication. van Kammen et al. (1990) examined the interaction of NE with negative symptoms in patients in the above-described protocol. They found that relapse following haloperidol withdrawal was associated with both increased negative symptoms and increased CSF NE and MHPG. Negative symptom scores significantly correlated with CSF NE and MHPG but not CSF HVA or 5-HIAA in the patients that relapsed. The authors concluded that increased NE activity in drug-free patients was associated with intensification of schizophrenic symptoms without necessarily causing the symptoms.

Taken together, the results of these studies appear to be inconsistent, but the inconsistencies may, at closer scrutiny, be understood. Zander et al. (1981) and Bagdy and Perenyi (1984) reported a decrease in CSF NE after a 2-week withdrawal from chronic neuroleptics in schizophrenic patients. This is similar to the results obtained by van Kammen et al. (1989a, 1990) in their group of patients that did not relapse after haloperidol withdrawal. In the Zander et al. study, patients that relapsed were unable to be tested, so one could speculate that the tested patients were stable and similar to the nonrelapsed patients in the van Kammen et al. study. Both of these groups were stable and showed decreases of CSF NE. On the other hand, the relapsed patients in the van Kammen et al. (1989a, 1990) studies were similar in condition to the patients in the study of Sternberg et al. (1981), in whom neuroleptic withdrawal was associated with an increase in psychosis ratings and increases in CSF NE. The van Kammen studies suggested that during maintenance drug treatment, CSF NE levels may increase when patients become biochemically unstable. The presence of increased NE levels may contribute to psychotic decompensation following the removal of the dopamine receptor blockade (i.e., the withdrawal of haloperidol). Future studies will have to test these speculations.

Conclusion

What is the role of NE in schizophrenia? Clearly there are differences in the levels of NE in the brains of patients with schizophrenia. This information comes from both autopsy studies and studies that measure the levels of NE and its metabolites in CSF, plasma, and urine. Antipsychotic agents affect not only dopamine, but also NE and other neurotransmitters. Animal studies have demonstrated that the NE and dopamine systems are interconnected so that changes or lesions in one system can have effects on the other system. These animal studies suggest that NE may regulate the dopamine system output so that lesions of the NE system may have their impact in effects on dopamine-controlled behaviors.

The dynamic nature of schizophrenia suggests that changes in the behavioral state of the patient are reflected in biochemical changes in neurotransmitter systems. These changes are indeed seen in NE and other neurotransmitter systems besides dopamine. One can reasonably speculate that these biochemical changes are the driving forces in the expression of the illness. Pharmacological interventions have behavioral effects, we presume, by altering neuronal transmission. The dopamine hypothesis rests on the fact that antipsychotic response was correlated most well with the dopamine receptor binding (Creese et al. 1976; Seeman et al. 1976; Snyder 1976), but the etiology of schizophrenia may be more complicated. As we progress in our understanding of how the brain functions in health and illness with animal studies and with studies of patients with schizophrenia, the view emerges that many systems in the brain are interconnected with parallel processing, and the idea of one neurotransmitter causing one psychiatric illness is gradually being replaced by the idea that an illness such as schizophrenia may come about because of imbalances or dysfunctions in the relations between various neurotransmitter systems.

In schizophrenia, changes in NE are seen as the patient's symptoms evolve and during different conditions of drug treatment or drug withdrawal, and there is some preliminary evidence that these changes may occur before the actual behavioral changes. This may take the form of a dysregulation of the NE system, causing or enhancing dysregulation in the dopamine system. Clearly there are many reports of alterations in the NE system in schizophrenia. Are these changes causal or are they interesting epiphenomena? What are the roles of other neurotransmitter systems besides dopamine and NE, such as glutamate, acetylcholine, serotonin, and the neuropeptides? Whatever the answers to these questions, a major neurotransmitter system such as NE, with the LC as its major cell group projecting to many areas of the brain, its involvement in the evaluation of the external and internal environments, and its demonstrated ability to modulate responses from other neurotransmitter systems (dopamine in particular) continue to make the exploration of the role of NE in schizophrenia of interest.

References

Albus M, Ackenheil M, Engel R, et al: Situational reactivity of autonomic functions in schizophrenic patients. Psychiatry Res 6:361–370, 1982

American Psychiatric Association: Diagnostic and Statistical Manual of Mental Disorders, 3rd Edition. Washington, DC, American Psychiatric Association, 1980

Antelman S, Caggiula A: Norepinephrine dopamine interactions and behavior. Science 195:646–653, 1977

Aston-Jones G, Bloom F: Norepinephrine containing locus coeruleus neurons in behaving rats exhibit pronounced responses to non-noxious environmental stimuli. J Neurosci 1:887–900, 1981

Bagdy G, Perenyi A: Decrease of CSF dopamine, its metabolites, and noradrenaline after withdrawal of chronic neuroleptic treatment in schizophrenia patients. Psychiatry Res 12:177–178, 1984

Bird E, Spokes E, Iversen L: Increased dopamine concentration in limbic areas of brains from patients dying with schizophrenia. Brain 102:347–360, 1979

Bowers M, Swigar M, Jatlow P, et al: Early neuroleptic response: Clinical profiles and plasma catecholamine metabolites. J Clin Psychopharmacol 7:83–86, 1987

Bridge T, Kleinman J, Karoum F, et al: Postmortem central catecholamines and antemortem cognitive impairment in elderly schizophrenics and controls. Neuropsychobiology 14:57–61, 1985

Chang W, Chen T, Lin S, et al: Plasma homovanillic acid levels and subtyping in schizophrenics: Evidence for the two-subtype concept. Biol Psychiatry 27:510–518, 1990

Chiodo L, Bunney B: Possible mechanisms by which repeated clozapine administration differentially affects the activity of two subpopulations of midbrain dopamine neurons. J Neurosci 5:2539–2544, 1985

Creese J, Burt DR, Snyder SH: Dopamine receptor binding predicts clinical and pharmacological potencies of antischizophrenic drugs. Science 192:481–483, 1976

Cross A, Crow T, Killpack W, et al: The activities of brain dopamine β-hydroxylase and catechol-O-methyltransferase in schizophrenic and controls. Psychopharmacology (Berl) 59:117–121, 1978

Crow T, Baker H, Cross A, et al: Monoamine mechanisms in chronic schizophrenia: postmortem neurochemical findings. Br J Psychiatry 134:249–256, 1979

Dajas F, Barbeito L, Martinez-Pesquera G, et al: Plasma noradrenaline and clinical psychopathology in schizophrenia. Neuropsychobiology 10:70–74, 1983

Dickinson S, Gadie B, Tulloch I: Alpha-1 and alpha-2 adrenoreceptor antagonists differentially influence locomotor and stereotyped behavior induced by D-amphetamine and apomorphine in the rat. Psychopharmacology (Berl) 96:521–527, 1988

Dinan T, Aston-Jones G: Chronic haloperidol inactivates brain noradrenergic neurons. Brain Res 325:385–388, 1985

Elizur A, Levy A, Favah M, et al: Preliminary trial of clonidine treatment in two patients suffering from chronic schizophrenia. Community Psychopharmacology (Berl) 4:507–514, 1980

Farley I, Price K, McCullough E, et al: Norepinephrine in chronic paranoid schizophrenia. Science 200:456–457, 1978

Freedman R: Interactions of antipsychotic drugs with norepinephrine and cerebellar neuronal circuitry: implications for the psychobiology of psychosis. Biol Psychiatry 12:181–196, 1977

Freedman R, Kirsch D, Bell J, et al: Clonidine treatment of schizophrenia: double blind comparison to placebo and neuroleptic drugs. Acta Psychiatr Scand 65:35–45, 1982

Gardos G, Cole J, Volcer I, et al: A dose response study of propranolol in chronic schizophrenics. Current Therapeutic Research 15:314–323, 1973

Gattaz W, Riederer P, Reynolds G, et al: Dopamine and noradrenaline in the cerebrospinal fluid of schizophrenic patients. Psychiatry Res 8:243–250, 1983

Glazer W, Charney D, Heninger G: Noradrenergic function in schizophrenia. Arch Gen Psychiatry 44:898–904, 1987

Glowinski J, Herve D, Tassin J: Heterologous regulation of receptors on target cells of dopamine neurons in the prefrontal cortex, nucleus accumbens and striatum. Ann N Y Acad Sci 537:112–123, 1988

Gomes U, Shanley B, Potgieter L, et al: Noradrenergic overactivity in chronic schizophrenia: evidence based on cerebrospinal fluid noradrenaline and cyclic nucleotide concentrations. Br J Psychiatry 137:346–351, 1980

Grenhoff J, Svensson T: Clonidine regularizes substantia nigra dopamine cell firing. Life Sci 42:2003–2009, 1988

Holmberg G, Gershon S: Autonomic and psychic effects of yohimbine hydrochloride. Psychopharmacologia 2:93–106, 1961

Hommer D, Zahn T, Pickar D, et al: Prazosin, a specific alpha-noradrenergic receptor antagonist, has not effect on symptoms but increases autonomic arousal in schizophrenic patients. Psychiatry Res 11:193–204, 1984

Hornykiewicz O: Brain catecholamines in schizophrenia: a good case for noradrenaline. Nature 299:484–486, 1982

Javitt D, Zukin S: Recent advances in the phencyclidine model of schizophrenia. Am J Psychiatry 148:1301–1308, 1991

Jimerson D, Post R, Stoddard F, et al: Preliminary trial of the noradrenergic agonist clonidine in psychiatric patients. Biol Psychiatry 15:45–57, 1980

Kafka M, Paul S: Platelet alpha 2-adrenergic receptors in depression. Arch Gen Psychiatry 43:91–95, 1986

Kafka M, van Kammen D: Alpha-adrenergic receptor function in schizophrenia. Arch Gen Psychiatry 40:264–270, 1983

Kafka M, van Kammen D, Kleinman J: Alpha adrenergic receptor function in schizophrenia. Psychopharmacol Bull 16:91–104, 1980

Kanof P, Davidson M, Johns C, et al: Clinical correlates of platelet prostaglandin receptor subsensitivity in schizophrenia. Am J Psychiatry 144:1556–1560, 1987

Karoum F, Karson C, Bigelow L, et al: Preliminary evidence of reduced combined output of dopamine and its metabolites in chronic schizophrenia. Arch Gen Psychiatry 44:604–607, 1987

Kemali D, Vecchio M, Maj M: Increased noradrenaline levels in CSF and plasma of schizophrenic patients. Biol Psychiatry 17:711–717, 1982

Kemali D, Maj M, Ariano M, et al: 24-hour plasma levels of prolactin, cortisol, growth hormone and catecholamines in schizophrenia patients. Biol Psychiatry 14:109–114, 1985

King D, Cooper S, Liddle J: The effect of propranolol on CSF amine metabolites in psychiatric patients. Br J Clin Pharmacol 15:331–337, 1983

Kirch D, Jaskiw G, Linnoila M, et al: Plasma amine metabolites before and after withdrawal from neuroleptic treatment in chronic schizophrenic inpatients. Psychiatry Res 25:233–242, 1988

Kleinman J, Karoum F, Rosenblatt J, et al: Postmortem neurochemical studies in chronic schizophrenia, in Biological Markers in Psychiatry and Neurology. Edited by Usdin E, Hanzn I. New York, Pergamon, 1982, pp 67–76

Ko G, Jimerson D, Wyatt R, et al: Plasma 3-methoxy-4-hydroxphenylglycol changes associated with clinical state and schizophrenic subtype. Arch Gen Psychiatry 45:842–846, 1988

Lake R, Sternberg D, van Kammen D, et al: Schizophrenia: elevated cerebrospinal fluid in norepinephrine. Science 207:331–333, 1980

Lane R, Blaha C, Rivet J: Selective inhibition of mesolimbic dopamine release following chronic administration of clozapine: involvement of alpha-1 noradrenergic receptors demonstrated by in vivo voltametry. Brain Res 460:398–401, 1988

Linnoila M, Ninan P, Scheinin M, et al: Reliability of norepinephrine and major monoamine metabolite measurements in CSF of schizophrenic patients. Arch Gen Psychiatry 40:1290–1294, 1983

Lohr J, Jeste D: Locus coeruleus morphometry in aging and schizophrenia. Acta Psychiatr Scand 77:689–697, 1988

Mueser K, Rosen A, Javaid J, et al: Urinary 3-methoxy-4-hydroxphenylglycol in affective and schizophrenic patients. Biol Psychiatry 15:57–61, 1986

Naber D, Albus M, Bürke H, et al: Neuroleptic withdrawal in chronic schizophrenia: CT and endocrine variables relating to psychopathology. Psychiatry Res 16:207–219, 1985

Peet M, Bethell M, Coates A, et al: Propranolol in schizophrenia: comparison of propranolol, chlorpromazine and placebo. Br J Psychiatry 139:105–111, 1981

Pickar D, Brier A, Hsiao J, et al: Cerebrospinal fluid and plasma monoamine metabolites and their relation to psychosis. Arch Gen Psychiatry 47:641–648, 1990

Pugh C, Steinert J, Priest R: Propranolol in schizophrenia: a placebo controlled trial of propranolol as an adjunct to neuroleptic medication. Br J Psychiatry 143:151–155, 1983

Roberts E: A hypothesis suggesting that there is a defect in the GABA system in schizophrenia: an essay. Neuroscience Research Program Bulletin 10:468–483, 1973

Rotrosen J, Miller A, Mandio D, et al: Prostaglandins, platelets, and schizophrenia. Arch Gen Psychiatry 37:1047–1054, 1980

Sedvall G, Alfredsson G, Bjerkenstedt L, et al: Central biochemical correlates to antipsychotic drug action in man, in The Impact of Biology on Modern Psychiatry. Edited by Gershon ES, Belmaker RH, Kety SS, et al. 1976, pp 41–54

Seeman P, Lee T, Chau-Wong M, et al: Antipsychotic drug doses and neuroleptic dopamine receptors. Nature 261:717–719, 1976

Simpson G, Dunz-Bartholini E, Watts T: A preliminary evaluation of the sedative effects of Catapres, a new antihypertensive agent in chronic schizophrenic patients. J Clin Pharmacol 7:221–225, 1967

Snyder S: The dopamine hypothesis of schizophrenia: focus on the dopamine receptor. Am J Psychiatry 133:197–202, 1976

Stein L, Wise C: Possible etiology of schizophrenia: progressive damage to the noradrenergic reward system by 5-hydroxydopamine. Science 171:1032–1036, 1971

Sternberg D, van Kammen D, Lake C, et al: The effect of pimozide on CSF norepinephrine in schizophrenia. Am J Psychiatry 138:1045–1051, 1981

Svensson T: Peripheral autonomic regulation of locus coeruleus noradrenergic neurons in the brain: putative implications for psychiatry and psychopharmacology. Psychopharmacology (Berl) 92:1–7, 1987

Taghzouti K, Simon H, Herve D, et al: Behavioral deficits induced by an electrolytic lesion of the rat ventral mesencephalic tegmentum are corrected by a superimposed lesion of the dorsal noradrenergic system. Brain Res 440:172–176, 1988

van Kammen DP, Gelernter J: Biochemical instability in schizophrenia, II: the serotonin and gamma-aminobutyric acid systems, in Psychopharmacology: The Third Generation of Progress. Edited by Meltzer HY. New York, Raven, 1987, pp 753–758 van Kammen D, Bunney W, Docherty J, et al: D-Amphetamine induced heterogeneous change in psychotic behavior in schizophrenia. Am J Psychiatry 139:991–997, 1982

van Kammen D, Peters J, van Kammen W, et al: Clonidine treatment of schizophrenia: can we predict treatment response? Psychiatry Res 27:297–311, 1989a

van Kammen D, Peters J, van Kammen W, et al: CSF norepinephrine in schizophrenia is elevated prior to relapse after haloperidol withdrawal. Biol Psychiatry 26:176–188, 1989b

van Kammen D, Peters J, Yao J, et al: Norepinephrine and relapse in chronic schizophrenia: negative symptoms revisited. Arch Gen Psychiatry 47:161–168, 1990

van Praag H, Kahn R, Asnis G: Denosologicalization of biological psychiatry or the specificity of 5HT disturbances in psychiatric disorders. J Affect Disord 13:1–8, 1987

Wise C, Stein L: Dopamine-beta-hydroxylase deficits in the brains of schizophrenic patients. Science 181:344–393, 1973

Wyatt R, Schwarz M, Erdelyi E, et al: Dopamine β-hydroxylase activity in brains of chronic schizophrenic patients. Science 187:368–370, 1975

Yamamoto K, Hoshino T, Takahashi Y, et al: Skin conductance activity after intraventricular administration of 6-hydroxydopa in rats. Biol Psychiatry 29:365–375, 1991

Yorkston N, Zaki S, Pitcher D, et al: Propranolol as an adjunct to the treatment of schizophrenia. Lancet 2:575–578, 1977

Zander K, Fischer B, Zimmer R, et al: Long-term neuroleptic treatment of chronic schizophrenic patients: clinical and biochemical effects of withdrawal. Psychopharmacology (Berl) 73:43–47, 1981

Zubenko G, Cohen B, Lipinski J, et al: Clonidine in the treatment of mania and mixed bipolar disorder. Am J Psychiatry 141:1617–1618, 1984

15

Affective Disorders and Adrenergic Function

HUSSEINI K. MANJI, M.D., F.R.C.P.C.
MATTHEW V. RUDORFER, M.D.
WILLIAM Z. POTTER, M.D., PH.D.

Originally proposed a quarter century ago, the catecholamine hypothesis of affective disorders posited a deficiency of norepinephrine (NE) at critical sites in the central nervous system (CNS) in depression and an excess of functional NE in mania (Bunney and Davis 1965; Schildkraut 1965). These hypotheses were generated from the clinical observation that drugs that deplete CNS monoamines (e.g., reserpine) produced depression in some individuals and from the in vitro monoamine reuptake blocking activity of the therapeutically effective tricyclic antidepressants. This postulate has been extensively investigated but has proven difficult to study experimentally due, at least in part, to the formidable methodological difficulties in assessing CNS noradrenergic function in humans.

Nevertheless, the possibility that NE function might specifically be involved in both the pathogenesis of affective disorders and the mechanism of action of antidepressant and antimanic drugs has been the focus of extensive research. The sheer volume of research data that have been generated has contributed to apparent controversies in the literature concerning not only the nature of noradrenergic dysfunction but also whether any such dysfunctions exist. In this chapter, we briefly review, critically appraise, and integrate the research findings in an attempt to present a coherent summary of the present status of findings on NE in affective illness.

Anatomical and Physiological Considerations

Although the details of the noradrenergic neuronal system and the sympathetic nervous system (SNS) have been described in detail elsewhere (Glover, Chapter 2, this volume), a brief outline may be useful here as a backdrop to the present discussion on affective disorders. The NE neurons in the brain stem and the locus coeruleus (LC) cell group in particular, although small in number, constitute a diffuse system of projections to widespread brain areas via highly branched axons. The extensive efferent innervation suggests that the LC plays a modulatory and integrative role, rather than having a role in specific sensory or motor processing (Foote et al. 1983; Jacobs 1990; Woodward et al. 1979). A number of physiological roles have been ascribed to the LC, notably in the control of vigilance and the initiation of adaptive behavioral response (Everitt et al. 1990; Foote et al. 1983; Svensson 1987). Considerable data support the hypothesis that noradrenergic neurons in the LC compose a CNS response or defense system, since the neurons are activated by "challenges" in both the behavioral-environmental and the physiological domains (Jacobs 1990).

Thus, although a variety of sensory stimuli are capable of increasing LC activity, noxious or stressful stimuli are particularly potent in this regard. Moreover, considerable evidence also supports a role of LC-NE neurons in the learning of aversively motivated tasks and in the conditioned response to stressful stimuli (Rasmussen and Jacobs 1986; Rasmussen et al. 1986), with obvious implications for a variety of psychiatric conditions. Indeed, tonic activation of the LC appears to occur preferentially in the response to stressful stimuli in contrast to stimuli limited to simply evoking activation or arousal (Rasmussen and Jacobs 1986; Rasmussen et al. 1986). Thus, dysregulation of the LC-NE system in affective illness has been postulated to be associated with both the prominent disturbances of vegetative function and the dysphoric hyperarousal frequently seen in these states (Siever 1987).

The relationship between the activity of central NE neurons and SNS activity is complex, with central NE serving both sympatho-excitatory and sympatho-inhibitory roles, depending on a variety of factors, including the brain region and receptor subtype(s) on which the NE acts. Moreover, the assumption that the SNS functions as a homogeneous whole, increasing or decreasing activity en masse, is no longer tenable. Thus, central circuits controlling sympathetic nerve discharge are capable of producing complex and highly differentiated response patterns. Nevertheless, it is possible to make some generalizations about the role of certain central noradrenergic nuclei in regulating sympathetic outflow.

Considerable evidence has accumulated that LC-NE projections can exert major *sympatho-excitatory* effects under experimental conditions. Thus, stimulation of the LC increases peripheral NE in blood. A variety of physiological and pharmacological manipulations that activate the LC also increase sympathetic activation (Reiner 1986; Svensson 1987), but whether this represents an ordered sequence of—stimulus → LC activation → SNS activation—is uncertain. Indeed, in animal experiments, one observes a higher degree of positive correlation between LC-NE unit activity and sympathetic activation following environmental challenges than after physiological challenges (Jacobs 1990). Interestingly, under certain physiological challenges (e.g., hemorrhage), the activity of the LC neurons is dissociated from sympathetic activation, demonstrating that pathways such as those involved in the baroreceptor response exist for sympathetic activation without stimulating the LC (Jacobs 1990). Thus, although LC-NE neurons are *generally* coactivated with both the neural and hormonal components of the sympathoadrenal system, sympathetic activation can occur in the absence of increased activity in this neuronal system.

The sympatho-excitatory effect of experimentally stimulating LC neurons appears to be exerted both via rostral projections from the brain stem to the posterior hypothalamus and even more indirectly via polysynaptic mechanisms involving rostral projections to a variety of limbic areas, including amygdala, hippocampus, and septum (Lechin et al. 1989). This strong correlation that generally exists between the activity of the LC-NE neurons and the SNS has led to the suggestion that they function in concert as a general defense system. Such a contention is supported by the electrophysiological evidence that NE increases the signal-to-noise ratio of afferent information transfer in the CNS (Woodward et al. 1979), characteristics that are ideal for a system hypothesized to integrate the activity of CNS structures responsible for sensory, motor, hormonal, autonomic, and cognitive responses to imperative or threatening environmental and physiological challenges.

In contrast to the sympatho-excitatory LC-NE neurons, NE neurons of the A1 nuclei are *sympatho-inhibitory* and play a role in cardiovascular regulation. Thus A1 noradrenergic neurons tonically inhibit nearby sympatho-excitatory C1 neurons in the rostral ventrolateral medulla, a major focus for the generation of vasomotor tone (Reis et al. 1988). A1 NE neurons also appear to exert sympatho-inhibitory effects via indirect inhibition of LC neurons (Lechin et al. 1989). Thus, although the relationship between the activity of "central NE neurons" and sympathetic activity is complex, it may be possible to make certain inferences from peripheral NE measures. Although caveats clearly are necessary, clinical investigators reason that peripheral sympathetic activity should be *positively* correlated with LC-NE activity and negatively correlated with A1 NE neuronal activity. It is clearly not possible to make assumptions about "brain NE" based on measures of peripheral NE activity, since the discrete major NE nuclei exert opposite effects on the SNS. Nonetheless, working on the assumption based on the reasoning described above that the LC-NE projections may be more implicated in the affective disorders than the medullary A1 neurons, *adequate* measures of peripheral NE activity (see Appendices 1 and 2 in this chapter) could yield useful information about the relative balance between central NE systems.

Studies on Norepinephrine and MHPG in Plasma

In the last decade, there has been an increasing number of investigations of plasma NE with the majority of the studies revealing some degree of elevation interpreted as evidence of increased peripheral SNS activity in patients with major depression (Barnes et al. 1983; Davis 1990; de Villiers et al. 1987; Koslow et al. 1983; Louis et al. 1975; Maas et al. 1987; Roy et al. 1985b, 1987, 1988; Rudorfer et al. 1985; Veith et al. 1988; Wyatt et al. 1971). Several studies using radioenzymatic and radiotracer assay techniques have demonstrated elevated plasma NE concentrations at presumed rest in patients with major depression, although there is considerable overlap. These elevations appear to be largely attributable to unipolar depressive patients, in particular those who meet criteria for melancholia (Roy et al. 1985b; Veith et al. 1985). Similarly, unipolar patients with dexamethasone nonsuppression have significantly higher lying plasma NE concentrations than patients with dexamethasone suppression (Barnes et al. 1983; Roy et al. 1985b). The variance in plasma NE has also been reported to be greater in depressed patients than control subjects in the absence of differences between the means (Siever et al. 1986).

However, since plasma NE is determined by a variety of factors, including release, reuptake, and degradation (Kopin 1985), NE "spillover" rates have been calculated. This technique, discussed elsewhere (Goldstein, Chapter 8, this volume), involves repeated blood sampling following an infusion of tritiated NE to a plateau concentration and avoids potential artifacts from interindividual differences in NE clearance. Esler et al. (1982) demonstrated that depressed patients had elevated spillover rates, theoretically representing the rate of entry of NE to plasma from peripheral sympathetic nerves. It is likely, however, that uptake-1 (the reuptake of the neurotransmitter into presynaptic nerve terminal) is much more important for the removal of endogenously released NE than for removal of circulating (and infused) NE, which may preclude accurate interpretation of the results.

This caveat notwithstanding, another study found that endogenously, but not nonendogenously, depressed patients showed a selective increase in NE spillover with no intergroup differences in clearance rate when compared with control subjects (Veith et al. 1985). Rosenblatt et al. (1969) had earlier infused labeled NE and measured the ratio of amines to oxidized metabolites. Depressed patients excreted relatively more NE and less oxidized metabolites. Attempting to control for a number of confounding variables, including period off prior medications, baseline activity level, and assay methodology, it appears that plasma NE may be heterogeneously abnormal in depressed patients under conditions of limited rest (supine for 30–60 minutes prior to blood drawing), with a significant subgroup (particularly unipolar, anxious, melancholic patients with dexamethasone nonsuppression) showing elevated plasma NE (see Table 15–1). A subgroup of bipolar depressed patients may, however, actually show reduced supine NE (Rudorfer et al. 1985).

Going beyond the study of plasma NE levels under resting conditions, studies of the responsiveness of plasma NE to various provocative challenge tests provide evidence for dysregulation of the NE system in depression. Thus, we have previously demonstrated, using an orthostatic challenge paradigm, that the increase in plasma NE produced by standing up is consistently greater in depressed unipolar or bipolar patients versus age- and sex-matched control subjects (Rudorfer et al. 1985). This has been

Table 15–1. Plasma norepinephrine concentrations across studies

		Norepinephrine (pmole/ml)				
		Control subjects			Depressed patients	
Study	*n*	**Supine**	**Upright**	*n*	**Supine**	**Upright**
Wyatt et al. 1971	22	1.21	—	1	2.18	—
Jones et al. 1978	12	1.42	3.19	—	—	—
Esler et al. 1982	17	1.39	—	11	2.56	—
Lake et al. 1982	22	1.70	3.03	15	3.21	4.72
Veith et al. 1983	8	2.31	3.45	14	1.8	3.4
Barnes et al. 1983	—	—	—	9	2.24	4.24
Siever et al. 1984	21	1.21	—	—	—	—
Cooper et al. 1985	—	—	—	9	5.41	—
Rudorfer et al. 1985	12	1.56	2.74	4	1.35	3.57
Roy et al. 1985b	10	0.96	1.89	17	2.05	4.06
de Villiers et al. 1987	20	1.2	—	15	3.1	—
Rothschild et al. 1987	6	1.87	—	22	2.5	—
Hauger et al. 1988	—	—	—	6	1.14	—
Sevy et al. 1989	14	0.51	—	14	0.61	—
Weighted means		**1.34**	**2.88**		**2.34**	**3.91**

demonstrated in several other studies (Roy et al. 1985b; Rubin et al. 1985; Veith et al. 1988). Depressed patients also show significantly higher plasma NE levels following a cold pressor test than age- and sex-matched control subjects, with the unipolar melancholic patients with dexamethasone nonsuppression showing a trend toward higher plasma NE than the patients with dexamethasone suppression (Roy et al. 1987). Taken together, the studies of plasma NE provide strong evidence for dysregulation of peripheral release of NE in affective illness, with a difference in the pattern of dysregulation between unipolar and bipolar patterns. Stated another way, bipolar patients appear to have reduced-to-average resting output of NE with a highly exaggerated NE response to standing; unipolar patients have an average-to-elevated resting NE with moderately exaggerated response to standing (and perhaps "stressors" in general).

Studies of plasma 3-methoxy-4-hydroxyphenylglycol (MHPG) have yielded variable results and do not generally support the concept of a unipolar-bipolar distinction (Goodwin and Jamison 1990). Plasma MHPG in unipolar depressive patients tends to be similar to that of control subjects, albeit with greater variance (Siever and Uhde 1984). Similar to the association with plasma NE, depressed patients with dexamethasone nonsuppression have higher levels of plasma MHPG (Jimerson et al. 1983; Roy et al. 1986). Plasma MHPG levels tend to be lower in bipolar than unipolar depressed patients (for a review, see Goodwin and Jamison 1990; Siever 1987) and interestingly are higher in bipolar patients when manic than when depressed (Halaris 1978; Jimerson et al. 1981). In a study of 18 manic patients, plasma MHPG correlated with their manic symptoms (in particular, grandiosity and elation) but not with anxiety, depression, motor behavior, acute psychosis, or severity of illness (Azorin et al. 1990). The mechanisms through which MHPG is elevated in mania remain to be determined.

Studies on Norepinephrine and MHPG in Cerebrospinal Fluid

For some years, measures of NE and its metabolites in CSF were thought to reflect brain NE "activity" directly. This assumption is, however, problematic since high correlations have been found between plasma and

CSF NE and MHPG (Goldstein et al. 1987; Kopin 1985). Pharmacological studies in dogs reveal parallel changes in plasma and CSF NE following ganglion blockade, suggesting that sympathetic outflow may determine (at least in part) NE concentrations in both compartments (Goldstein et al. 1987). Plasma MHPG is the major source of CSF MHPG, which readily crosses the blood-brain barrier (Kopin 1985). Using the high correlation between plasma and CSF MHPG observed in comparisons of high and low catecholamine output states, an equation has been derived to "correct" for the contribution of plasma MHPG to that of CSF (Goldstein et al. 1987; Kopin 1985). However, in the relatively narrow range of values observed in depressed patients and control subjects, the utility of such an equation to identify "brain MHPG" is questionable (Linnoila et al. 1986) and, to date, has not revealed new insights in studies of depression.

Certain other methodological problems are unique to CSF measures. Standards for obtaining spinal fluid such as elapsed time between needle insertion and sample collection have not been established. Therefore, subjects may not have the same degree of accommodation to the stress of the needle stick. Moreover, sampling at a single point in time may not reflect the biochemical process of depression or mania but rather a state-dependent fluctuation from a recent external or internal stress.

Four of the seven studies of CSF concentrations of NE that compared depressed patients and control subjects in the same study reported elevations (Christensen et al. 1980; Dencker et al. 1966; Post et al. 1989; Roy et al. 1988); one study reported a decrease (Post et al. 1978), and the other two no significant change (Berrettini et al. 1984, 1985) (Table 15–2). One noted an overall decrease in reported values of NE after investigators began to use high-performance liquid chromatography with electrochemical detection in the 1980s (Davis 1990). CSF concentrations of NE have been reported to be elevated in depressed patients with an atypical presentation, higher scores for nurse-rated anxiety, and a longer duration of hospitalization (Post et al. 1984). Earlier investigations had shown that CSF NE was higher in mania than in depression (Post et al. 1978). Perhaps the most novel finding has been that in dysphoric mania, defined by the coexistence of high depression ratings, NE in CSF correlates modestly but significantly ($r \approx .5$) with ratings of dysphoria and anxiety but not with ratings of mania (Post et al. 1989). These authors suggested that CSF NE may be positively correlated with the degree of anxiety across a variety of psychopathological syndromes, including depression, mania, and perhaps anxiety.

CSF MHPG in depressed patients is consistently reported as not different from that in control subjects, except for higher levels in postmenopausal women (Koslow et al. 1983). Within the depressed group, however, signif-

Table 15–2. Cerebrospinal fluid norepinephrine concentrations across studies

	Control subjects		Depressed patients	
Study	*n*	**pg/ml**	*n*	**pg/ml**
Before high-performance liquid chromatography				
Dencker et al. 1966	7	200	13 (depressed)	700
			3 (manic)	500
Post et al. 1978	17	225	13 (depressed)	210
			6 (manic)	460
After high-performance liquid chromatography				
Christensen et al. 1980	18	102	15 (unipolar and bipolar)	132
Linnoila et al. 1983			16	95
Berrettini et al. 1984	20	54	10 (bipolar)	57
Siever et al. 1984a			12	319
Berrettini et al. 1985	23	67	12	76
Roy et al. 1985a			15 (melancholia)	123
			8 (episode)	118
			5 (dysthymic disorder)	71
Roy et al. 1988	25	100	28	131
Post et al. 1989	48	95	26	152
Weighted means		**84.2**		**129.4**

icant subgroup differences emerged in some reports, suggesting lower CSF MHPG in bipolar I than in unipolar patients (Potter et al. 1987). As was the case for CSF NE, variables other than overall diagnosis may influence values. For instance, within a group of depressed patients, those with increased anxiety, agitation, somatization, and sleep disturbance had significantly elevated levels of CSF MHPG (Redmond et al. 1986). CSF MHPG has also been reported to be elevated in manic patients compared with control subjects (Post et al. 1984; Redmond et al. 1986; Swann et al. 1983, 1986). Swann et al. (1987) also noted that CSF MHPG concentration correlated with several dysphoric elements of the manic syndrome (i.e., total manic severity and hostility). Moreover, this study found a significant reduction in CSF MHPG with lithium treatment, even when treatment was unsuccessful.

Taken together, the CSF studies of NE and its metabolite MHPG suggest that NE output is higher in mania than in depression and that there may be relatively higher values in unipolar versus bipolar depression. Relative elevation within patient groups may, in turn, be related to anxiety or the overall severity of the condition. As noted previously, it is possible that the CSF findings reflect events occurring in the SNS as much as events occurring in the brain, and it is therefore not surprising that the pattern of findings is similar in CSF and plasma. In a broad sense, the distinction separating brain activity and sympathetic outflow is somewhat artificial since, as noted at the outset, there are intimate links between CNS function and peripheral NE output. Nevertheless, CSF NE or MHPG was correlated with dysphoric elements of the manic syndrome, whereas urinary measures tended to be associated with euphoric components (Post et al. 1989; Swann et al. 1986), suggesting that studies examining multiple measures and components of the NE system(s) allow for the most meaningful interpretations.

Studies on Norepinephrine and Its Metabolites in Urine

Since the catecholamine hypothesis of affective disorders was proposed, attempts to characterize output of the NE system in depressed patients have focused on measurements of MHPG in urine more than any other single parameter (see Filser et al. 1988; Potter and Linnoila 1989; Potter et al. 1987; Schatzberg et al. 1980; Schildkraut et al. 1978a). These studies were fueled (at least to some extent) by the early interpretations of data suggesting that about 50% of MHPG in the urine was derived from the CNS. However, subsequent work indicated that MHPG readily crosses the blood-brain barrier to the CSF; earlier estimates of CNS contribution to urinary MHPG have thus been revised to approximately one-third (reviewed in Kopin 1985). Nevertheless, as discussed, peripheral and central NE systems interact, and thus even a "primarily peripheral" measure such as urinary MHPG can theoretically provide useful information.

Detailed reviews of the literature differ in their conclusions, but some emphasize that there is decreased (albeit modestly) 24-hour urinary excretion of MHPG in depressed patients compared with control subjects. When one examines the studies in more detail, it appears that the MHPG reduction is accounted for exclusively by bipolar patients, although a study of subsequent patients in the same center does not replicate this finding, perhaps suggesting a change in the patient population (Potter et al. 1993). Reduced urinary MHPG may be present only in bipolar I and not in bipolar II patients (Beckman and Goodwin 1980; Muscettola et al. 1984). Average urinary MHPG is not reduced in unipolar populations taken as a whole; rather, there may be a subgroup of unipolar patients who have elevated MHPG as compared with control subjects and bipolar subjects (Schatzberg et al. 1982). Investigators have also attempted to use this variance to subtype patients within the unipolar group and to attempt to predict treatment response based on pretreatment urinary MHPG values (Goodwin et al. 1978; Hollister et al. 1980; Schildkraut et al. 1978b). Indeed, several reports supported the notion that in unipolar depression, lower urinary MHPG predicts favorable responses to antidepressants that primarily block NE reuptake (imipramine, desipramine, nortriptyline, maprotiline) (Faucett et al. 1972; Maas et al. 1972, 1982; Rosenbaum et al. 1980). Others, however, could find no evidence for such a prediction (Janicak et al. 1986; Muscettola et al. 1984). Again, these different findings may represent the shifting patient profile studied at tertiary referral centers.

Lower levels of urinary MHPG in depressed patients have been reported to approach normal values with clinical improvement, suggesting that this measure may be state dependent (Pickar et al. 1978). Similarly, longitudinal studies of bipolar patients suggest increased MHPG excretion in the manic state compared with the depressed state (Post et al. 1984; Potter et al. 1987). Swann et al. (1987) noted that manic patients responding to lithium had decreased MHPG and increased NE excretion relative to the total excretion of NE and its metabolites, suggesting that lithium response was associated with an alteration of catecholamine metabolism pathways. Overall, urinary MHPG by itself has not proven to be a sufficiently robust and consistent measure to be generally accepted as a useful tool in diagnosis or prediction of treatment response (Davis and Bresnahan 1987).

In recent years, investigators have attempted to go beyond the "too little or too much" hypothesis of affective disorders; one approach has been to measure the 24-hour concentrations of urinary catecholamines and their metabolites to identify possible abnormalities in the relative activity of NE metabolic pathways in depression and mania. As discussed in Appendices 1 and 2, one can obtain reliable information about the average output of NE and its metabolites within an individual in a stable mood state from only two consecutive 24-hour urine collections. Furthermore, since all major metabolites are measured, 24-hour urinary measures of NE and of its metabolites can account for interindividual differences in the relative metabolism of NE as well as its turnover (amount formed and excreted per 24 hours at steady state).

Consistent with the findings of elevated basal or stress-induced plasma NE, several investigators observed elevated urinary excretion of NE and of its major extraneuronal metabolite, normetanephrine, in depressed patients (Davis et al. 1988; Maas et al. 1987; Roy et al. 1985b, 1986, 1988). Moreover, the finding appears more impressive when the excretion of NE and normetanephrine in depression is examined *relative* to total NE excretion. Thus, in the study of Maas et al., a modest increase in total urinary catecholamine excretion (16%) was accompanied by marked increases in urinary NE (57%) and normetanephrine (42%) in depressed patients. These results suggest a shift toward extraneuronal metabolic pathways and are consistent with the suggestions of increased NE release and "spillover" in depression. Indeed, experimental data in humans indicate that a greater pulsatile release of NE is associated with a relative increase in NE excretion and a relative decrease in MHPG excretion (Maas et al. 1970, 1971).

Our group has recently analyzed data collected from more than a decade of studies on an inpatient unit at the National Institute of Mental Health. We compared baseline excretion of urinary NE and its metabolites and the fractional extraneuronal concentration of NE (NE + normetanephrine/ΣNE) in unmedicated unipolar and bipolar depressed patients and healthy volunteer subjects. Consistent with previously reported results, urinary NE as well as the urinary NE and normetanephrine relative to total turnover were elevated in the depressed patients as a whole compared with healthy volunteer subjects. In contrast to previous studies, we found significantly elevated concentrations of urinary NE in bipolar, but not unipolar, depressed patients compared with healthy volunteer subjects and a trend toward decreased MHPG in both patient groups (Potter et al. 1993). Several possible methodological factors that may account for these differences, including differences in assay and length of inpatient drug-free duration (a minimum of 4–8 weeks at our Center). Interestingly, total turnover of NE (NE + normetanephrine + MHPG + vanillylmandelic acid) was significantly lower in unipolar and bipolar patients compared with healthy volunteer subjects, suggesting a reduction of tyrosine hydroxylase activity in sympathetic neurons.

Clinical Studies of Adrenergic Receptors

The future development of selective receptor ligands for positron-emission tomography (PET) studies may soon permit the direct assessment of CNS adrenergic receptors in humans. To date, studies of NE receptors in affective disorders have been limited to indirect research strategies. The most commonly utilized strategies include 1) characterization of receptor number and function in readily accessible blood elements and 2) pharmacological "challenge" strategies whereby alterations in biochemical neuroendocrine, cardiovascular, or behavioral parameters in response to various receptor agonists and/or antagonists are measured.

Receptor Studies on Blood Cells

Due to the accessibility of platelets and lymphocytes, both α_2- and β_2-adrenergic receptors have been extensively studied in affective disorders, frequently generating elaborate hypotheses about CNS adrenergic receptor dysfunction based solely on studies of these tissues. There are several problems with the assumption that changes in adrenergic receptors on peripheral cells reflect similar alterations in the CNS. To begin with, receptors on blood cells are, by definition, noninnervated, exist in a markedly different environment, and may therefore poorly reflect central, innervated adrenergic receptors.

Another major (often overlooked) consideration when interpreting studies of dynamic receptor regulation in blood cells is that white blood cell counts and the relative proportions of subsets of lymphocytes may vary. Recruitment of cells with different characteristics into the circulation may frequently explain altered receptor function. Precisely such a mechanism appears to be operative in the studies demonstrating the seemingly paradoxical *increase* in lymphocyte β-adrenergic receptor density and responsiveness during short-term isoproterenol infusion, mental arithmetic, and dynamic exercise (procedures that stimulate the SNS) (Maisel et al. 1990a, 1990b; Van Tits et al. 1990). The mechanism appears to be a release of subsets of "fresh" lymphocytes from the spleen into the circulation (see Van Tits et al. 1990; Werstiuk et al. 1990). These fresh, activated lymphocytes express enhanced β-adrenergic receptor responsiveness, and this probably accounts for the exercise- and catecholamine-induced increases in β-adrenergic receptor responsiveness. Indeed, one study reported a "normalization" of previously blunted lymphocyte β-adrenergic receptor responsiveness in depressed patients following electroconvulsive therapy (Mann et al. 1990b). Given the increase in plasma catecholamine produced acutely by electroconvulsive therapy (Mann et al. 1990a) and the intermittent "phasic" nature in which electroconvulsive therapy is administered (e.g., three times per week), the exchange of lymphocyte subsets from the spleen to the circulation may explain the apparent "resensitization" of circulating β-receptors.

Platelet α_2-Adrenergic Receptors

Based on the assumption that postulated alterations in CNS α_2-adrenergic receptors may also be reflected in peripheral tissues, numerous studies have measured the binding of α_2-agonist or -antagonist ligands to platelets obtained from patients with affective illness and from control subjects. However, although human platelets and cerebral cortex contain homogeneous populations of the same α_{2A} receptors (Bylund et al. 1988), radioligand studies across groups are confounded by numerous methodological problems, such as varying patient populations, sex, age, clinical state, drug washout period, and assay technique (see review by

Piletz et al. 1986). A review of 13 studies using yohimbine-alkaloid radioligands revealed no significant differences in the B_{max} of platelet α_2-receptors between depressed patients and control subjects (see Doyle et al. 1985; Georgotas et al. 1987; Kafka and Paul 1986; Kafka et al. 1986; Katona et al. 1987; Lenox et al. 1983; Piletz et al. 1986; Wolfe et al. 1986, 1989). Most of the studies using partial or full agonists, however, have observed increased B_{max} in the platelets of depressed patients as compared with control subjects (Garcia-Sevilla et al. 1983, 1986, 1990; Pandey et al. 1989; Piletz et al. 1986; Smith et al. 1983; Stahl et al. 1983). These results have frequently been interpreted as evidence for the α_2-hypersensitivity theory of depression (Garcia-Sevilla et al. 1986, 1990; Piletz et al. 1986).

These studies, however, have used the imidazoline compounds clonidine, *p*-aminoclonidine, and UK-14304 as radioligands, and there are problems with the use of these radioligands that go beyond reproducibility of measures across groups. Thus, it is now clear that a variety of compounds (e.g., clonidine, *p*-aminoclonidine, idazoxan, UK-14304) previously considered to be selective α_2-ligands bind also with high affinity to non-adrenergic catecholamine-insensitive sites (Boyajian and Leslie 1987; Bricca et al. 1988a, 1988b; Ernsberger et al. 1987; Meeley et al. 1986). These sites, termed imidazoline sites, appear to mediate the hypotensive effects of clonidine (Bousquet and Feldman 1987) and are also present on human platelets (Michel et al. 1990). Furthermore, the ratio of imidazoline sites to α_2-receptors in human platelets varies widely among healthy volunteer subjects (Michel et al. 1990), making it difficult to arrive at any meaningful inferences about α_2-adrenergic receptors on the basis of these agonist binding studies. Indeed, a report even suggests that the elevated platelet *p*-aminoclonidine binding in depressed subjects is due solely to increases in the catecholamine-insensitive (imidazoline) sites (Piletz et al. 1990).

Since these imidazoline receptors may also be present on noradrenergic neurons and regulate NE release (Starke et al. 1989), these findings, if replicated, could nonetheless have important implications for the noradrenergic hypotheses of depression; however, the present evidence fails to support the suggestion of α_2-supersensitivity in depression. If anything, the results of some (but not all) studies examining α_2-inhibition of platelet adenylate cyclase (AC) activity are more supportive of subsensitive α_2-receptors in depression (reviewed in Kafka and Paul 1986; Pandey et al. 1990). As discussed, however, receptor number and sensitivity on circulating blood cells may be significantly regulated by plasma catecholamines (Sacchetti et al. 1985). In support of this, Freedman et al. (1990) demonstrated stress-induced desensitization of α_2-adrenergic receptors in human platelets, accompanied by significant increases in plasma catecholamines and subjective anxiety.

Mononuclear Cell β-Adrenergic Receptors

The well-documented decrease in the number of cortical β-adrenergic receptors and reduced sensitivity of β-adrenergic receptor–stimulated AC activity found in rodent brain following chronic administration of all classes of antidepressant drugs and repeated electroconvulsive shocks (Banerjee et al. 1977; Bergstrom and Kellar 1979; Deakin et al. 1981; Sulser et al. 1978) suggest that such changes may be related to the therapeutic action of antidepressant treatments. Peripheral β-receptors, which mirror changes in central β-adrenergic receptors, would clearly represent useful tools for defining the role of β-adrenergic receptors in depressive illness and in the effects of treatment. However, alterations in β-adrenergic receptor density in the rat brain induced by antidepressants appear to be restricted to the β_1-subtype (Minneman et al. 1979), whereas human mononuclear cells contain only the β_2-subtypes (Meurs et al. 1982).

Despite these caveats, several groups have investigated the density of β-adrenergic receptors in untransformed lymphocytes or leukocytes of untreated depressed patients, yielding fairly inconsistent results. Many groups reported a decrease in β-adrenergic receptor number (Carstens et al. 1988; Extein et al. 1979; Magliozzi et al. 1989; Pandey and Davis 1986; Pandey et al. 1990); others described an increase or no change when depressed patients were compared with healthy volunteer subjects (Cooper et al. 1985; Healy et al. 1985; Mann et al. 1985; Pandey et al. 1985; Sarai et al. 1982; Zohar et al. 1983). These studies and possible methodological sources of the differences in results (e.g., methods of tissue preparation, type of ligand used, subtypes of patient populations, length of drug-free interval) have been discussed in a very thorough critical appraisal (Werstiuk et al. 1990).

In contrast to the inconsistent results from binding studies described earlier, most studies measuring mononuclear cell β-adrenergic receptor-stimulated AC activity reported decreased responsiveness in depressed patients compared with healthy volunteer subjects (Ebstein et al. 1988; Extein et al. 1979; Halper et al. 1988; Healy et al. 1983; Klysner et al. 1987; Mann et al. 1985; Pandey et al. 1979). The consistently observed decrease in leukocyte β-adrenergic receptor function in depression could reflect an inherited abnormality of the β-adrenergic receptor/G protein/AC complex, as suggested by the findings of Wright et al. (1984), using Epstein-Barr virus–transformed lymphocytes from manic-depressive patients and control subjects. However, these findings need to be replicated, and a number of additional confounding factors need to be considered. In this context, twin studies do not show a significant variability of isoproterenol-stimulated cyclic adenosine monophosphate (cAMP) production, suggesting that variations in isoproterenol-stimulated cAMP production are most likely due to "environmental" effects on the number of sensitivity of β-adrenergic receptors (Ebstein et al. 1986). As discussed, elevated catecholamine-induced changes in mononuclear cell β-adrenergic receptor sensitivity may be due to alterations in the subsets of lymphocytes. Additionally, elevated circulating catecholamines and glucocorticoids may exert significant effects on β-adrenergic receptor sensitivity (Buckholtz et al. 1988; Werstiuk et al. 1990), perhaps via their effects at the level of the G protein–stimulating AC, Gs (Malbon et al. 1990; Manji 1992). Thus, additional studies are clearly needed to ascertain if the consistent differences in mononuclear cell β-adrenergic receptor sensitivity in depression truly reflect comparable differences in the brain.

Pharmacological Challenge Strategies

Pharmacological challenge paradigms, which utilize agents known to stimulate receptor sites directly or indirectly, have been extensively utilized for testing pathophysiological hypotheses about noradrenergic dysfunction in affective illness (Horton et al. 1986; Siever 1987). The α_2-adrenergic agonist clonidine (which may also exert effects at imidazoline sites) has been administered to depressed patients, and the response of plasma MHPG, blood pressure, heart rate, sedation, growth hormone, and cortisol has been measured. Clonidine-induced decreases of plasma MHPG have been found to be somewhat more marked (Leckman et al. 1980; Siever et al. 1984a, 1984b, 1984c) or unchanged (Charney et al. 1983), when using either oral or intravenous clonidine, respectively. Similarly, the plasma MHPG response to yohimbine is unchanged in depressed patients (Heninger et al. 1988). Clonidine-induced decreases in blood pressure and increases in sedation have been found to be not significantly different in depressive patients compared with control subjects (Charney et al. 1982; Checkley et al. 1981). Similarly, responses of cortisol and adrenocorticotropic hormone to acute clonidine administration in depressed patients vary, with levels increased, decreased, or unchanged (Siever 1987).

In contrast, a series of studies have consistently shown a significantly reduced growth hormone response to clonidine (presumably mediated by postsynaptic hypothalamic α_2-receptors) in depressed patients (Amsterdam et al. 1989; Ansseau et al. 1988; Boyer et al. 1986; Carstens et al. 1987; Charney et al. 1982; Checkley et al. 1981, 1984, 1985; Hoehe et al. 1986; Lechin et al. 1985; Matussek et al. 1980; Mitchell et al. 1978; Siever et al. 1982; Uhde et al. 1986). These findings generally have been interpreted as evidence for subsensitive central postsynaptic α_2-receptors in depression, perhaps secondary to elevations in NE. However, it is clear that this response is not specific for depression, since a blunted growth hormone response to clonidine has been reported in patients with panic disorder (Charney and Heninger 1986; Nutt 1989; Uhde et al. 1986), generalized anxiety symptoms (Abelson et al. 1991), obsessive-compulsive disorder (Siever et al. 1983), and even in mania (Dinan et al. 1991). Thus, a blunted α_2-response may be observed in any condition characterized by tonic or episodic abnormally elevated central NE. Moreover, it is possible that factors distal to the α_2-receptors in the hypothalamus, such as growth hormone–releasing hormone or somatostatin may underlie the blunted growth hormone response to clonidine. Attempts to address this by the use of growth hormone–releasing hormone challenge tests have thus far proved unsuccessful, with conflicting results reported (Laakmann et al. 1990). Adding to the difficulty in interpreting the blunted growth hormone responses to clonidine are the findings suggesting that the major growth hormone-releasing effects of clonidine are exerted via inhibition of hypothalamic release of somatostatin rather than by stimulating growth hormone-releasing hormone secretion (Devesa et al. 1990).

A Synthesis

Efforts to study NE output and function in affective disorders have evolved for almost two decades and yielded a cumulative data base pointing to some overall degree of abnormality. Considerable evidence suggests that depressed patients excrete disproportionately greater amounts of NE and its major extraneuronal metabolite, normetanephrine, relative to total catecholamine synthesis compared with control subjects. This is particularly true of melancholic, unipolar depressed subjects, but our recent data suggest that under adequately controlled drug-free (> 4 weeks) study, bipolar depressed subjects may exhibit a similar dysregulation of the noradrenergic system. At least with regard to mania, the original catecholamine hypothesis has withstood the test of time, with increased noradrenergic function consistently observed in mania (although this finding may ultimately reflect a secondary effect). The intriguing recent findings that CSF and urinary NE measures may be associated with the dysphoric and euphoric components of the manic syndrome, respectively, deserve further investigation and suggest that something other than mania per se may be producing the changes.

Findings of increased fractional urinary output of urinary NE and normetanephrine and of an exaggerated raise in plasma NE on orthostatic challenge in depressed patients are compatible with those of Esler et al. (1982) and Veith et al. (1985) of increased "leakiness" of presynaptic NE terminals. The mechanisms for this increased "leakiness" remain unclear, but may involve both central and peripheral sites.

One possible mechanism is a subsensitivity of nerve terminal α_2-autoreceptors (Siever et al. 1985). These receptors operate as "thermostats" and depress NE release on activation (Starke et al. 1989). They do not directly influence the firing rate of the neuron but attenuate the release of neurotransmitters when an action potential depolarizes the varicosity or terminal or may even prevent the action potential from "invading" the nerve terminal (Starke et al. 1989). Indeed, preclinical studies have demonstrated that blockade of presynaptic α_2-receptors on sympathetic neurons markedly increases the amount of neurotransmitters released per impulse (Starke et al. 1989). Subsensitive peripheral α_2-autoreceptors would be expected to result in a greater fractional excretion of NE and an exaggerated NE release on any activation of the SNS (e.g., orthostatic stress, cold stress, "early hospitalization stress"). The suggestion of a subsensitivity of α_2-receptors receives additional support from the blunted growth hormone responses to clonidine and attenuated platelet α_2-function (as assessed by inhibition of AC activity). One might justifiably argue that the subsensitive α_2-receptors may be the sequelae, not the cause, of increased NE release in depression. Although this is likely true of the platelet (circulating) α_2-receptors, such a mechanism is difficult to reconcile teleologically for the nerve terminal α_2-receptors. If these autoreceptors do downregulate in the presence of increased NE, it would suggest the potential for an escalating feed-forward cycle, generally not seen in biological systems.

Another putative peripheral mechanism to explain the findings seen in depressed patients is that of an attenuation of the reuptake mechanism. The pattern of the relative excretion of NE and its metabolites in urine from depressed patients compared with that from control subjects is strikingly similar to the further shift in the pattern observed during treatment with NE reuptake blockers. Although considerable evidence suggests that depression is associated with reduced platelet *serotonin* reuptake sites, this has not been adequately investigated for NE due to the absence of suitable peripheral tissues. Nevertheless, there is some evidence that there may be physiological regulation of the reuptake process (Lee et al. 1983), and this requires further study in animal models or appropriate cultured cells.

As discussed at the beginning of the chapter, it is difficult to make inferences about "brain NE" based on peripheral measures. Nevertheless, the increased sympathetic activity consistently observed in depression would be compatible with *increased* activity of the A6 LC noradrenergic projections and/or *decreased* activity of the A1 "vasomotor" NE projections. The increased LC activity would also be compatible with the observation of increased CSF NE in depression and dysphoric mania (Post et al. 1984, 1989). Additionally, if one uses a "pharmacological bridge," the decreased firing of LC neurons, the decreased levels of tyrosine hydroxylase and of its messenger RNA in LC neurons (Nestler et al. 1990), and the reduced turnover in rat brain following chronic administration of antidepressants would support such a contention. Moreover, the observations by several independent laboratories that antidepressants downregulate β-adrenergic receptors in vitro (in the absence of presynaptic inputs) (Fishman and Finberg 1987; Honegger et al. 1986; Manji et al. 1991a) suggest that the aforementioned biochemical and electrophysiological changes need not simply be the "compensatory adaptations" secondary to elevated intrasynaptic NE.

Conclusion

Studies of the noradrenergic system in affective disorders clearly need to move beyond the "too little or too much" phase. It is clear the field needs to adopt new strategies to address the abnormalities of the NE system in these disorders. Our recent data indicate that it may be impossible to pick up subtle "signals" in this stress-responsive system by examining relatively heterogenous populations under poorly controlled conditions and in the presence of varying degrees of drug (e.g., tricyclic antidepressants, benzodiazepines) withdrawal. It may be considerably more fruitful to focus resources on examining smaller numbers of patients longitudinally under rigorous conditions.

There are also an increasing number of research strategies that are becoming feasible in clinical studies that have not yet been systematically applied to the study of affective illness. These include imaging techniques (in particular, PET), utilization of cell cultures from human subjects, signal transduction measures in circulating blood cells, and newer molecular biological techniques. The development of PET in recent years has made it theoretically possible to examine receptors in the living brain. Selective PET ligands for various receptors, as well as for reuptake sites, are currently being developed and will undoubtedly yield information about central noradrenergic systems hitherto unquantifiable in humans.

Another exciting approach that has already been applied to the study of alcoholism (Nagy et al. 1988) is the use of cultured, immortalized cells from patients with various clinical diagnoses. The unique property of immortalized cells is that they continue to divide for many generations and the progeny are genotypically identical and retain a stable phenotype. This allows for the examination of potential receptors and postreceptor defects in these illnesses in the absence of the confounding (frequently considerable) effects of circulating catecholamines or stress hormones (which are also abnormal in these populations).

There is increasing evidence that a variety of psychotropic drugs (including antidepressants and lithium) exert significant effects as postreceptor sites (Avissar et al. 1988; Manji 1992; Manji et al. 1991a, 1991b; Risby et al. 1991; Yamaoka et al. 1988). These include various G proteins and second messengers (e.g., cAMP, protein kinase C). These postreceptor targets for the psychotropic drugs suggest that they may represent the biochemical-molecular factors predisposing individuals to affective illness. Indeed, it is noteworthy that there are a growing number of illnesses in which G protein abnormalities have been implicated, whereas decades of studies of neurotransmitters and receptors have yielded less consistent results. Given their crucial role in signal amplification and integration, it is not surprising that G protein dysfunction has been identified in a variety of pathophysiological states. Furthermore, G proteins are far more selective with respect to their functions and localization than previously believed (Birnbaumer et al. 1990) and could very well be implicated in the pathophysiology of affective illness as preliminary results indicate (Manji 1992; Schreiber et al. 1991). Although speculative, these possibilities remain of interest given the cumulative experiences from studies of NE in depression and the frequency with which the noradrenergic system is implicated in the action of antidepressant treatments (Potter et al. 1988).

Appendix 1: NE Metabolites as an Index of Sympathetic Tone

Plasma NE is derived largely from the vast array of sympathetic nerves innervating blood vessels (Esler et al. 1982). However, most of the NE released by sympathetic nerves never reaches the circulation but is cleared by a variety of processes, particularly reuptake into presynaptic sympathetic nerve terminals (Kopin 1985). Intraneuronally, NE is deaminated by monoamine oxidase to dihydroxyphenylglycol (DHPG), which diffuses out of the neurons and is *O*-methylated by catechol-*O*-methyltransferase to MHPG. MHPG can be further oxidized (primarily in the liver) to vanillylmandelic acid. Ultimately, all NE produced must be excreted (primarily in urine), either as unchanged amine or as one of several metabolites, and the sum of NE and its metabolites in urine (ΣNE) should provide a good index of whole-body NE production. This may not be the best index of NE release by sympathetic neurons, however, since some amount of MHPG and vanillylmandelic acid derives from intraneuronal NE that was never released but that leaked from storage vesicles into the cytoplasm where it was oxidized by monoamine oxidase (Goldstein et al. 1988). Eisenhofer et al. (1988) estimated that under resting conditions, approximately 64% of DHPG is derived from NE leakage from storage vesicles. In contrast to MHPG and vanillylmandelic acid, normetanephrine is formed exclusively by extraneuronal *O*-methylation from NE that escaped reuptake (Kopin 1985). In patients with multiple system atrophy, it has been argued that NE and normetanephrine may better reflect NE release by sympathetic neurons, although the deaminated metabolites reflect total NE metabolism (Kopin 1985). Therefore, the ratio (NE + normetanephrine)/ΣNE is probably a better reflection of actual NE release than is ΣNE. Since the amount of MHPG and vanillylmandelic acid excreted per day is some 20–30 times greater than the amount of NE and normetanephrine, a small increase in NE release might be reflected in increases in NE and normetanephrine excretion without concomitant increases in MHPG and/or vanillylmandelic acid.

Appendix 2: Plasma NE Measurements—Additional Caveats

Of the various clinical indices of sympathetic nervous function, the measurements of the plasma concentration of NE have been used most extensively. Measuring NE (and MHPG) in plasma has the advantages of convenience, accessibility, and control. It allows frequent sampling under a variety of conditions such as controlled physical and mental stress. Indeed, under some conditions, venous plasma NE measures do appear to provide an adequate estimation of average sympathetic outflow. However, plasma NE measurements do have limitations when used as the sole indices of sympathetic activity under a variety of conditions. The plasma NE concentration is determined by the release, reuptake, degradation, redistribution, and partitioning into more than one physiological space (Linares et al. 1987). A variety of factors alter the rate at which NE is removed from plasma in humans (Esler et al. 1982), thereby rendering plasma NE less useful as a valid index of NE release and sympathetic tone (Ludwig et al. 1988). In this context, several investigators (including ourselves) have utilized the rise in plasma NE with upright posture as a clinical index of reflex sympathetic nervous responsiveness. Indeed, one study documented parallel increases in sympathetic activity and plasma NE (albeit with several minutes time lag), following nitroprusside-induced hypotension in healthy volunteer subjects (Rea et al. 1990). However, plasma NE clearance also falls with upright posture (Esler et al. 1988), presumably due to reduction in cardiac output and organ blood flows. The relative contribution of these two processes to the observed elevation of plasma NE is presently unclear, but highlights the need to consider alterations in clearance when interpreting changes in plasma catecholamine levels. One approach to overcome the confounding factors of clearance has been to determine NE "spillover" calculated by measuring the ratio of the infusion rate of ^{3}H-NE to the specific activity of plasma NE with repeated blood sampling following an infusion of ^{3}H-NE to a plateau concentration (Esler et al. 1982).

An additional concern is that circulating NE, probably consisting for the most part of spillover from the synaptic clefts of sympathetic innervation to arterial vascular walls, reflects peripheral, *regionalized* sympathetic outflow. Thus, the site of catecholamine sampling is an important factor, and several studies have outlined the limitations of antecubital venous measurements (Goldstein et al. 1987). Since the forearm is a site of net NE production, antecubital venous NE measurements tend to reflect forearm sympathetic nerve activity disproportionately.

Finally, unlike 24-hour urine collection for 2–3 consecutive days, there is greater intraindividual variability over time, at least for plasma samples at single time points. Given the very short half-life of NE (< 10 minutes), even with repeated plasma measuring, fluctuations may go undetected.

References

Abelson JL, Glitz D, Cameron OG, et al: Blunted growth hormone response to clonidine in patients with generalized anxiety disorder. Arch Gen Psychiatry 48:157–162, 1991

Amsterdam JD, Maislin G, Skolnick GB, et al: Multiple hormone responses to clonidine administration in depressed patients and healthy volunteers. Biol Psychiatry 26:265–278, 1989

Ansseau M, von Frenckell R, Certontaine JL, et al: Blunted response of growth hormone to clonidine and apomorphine in endogenous depression. Br J Psychiatry 153:65–71, 1988

Avissar S, Schreiber G, Danon A, et al: Lithium inhibits adrenergic and cholinergic increases in GTP binding in rat cortex. Nature 331:440–442, 1988

Azorin JM, Pupeschi G, Valli M, et al: Plasma 3-methoxy-4-hydroxyphenylglycol in manic patients: relationships with clinical variables. Acta Psychiatr Scand 81:14–18, 1990

Banerjee SD, Kung LS, Riggi SS, et al: Development of beta-adrenergic receptor subsensitivity by antidepressants. Nature 268:455–456, 1977

Barnes RF, Vieth RC, Borson S, et al: High levels of plasma catecholamines in dexamethasone-resistant depressed patients. Am J Psychiatry 140:1623–1625, 1983

Beckmann H, Goodwin FK: Urinary MHPG in subgroups of depressed patients and normal controls. Neuropsychobiology 6:91–100, 1980

Bergstrom DA, Kellar KJ: Effect of electroconvulsive shock on monoaminergic receptor binding sites in rat brain. Nature 278:464–466, 1979

Berrettini WH, Goldin LR, Nurnberger JI, et al: Genetic factors in affective illness. J Psychiatr Res 18:329–335, 1984

Berrettini WH, Nurnberger JI, Scheinin M, et al: Cerebrospinal fluid and plasma monoamines and their metabolites in euthymic bipolar patients. Biol Psychiatry 20:257–269, 1985

Birnbaumer L, Abramowitz J, Brown AM: Receptor-effector coupling by G proteins. Biochim Biophys Acta 1031:163–224, 1990

Bousquet P, Feldman J: The blood pressure effects of alpha-adrenoceptor antagonists injected in the medullary site of action of clonidine: the nucleus reticularis lateralis. Life Sci 40:1045–1052, 1987

Boyajian CL, Leslie FM: Pharmacological evidence for alpha$_2$-adrenoceptor heterogeneity: differential binding properties of [^{3}H]rauwolscine and [^{3}H]idazoxan in rat brain. J Pharmacol Exp Ther 241:1092–1098, 1987

Boyer P, Davila M, Schaub C, et al: Growth hormone response to clonidine stimulation in depressive states, part I. Psychiatric Psychobiology 1:189–195, 1986

Bricca G, Dontenwill M, Molines A, et al: Evidence for the existence of a homogeneous population of imidazoline receptors in the human brainstem. Eur J Pharmacol 150:401–402, 1988a

Bricca G, Dontenwill M, Molines A, et al: The imidazoline preferring receptor: binding studies in bovine, rat, and human brainstem. Eur J Pharmacol 162:1–9, 1988b

Buckholtz NS, Davies AO, Rudorfer MV, et al: Lymphocyte beta-adrenergic function versus catecholamines in depression. Biol Psychiatry 24:451–457, 1988

Bunney WE Jr, Davis JM: Norepinephrine in depressive reactions. Arch Gen Psychiatry 13:483–494, 1965

Bylund DB, Ray-Prenger C, Murphy TJ: Alpha$_{2A}$, and alpha$_{2B}$ adrenergic receptor subtypes: antagonist binding in tissues and cell lines containing only one subtype. J Pharmacol Exp Ther 245:600–607, 1988

Carstens ME, Engelbrecht AH, Russell VA, et al: Beta-adrenoreceptors on lymphocytes of patients with major depressive disorders. Psychiatry Res 20:239–248, 1987

Carstens ME, Engelbrecht AH, Russell VA, et al: Biological markers in juvenile depression. Psychiatry Res 23:77–88, 1988

Charney DS, Heninger GR: Abnormal regulation of noradrenergic function in panic disorders. Arch Gen Psychiatry 43:1041–1054, 1986

Charney DS, Heninger GR, Sternberg DE, et al: Adrenergic receptor sensitivity in depression. Arch Gen Psychiatry 39:290–294, 1982

Charney DS, Heninger GR, Sternberg DE: Alpha$_2$ adrenergic receptor sensitivity and the mechanism of action of antidepressant therapy. Br J Psychiatry 142:265–275, 1983

Checkley SA, Slade AP, Shur E: Growth hormone and other responses to clonidine in patients with endogenous depression. Br J Psychiatry 138:51–55, 1981

Checkley SA, Glass TB, Thompson C, et al: The growth hormone response to clonidine in endogenous as compared to reactive depression. Psychol Med 14:773–777, 1984

Checkley SA, Corn TH, Glass TB, et al: The responsiveness of central alpha adrenoceptors in depression, in The Biology of Depression. Edited by Deakin JFW. London, England, Gaskell, 1985, pp 110–119

Christensen NJ, Vestergaard P, Sorensen T, et al: Cerebrospinal fluid adrenaline and noradrenaline in depressed patients. Acta Psychiatr Scand 62:178–186, 1980

Cooper SJ, Kelly JG, King DJ: Adrenergic receptors in depression: effects of electroconvulsive therapy. Br J Psychiatry 147:23–29, 1985

Davis BA: Biogenic Monoamines and Their Metabolites in the Urine, Plasma, and Cerebrospinal Fluid of Normal, Psychiatric, and Neurological Subjects. Boca Raton, FL, CRC Press, 1990

Davis JM, Bresnahan DB: Psychopharmacology in clinical psychiatry, in Psychiatry Update: American Psychiatric Association Annual Review, Vol 6. Edited by Hales RE, Frances AJ. Washington, DC, American Psychiatric Press, 1987, pp 159–188

Davis JM, Koslow SH, Gibbons RD, et al: Cerebrospinal fluid and urinary biogenic amines in depressed patients and healthy controls. Arch Gen Psychiatry 45:705–717, 1988

Deakin JFW, Owen F, Cross AJ, et al: Studies on possible mechanisms of action of electroconvulsive therapy: effects of repeated electrically induced seizures on rat brain receptors for monoamines and other neurotransmitters. Psychopharmacology (Berl) 73:345–349, 1981

Dencker SJ, Häggendal J, Malm U: Noradrenaline content of cerebrospinal fluid in mental disease. Lancet 2:754–755, 1966

Devesa J, Arce V, Lois N, et al: α_2-Adrenergic agonism enhances the growth hormone (GH) response to GH-releasing hormone through an inhibition of hypothalamic somatostatin release in normal men. J Clin Endocrinol Metab 71:1581–1588, 1990

de Villiers AS, Russel VA, Carstens ME, et al: Noradrenergic function and hypothalamic-pituitary-adrenal axis activity in primary unipolar major depressive disorder. Psychiatry Res 22:127–140, 1987

Dinan TG, Lakshmi N, Yatham MB, et al: Blunting of noradrenergic-stimulated growth hormone release in mania. Am J Psychiatry 148:936–938, 1991

Doyle MC, George AJ, Ravindran AV, et al: Platelet α_2-adrenoreceptor binding in elderly depressed patients. Am J Psychiatry 142:1489–1490, 1985

Ebstein R, Oppenheim G, Ebstein BS, et al: The cyclic AMP second messenger system in man: the effects of heredity, hormones, drugs, aluminum, age and disease on signal amplification. Prog Neuropsychopharmacol Biol Psychiatry 10:323–353, 1986

Ebstein R, Lerer B, Shapira B, et al: Cyclic AMP second messenger signal amplification in depression. Br J Psychiatry 152:665–669, 1988

Eisenhofer G, Goldstein DS, Ropchak TG, et al: Source and physiological significance of plasma 3,4-dihydroxyphenylglycol and 3-methoxy-4-hydroxyphenylglycol. J Auton Nerv Syst 24:1–14, 1988

Ernsberger P, Meeley MP, Mann JJ, et al: Clonidine binds to imidazole binding sites as well as α_2-adrenoceptors in the ventrolateral medulla. Eur J Pharmacol 134:1–13, 1987

Esler M, Turbott JT, Schwartz R, et al: The peripheral kinetics of norepinephrine in depressive illness. Arch Gen Psychiatry 39:295–300, 1982

Esler M, Jennings G, Korner P, et al: Assessment of human sympathetic nervous system activity from measurements of norepinephrine turnover. Hypertension 11:3–20, 1988

Everitt RJ, Robbins TW, Selden NRW: Function of the locus coeruleus noradrenergic system: a neurobiological and behavioral synthesis, in The Pharmacology of Noradrenaline in the Central Nervous System. Edited by Heal DJ, Marsden CA. Oxford, England, Oxford University Press, 1990, pp 349–378

Extein I, Tallman J, Smith CC, et al: Changes in lymphocyte beta-adrenergic receptors in depression and mania. Psychiatry Res 1:191–197, 1979

Fawcett JA, Maas JW, Dekirmenjian H: Depression and MHPG excretion: response to dextroamphetamines and tricycle antidepressants. Arch Gen Psychiatry 26:246–251, 1972

Filser JG, Spira J, Fischer M, et al: The evolution of 4-hydroxy-3-methoxy-phenylglycol sulfate as a possible marker of control norepinephrine turnover studies in healthy volunteers and depressed patients. J Psychiatr Res 22:171–181, 1988

Fishman PH, Finberg JPM: Effect of the tricyclic antidepressant desipramine on β-adrenergic receptors in cultured rat glioma C6 cells. J Neurochem 49:282–287, 1987

Foote SL, Bloom FE, Aston-Jones G: The nucleus locus coeruleus: new evidence of anatomical and physiological specificity. Physiol Rev 69:844–914, 1983

Freedman RR, Embury J, Migaly P, et al: Stress-induced densensitization of alpha$_2$-adrenergic receptors in human platelets. Psychosom Med 52:624–630, 1990

Garcia-Sevilla JA, Garcia-Vallejo P, Guimon J: Enhanced α_2-adrenoceptor-mediated platelet aggregation in patients with major depressive disorder. Eur J Pharmacol 94:359–360, 1983

Garcia-Sevilla JA, Guimon J, Garcia-Vallejo P, et al: Biochemical and functional evidence of supersensitive platelet α_2-adrenoreceptors in major affective disorder: effect of long-term lithium carbonate treatment. Arch Gen Psychiatry 43:51–57, 1986

Garcia-Sevilla JA, Padro D, Giralt MT, et al: α_2-Adrenoceptor-mediated inhibition of platelet adenylate cyclase and induction of aggregation in major depression: effect of long-term antidepressant drug treatment. Arch Gen Psychiatry 47:125–132, 1990

Georgotas A, Schweitzer J, McCue RE, et al: Clinical and treatment effects on ^{3}H-clonidine and ^{3}H-imipramine binding in elderly depressed patients. Life Sci 40:2137–2143, 1987

Goldstein DS, Zimlichman R, Kelly GD, et al: Effect of ganglion blockade on cerebrospinal fluid norepinephrine. J Neurochem 49:1484–1490, 1987

Goldstein DS, Eisenhofer G, Stull R, et al: Plasma dihydroxyphenylglycol and the intraneuronal disposition of norepinephrine in humans. J Clin Invest 81:213–220, 1988

Goodwin FK, Jamison KR: Biochemical and pharmacologic studies, in Manic Depressive Illness. Edited by Goodwin FK, Jamison KR. New York, Oxford University Press, 1990, pp 416–502

Goodwin FK, Cowdry RW, Webster MH: Predictors of drug response in the affective disorders: toward an integrated approach, in Psychopharmacology: A Generation of Progress. Edited by Lipton MA, DiMaschio AD, Killam KF. New York, Raven, 1978, pp 1277–1288

Halaris AE: Plasma 3-methoxy-4-hydroxyphenylglycol in manic psychosis. Am J Psychiatry 135:493–494, 1978

Halper JP, Brown RP, Sweeney JA, et al: Blunted beta-adrenergic responsivity of peripheral blood mononuclear cells in endogenous depression: isoproterenol dose response studies. Arch Gen Psychiatry 45:241–244, 1988

Hauger RL, Scheinin M, Siever LJ, et al: Dissociation of norepinephrine turnover from alpha-2 responses after clorgyline. Clin Pharmacol Ther 43:32–38, 1988

Healy D, Carney PA, Leonard BE: Monoamine-related markers of depression: changes following treatment. J Psychiatr Res 17:251–260, 1983

Healy D, Halloran AO, Carney PA, et al: Peripheral adrenoreceptors and serotonin receptors in depression: changes associated with response to treatment with trazodone on amitriptyline. J Affective Disord 9:285–296, 1985

Heninger GR, Charney DS, Price LH: Alpha$_2$-adrenergic receptor sensitivity in depression: the plasma MHPG, behavioral, and cardiovascular responses to yohimbine. Arch Gen Psychiatry 45:718–726, 1988

Hoehe M, Valido G, Matussek N: Growth hormone response to clonidine in endogenous depressive patients: evidence for a trait marker in depression, in Biological Psychiatry Developments in Psychiatry. Edited by Shagass C, Josiassen EC, Bridger WH, et al. Amsterdam, Elsevier, 1986, pp 862–864

Hollister LE, Davis KL, Berger PA: Subtypes of depression based on excretion of MHPG and response to nortriptyline. Arch Gen Psychiatry 37:1107–1110, 1980

Honegger UE, Disler B, Wiesmann UN: Chronic exposure of human cells in culture to the tricyclic antidepressant desipramine reduces the number of beta-adrenoceptors. Biochem Pharmacol 35:1899–1903, 1986

Horton RW, Katona CLE, Theodorou AE, et al: Platelet radioligand binding and neuroendocrine challenge tests in depression: antidepressant and receptor function, in Ciba Foundation Symposium. Chichester, Wiley, 1986, pp 84–105

Jacobs BL: Locus coeruleus neuronal activity in behaving animals, in The Pharmacology of Noradrenaline in the Central Nervous System. Edited by Heal DJ, Marsden CA. Oxford, England, Oxford University Press, 1990, pp 248–265

Janicak PG, Davis JM, Chan C, et al: Failure of urinary MHPG levels to predict treatment response in patients with unipolar depression. Am J Psychiatry 143:1398–1402, 1986

Jimerson DC, Post RM, van Kammen DP, et al: Cerebrospinal fluid cortisol levels in depression and schizophrenia. Am J Psychiatry 137:979–980, 1980

Jimerson DC, Nurnberger JI Jr, Post RM, et al: Plasma MHPG in rapid cyclers and healthy twins. Arch Gen Psychiatry 38:1287–1290, 1981

Jimerson DC, Insel TR, Reus VI, et al: Increased plasma MHPG in dexamethasone-resistant depressed patients. Arch Gen Psychiatry 40:173–176, 1983

Jones DH, Hamilton CA, Reid JL: Plasma noradrenaline, age, and blood pressure: a population study. Clinical Science and Molecular Medicine 55:735–755, 1978

Kafka MS, Paul SM: Platelet α_2-adrenergic receptors in depression. Arch Gen Psychiatry 43:91–95, 1986

Kafka MS, Nurnberger JI, Siever LJ, et al: Alpha$_2$-adrenergic receptor function in patients with unipolar and bipolar affective disorder. J Affect Disord 10:163–169, 1986

Katona CLE, Theodorou AE, Davies SL, et al: [^{3}H]Yohimbine binding to platelet α_2-adrenoreceptors in depression. J Affect Disord 17:219–228, 1987

Klysner R, Geisler A, Rosenberg R: Enhanced histamine and β-adrenoreceptor mediated cyclic AMP formation in leukocytes from patients with endogenous depression. J Affective Disord 13:227–232, 1987

Kopin IJ: Catecholamine metabolism: basic aspects and clinical significance. Pharmacol Rev 37:333–364, 1985

Koslow SH, Maas JW, Bowden CL, et al: CSF and urinary biogenic amines and metabolites in depression and mania: a controlled, univariate analysis. Arch Gen Psychiatry 40:999–1010, 1983

Laakmann G, Hinz A, Voderholzer U, et al: The influence of psychotropic drugs and releasing hormones on anterior pituitary hormone secretion in healthy subjects and depressed patients. Pharmacopsychiatria 23:18–26, 1990

Lake CR, Pickar D, Ziegler MG, et al: High plasma norepinephrine levels in patients with major affective disorder. Am J Psychiatry 139:1315–1318, 1982

Lechin F, van der Dijs B, Jakubowicz D, et al: Effects of clonidine on blood pressure, noradrenaline, cortisol, growth hormone and prolactin plasma levels in high and low intestinal tone depressed patients. Neuroendocrinology 41:156–162, 1985

Lechin F, Van Der Dijs B, Amat J, et al: Neuroanatomical basis, in Neurochemistry and Clinical Disorders: Circuitry of Some Psychiatric and Psychosomatic Syndromes. Edited by Lechin F, Van Der Dijs F. Boca Raton, FL, CRC Press, 1989, pp 2–48

Leckman JF, Maas JW, Redmond DE Jr, et al: Effects of oral clonidine on plasma-3-methoxy-4-hydroxy-phenethyleneglycol (MHPG) in man: preliminary report. Life Sci 26:2179–2185, 1980

Lee CM, Javitch JA, Snyder A: Recognition sites for norepinephrine uptake: regulation by neurotransmitter. Science 220:626–629, 1983

Lenox RH, Ellis JE, Van Riper DA, et al: Platelet α_2-adrenergic receptor activity in clinical studies of depression, in Frontiers in Neuropsychiatric Research. Edited by Usdin E. London, Macmillan, 1983, pp 331–355

Linares OA, Jacquez JA, Zech LA, et al: Norepinephrine metabolism in humans: kinetic analysis and model. J Clin Invest 80:1332–1341, 1987

Linnoila M, Ninan PT, Scheinin M, et al: Reliability of norepinephrine and major monoamine metabolite measurements in CSF of schizophrenic patients. Arch Gen Psychiatry 40:1290–1296, 1983

Linnoila M, Guthrie S, Lane EA, et al: Clinical studies on norepinephrine metabolism: how to interpret the number. Psychiatry Res 17:229–239, 1986

Louis WJ, Doyle AE, Anavekar SN: Plasma noradrenaline concentration and blood pressure in essential hypertension, phaeochromocytoma and depression. Clinical Science and Molecular Medicine 48:239S–242S, 1975

Ludwig J, Gerhardt T, Halbrugge T, et al: Plasma concentrations of noradrenaline and 3,4-dihydroxyphenylglycol under conditions of enhanced sympathetic activity. Eur J Clin Pharmacol 35:261–267, 1988

Maas JW, Benensohn H, Landis DH: A kinetic study of the disposition of circulating norepinephrine in normal male subjects. J Pharmacol Exp Ther 174:381–387, 1970

Maas JW, Dekirmenjian H, Fawcett J: Catecholamine metabolism, depression and stress. Nature 230:330–331, 1971

Maas JW, Fawcett JA, Dekirmenjian H: Catecholamine metabolism, depressive illness, and drug response. Arch Gen Psychiatry 26:252–262, 1972

Maas JW, Kocsis JH, Bowden CL, et al: Pretreatment neurotransmitter metabolites and response to imipramine or amitriptyline treatment. Psychol Med 12:37–43, 1982

Maas JW, Koslow SH, Davis J, et al: Catecholamine metabolism and disposition in healthy and depressed subjects. Arch Gen Psychiatry 44:337–344, 1987

Magliozzi JR, Gietzen D, Maddock RJ, et al: Lymphocyte beta-adrenoreceptor density in patients with unipolar depression and normal controls. Biol Psychiatry 26:15–25, 1989

Maisel AS, Knowlton KU, Fowler P, et al: Adrenergic control of circulating lymphocyte subpopulations: effects of congestive heart failure, dynamic exercise, and terbutaline treatment. J Clin Invest 85:462–467, 1990a

Maisel AS, Harris T, Rearden CA, et al: Beta-adrenergic receptors in lymphocyte subsets after exercise: alterations in normal individuals and patients with congestive heart failure. Circulation 82:2003–2010, 1990b

Malbon CC, Hadcock JR, Rapiejko PJ, et al: Regulation of transmembrane signaling elements: transcriptional, post-transcriptional and post-translational controls. Biochem Soc Symp 56:155–164, 1990

Manji HK: G proteins: implications for psychiatry. Am J Psychiatry 149:746–760, 1992

Manji HK, Chen G, Bitran JA, et al: Chronic exposure of C6 glioma cells to desipramine desensitizes β-adrenoceptors, but increases K_L/K_H ratio. Eur J Pharmacol 206:159–162, 1991a

Manji HK, Hsiao JK, Risby ED, et al: The mechanisms of action of lithium: effects on serotonergic and noradrenergic systems in normal subjects. Arch Gen Psychiatry 48:505–512, 1991b

Mann JJ, Brown RP, Halper JP, et al: Reduced sensitivity of lymphocyte beta-adrenergic receptors in patients with endogenous depression and psychomotor agitation. N Engl J Med 313:715–720, 1985

Mann JJ, Manevitz AZ, Chen JS, et al: Acute effects of single and repeated electroconvulsive therapy on plasma catecholamines and blood pressure in major depressive disorder. Psychiatry Res 34:127–137, 1990a

Mann JJ, Mahler JC, Wilner PJ, et al: Normalization of blunted lymphocyte beta-adrenergic responsivity in melancholic inpatients by a course of electroconvulsive therapy. Arch Gen Psychiatry 47:461–464, 1990b

Matussek N, Ackenheil M, Hippius H, et al: Effect of clonidine on growth hormone release in psychiatric patients and controls. Psychiatry Res 2:25–36, 1980

Meeley MP, Ernsberger PR, Granata AR, et al: An endogenous clonidine-displacing substance from bovine brain: receptor binding and hypotensive actions in the ventrolateral medulla. Life Sci 38:1119–1126, 1986

Meurs H, Van Den Bogaard W, Kauffman HF, et al: Characterization of (-)-[^{3}H]dihydroalprenolol binding to intact and broken cell preparations of human peripheral blood lymphocytes. Eur J Pharmacol 85:185–194, 1982

Michel MC, Regan JW, Gerhardt MA, et al: Nonadrenergic [^{3}H]idazoxan binding sites are physically distinct from alpha$_2$-adrenergic receptors. Mol Pharmacol 37:65–68, 1990

Minneman KP, Dibner MD, Wolfe BB, et al: β_1- and β_2-adrenergic receptors in rat cerebral cortex are independently regulated. Science 204:866–868, 1979

Mitchell PB, Bearn JA, Corn TH, et al: Growth hormone response to clonidine after recovery in patients from endogenous depression. Br J Psychiatry 152:34–38, 1988

Muscettola G, Potter WZ, Pickar D, et al: Urinary 3-methoxy-4-hydroxyphenylglycol and major affective disorders. Arch Gen Psychiatry 41:337–342, 1984

Nagy LE, Diamond I, Gordon A: Cultured lymphocytes from alcoholic have altered cAMP signal transduction. Proc Natl Acad Sci U S A 8[illegible]–6976, 1988

Nestler EJ, McMahon A, Sabban EL, et al: Chronic antidepressant administration decreases the expression of tyrosine hydroxylase in the rat locus coeruleus. Proc Natl Acad Sci U S A 87:7522–7566, 1990

Nutt DJ: Altered central alpha$_2$-adrenoreceptor sensitivity in panic disorder. Arch Gen Psychiatry 46:165–169, 1989

Pandey GN, Davis JM: Leucocyte beta-adrenergic receptors: a marker for central beta-adrenergic receptor function in depression. Clin Neuropharmacol 9 (suppl 14):353–355, 1986

Pandey GN, Dysken MW, Garver DL, et al: Beta-adrenergic receptor function in affective illness. Am J Psychiatry 136:675–678, 1979

Pandey GN, Janicak P, Davis JM: Studies of beta-adrenergic receptors in leukocytes of patients with affective illness and effects of antidepressant drugs. Psychopharmacol Bull 21:603–609, 1985

Pandey GN, Janicak PG, Javaid JI, et al: Increased ^{3}H-clonidine binding in the platelets of patients with depressive and schizophrenic disorders. Psychiatry Res 28:73–88, 1989

Pandey GN, Pandey SC, Davis JM: Peripheral adrenergic receptors in affective illness and schizophrenia. Pharmacol Toxicol 66 (suppl 3):313–336, 1990

Pickar D, Sweeney DR, Maas JW, et al: Primary affective disorder, clinical state change and MHPG excretion: a longitudinal study. Arch Gen Psychiatry 35:1378–1383, 1978

Piletz JE, Schubert DSP, Halaris A: Evaluation of studies on platelet alpha$_2$-adrenoreceptors in depressive illness. Life Sci 39:1589–1616, 1986

Piletz JE, Halaris A, Saran A, et al: Elevated ^{3}H-para-aminoclonidine binding to platelet purified plasma membranes from depressed patients. Neuropsychopharmacology 3:201–210, 1990

Post RM, Lake CR, Jimerson DC, et al: Cerebrospinal fluid norepinephrine in affective illness. Am J Psychiatry 135:907–912, 1978

Post RM, Jimerson DC, Ballenger JC, et al: Cerebrospinal fluid norepinephrine and its metabolites in manic-depressive illness, in Neurobiology of Mood Disorders. Edited by Post RM, Ballenger JC. Baltimore, MD, Williams & Wilkins, 1984, pp 539–554

Post RM, Rubinow DR, Uhde TW, et al: Dysphoric mania. Arch Gen Psychiatry 46:353–358, 1989

Potter WZ, Linnoila M: Biochemical classifications of diagnostic subgroups and D-type scores. Arch Gen Psychiatry 46:269–271, 1989

Potter WZ, Rudorfer MV, Goodwin FK: Biological findings in bipolar disorders, in Psychiatry Update: American Psychiatric Association Annual Review, Vol 6. Edited by Hales RE, Francis AJ. Washington, DC, American Psychiatric Press, 1987, pp 32–60

Potter WZ, Rudorfer MV, Linnoila M: Effects of antidepressants in norepinephrine and its metabolites in cerebrospinal fluid, plasma, and urine. Progress in Catecholamine Research Part C: Clinical Aspects 301–306, 1988

Rasmussen K, Jacobs BL: Single unit activity of locus coeruleus neurons in the freely moving cat, II: conditioning and pharmacologic studies. Brain Res 371:335–344, 1986

Rasmussen K, Morilak DA, Jacobs BL, et al: Single unit activity of locus coeruleus neurons in the freely moving cat, I: during naturalistic behaviors and in response to simple and complex stimuli. Brain Res 371:324–334, 1986

Rea RF, Eckberg DL, Fritsch JM, et al: Relation of plasma norepinephrine and sympathetic traffic during hypotension in humans. Am J Physiol 258:R982–R986, 1990

Redmond DE Jr, Katz MM, Maas JW, et al: Cerebrospinal fluid amine metabolites: relationships with behavioral measurements in depressed, manic, and healthy control subjects. Arch Gen Psychiatry 43:938–947, 1986

Reiner PB: Correlational analysis of central noradrenergic neuronal activity and sympathetic tone in behaving cats. Brain Res 378:86–96, 1986

Reis DJ, Morrison S, Ruggiero DA: The C1 area of the brainstem in tonic and reflex control of blood pressure: state of the art lecture. Hypertension 2:18–23, 1988

Risby ED, Hsiao JK, Manji HK, et al: The mechanisms of action of lithium: effects on adenylate cyclase activity and β-adrenergic receptor binding in normal subjects. Arch Gen Psychiatry 48:513–524, 1991

Rosenbaum AH, Schatzberg AF, Maruter T, et al: MHPG as a predictor of antidepressant responses to imipramine and maprotiline. Am J Psychiatry 137:1090–1092, 1980

Rosenblatt S, Chanley JD, Leighton WP: The investigation of adrenergic metabolism with ^{3}H-norepinephrine in psychiatric disorders, II: temporal changes in the distribution of urinary tritiated metabolites in affective disorders. J Psychiatr Res 6:321–333, 1969

Rothschild AJ, Schatzberg AF, Langlais PJ, et al: Psychotic and nonpsychotic depressions, I: comparison of plasma catecholamines and cortisol measures. Psychiatry Res 20:143–153, 1987

Roy A, Pickar D, Linnoila M, et al: Cerebrospinal fluid monoamine and monoamine metabolite concentrations in melancholia. Psychiatry Res 15:281–292, 1985a

Roy A, Pickar D, Linnoila M, et al: Plasma norepinephrine level in affective disorders: relationship to melancholia. Arch Gen Psychiatry 42:1181–1185, 1985b

Roy A, Jimerson DC, Pickar D: Plasma MHPG in depressive disorders and relationship to the dexamethasone suppression test. Am J Psychiatry 143:846–851, 1986

Roy A, Guthrie S, Pickar D, et al: Plasma norepinephrine responses to cold challenge in depressed patients and normal controls. Psychiatry Res 21:161–168, 1987

Roy A, Pickar D, De Jong J, et al: Norepinephrine and its metabolites in cerebrospinal fluid, plasma and urine: relationship to hypothalamic pituitary-adrenal axis function in depression. Arch Gen Psychiatry 45:849–856, 1988

Rubin AL, Price LH, Charney DS, et al: Noradrenergic function and the cortisol response to dexamethasone in depression. Psychiatry Res 15:5–15, 1985

Rudorfer MV, Ross RS, Linnoila M, et al: Exaggerated orthostatic responsivity of plasma norepinephrine in depression. Arch Gen Psychiatry 42:1186–1192, 1985

Sacchetti E, Conte G, Pennati A, et al: Platelet α_2-adrenoceptors in major depression: relationship with urinary 4-hydroxy-3-methoxy-phenylglycol and age at onset. J Psychiatr Res 19:579–586, 1985

Sarai K, Nakahara T, Kanagana K, et al: Lymphocyte beta-adrenergic receptor function in affective disorders. Advances in Bioscience 40:161–165, 1982

Schatzberg AF, Orsulak PJ, Rosenbaum AH, et al: Catecholamine measures for diagnosis and treatment of patients with depressive disorders. J Clin Psychiatry 41:35–39, 1980

Schatzberg AF, Orsulak PJ, Rosenbaum AH, et al: Toward a biochemical classification of depressive disorders, V: heterogeneity of unipolar depressions. Am J Psychiatry 139:471–474, 1982

Schatzberg AF, Samson JA, Bloomingdale KC, et al: Toward a biochemical classification of depressive disorders: urinary catecholamines, their metabolites and D-type scores in subgroups of depressive disorders. Arch Gen Psychiatry 46:260–268, 1989

Schildkraut JJ: The catecholamine hypothesis of affective disorders: a review of supporting evidence. Am J Psychiatry 122:509–522, 1965

Schildkraut JJ, Orsulak PJ, Schatzberg AF, et al: Biochemical discrimination of subgroups of depression disorders based on differences in catecholamine metabolism, in Biological Markers in Psychiatry and Neurology. Edited by Usdin E, Hamin I. New York, Pergamon, 1978a, pp 23–33

Schildkraut JJ, Orsulak PJ, Schatzberg AF, et al: Toward a biochemical classification of depressive disorders: differences in urinary excretion of MHPG and other catecholamine metabolites in clinically defined subtypes of depression. Arch Gen Psychiatry 35:1427–1433, 1978b

Schreiber G, Avissar S, Danon A, et al: Hyperfunctional G proteins in mononuclear leukocytes of patients with mania. Biol Psychiatry 29:273–280, 1991

Sevy S, Papadimitriou GN, Surmont DW, et al: Noradrenergic function in generalized anxiety disorder, major depressive disorder, and health subjects. Biol Psychiatry 25:141–147, 1989

Siever LJ: Role of noradrenergic mechanisms in the etiology of the affective disorders, in Psychopharmacology: The Third Generation of Progress. Edited by Meltzer HY. New York, Raven, 1987, pp 493–504

Siever LJ, Uhde TW: New studies and perspectives on the noradrenergic receptor system in depression: effects of the alpha 2-adrenergic agonist clonidine. Biol Psychiatry 19:131–156, 1984

Siever LJ, Uhde TW, Silberman EK, et al: Growth hormone response to clonidine as a probe of noradrenergic receptor responsiveness in affective disorder patients and controls. Psychiatry Res 6:171–183, 1982

Siever LJ, Insel TR, Jimerson DC, et al: Growth hormone response to clonidine in obsessive-compulsive patients. Br J Psychiatry 142:184–187, 1983

Siever LJ, Uhde TW, Jimerson DC, et al: Differential inhibitory noradrenergic responses to clonidine in 25 depressed patients and 25 normal control subjects. Am J Psychiatry 141:733–741, 1984a

Siever LJ, Uhde TW, Jimerson DC, et al: Plasma cortisol responses to clonidine in depressed patients and controls: evidence for possible alteration in noradrenergic-neuroendocrine relationships. Arch Gen Psychiatry 41:63–68, 1984b

Siever LJ, Kafka MS, Targum S, et al: Platelet alpha-adrenergic binding and biochemical responsiveness in depressed patients and controls. Psychiatry Res 11:287–302, 1984c

Siever LJ, Pickar D, Lake R, et al: Extreme elevations in plasma norepinephrine associated with decreased alpha-adrenergic responsivity in major depressive disorder: two case reports. J Clin Psychopharmacol 3:39–41, 1985

Siever LJ, Uhde TW, Jimerson DC, et al: Indices of noradrenergic system in depression. Psychiatry Res 19:59–73, 1986

Smith CB, Hollingsworth PJ, Garcia-Sevilla JA, et al: Platelet alpha$_2$-adrenoreceptors are decreased in number after antidepressant therapy. Prog Neuropsychopharmacol Biol Psychiatry 7:241–247, 1983

Stahl SM, Lemoine PM, Ciaranello RD, et al: Platelet alpha2-adrenergic receptor sensitivity in major depressive disorder. Psychiatry Res 10:157–164, 1983

Starke CK, Gothert M, Kilbinger H: Modulation of neurotransmitter release by presynaptic autoreceptors. Physiol Rev 69:864–989, 1989

Sulser F, Vetulani J, Mobley PL: Mode of action of antidepressant drugs. Biochem Pharmacol 27:257–271, 1978

Svensson TH: Peripheral, autonomic regulation of locus coeruleus noradrenergic neurons in brain: putative implications for psychiatry and psychopharmacology. Psychopharmacology (Berl) 92:1–7, 1987

Swann AC, Secunda S, Davis JM, et al: CSF monoamine metabolites in mania. Am J Psychiatry 140:396–400, 1983

Swann AC, Secunda SK, Katz MM, et al: Lithium treatment of mania: clinical characteristics, specificity of symptom change, and outcome. Psychiatry Res 18:127–141, 1986

Swann AC, Koslow SH, Katz MM, et al: Lithium carbonate treatment of mania: cerebrospinal fluid and urinary monoamine metabolites and treatment outcome. Arch Gen Psychiatry 44:345–354, 1987

Uhde TW, Vittone BJ, Siever LJ, et al: Blunted growth hormone response to clonidine in panic disorder patients. Biol Psychiatry 21:1081–1085, 1986

Van Tits LJ, Michel MC, Grosse-Wilde H, et al: Catecholamines increase lymphocyte beta 2-adrenergic receptors via a beta 2-adrenergic, spleen-dependent process. Am J Physiol 258:E191–202, 1990

Veith RC, Raskind MA, Barnes RF, et al: Tricyclic antidepressants and supine, standing, and exercise plasma norepinephrine levels. Clin Pharmacol Ther 33:763–769, 1983

Veith RC, Halter JB, Murburg MM, et al: Increased plasma NE appearance rate in dexamethasone resistant depression. Fourth World Congress of Biological Psychiatry, Athens, Greece, October 13–17, 1985

Veith RC, Barnes RF, Villacres RC, et al: Plasma catecholamines and norepinephrine kinetics in depression and panic disorder, in Catecholamines: Clinical Aspects. Edited by Belmaker R. New York, Alan R Liss, 1988, pp 197–202

Werstiuk ES, Steiner M, Burns T: Studies on leukocyte β-adrenergic receptors in depression: a critical appraisal. Life Sci 47:85–105, 1990

Wolfe N, Cohen BM, Gelenberg AJ: Alpha2-adrenergic receptors in platelet membranes of depressed patients: increased affinity for ^{3}H-yohimbine. Psychiatry Res 20:107–116, 1986

Wolfe N, Gelenberg AJ, Lydiard RB: Alpha2-adrenergic receptor sensitivity in depressed patients: relation between ^{3}H-yohimbine binding to platelet membranes and clonidine-induced hypotension. Biol Psychiatry 25:382–392, 1989

Woodward DJ, Moises HC, Waterhouse BD, et al: Modulatory actions of norepinephrine in the mammalian central nervous system. Federation Proceedings 38:59–66, 1979

Wright AF, Crichton DN, Loudon JB, et al: β-Adrenoreceptor binding defects in cell lines from families with manic-depressive disorder. Ann Hum Genet 48:201–214, 1984

Wyatt RJ, Portnoy B, Kupfer DJ, et al: Resting plasma catecholamine concentrations in patients with depression and anxiety. Arch Gen Psychiatry 24:65–70, 1971

Yamaoka K, Nanba T, Nomura S: Direct influence of antidepressants on GTP binding protein of adenylate cyclase in cell membranes of the cerebral cortex of rats. J Neural Transm 71:165–175, 1988

Zohar J, Bannet J, Drummer D, et al: The responses of lymphocyte beta-adrenergic receptors to chronic propranolol treatment in depressed patients, schizophrenic patients, and normal controls. Biol Psychiatry 18:553–560, 1983

16

Adrenergic Dysfunction in Anxiety Disorders

JAMES L. ABELSON, M.D., PH.D.
OLIVER G. CAMERON, M.D., PH.D.

Be not afraid of sudden fear.

—Proverbs 3:25

The only thing we have to fear is fear itself.

—F. D. Roosevelt, First Inaugural Address, March 1933

The relationship between stress and the fight-or-flight reaction with activation of the sympathoadrenal endocrine system has been recognized since early in this century (for details, see later in this chapter and Cameron, Chapter 9, this volume; Goldstein, Chapter 8, this volume). Since the symptoms of stress and the fight-or-flight reaction, as well as symptoms often reported in response to administration of sympathomimetic substances, are similar to symptoms often referred to as anxiety, it has been widely assumed that anxiety is associated with activation of the sympathoadrenal system and increases in systemic catecholamine—

epinephrine and norepinephrine (NE)—levels. Furthermore, activation of the central nervous system (CNS) noradrenergic neurotransmitter system, either by administration of specific pharmacological agents or by means of electrical stimulation of the locus coeruleus, produces a syndrome in monkeys that has been proposed as a model of human anxiety (Charney and Redmond 1983; Redmond 1987).

An immediate problem in understanding the potential relationship between anxiety and adrenergic changes is that the term *anxiety* is used in different ways. Anxiety can refer to the normal subjective emotional response to a threatening situation. It can also refer to a syndrome involving not only the subjective psychological response but specific physiological changes. Finally, it can be a disorder that includes not only characteristic signs and symptoms but also other specific aspects such as a typical age at onset, gender distribution, prognosis, response to treatment, and familial pattern. It is not clear to what extent the different meanings of anxiety refer to overlapping versus distinct psychobiological states.

The proposed relationship between anxiety and the stress or fight-or-flight reaction refers to the second meaning of anxiety: as a normal psychological and physiological reaction to a threatening situation. Much of the research prior to the last 10–15 years concerning anxiety and sympathoadrenal activation involves this second meaning of anxiety. Although some of these studies undoubtedly involve people with pathological anxiety states, these states typically are not well characterized and are very heterogeneous. In addition, some studies appear to include a mixture of individuals without pathology and individuals with pathological anxiety. Thus, it is not clear to what extent this body of research has applicability to the question of whether there is evidence of adrenergic dysfunction associated with specific abnormal anxiety states. The study of adrenergic function in abnormal anxiety states has been facilitated greatly by the development of DSM-III (American Psychiatric Association 1980), DSM-III-R (American Psychiatric Association 1987), and DSM-IV (American Psychiatric Association 1994), which have allowed more precise delineations of study populations.

In this chapter, we review studies that have evaluated adrenergic function in persons with anxiety disorders (for prior reviews, see Cameron et al. 1990; Charney et al. 1990). Emphasis is given to studies that have appeared in the last 5–7 years and that have used DSM-III or DSM-III-R diagnostic criteria. Results of challenge studies receive particular attention. The review covers panic disorder, generalized anxiety disorder (GAD), obsessive-compulsive disorder (OCD), posttraumatic stress disorder (PTSD), and other anxiety and related disorders. Adrenergic symptomatology and results of 1) basal endocrine and physiological studies, 2) challenge studies that influence adrenergic mechanisms, and 3) adrenergic receptor binding studies for each specific disorder are described.

Adrenergic Symptomatology

Since the development of DSM-III, psychiatric diagnostic nosology has increasingly focused on specific symptom profiles. The presence of an anxiety disorder has been defined by the presence of symptoms of anxiety or worry, along with avoidance behavior and/or increased arousal. Very specific symptom lists are provided for most of the anxiety disorders and many of the listed symptoms may involve adrenergic control. For example, panic attacks are defined in the DSM by the presence of 4 of 13 specific symptoms, such as a pounding heart, sweating, or shaking. GAD requires the presence of 3 of 6 symptoms, such as muscle tension, irritability, or restlessness. A PTSD diagnosis requires persistent symptoms of increased arousal.

Perhaps because the presence of specific symptoms is used to define these disorders, the symptoms themselves have not been the explicit focus of much research (i.e., as dependent variables), despite the fact that it was the association of "adrenergic symptomatology" with anxiety that initially stimulated the development of adrenergic hypotheses about anxiety disorders. Although the occurrence and severity of symptoms are often recorded, the symptoms are usually used only to verify that an anxiety reaction (e.g., a panic attack) occurred or to correlate the occurrence and severity of the anxiety reaction with other variables (e.g., physiological changes, response to treatment). Thus, since symptoms in general have not been studied very intensively, adrenergic symptoms (as a subclass of all anxiety symptoms) have not been studied to any significant extent. Furthermore, as noted by Cameron (Chapter 9, this volume), it is often not clear to what extent a particular symptom is associated with physiological changes that are actually mediated completely or in large part by adrenergic mechanisms. Nevertheless, some research has been reported and is described briefly. Additionally, symptoms are referred to when considered germane under the categories of basal, challenge, and binding studies.

Cameron et al. (1986) studied a large group of anxious patients, with all diagnoses except PTSD represented. Based on factor analysis, they concluded that symptoms clustered into psychological, sympathetic, and parasympathetic groups; the sympathetic factor included cardiorespiratory symptoms and sweating, whereas the parasympathetic factor included gastrointestinal and urinary symptoms. Hibbert (1984) and Bruce and Lader (1991) reported a specific increase in frequency of cardiorespiratory symptoms and sweating in panic patients, and Katerndahl (1988) reported that cardiorespiratory symptoms typically occurred early in a panic attack. Thus, these studies supported the finding of Cameron et al. (1986) that cardiorespiratory symptoms and sweating tend to occur together as a cluster.

Tyrer et al. (1980) and McLeod et al. (1986) reported that, under some circumstances, there are significant correlations between reports of cardiac symptoms and sweating with actual physiological changes. This awareness might not be specific to panic attacks per se, however; persons with both panic and alcoholism describe alcohol withdrawal as very similar in somatic (and other) symptoms to a panic attack (George et al. 1988). Furthermore, although anxious patients appear to have increased awareness of cardiac activity (Barsky 1992; Ehlers and Breuer 1992; McLeod and Hoehn-Saric 1993), there is some dissociation between symptoms and physiological changes. Of particular interest is the question of whether patients report cardiac symptoms when no changes in heart action are observed, and, if so, what is the source of this subjective sensation. Finally, Goetz et al. (1989) reported that panic patients do not indiscriminately endorse somatic symptoms (although prior studies had suggested that they do).

In summary, the evidence is rather limited, but it suggests that patients with anxiety disorders do exhibit a cluster of symptoms (cardiac, respiratory, and sweating) that are mediated at least in part by sympathetic activation. This evidence comes mainly from studies of panic patients. It supports the hypothesized link between adrenergic mechanisms and human anxiety.

Panic Disorder

Many studies of anxiety prior to the 1980s most likely included individuals with a mixture of anxiety syndromes. Since the division of pathological anxiety into specific disorders by DSM-III, much of the research concerning the psychobiology of pathological anxiety has focused on panic disorder, probably because panic disorder appears to be a well-defined syndrome that is reliably diagnosable and is responsive to specific pharmacological (and nonpharmacological) treatments.

Basal Studies

A number of studies have determined resting plasma epinephrine and NE levels in panic patients (see reviews, Cameron and Nesse 1988; Cameron et al. 1990; see also, Stein et al. 1992b). Although several studies reported elevations, others did not find any difference. Based on all studies reported to date, it appears that if a patient is in a truly basal state, there is either no increase in catecholamines or only small catecholamine elevations in panic patients. Patients, however, may be more reactive to various "stressful" circumstances (e.g., venipuncture, postural change, and ambulation) and may have a response that exceeds the response of control subjects (Albus et al. 1988; Hoehn-Saric et al. 1991), although not all studies have reported this (Roth et al. 1992).

3-Methoxy-4-hydroxyphenylglycol (MHPG) is a metabolite of NE that predominates in the CNS. Early basal and ambulatory studies of plasma MHPG in panic patients were inconclusive (Cameron and Nesse 1988); MHPG levels did not change during "spontaneous" panic attacks (Cameron et al. 1987). Gurguis et al. (1991) reported normal MHPG levels in panic patients. No relationship was observed between anxiety severity and MHPG in depressed patients (Garvey et al. 1991). A study of yohimbine effects in panic patients found no preinfusion difference between patients and control subjects (Gurguis and Uhde 1990), whereas another found a trend for preinfusion levels to be higher in patients (Charney et al. 1987). Another group found that higher preinfusion MHPG levels predicted the occurrence of a panic response to lactate (den Boer et al. 1989). (This group also reported that 5-hydroxytryptophan administration had no effect on MHPG in panic patients [den Boer and Westenberg 1990].) Flumazenil, a benzodiazepine receptor antagonist, did not affect MHPG levels in either controls or patients who panicked in response to flumazenil administration (Woods et al. 1991). Finally, in a group of panic patients, low MHPG predicted entry into, and continuation of, treatment (Edlund and Swann 1989).

Several studies have also evaluated MHPG levels in urine and cerebrospinal fluid. One group found that 24-hour urinary levels were only slightly increased (14%) in patients (Garvey et al. 1989), and patients with *low* levels were more likely to have experienced spontaneous panic attacks in the preceding month (Garvey et al. 1991). Another group reported more substantial elevations (Lima et al. 1992). Cerebrospinal fluid MHPG levels do not appear to be elevated in panic patients (Eriksson et al. 1991; George et al. 1990). In summary, as with to catecholamines, it appears that basal MHPG levels are either normal or only modestly increased in panic patients.

Melatonin is controlled in part by β-adrenergic mechanisms. Two earlier studies (Cameron et al. 1987; McIntyre et al. 1987) reported reductions in normal nighttime melatonin levels. A more recent study found decreases in recently medicated patients, but increases in never-medicated patients (McIntyre et al. 1990). Melatonin increased in response to 5-hydroxytryptophan, but no difference between panic patients and control subjects was reported (den Boer and Westenberg 1990).

Monoamine oxidase is an intraneuronal degradative enzyme that can affect levels and activity of catecholamines. Monoamine oxidase activity in panic patients has been reported to be increased (Flaskos et al. 1989), decreased (Balon et al. 1987), or normal (Norman et al. 1988). Thus, no conclusions can be drawn, and further research is needed. Phenylethylamine, a substance involved in the metabolic pathways of the catecholamines, was not abnormal in agoraphobic subjects (Theofilopoulos et al. 1991).

Physiological as well as endocrinological studies have been carried out in panic patients. As in the endocrinological studies, habituation, reactivity, and "spontaneous" variability are potentially important reasons for the differences sometimes observed between patients and control subjects. Adrenergic mechanisms play a role in control of heart rate, blood pressure, respiration, other cardiac abnormalities (possibly), peripheral blood flow, sweating (skin conductance/galvanic skin response, which is "sympathetic" but mainly cholinergically rather than adrenergically mediated), and electromyographic activity (possibly). Earlier studies of skin conductance and electromyographic activity (see Cameron and Nesse 1988) indicated that both were increased in anxious patients. Hoehn-Saric et al. (1991) also found increased electromyographic activity in patients (specifically in panic patients), but a smaller skin conductance response. However, Roth et al. (1988) found greater skin conductance in patients.

Heart rate is controlled mainly by a balance between adrenergic input (activating) and vagal input (cholinergic and inhibitory); thus, an increase in heart rate can be due to increased adrenergic activity and/or decreased vagal tone. There are means to separate these effects (e.g., use of specific adrenergic and cholinergic pharmacological agonists and antagonists); some studies have used such strategies, but many have not. Therefore, it is not possible to determine in many studies whether a change is due to adrenergic or cholinergic changes, or to some combination. Respiration is also controlled by both adrenergic and cholinergic inputs. Blood pressure is less influenced by cholinergic effects, but other nonadrenergic effects are important. Thus, for each of these variables (heart rate, blood pressure, respiration), differences between patients and control subjects might or might not represent differences in adrenergic control. All findings that do not involve methods to identify adrenergic effects specifically must be understood and interpreted with this caveat in mind.

Most (but not all) early studies found significant elevations of heart rate in anxious individuals (see Cameron and Nesse 1988), and anxious persons seem to be unusually aware of their own heartbeat (Ehlers and Breuer 1992; Pyke and Greenberg 1986; Tyrer et al. 1980). Among more recent studies, several found increases in resting heart rate (Aronson et al. 1989a, 1989b; Gorman et al. 1990; Hoehn-Saric et al. 1991; Nutt 1989; Roth et al. 1988; Uhde et al. 1989), but several others did not (Cameron et al. 1990; Clark et al. 1990; Rapaport et al. 1991; Tancer et al. 1990a; Yeragani et al. 1987a). Two studies found no difference in heart rate in response to postural challenge (Cameron et al. 1990; Yeragani et al. 1990b), whereas one study noted an abnormal increase (Stein et al. 1992b). Heart rate is elevated in some, but not all, panic patients during "spontaneous" panic attacks (Cameron et al. 1987; Freedman et al. 1985; Taylor et al. 1983). There are no obvious systematic differences among these studies that explain the discrepancy; differences, when they were observed, were typically not large (usually less than 10 beats per minute). Finally, King et al. (1988) evaluated the relationship between heart rate and personality variables in panic patients; they found a negative relationship between heart rate and "sociability."

As already noted, vagal, as well as adrenergic, effects influence heart action; increases in heart rate, suggestive of adrenergic activation, could actually reflect not adrenergic changes but rather decreases in vagal tone.

For example, fluctuations in respiration, through variations in intrathoracic pressure, modify vagal tone. Specifically, beat-to-beat changes in heart rate, as determined by fluctuation in R-R interval (R-R variance), are a measure of vagal tone; greater fluctuation is associated with more vagal input.

Several studies have evaluated the R-R interval in panic patients (Gorman et al. 1987; Middleton 1990; Roth et al. 1986; Weissman et al. 1987; Yeragani et al. 1990a, 1993). Yeragani et al. (1990a, 1993) found that R-R variance was lower in patients than in control subjects, especially on standing, suggestive of less vagal activity in patients. Middleton reported less heart rate variability in recovered panic patients. Roth et al. (1986) also reported differences between agoraphobic patients with panic attacks, but this difference disappeared when covaried with baseline heart rate. Weissman et al. did not find any R-R variability differences among control subjects and panic patients with and without mitral valve prolapse in response to deep breathing. Gorman et al. (1987) found that heart rate variability did not change during a lactate-induced panic attack. However, George et al. (1989) did report that lactate infusion and hyperventilation, procedures known to produce panic attacks in panic patients, decreased R-R interval variance in control subjects.

Blood pressure has also been studied in panic patients; earlier studies often (but not always) reported small elevations, especially of systolic blood pressure (Cameron and Nesse 1988). Some studies found a tendency for systolic and/or diastolic blood pressure to be higher in panic patients (Gorman et al. 1990; Hoehn-Saric et al. 1991; Rapaport et al. 1991; Uhde et al. 1989; Yeragani et al. 1990b), whereas others did not (Cameron et al. 1990; Nutt 1989; Tancer et al. 1990a). There was no difference between patients and control subjects on systolic or diastolic blood pressure in response to flumazenil (Woods et al. 1991). Systolic blood pressure was correlated with a report of "palpitations" (McLeod et al. 1986). Interestingly, Yeragani et al. (1990b) reported that the elevations of blood pressure and heart rate observed in a sample of panic patients were due only to those panic patients who were smokers, suggesting that it was tobacco use rather than panic that produced these elevations.

Respiratory changes, including hyperventilation, are associated with pathological anxiety (see Cameron and Nesse 1988); whether the "hyperventilation syndrome" is an entity distinct from panic remains to be determined (Cowley and Roy-Byrne 1987). The potential relationship between hyperventilation and adrenergic function has not been studied in detail. Gorman et al. (1988) reported that panic in response to inhalation of carbon dioxide was associated with increases in plasma NE and diastolic blood pressure.

Mitral valve prolapse is a cardiac abnormality that is observed with increased frequency in panic disorder (Margraf et al. 1988). Mitral valve prolapse has been associated with abnormalities of adrenergic function, including augmented heart rate and symptom responses to isoproterenol infusion and more orthostasis and syncope; 24-hour catecholamine excretion of catecholamine has been reported to be increased or decreased (Boudoulas et al. 1984; Weissman et al. 1987). Although it has been reported that panic patients have a different pattern of physiological and symptom abnormalities than do patients with mitral valve prolapse (Weissman et al. 1987), two studies found no difference between panic patients with versus without mitral valve prolapse on several markers of adrenergic functioning, including plasma epinephrine and NE, urinary MHPG, heart rate and specific electrocardiographic parameters, and blood pressure (Dager et al. 1987; Gorman et al. 1988). It has also been suggested that idiopathic cardiomyopathy is a specific cardiac abnormality associated with panic disorder and that the association might be mediated at least in part by adrenergic mechanisms (Kahn et al. 1987).

Challenge Studies

Panic disorder has been extensively studied utilizing pharmacological challenges that interact specifically with noradrenergic neurons. Clonidine has been the most extensively studied probe. Blunting of the growth hormone (GH) response to clonidine is the most robust psychobiological abnormality thus far identified in panic disorder. Four of six studies have demonstrated significantly smaller GH responses to clonidine in panic patients compared with control subjects (Abelson et al. 1992; Charney and Heninger 1986; Nutt 1989; Uhde et al. 1986). At least 35 of the 49 patients (71%) studied in these four studies had blunted responses (defined as less than a 5 ng/ml increase in GH after clonidine infusion). The only negative studies have come from a single research group (Schittecatte et al. 1988, 1992). The first study involved only seven patients, and two of the three male patients did have blunted responses as defined here. None of the four female patients' responses were blunted. Blunting can be obscured by alcohol use, birth control pills, and/or the preovulatory estrogen surge. In this study, the females were studied within 10 days of onset of menses, which should keep most women out of the phase of estrogen increase, but perhaps not all; no mention is made of the use of alcohol or birth control pills. The second study, of 10 male patients, reported a nonsignificantly higher GH response for patients compared with control subjects. This result is very difficult to rectify with the rest of the literature, although it should be noted that the rate of blunted responses within the control group was also lower than the rate seen in control groups in other studies of panic patients. In both of the negative studies, patients were free of benzodiazepines for only a minimum of 5 days before study, but neither study addressed the possibility that benzodiazepine withdrawal could acutely activate adrenergic systems and obscure GH blunting.

Although the blunting of the GH response to clonidine in patients with panic disorder is a solidly established finding, its implications are less clear. GH blunting has generally been interpreted, based on animal data, as indicating subsensitivity of postsynaptic α_2-adrenoreceptors in the hypothalamus (Siever et al. 1982). However, there is now evidence that some patient groups showing blunted GH responses to clonidine also have blunted GH responses to stimulation with GH-releasing hormone (GHRH) (Lesch et al. 1987; Rapaport et al. 1989; Tancer et al. 1993), which acts at the level of the pituitary. Blunted GH responses to clonidine and GHRH could be due to abnormalities within the pituitary (e.g., changes in GHRH receptor function or number) or in other neurotransmitter systems (e.g., decreased cholinergic activity leading to increased somatostatin release and decreased GH responsivity). However, even though the GHRH-stimulated release of GH does not involve the α_2-adrenoreceptor directly, it remains quite plausible that a hypothalamic adrenergic abnormality could underlie the blunting of the GH response. Chronic excessive adrenergic input to the hypothalamus could lead to down-regulation of both the postsynaptic α_2-adrenoreceptor and the GHRH receptor, leading to blunting of the GH response to both clonidine and exogenous GHRH. Excessive adrenergic input could also activate counterregulatory responses further "downstream," such as increases in somatomedin-C levels, which would also blunt GH responses to both clonidine and GHRH (even in the absence of primary changes in α_2-adrenoreceptor sensitivity). Alternatively, since the adrenergic control of GH in humans may be mediated primarily by somatostatin (Devesa et al. 1992), the blunted response to GHRH could actually be a separate abnormality. However, data from depressed and control subjects suggest that blunted GH responses to clonidine and GHRH in these subjects are highly correlated with each other (Lesch et al. 1988; Tancer et al. 1990b) and are also correlated with somatomedin-C levels, and likely derive from abnormalities at or above the level of the hypothalamus (Lesch et al. 1988). Further research is needed to determine if this abnormality involves adrenergic input in a primary fashion. This finding, despite complications in its interpretation, remains consistent with the hypothesis that chronic excessive noradrenergic input reaches the hypothalamus, presumably from the locus coeruleus, and leads to down-regulation of postsynaptic α_2-adrenoreceptors. Clarification of the precise mechanisms leading to blunted GH responses to clonidine in panic patients must await more detailed studies that combine the use of multiple adrenergic probes in the same subjects. Further study of the robust inhibition of GH release seen in panic in response to the adrenergic agonist clonidine is likely to be a fruitful avenue for clarifying the role of adrenergic mechanisms in this disorder.

Adrenergic regulation in panic disorder has also been studied by examining MHPG and cardiovascular responses to clonidine infusions. Normally, clonidine produces a decrease in MHPG. Panic patients showed a greater clonidine-induced decrease in MHPG than did control subjects in two studies (Charney and Heninger 1986; Nutt 1989), but not in a third (Abelson et al. 1992). In one of the positive studies, the group differences in response may have been due to elevation in baseline levels (Nutt 1989). In the nonreplicating study, group differences were in the same direction as in the other two studies (greater fall in patients) but did not reach significance; a trend toward elevated baseline levels was also present, which would complicate interpretation of response differences. The available data therefore suggest that panic patients may have greater declines in MHPG following clonidine than control subjects, but this is not yet clearly established and is much less robust than the GH findings. There does not appear to be any association between the GH and the MHPG abnormalities.

Conclusions about cardiovascular responses to clonidine in panic disorder are similarly tenuous. Normally, clonidine lowers blood pressure and heart rate. Four studies have compared panic patients with control subjects on these variables. One study demonstrated a greater decrease in systolic blood pressure after clonidine infusion in patients (Nutt 1989), but the other three studies did not report this finding (Abelson et al. 1992; Charney and Heninger 1986; Uhde et al. 1989). Two studies reported greater declines in diastolic blood pressure in patients (Charney and Heninger 1986; Nutt 1989), but two others found no differences (Abelson et al. 1992; Uhde et al. 1989). Heart rate responses to clonidine were normal in all studies. Differences in dosing schedules (fixed-dose versus per-kilogram dose) and in methods of data analysis ("peak" responses versus averaged values) may explain some of the discrepancies in the blood pressure findings, but it does not appear that panic disorder is associated with reliably reproducible abnormalities in cardiovascular responses to clonidine infusion.

Panic patients have also been compared with control subjects in the responses of MHPG and cardiovascular variables to oral yohimbine. Yohimbine produces significant rises in subjective anxiety in patients with panic disorder and stimulates panic attacks in somewhat more than 50% of such patients, but has little effect on anxiety or panic states in control subjects (Charney et al. 1987; Gurguis and Uhde 1990). An initial study demonstrated that patients with frequent panic attacks had increased MHPG and cardiovascular responses to yohimbine when compared with control subjects or patients with less frequent attacks (Charney et al. 1984). However, when the sample from this study was expanded and reanalyzed, it was found that it was only those patients who panicked in response to the yohimbine who had increased MHPG, systolic blood pressure, and heart rate responses when compared with control subjects; nonpanicking patients did not differ from control subjects on these variables (Charney et al. 1987). A more recent replication study (Gurguis and Uhde 1990) confirmed a strong relationship between "panicky" feelings in patients after yohimbine infusion and their MHPG response. This study reported a "trend" toward greater MHPG rises in panic patients compared with control subjects. A third study reported augmented increases in NE and heart rate and a greater decrease in skin conductance in response to yohimbine in panic patients (Albus et al. 1992). Abnormal changes in regional cerebral blood flow following yohimbine infusions have also been noted in panic patients (Woods et al. 1988b).

Because it is now known that an interaction between cognitive factors and somatic symptoms may play a role in panic attacks (Sanderson et al. 1989), it has become very difficult to determine whether the "abnormal" responses of panic patients to probes like yohimbine are due to nonspecific somatic cues and cognitive factors that elicit secondary neuroendocrine-cardiovascular effects or to direct pharmacological effects that provide information about the functioning of α_2-adrenoreceptors in these patients. The yohimbine data suggest at a minimum that subjective anxiety and "panicky" feelings may be associated with alterations in noradrenergic activity and that panic patients have a lower threshold for eliciting subjective anxiety. Whether panic anxiety involves abnormal sensitivity of the α_2-adrenergic receptor remains unclear. Further clarification could be achieved by studies that combine the use of clonidine and yohimbine in the same subjects (Charney and Heninger 1986).

Adrenergic responses to other substances have also been studied in panic patients. Infusions of isoproterenol (a peripheral β-adrenergic agonist) produced dose-related panic attacks (Pohl et al. 1988a) that were symptomatically similar to natural attacks (Balon et al. 1988). Isoproterenol infusions also have demonstrated cardiac β-receptor "down-regulation" in panic patients (Nesse et al. 1984). The symptoms most associated with isoproterenol-induced attacks were "shortness of breath" and "fear of going crazy" (Balon et al. 1990). If peripheral β-receptors are "down-regulated" in panic, then the panic attacks produced by isoproterenol are likely to be secondary to nonspecific factors (e.g., somatic symptoms and cognitive responses). Isoproterenol-induced panic attacks are not associated with elevations of MHPG (Pohl et al. 1987) or heart rate (Freedman et al. 1984). NE infusions in panic patients may induce panic attacks that are associated with increased blood pressure and decreased heart rate (Pyke and Greenberg 1986), but the lack of controlled studies utilizing this model precludes meaningful interpretation. Based on cortisol responses to desipramine and m-chlorophenylpiperazine (mCPP), one group (Asnis et al. 1992) concluded that there were interrelated abnormalities of adrenergic and serotonergic functioning in panic disorder.

Lactate- and CO_2-induced panic attacks have been extensively studied, but the mechanism of panic induction in these two models remains unknown. Both have been hypothesized to induce panic through central noradrenergic mechanisms (Liebowitz et al. 1985; Woods et al. 1986), but support for these hypotheses remains weak. Panic in these models is not associated with elevations of MHPG (Carr et al. 1986; Woods et al. 1988a). It may be associated with increases in NE, heart rate, respiration, and blood pressure, but these changes tend to be small and inconsistent (Bystritsky and Shapiro 1992; Carr et al. 1986; Gaffney et al. 1988; Gorman et al. 1988; Liebowitz et al. 1985; Woods et al. 1988a; Yeragani et al. 1989). Preinfusion changes in heart rate do not predict who will panic in response to lactate (Yeragani et al. 1987b), and pretreatment with clonidine only partially attenuates panic responses to lactate (Coplan et al. 1992). Increases in preinfusion heart rate may or may not predict who will panic in response to a placebo (Goetz et al. 1993; Yeragani et al. 1987c). McCann et al. (1991) reported that depletion of central catecholamine levels can produce panic attacks in a minority of control subjects.

Caffeine-induced panic also fails to produce elevations in plasma MHPG (Charney et al. 1985; Uhde et al. 1984), plasma NE, or urinary measures of MHPG, normetanephrine, and metanephrine (T. W. Uhde, personal communication, March 1993). Although never used specifically as pharmacological challenge agents, tricyclic antidepressants are known to worsen anxiety symptoms and panic attacks early in treatment in a dose-dependent fashion in some patients (Nutt 1990; Pohl et al. 1988b). This may represent a specific adrenergic challenge in that tricyclic antidepressants are noradrenergic uptake blockers and may thus increase noradrenergic transmission early in treatment. Further study of this phenomenon may prove fruitful.

Adrenergic function in panic disorder has also been studied utilizing nonpharmacological challenge models, including exposure to phobic stimuli, the cold pressor test, exercise, and postural challenge. Anxiety induced in panic patients by exposure to phobic situations did produce increases in MHPG in one study (Ko et al. 1983), but not in another (Woods et al. 1987). Panic patients did not differ from control subjects in blood pressure responses to the cold pressor test (Grunhaus et al. 1983), nor in heart rate, blood pressure, or catecholamine responses to the postural challenge of standing for 15 minutes (Cameron et al. 1990). When measured in a second study after 5 minutes of standing, panic patients did have greater heart rate responses than did control subjects, but NE responses remained normal (Stein et al. 1992b). In this study, the relations among postural changes in heart rate and NE also differed between groups; this was interpreted as representing decreased vagal functioning in the panic patients. Panic patients do not appear to have an abnormal physiological response to exercise (Sandler et al. 1992; Stein et al. 1992a).

Receptor Binding and Related Studies

Several studies have evaluated adrenergic receptor binding in panic patients. Two of three studies demonstrated that lymphocyte β-adrenergic receptor number was lower in panic patients than in control subjects (Albus et al. 1986; Aronson et al. 1989a; Brown et al. 1988); one reported an elevation (Butler et al. 1992). Two others demonstrated a reduction in adrenergically mediated lymphocyte cyclic adenosine monophosphate production (Charney et al. 1989b; Lima and Turner 1983).

Platelet binding of a nonselective α-antagonist (dihydroergocryptine) was increased in panic patients (Roy-Byrne and Uhde 1985); binding of α_2-adrenoreceptor selective ligands (the partial agonist clonidine and the antagonists yohimbine and rauwolscine) have been reported to be decreased (Albus et al. 1986; Cameron et al. 1984, 1990), normal (Charney et al. 1989b; Norman et al. 1987; Nutt and Fraser 1987), or increased (Butler et al. 1992). Grunhaus et al. (1990) demonstrated that patients with comorbid panic disorder and major depression had decreases in platelet α_2-adrenergic receptor binding similar to panic patients and different from depressed patients. Correlations between catecholamines and receptor binding parameters have not yielded consistent significant associations (Albus et al. 1986; Cameron et al. 1984, 1990).

Several of the above studies evaluated potential relationships among binding parameters with symptom and demographic parameters. Three reported no relationship (Albus et al. 1986; Cameron et al. 1984; Nutt and Fraser 1987). Aronson et al. (1989a) reported a significant negative correlation between β-adrenergic binding and trait anxiety scores, and Charney et al. (1989b) reported that panic attack frequency was positively correlated with parameters of adenylate cyclase activity. Cameron and colleagues reported that clonidine binding was correlated with panic attack severity (Cameron et al. 1990), and that the reduction in symptom severity in response to imipramine treatment was positively correlated with changes in binding parameters (Cameron et al. 1984). In all of these studies, many other correlations were not significant, and thus those that were might have occurred by chance.

Conclusions

Although there is evidence that abnormalities in noradrenergic neuroendocrine pathways play a role in the pathophysiology of panic disorder, the noradrenergic hypothesis of panic has not been unequivocally demonstrated. Data from studies of panic patients in basal states provide conflicting and inconclusive evidence. It appears more likely that any adrenergic abnormality in panic involves regulation of the system's responsivity rather than its resting tone. As a result, challenge studies have proven more informative.

The best evidence for noradrenergic abnormalities in panic disorder has come from challenge probes that have specific pharmacological effects on the noradrenergic system. Further study of these probes is needed. Questions have been raised about the relevance of platelet α_2-adrenoreceptor binding assays to central noradrenergic receptors (Nutt 1987), but the binding study results in panic patients are thus far fairly consistent with the noradrenergic challenge study results. Application of both paradigms to the same subjects will help clarify the relationship between the peripheral and central probes and could strengthen the evidence for adrenergic dysfunction in panic.

The provocation of panic attacks by the β-adrenergic agonist isoproterenol appears to implicate the β-adrenergic receptor in the pathophysiology of panic as well. However, since isoproterenol does not cross the blood-brain barrier, it appears likely that it provokes the full panic symptom constellation as a secondary response to the peripheral somatic symptoms induced. Although abnormal lymphocyte β-adrenoreceptor binding also implicates the β-adrenergic receptor in panic, propranolol (a β-adrenergic receptor antagonist that does cross the blood-brain barrier) does not reliably block panic (see Cameron, Chapter 9, this volume). Much further research is needed to understand the role of the β-adrenergic receptor in panic disorder.

Panic attacks induced by adrenergically nonspecific challenges (e.g., caffeine, lactate, CO_2) do not necessarily involve adrenergic mechanisms. Whether this reflects weaknesses in the models, insensitivity of the adrenergic variables monitored, or evidence against the noradrenergic hypothesis of panic is not yet known. Better understanding of the role of nonspecific somatic cues and cognitive factors in laboratory models of panic is also needed to clarify the results of studies using challenges that can produce somatic cues that are similar to panic attack symptoms. Of the challenge paradigms reviewed here, only the clonidine model is uncontaminated by elicitation of somatic cues that may induce panic by associative learning or cognitive processes. Clonidine is in fact anxiolytic in acute administration (Uhde et al. 1989). Further understanding of the role of adrenergic mechanisms in panic anxiety will require combined utilization of, for example, platelet binding assays with infusions of clonidine, GHRH, and yohimbine, as well as the development of new models.

Generalized Anxiety Disorder

As noted, many studies of anxiety prior to the 1980s most likely included subjects with a variety of different anxiety disorders, many of whom would now be diagnosed as having GAD. As reviewed previously, there was some evidence from both endocrine and physiological studies that there was increased adrenergic function in these patients (see Cameron and Nesse 1988). There is, however, substantially less information on adrenergic function in those with specifically diagnosed GAD than in individuals with panic attacks.

Basal Studies

An initial study by Mathew et al. (1980) suggested that patients with GAD had elevated plasma levels of epinephrine and NE. However, the elevations noted may have actually represented responses to the challenge of venipuncture since blood was drawn immediately after the venipuncture was performed. Cameron et al. (1990) reported normal baseline NE, heart rate, and blood pressure in GAD. Hoehn-Saric et al. (1989) reported normal baseline electrodermal activity, "heart period," and blood pressure; they also reported that cardiac symptoms were associated with lability of cardiac action (Hoehn-Saric et al. 1991). McLeod et al. (1990), in a study that did not include a control group, found pretreatment levels of epinephrine and NE that were comparable to published norms for control subjects; after 6 weeks of treatment, the study found that a benzodiazepine (alprazolam) decreased epinephrine and decreased the NE response to "stress," whereas imipramine increased epinephrine nonsignificantly and significantly increased the NE response to "stress."

Munjack et al. (1990) examined plasma MHPG levels in resting GAD patients and, although initial analysis suggested elevated levels relative to control subjects, it was concluded, based on discriminant analysis, that the group differences were not meaningful. One study did find elevated plasma levels of NE and MHPG in GAD patients compared with control subjects, but this report did not describe a delay between venipuncture and blood drawing, so the elevations may again reflect greater reactivity to the procedure in the patients (Sevy et al. 1989). Two reports supported the conclusion that there are no baseline MHPG differences between GAD patients and control subjects prior to challenge studies (Abelson et al. 1991; Charney et al. 1989a).

In summary, it does not appear that patients with GAD have elevated basal levels of catecholamines, MHPG, or autonomic activity. Similar to panic patients, patients with GAD may be more adrenergically reactive to a physical stressor such as venipuncture, although not necessarily to all "stressors" (Hubert and de Jong-Meyer 1990). Uncontrolled differential reactivity of GAD patients (or "impure" diagnostic groups) might account for the positive findings in earlier studies.

Challenge Studies

Only two pharmacological challenge studies of adrenergic activity in patients with GAD have been published. Charney et al. (1989a) reported that GAD patients do not differ from control subjects in their behavioral, cortisol, and cardiovascular responses to yohimbine challenge. The GAD patients did not show any greater increase in anxiety or a higher frequency of panic-like reactions in response to yohimbine than did control subjects. Yohimbine increased blood pressure equally in the two groups and had no effect on heart rate in either. Yohimbine produced increases in MHPG levels in patients and control subjects; the GAD patients tended to have smaller increases in MHPG following yohimbine than control subjects. Charney et al. concluded that patients with GAD do not display increased noradrenergic neuronal activity.

In contrast, Abelson et al. (1991) reported blunted GH responses to clonidine infusion. This finding has usually been interpreted as reflecting subsensitivity of postsynaptic α_2-adrenoreceptors. Although recent work with GHRH has complicated this interpretation (see earlier on panic disorder), the finding is consistent with the hypothesis that patients with GAD do have chronic central noradrenergic overactivity, which leads to down-regulation of α_2-adrenoreceptors. In this study, there were no significant differences between groups in cardiovascular, MHPG, or subjective (e.g., anxiety, sedation) responses to clonidine. GAD patients and control subjects also did not differ in epinephrine, NE, heart rate, or blood pressure responses to postural challenge (Cameron et al. 1990). During the challenge of performance stress, patients with GAD may show inhibition of sympathetic activity, reflected in a weaker electrodermal response and a narrower heart interbeat interval range than control subjects (Hoehn-Saric et al. 1989).

Receptor Binding Studies

Apparently, there are no studies of β-adrenergic receptor binding studies in persons with GAD. There are two reports of decreased numbers of platelet α_2-adrenoreceptors in patients with GAD. Sevy et al. (1989) reported a decrease in binding of the antagonist tritiated yohimbine; Cameron et al. (1990) reported decreased tritiated clonidine binding. Tritiated yohimbine binding was also lower in GAD subjects in the study by Cameron et al., but this difference did not reach statistical significance. Reduced platelet α_2-adrenoreceptors is consistent with the hypothesis of reduced adrenergic receptor function in these patients.

Conclusions

Patients with GAD do not appear to differ from control subjects in basal state measures of noradrenergic activity. However, like panic patients, GAD patients may differ from control subjects in the *regulation* of noradrenergic systems, reflected in a greater noradrenergic reactivity to some stressors. Chronic excessive noradrenergic reactivity to certain kinds of stimuli could conceivably produce down-regulatory responses within the noradrenergic system and produce the blunted GH response to clonidine and reduced platelet α_2-adrenoreceptor number reported in GAD. Considerably more research will be needed to explore this hypothesis, especially in light of the preliminary evidence that patients with

GAD may show sympathetic *inhibition* in response to some types of challenges.

Patients with GAD appear to have normal responses to the α_2-adrenoreceptor antagonist yohimbine but abnormal responses to the agonist clonidine. Thus far, there is very little knowledge of the relationships between various probes of adrenergic function in humans because multiple probes have not been applied to a single group of subjects. The data indicate that some GAD patients may have elevated NE or MHPG levels in some circumstances, some have reduced platelet α_2-adrenoreceptor number, some have blunted GH but normal MHPG responses to clonidine, some have normal or reduced MHPG responses to yohimbine, and some have inhibited sympathetic activity under performance stress. It is not clear if the abnormalities are identifying separate groups of patients or whether the pattern of adrenergic abnormality reflects a single underlying defect that is relevant to the pathophysiology of the disorder. Studies examining the relationships among the known abnormalities in the same patients are needed. It will remain difficult to determine, using currently available models, whether identified abnormalities are due either to an underlying biological abnormality in central noradrenergic systems that is genetically or developmentally determined, or to secondary physiological responses to other, non-noradrenergic, cognitive (cortical?) or emotional (limbic?) abnormalities.

Despite the attempts in DSM-III-R and DSM-IV to make the diagnosis of GAD a less residual and more homogeneous category (Barlow et al. 1986), it has still attracted relatively little scientific attention and remains subject to controversy (Abelson et al. 1991; Breier et al. 1985). However, this diagnosis does describe a group of patients, however heterogeneous, who do have abnormal levels of worry and anxiety; these patients do appear to have some abnormalities in adrenergic function. Further study of patients with GAD can enhance our understanding of the relationship between adrenergic neurotransmission and anxiety.

Obsessive-Compulsive Disorder

Adrenergic mechanisms have received less attention in studies of OCD than in studies of GAD and, especially, panic, perhaps because of the dominance of serotonergic hypotheses in the literature on OCD.

Basal Studies

Baseline elevations of NE and MHPG have been reported in patients with OCD (Insel et al. 1984; Rasmussen et al. 1987; Siever et al. 1983), but these findings have not been uniformly replicated (Benkelfat et al. 1991; Lee et al. 1990). One study (Benkelfat et al. 1991) also reported normal levels of epinephrine; this study did report increased resting heart rates. Urinary excretion of phenylethylamine, a substance related to the metabolic pathways of the catecholamines, has also been reported to be normal in patients with OCD (Theofilopoulos et al. 1991).

Challenge Studies

One early report described blunted GH responses to clonidine infusion in patients with OCD (Siever et al. 1983), but subsequent attempts to replicate this finding have failed (Hollander et al. 1991; Lee et al. 1990). MHPG, blood pressure, and heart rate responses to clonidine infusion are also normal in patients with OCD (Hollander et al. 1991; Lee et al. 1990). These patients do not differ from healthy control subjects in their behavioral and MHPG responses to oral yohimbine either (Rasmussen et al. 1987). Although the α_2-adrenoreceptor antagonist yohimbine does not appear to exacerbate obsessive-compulsive symptoms, the agonist clonidine does appear to reduce such symptoms (Hollander et al. 1988); whether this is due to a specific adrenergic effect or is secondary to the general calming and sedating effects of clonidine is not clear.

Receptor Binding Studies

One small study (Lee et al. 1990) evaluated α_2-adrenoreceptor binding in patients with OCD. In contrast to patients with panic or GAD, these patients did not have any reduction in receptor binding parameters. In fact, the number of binding sites for tritiated clonidine was somewhat increased in these individuals.

Conclusions

The little data available concerning adrenergic function in OCD are perhaps most useful in the contrast provided to the panic and GAD results. The data suggest that the abnormalities seen in panic and GAD have some specificity to those disorders; they are not seen in patients with every anxiety disorder or in every patient with a high level of anxiety. Patients with OCD score quite comparably to panic and GAD patients on state-trait anxiety scales (J. L. Abelson, O. G. Cameron, G. C. Curtis, unpublished data, April 1992), but they differ from the latter two groups in the GH response to clonidine and in their platelet α_2-adrenoreceptor binding. There is a noteworthy concordance between results of platelet binding studies and the challenge studies across the three diagnostic groups, with panic and GAD patients showing blunted GH responses to clonidine and reduced binding, whereas OCD patients show normal GH responses and normal or increased binding.

Posttraumatic Stress Disorder

Although PTSD is by definition a disorder, it is not clear to what extent the disorder is due strictly to the occurrence and severity of the stressful event, or whether some individuals are more susceptible to developing a posttraumatic reaction than others. To the extent that the disorder is due primarily to the occurrence of the stressor, it might be viewed as an example of the second meaning of anxiety described above—that is, a normal psychological and physiological reaction to a threatening situation. Conversely, if some individuals are constitutionally more susceptible than others, then abnormalities of adrenergic functioning in these persons might represent changes *qualitatively* different than a "normal" stress reaction.

Basal Studies

Vietnam veterans with PTSD have been reported to have elevated plasma (Blanchard et al. 1991) and urinary (Kosten et al. 1987; Yehuda et al. 1992) levels of epinephrine and NE compared with patients with a number of other psychiatric disorders. Relatively low levels of urinary free cortisol have led to the proposal of the NE/cortisol ratio as a relatively powerful biological marker of PTSD, discriminating this disorder from other psychiatric disorders, with a sensitivity of 78% and a specificity of 89% (Mason et al. 1988). Pitman and Orr (1990), however, failed in their attempt to replicate the earlier reports of elevated urinary catecholamines and elevated NE/cortisol ratio in Vietnam veterans with

PTSD. McFall et al. (1992) also reported no difference between PTSD patients and control subjects in tonic levels of heart rate, blood pressure, and plasma catecholamines. Although not specifically diagnosed with PTSD, American hostages freed from extended captivity in Iran did have elevated levels of urinary catecholamines (Rahe et al. 1990). Plasma NE levels have also been reported to be elevated in PTSD (Blanchard et al. 1991), whereas levels of the second messenger adenylate cyclase have been reported to be reduced (Lerer et al. 1990).

Baseline levels of autonomic arousal in patients with PTSD have been reported in a variety of psychophysiological challenge studies. A comprehensive review of these studies (McFall et al. 1989) concluded that patients with PTSD have elevated resting heart rates relative to stressed or nonstressed control subjects without PTSD, but not relative to other psychiatric control subjects. They do not show elevations in resting blood pressure. They do appear to be "more autonomically arousable," as indicated by heart rate and event-related brain potential responses to randomly presented tones (Paige et al. 1990).

Challenge Studies

A predominant theory of the pathophysiology of PTSD has postulated that it involves the conditioning of emotional, behavioral, and physiological responses to traumatic stress such that these responses are repeatedly and inappropriately elicited by previously nontraumatic internal and external stimuli (McFall et al. 1989). This phenomenon is reflected in the DSM-IV diagnostic criterion of "physiologic reactivity on exposure to internal or external cues that symbolize or resemble an aspect of the traumatic event" (p. 428). It has spawned a series of behavioral challenge studies in which patients with PTSD are compared with various control groups in their emotional and physiological responses to trauma-related and unrelated stimuli. Dobbs and Wilson (1960) were the first to demonstrate that combat-exposed veterans had greater heart-rate responses to combat-related stimuli than did control subjects. A methodologically more sophisticated series of studies in the 1980s confirmed that combat veterans with PTSD responded with significantly greater heart rate increases on exposure to combat-related stimuli, but not to combat-unrelated stimuli, when compared with combat-exposed veterans without PTSD, veterans not exposed to combat, and patients with other psychiatric disorders (Blanchard et al. 1982, 1986; Pallmeyer et al. 1986). Malloy et al. (1983) provided further confirmation of these findings and extended them to include increased electrodermal responsiveness to combat-related stimuli in the patients with PTSD. All of the above challenge studies utilized recorded sights and sounds of combat as a standard combat-related stimulus for all subjects. Pitman et al. (1987) utilized individualized imagery scripts to compare combat-exposed veterans with and without PTSD on heart rate, skin conductance, and facial muscle tension responses to combat-related and -unrelated stimuli. This study further demonstrated that PTSD is associated with excessive autonomic arousal in response to trauma-related stimuli, but not to neutral stimuli, nor to stressful but nontrauma-related stimuli.

Despite this solid body of evidence for heightened autonomic arousal in patients with PTSD, and the extensive theoretical writings that invoke noradrenergic mechanisms in the pathophysiology of this disorder (Kolb 1987; McFall et al. 1989; van der Kolk et al. 1985), there have been only a few studies using noradrenergic neuroendocrine challenge models in patients with PTSD. McFall et al. (1990) examined arterialized plasma catecholamine levels (epinephrine and NE) in patients with PTSD in a standardized, recorded, combat-stimuli challenge paradigm similar to those used to demonstrate autonomic hyperreactivity. They did not detect any baseline differences between PTSD patients and a mixed control group in heart rate, blood pressure, or catecholamine levels. The PTSD patients responded to a stressful, but nontrauma-related stimulus, with significant increases in blood pressure and emotional arousal. They responded to the trauma-related stimulus with significant increases in heart rate, blood pressure, emotional arousal, and epinephrine. They were more reactive to the trauma-related than the nonrelated stimulus. Control subjects did not show significant increases on cardiovascular or neuroendocrine variables during exposure to either stimulus. Direct comparisons of the responsivity of the patient and control groups showed that during or immediately following the trauma-related stimulus, the patients had greater increases from baseline in heart rate, blood pressure, emotional arousal, and epinephrine.

One noradrenergic neuroendocrine challenge study of PTSD utilized the GH response to desipramine challenge as a probe for α_2-adrenoreceptor functioning (Dinan et al. 1990). Chronic excessive noradrenergic activity should lead to down-regulation of the α_2-adrenoreceptor, which would be reflected in blunted GH responses to desipramine challenge. Dinan et al. report no differences between patients with PTSD and a healthy control group on this measure. However, weaknesses in the study do not allow any clear conclusions about adrenoreceptor status in PTSD. The sample size was small ($n = 8$ in each group), and the patients were females who had suffered violent sexual assaults, whereas combat veterans were used in all of the studies cited earlier. Most importantly, all of the subjects were tested during the third week of their menstrual cycle, which is a time at which the high estrogen levels that are present serve to sensitize the hypothalamic-pituitary-somatotroph axis to stimulation and might thus obscure blunted GH responses that might be seen earlier in the cycle. In contrast to this study, Hansenne et al. (1991) found a blunted GH response to clonidine in one male PTSD patient; after treatment, the GH response was normal. Further, PTSD patients who had panic attacks or flashbacks in response to yohimbine showed abnormal increases in heart rate, blood pressure, and MHPG levels (Southwick et al. 1993).

Receptor Binding Studies

Perry et al. (1987) reported a decreased number of platelet α_2-adrenergic receptor binding sites in patients with PTSD, as measured with tritiated rauwolscine. This decrease is similar to the findings in panic disorder and GAD.

Conclusions

Patients with PTSD do show heightened autonomic arousal when challenged with trauma-related stimuli. Noradrenergic mechanisms are presumed to play a role in this arousal, but direct documentation of noradrenergic responsivity and receptor status in PTSD remains relatively sparse.

Other Anxiety and Related Disorders

Studies of adrenergic function are limited in other anxiety disorders (see Cameron and Nesse 1988). In persons with public speaking anxiety (social phobia), heart rate and blood pressure are increased both in anticipation of and during public speaking (Fredrikson et al. 1990). Similar to panic patients, they might have an increased incidence of mitral valve prolapse (Chaleby and Ziady 1988). Epinephrine infusion does not produce symptoms of social anxiety in patients with social phobia (Papp et al. 1988). Patients with social phobia, like those with panic disorder and GAD, appear to have blunted GH responses to clonidine infusion (Uhde et al. 1991), as well as increased lying and standing plasma NE levels (Stein et al. 1992b). One case report suggests that clonidine may be an effective treatment in some cases of social phobia (Goldstein 1987). Persons with social phobias had greater increases in systolic blood pressure than control subjects in response to thyrotropin-releasing hormone (Tancer et al. 1990a). In contrast, Gorman et al. (1990) did not find any difference between social phobic persons and control subjects in the heart rate or blood pressure responses to CO_2 inhalation.

Pharmacological challenge studies of patients with simple phobias have not been reported. Exposure to feared stimuli represents a specific behavioral challenge for such patients, and catecholamine increases have been seen in association with phobic anxiety (Abelson and Curtis 1989; Nesse et al. 1985). Children with separation anxiety have higher heart rate and blood pressure than do control subjects (Rogeness et al. 1990). Furthermore, the presence of anxiety symptoms in individuals with affective disorders (Abramson et al. 1989), or at risk for affective disorders (Zahn et al. 1991), is associated with elevated heart rates. Finally, Southwick et al. (1990) reported a decrease in platelet α_2-adrenoreceptors (similar to panic and GAD patients) in anxious, but not in nonanxious, persons with borderline personality disorder.

Overall Conclusions

The association of putative adrenergic symptoms with subjective anxiety has long implicated adrenergic dysfunction as an important part of the psychobiology of both normal and pathological anxiety. Furthermore, the idea that abnormalities in central noradrenergic centers such as the locus coeruleus underlie abnormal anxiety states—initially based primarily on results from animal studies—remains one of the dominant hypotheses shaping efforts to define the pathophysiology of panic disorder. These observations have motivated a great deal of research into potential adrenergic abnormalities in persons with anxiety disorders.

Prior to the 1980s, studies of potential psychobiological abnormalities associated with abnormal anxiety generally included a mixture of what are now defined as discrete anxiety disorders, with panic disorder and GAD most heavily represented. More recent research has generally studied these disorders separately. Some recent studies, as well as some earlier studies, have suggested that in the basal state catecholamines, catecholamine metabolites (mainly MHPG), other biochemical markers of adrenergic function, and physiological markers such as heart rate and blood pressure are elevated in at least some of these patient groups. However, many studies have not found such elevations. Some of these studies may not have evaluated the patients under truly basal conditions, and some but not all investigators have reported (for panic at least) that patients are more physiologically reactive to stress than are control subjects. Thus, any reported abnormalities might be due to study under nonbasal conditions. It remains unresolved whether these peripheral adrenergic markers are abnormal in patients with anxiety disorder when the patients are in truly basal conditions.

Results of challenge studies have indicated that there is an abnormality involving the α_2-adrenoreceptor in panic patients and possibly in GAD patients, but probably not in patients with OCD. The evidence is strongest in the panic patients, as indicated by a blunted GH response to the α_2-adrenoreceptor partial agonist clonidine. Abnormalities in the MHPG response to clonidine, as well as to the α_2-adrenoreceptor antagonist yohimbine, are not as robust or consistent as the GH findings, but they also support an abnormality of the adrenergic functioning involving the α_2-adrenoreceptor in panic. It is not clear, however, whether this abnormality lies in the α_2-adrenoreceptor per se, or somewhere else in the pathways controlling GH and MHPG (see later). The status of the response to β-adrenergic challenge in these patients has been studied, but is less well characterized; the relative lack of therapeutic efficacy of β-adrenergic antagonist drugs suggests that any abnormality of this receptor may not be of fundamental importance in panic disorder.

Receptor binding studies have indicated that the platelet α_2-adrenoreceptor number may be decreased in panic disorder (although not all studies of panic have found this), GAD, and possibly posttraumatic stress, but not in OCD. The general concordance of results between α_2-adrenoreceptor binding data and GH responses to clonidine across diagnoses—both markers support reduced α_2-adrenoreceptor status in panic and GAD, whereas both are normal in OCD—tends to support the validity of the findings. There is evidence for decreased numbers of β-adrenergic receptors on lymphocytes in panic disorder as well.

Regulation and Dysregulation

Regulation means the control, adjustment, or feedback in a system that attempts to maintain a functionally adaptive endpoint or output; dysregulation is said to occur either when the endpoint or output is not maintained or when the steps in the system maintaining this endpoint or output are not functioning the same as in the general population. Adrenergic dysfunction is more readily apparent in patients with anxiety disorders when they are pharmacologically challenged than when they are observed in the basal resting state; this suggests that the adrenergic abnormality in these disorders is in regulatory processes. The system appears able to maintain endpoint physiological parameters under adrenergic regulatory control (e.g., basal plasma catecholamines and heart rate) within normal limits; nevertheless abnormal responses to challenges that evaluate function at some point in the system other than at an endpoint suggest that some regulatory steps or set points are not normal. (If confirmed, the finding that the stress response in persons with pathological anxiety is abnormal might be viewed as evidence against the hypothesis that the system is maintaining a normal endpoint; on the other hand, however, it could equally legitimately be viewed as another type of challenge that also provides evidence of dysregulation.)

The endpoints or outputs of interest in this context are the various psychobiological functions (e.g., neuronal pathways and activity; neurotransmitter release, reuptake, and metabolism; number and responsivity of receptors) in the central and peripheral nervous systems that involve adrenergic regulatory mechanisms. Examples of markers of these functions that have been studied in anxiety disorders include GH release, production of the NE metabolite MHPG, and maintenance of blood pressure in the normal range. Are there any other disorders associated with adrenergic functioning that demonstrate dysregulation and, as such, support the idea that dysregulation in this system could also occur in patients with anxiety disorder? The development of essential hypertension might provide one example.

There is evidence for sympathetic dysfunction early in the course of essential hypertension—that is, either in hypertensive patients when they are young or in the subset of hypertensive patients who develop the illness at an earlier age; for example, resting levels of plasma catecholamines are abnormally elevated (Abboud 1982; Goldstein 1983; Nelesen and Dimsdale, Chapter 10, this volume; Tuck 1986). Later in the course of the disorder the plasma catecholamine abnormality is no longer apparent, suggesting that compensatory changes in the adrenergic systems occur in longer-standing hypertension. Nonetheless, although regulatory control mechanisms have restored the basal endpoint measure (the plasma catecholamine levels) to normal, abnormalities in adrenergic regulatory processes are still apparent as abnormal responses to various pharmacological challenges and to stress. Acute responses to adrenergic stimulation with adrenergically active substances are abnormal in hypertension (Bertel et al. 1980; Esler et al. 1977; Kotchen et al. 1982; Trimarco et al. 1983), and hypertensive patients are more responsive to stress than are nonhypertensive subjects (Goldstein 1981; Horikoshi et al. 1985; Julius 1991; Tuck 1986). The effectiveness of various adrenergically active substances in the treatment of hypertension, even if long standing, also indicates that abnormalities in adrenergic functioning are involved in the blood pressure elevations. Conceptually, it appears that in hypertension some perturbation to the system has forced it to accept an altered, or "abnormal," set point for one adrenergically regulated endpoint (blood pressure), whereas other endpoints (e.g., plasma catecholamines) have returned to the normal range. The regulatory processes themselves, however, are no longer "normal," as revealed by their inability to maintain normal responsivity to adrenergic challenges.

Abnormalities in adrenergic challenge studies and in response to stress, indicative of adrenergic dysregulation, suggest that adrenergic dysregulation in pathological anxiety states might be similar to regulatory dysfunction in hypertension. Thus, the pathophysiology of hypertension appears to provide supportive evidence that dysregulation of adrenergic systems does occur, and could be a contributing factor in the physiology (and etiology) of pathological anxiety states. Applying the conceptual model of hypertension to anxiety disorders, it appears that some as-yet-unidentified perturbation has forced the adrenergic system to accept an altered set point for one of its regulated endpoints (anxiety), but other endpoints are maintained in the normal range. Abnormal responses to challenges, such as clonidine or stress, however, reveal that the perturbation, whatever its origin, has produced a dysregulation of adrenergic control mechanisms.

CNS Adrenergic Function and Anxiety

If adrenergic control mechanisms are dysregulated in patients with anxiety disorders, it seems logical to wonder what role this adrenergic dysregulation is likely to play in the overall presentation of these disorders. To address that question, it is necessary to know what function CNS adrenergic mechanisms normally serve. A substantial amount of research indicates that the locus coeruleus, and CNS adrenergic functioning in general, mediate attention, arousal (Aston-Jones et al. 1991; Foote et al. 1991; Robbins 1988; Sara et al. 1988), and stress and defense (threat) reactions (Jacobs 1990; Stanford 1990). The abnormal reactivity of the adrenergic system in individuals with abnormal anxiety (documented in this chapter) thus may reflect their propensity to one or more of the following tendencies: 1) to pay excessive attention to the potential of stress or threat in a variety of situations, 2) to respond to truly threatening situations with excessive arousal, or 3) to mount excessive defense reactions after appropriate attention and arousal responses have been made to a perceived threat. In other words, anxious patients may be excessively vigilant in scanning the environment for threat, may get overly aroused once threat is perceived, and/or may get overly "defensive" (avoidant) once aroused. All of these response tendencies may be at least partially mediated by excessive activity in the locus coeruleus-adrenergic system.

The types of situations perceived as threatening vary widely across individuals and across anxiety disorders. Some anxious patients appear to be tonically anxious, perhaps due to near-constant anticipation of threat. Although adrenergic dysregulation may mediate attention, arousal, and defense aspects of their responses to threat, it may not account either for the qualitative differences in anxiety-provoking stimuli across individuals and disorders, nor for the fact that a particular stimulus would not be perceived as threatening by an individual who was not abnormally anxious. Although an altered set point within the adrenergic system could account for a generally lowered threshold for the perception of threat, a generally altered threshold cannot account for the specificity of the types of fears that develop in patients with panic-agoraphobia, OCD, social phobia, or PTSD. Adrenergic mechanisms, however, can account for the form of psychophysiological reaction that occurs when a particular threatening stimulus is present and also for the apparently excessive (dysfunctional) magnitude of reaction. In other words, adrenergic dysregulation probably does not determine the characteristics of the particular stimulus that is perceived as threatening, but does determine that a reaction of a particular kind and of excessive intensity will occur when any stimulus perceived as threatening is present. As referred to at the beginning of this chapter, this is the so-called fight-or-flight reaction; in these individuals, flight (avoidance behavior) is the response seen. This perspective, if accurate, suggests that adrenergic dysregulation could play an important role in the pathophysiology of many of the anxiety disorders; this does not, however, necessarily directly implicate it in the etiology of these disorders.

Future Directions

There are several questions that follow directly from the research reviewed in this chapter and are basic to the issue of how adrenergic functioning might be abnormal in pathological anxiety states.

1. Although some studies have demonstrated abnormalities in adrenergic markers under "basal" or "resting" conditions, many have found no differences; a few studies have suggested that anxious patients might be more responsive to stress than are control subjects. Is basal adrenergic functioning actually normal and are differences when reported always due to measurements under nonbasal conditions? Is anxiety-related adrenergic dysfunction always due to excessive reactivity?
2. What is the relationship between adrenergic functioning in the periphery and functioning in the CNS? Specifically, how are findings of abnormalities in the periphery (such as catecholamine levels and receptor binding parameters) related to abnormalities in the CNS (such as the blunted GH response to clonidine)?
3. How are the different anxiety disorders similar to, or different from, each other in adrenergic functioning, and as a group, do the anxiety disorders demonstrate any common adrenergic abnormality that is distinctive from other psychiatric disorders (e.g., depressive disorders)?

4. Is the blunted GH response to clonidine, which has been demonstrated convincingly in panic disorder and also appears to be present in GAD but not in OCD, actually due to an abnormality of adrenergic functioning or is the source of the abnormality "upstream" or "downstream" from the (α_2-) adrenergic receptor?

Other issues and questions for future research and consideration include the following:

1. Is the dysregulation model generally a useful way to understand any adrenergic abnormality present in pathological anxiety?
2. What are the relationships among pathological anxiety, stress, and putative animal models of normal and pathological anxiety states?
3. What are the relationships between pathological anxiety states and medical illnesses such as hypertension and cardiovascular disease (Starkman et al. 1990)?
4. Are there any long-term reversible or irreversible psychobiological consequences associated with the persistence of untreated or partially treated anxiety (Lee and Cameron 1986)?
5. What information might cerebral blood flow or cerebral glucose metabolism changes in response to adrenergic challenge studies, or the development of adrenergic imaging ligands for physiological imaging techniques (such as single photon emission computed tomography or positron-emission tomography), provide about CNS dysfunction in pathological anxiety?
6. What are the physiological mechanisms by which the physical symptoms associated with anxiety are perceived; in other words, what and where are the "interoceptors" that are activated during subjective anxiety?

Further research using the experimental paradigms discussed in this chapter and addressing the questions posed should help to locate more precisely the adrenergic regulatory abnormalities in patients with anxiety disorders. However, new paradigms also will be needed to determine whether the initial perturbation that produces adrenergic dysregulation is itself within the adrenergic nervous system or whether it involves in a primary fashion other neurotransmitter systems with which adrenergic neurons interact. In other words, what is the etiology as well as the physiology of pathological anxiety?

References

Abboud FM: The sympathetic system in hypertension. Hypertension 4 (suppl 2):208–225, 1982

Abelson JL, Curtis GC: Cardiac and neuroendocrine responses to exposure therapy in height phobics: desynchrony within the "physiological response system." Behav Res Ther 27:561–567, 1989

Abelson JL, Glitz D, Cameron OG, et al: Blunted growth hormone response to clonidine in patients with generalized anxiety disorder. Arch Gen Psychiatry 48:157–162, 1991

Abelson JL, Glitz D, Cameron OG, et al: Endocrine, cardiovascular, and behavioral responses to clonidine in patients with panic disorder. Biol Psychiatry 32:18–25, 1992

Abramson L, Keshavan MS, Sitaram N: Elevated sleep and waking heart rates in anxious depressions. Biol Psychiatry 26:496–499, 1989

Albus M, Bondy B, Ackenheil M: Adrenergic receptors on blood cells: relation to the pathophysiology of anxiety. Clin Neuropharmacol 9 (suppl 4):359–361, 1986

Albus M, Braune S, Hohn T, et al: Do anxiety patients with or without frequent panic attacks differ in response to stress? Stress Medicine 3:189–194, 1988

Albus M, Zahn TP, Breier A: Anxiogenic properties of yohimbine, I: behavioral, physiological and biochemical measures. Eur Arch Psychiatry Clin Neurosci 241:337–344, 1992

American Psychiatric Association: Diagnostic and Statistical Manual of Mental Disorders, 3rd Edition. Washington, DC, American Psychiatric Association, 1980

American Psychiatric Association: Diagnostic and Statistical Manual of Mental Disorders, 3rd Edition, Revised. Washington, DC, American Psychiatric Association, 1987

American Psychiatric Association: Diagnostic and Statistical Manual of Mental Disorders, 4th Edition. Washington, DC, American Psychiatric Association, 1994

Aronson TA, Carasiti I, McBane D, et al: Biological correlates of lactate sensitivity in panic disorder. Biol Psychiatry 26:463–477, 1989a

Aronson TA, Whitaker-Azmitia P, Caraseti I: Differential reactivity to lactate infusions: the relative role of biological, psychological, and conditioning variables. Biol Psychiatry 25:469–481, 1989b

Asnis GM, Wetzler S, Sanderson WC, et al: Functional interrelationship of serotonin and norepinephrine: cortisol response to MCPP and DMI in patients with panic disorder, patients with depression, and normal control subjects. Psychiatry Res 43:65–76, 1992

Aston-Jones G, Chiang C, Alexinsky T: Discharge of noradrenergic locus coeruleus neurons in behaving rats and monkeys suggests a role in vigilance. Prog Brain Res 88:501–520, 1991

Balon R, Rainey JM, Pohl R, et al: Platelet monoamine oxidase activity in panic disorder. Psychiatry Res 22:37–41, 1987

Balon R, Yeragani VK, Pohl R: Phenomenological comparison of dextrose, lactate, and isoproterenol associated panic attacks. Psychiatry Res 26:43–50, 1988

Balon R, Yeragani VK, Pohl R, et al: Somatic and psychological symptoms during isoproterenol-induced attacks. Psychiatry Res 32:103–112, 1990

Barlow DH, Blanchard EB, Vermilyea JA, et al: Generalized anxiety and generalized anxiety disorder: description and reconceptualization. Am J Psychiatry 143:40–44, 1986

Barsky AJ: Palpitations, cardiac awareness, and panic disorder. Am J Med 92 (suppl 1A):31–34, 1992

Benkelfat C, Mefford IN, Masters CF, et al: Plasma catecholamines and their metabolites in obsessive-compulsive disorder. Psychiatry Res 37:321–331, 1991

Bertel O, Buhler R, Kiowski W, et al: Decreased beta-adrenoreceptor responsiveness as related to age, blood pressure, and plasma catecholamines in patients with essential hypertension. Hypertension 2:130–138, 1980

Blanchard EB, Kolb LC, Pallmeyer TP, et al: Psychophysiological study of post traumatic stress disorder in Vietnam veterans. Psychiatr Q 54:220–229, 1982

Blanchard EB, Kolb LC, Gerardi RJ, et al: Cardiac response to relevant stimuli as an adjunctive tool for diagnosing post-traumatic stress disorder in Vietnam veterans. Behavior Therapy 24:645–652, 1986

Blanchard EB, Kolb LC, Prins A, et al: Changes in plasma norepinephrine to combat-related stimuli among Vietnam veterans with posttraumatic stress disorder. J Nerv Ment Dis 179:371–373, 1991

Boudoulas H, Kind BD, Wooley CF: Mitral valve prolapse: a marker for anxiety or overlapping phenomenon? Psychopathology 17:98–106, 1984

Breier A, Charney DS, Heninger GR: The diagnostic validity of anxiety disorders and their relationship to depressive illness. Am J Psychiatry 142:787–797, 1985

Brown SL, Charney DS, Woods SW, et al: Lymphocyte beta-adrenergic receptor binding in panic disorder. Psychopharmacology (Berl) 94:24–28, 1988

Bruce M, Lader M: DSM-III-R state anxiety symptoms in anxiety disorder patients. Biol Psychiatry 30:519–522, 1991

Butler J, O'Halloran A, Leonard BE: The Galway study of panic disorder, II: changes in some peripheral markers of noradrenergic and serotonergic function in DSM-III-R panic disorder. J Affect Disord 26:89–100, 1992

Bystritsky A, Shapiro D: Continuous physiological changes and subjective reports in panic patients: a preliminary methodological report. Biol Psychiatry 32:766–777, 1992

Cameron OG, Nesse RM: Systemic hormonal and physiological abnormalities in anxiety disorders. Psychoneuroendocrinology 13:287–307, 1988

Cameron OG, Smith CB, Hollingsworth PJ, et al: Platelet α_2-adrenergic receptor binding and plasma catecholamines. Arch Gen Psychiatry 41:1144–1148, 1984

Cameron OG, Thyer BA, Nesse RM, et al: Symptom profiles of patients with DSM-III anxiety disorders. Am J Psychiatry 143:132–137, 1986

Cameron OG, Lee MA, Curtis GC, et al: Endocrine and physiological changes during "spontaneous" panic attacks. Psychoneuroendocrinology 12:321–331, 1987

Cameron OG, Smith CB, Lee MA, et al: Adrenergic status in anxiety disorders: platelet α_2-adrenergic receptor binding, blood pressure, and plasma catecholamines in panic and generalized anxiety disorder patients and normal subjects. Biol Psychiatry 28:3–20, 1990

Carr DB, Sheehan DV, Surman OS, et al: Neuroendocrine correlates of lactate-induced anxiety and their response to chronic alprazolam therapy. Am J Psychiatry 143:483–494, 1986

Chaleby J, Ziady G: Mitral valve prolapse and social phobia. Br J Psychiatry 152:280–281, 1988

Charney DS, Heninger GR: Abnormal regulation of noradrenergic function in panic disorder. Arch Gen Psychiatry 43:1042–1054, 1986

Charney DS, Redmond DE: Neurobiological mechanisms in human anxiety: evidence supporting central noradrenergic hyperactivity. Neuropharmacology 12B:1531–1536, 1983

Charney DS, Heninger GR, Breier A: Noradrenergic function in panic anxiety: effects of yohimbine in healthy subjects and patients with agoraphobia and panic disorder. Arch Gen Psychiatry 41:751–763, 1984

Charney DS, Heninger GR, Jatlow PI: Increased anxiogenic effects of caffeine in panic disorders. Arch Gen Psychiatry 42:233–243, 1985

Charney DS, Woods SW, Goodman WK, et al: Neurobiological mechanisms of panic anxiety: biochemical and behavioral correlates of yohimbine-induced panic attacks. Am J Psychiatry 144:1030–1036, 1987

Charney DS, Woods SW, Heninger GR: Noradrenergic function in generalized anxiety disorder: effects of yohimbine in healthy subjects and patients with generalized anxiety disorder. Psychiatry Res 27:173–182, 1989a

Charney DS, Innis RB, Duman RS, et al: Platelet alpha-2-receptor binding and adenylate cyclase activity in panic disorder. Psychopharmacology (Berl) 98:102–107, 1989b

Charney DS, Woods SW, Nagy LM, et al: Noradrenergic function in panic disorder. J Clin Psychiatry 51:5–11, 1990

Clark DB, Taylor CB, Hayward C, et al: Motor activity and tonic heart rate in panic disorder. Psychiatry Res 32:45–53, 1990

Coplan JD, Liebowitz MR, Gorman JM, et al: Noradrenergic function in panic disorder: effects of intravenous clonidine pretreatment on lactate induced panic. Biol Psychiatry 31:135–146, 1992

Cowley DS, Roy-Byrne PP: Hyperventilation and panic disorder. Am J Med 83:929–937, 1987

Dager SR, Cowley DS, Dunner DL: Biological markers in panic states: lactate-induced panic and mitral valve prolapse. Biol Psychiatry 22:339–359, 1987

den Boer JA, Westenberg HGM: Behavioral, neuroendocrine, and biochemical effects of 5-hydroxytryptophan administration in panic disorder. Psychiatry Res 31:267–278, 1990

den Boer JA, Westenberg HGM, Klompmakers AA, et al: Behavioral biochemical and neuroendocrine concomitants of lactate-induced panic anxiety. Biol Psychiatry 26:612–622, 1989

Devesa J, Lima L, Tresquerres JAF: Neuroendocrine control of growth hormone secretion in humans. Trends in Endocrinology and Metabolism 3:175–183, 1992

Dinan TG, Barry S, Yatham LN, et al: A pilot study of a neuroendocrine test battery in posttraumatic stress disorder. Biol Psychiatry 28:665–672, 1990

Dobbs D, Wilson WP: Observations on persistence of war neurosis. Diseases of the Nervous System 21:686–691, 1960

Edlund MJ, Swann AC: Low MHPG and continuing treatment in panic disorder. Prog Neuropsychopharmacol Biol Psychiatry 13:701–707, 1989

Ehlers A, Breuer P: Increased cardiac awareness in panic disorder. J Abnorm Psychol 101:371–382, 1992

Eriksson E, Westberg P, Alling C, et al: Cerebrospinal fluid levels of monoamine metabolites in panic disorder. Psychiatry Res 36:243–251, 1991

Esler M, Julius S, Zweifler A, et al: Mild high-renin essential hypertension: neurogenic human hypertension? N Engl J Med 296:405–411, 1977

Flaskos J, Theophilopoulos N, George AJ: Platelet monoamine oxidase activity and 5-hydroxytryptamine uptake in agoraphobic patients. Br J Psychiatry 155:680–685, 1989

Foote SL, Berridge CW, Adams LM, et al: Electrophysiological evidence for the involvement of the locus coeruleus in alerting, orienting, and attending. Prog Brain Res 88:521–532, 1991

Fredrikson M, Klein K, Ohman A: Do instructions modify effects of beta-adrenoceptor blockage on anxiety? Psychophysiology 27:309–317, 1990

Freedman RR, Ianni P, Ettedgui E: Psychophysiologic factors in panic disorder. Psychopathology 17 (suppl 1):67–73, 1984

Freedman RR, Ianni P, Ettedgui E, et al: Ambulatory monitoring of panic disorder. Arch Gen Psychiatry 42:244–248, 1985

Gaffney FA, Fenton BJ, Lane LD, et al: Hemodynamic, ventilatory, and biochemical responses of panic patients and normal controls with sodium lactate infusion and spontaneous panic attacks. Arch Gen Psychiatry 45:53–60, 1988

Garvey M, Noyes JR, Cook B, et al: The relationship of panic disorder and its treatment outcome to 24-hour urinary MHPG levels. Psychiatry Res 30:53–61, 1989

Garvey M, Hollon S, DeRubeis R: Is there a relationship between MHPG and anxiety in depressed patients? Psychiatry Res 38:187–195, 1991

George DT, Zerby A, Noble S, et al: Panic attacks and alcohol withdrawal: can subjects differentiate the symptoms. Biol Psychiatry 24:240–243, 1988

George DT, Nutt DJ, Walker WV, et al: Lactate and hyperventilation substantially attenuate vagal tone in normal volunteers. Arch Gen Psychiatry 46:153–156, 1989

George DT, Adinoff B, Ravotz B, et al: A cerebrospinal fluid study of the pathophysiology of panic disorder associated with alcoholism. Acta Psychiatr Scand 82:1–7, 1990

Goetz RR, Gorman JM, Dillon DJ, et al: Do panic disorder patients indiscriminately endorse somatic complaints? Psychiatry Res 29:207–213, 1989

Goetz RR, Klein DF, Gully R, et al: Panic attacks during placebo procedures in the laboratory. Arch Gen Psychiatry 50:280–285, 1993

Goldstein DS: Plasma norepinephrine during stress in essential hypertension. Hypertension 3:551–556, 1981

Goldstein DS: Plasma catecholamines and essential hypertension: an analytical review. Hypertension 5:86–99, 1983

Goldstein DS: Treatment of social phobia with clonidine. Biol Psychiatry 22:369–372, 1987

Gorman JM, Davies M, Steinman R, et al: An objective marker of lactate-induced panic. Psychiatry Res 22:341–348, 1987

Gorman JM, Fyer MR, Goetz R, et al: Ventilatory physiology of patients with panic disorder. Arch Gen Psychiatry 45:31–39, 1988

Gorman JM, Papp LA, Martinez J, et al: High-dose carbon dioxide challenge test in anxiety disorder patients. Biol Psychiatry 28:743–757, 1990

Grunhaus L, Gloger S, Birmacher B, et al: Prolactin response to the cold pressor test in patients with panic attacks. Psychiatry Res 8:171–177, 1983

Grunhaus LJ, Cameron O, Pande AC, et al: Comorbidity of panic disorder and major depressive disorder: effects on platelet alpha2 adrenergic receptors. Acta Psychiatr Scand 81:216–219, 1990

Gurguis GNM, Uhde TW: Plasma 3-methoxy-4-hydroxyphenylethylene glycol (MHPG) and growth hormone responses to yohimbine in panic disorder patients and normal controls. Psychoneuroendocrinology 15:217–224, 1990

Gurguis GNM, Mefford IN, Uhde TW: Hypothalamic-pituitary-adrenocortical activity in panic disorder: relationship to plasma catecholamine metabolites. Biol Psychiatry 30:502–506, 1991

Hansenne M, Pitchot W, Ansseau M: The clonidine test in posttraumatic stress disorder. Am J Psychiatry 148:810–811, 1991

Hibbert GA: Ideational components of anxiety: their origin and content. Br J Psychiatry 144:618–624, 1984

Hoehn-Saric R, McLeod DR, Zimmerli WD: Somatic manifestations in women with generalized anxiety disorder. Arch Gen Psychiatry 46:1113–1119, 1989

Hoehn-Saric R, McLeod DR, Zimmerli WD: Psychophysiological response patterns in panic disorder. Acta Psychiatr Scand 83:4–11, 1991

Hollander E, Fay M, Cohen B, et al: Serotonergic and noradrenergic sensitivity in obsessive-compulsive disorder: behavioral findings. Am J Psychiatry 145: 1015–1017, 1988

Hollander E, DeCaria C, Nitescu A, et al: Noradrenergic function in obsessive-compulsive disorder: behavioral and neuroendocrine responses to clonidine and comparison to healthy controls. Psychiatry Res 37:161–177, 1991

Horikoshi Y, Igarashi ITH, Inui M, et al: The adreno-sympathetic system, the genetic predisposition to hypertension, and stress. Am J Med Sci 289:186–191, 1985

Hubert W, de Jong-Meyer R: Psychophysiological response patterns to positive and negative film stimuli. Biol Psychology 31:73–93, 1990

Insel TR, Mueller EA, Gillin JC, et al: Biological markers in obsessive-compulsive and affective disorders. J Psychiatr Res 4:407–423, 1984

Jacobs BL: Locus coeruleus neuronal activity in behaving animals, in The Pharmacology of Noradrenaline in the Central Nervous System. Edited by Heal DJ, Marsden CA. New York, Oxford University Press, 1990, pp 248–265

Julius S: Autonomic nervous dysfunction in essential hypertension. Diabetes Care 14:249–259, 1991

Kahn JP, Drusin RE, Klein DF: Idiopathic cardiomyopathy and panic disorder: clinical association in cardiac transplant candidates. Am J Psychiatry 144: 1327–1330, 1987

Katerndahl DA: The sequence of panic symptoms. J Fam Pract 26:49–52, 1988

King RJ, Bayon EP, Clark DB, et al: Tonic arousal and activity: relationships to personality and personality disorder traits in panic patients. Psychiatry Res 25:65–72, 1988

Ko GN, Elsworth JP, Roth RH, et al: Panic-induced elevations of plasma MHPG levels in phobic-anxious patients. Arch Gen Psychiatry 40:425–430, 1983

Kolb LC: A neuropsychological hypothesis explaining posttraumatic stress disorders. Am J Psychiatry 144:989–995, 1987

Kosten TR, Mason JW, Giller EL, et al: Sustained urinary norepinephrine and epinephrine elevation in post-traumatic stress disorder. Psychoneuroendocrinology 12:13–20, 1987

Kotchen TA, Guthrie GP, McKean H, et al: Adrenergic responsiveness in prehypertensive subjects. Circulation 65:285–290, 1982

Lee MA, Cameron OG: Anxiety, type A behavior, and cardiovascular disease. Int J Psychiatry Med 16:123–129, 1986

Lee MA, Cameron OG, Gurguis GNM, et al: Alpha 2-adrenoreceptor status in obsessive-compulsive disorder. Biol Psychiatry 27:1083–1093, 1990

Lerer B, Bleich A, Bennett ER, et al: Platelet adenylate cyclase and phospholipase C activity in posttraumatic stress disorder. Biol Psychiatry 27:735–740, 1990

Lesch KP, Laux G, Erb A, et al: Attenuated growth hormone response to growth hormone-releasing hormone in major depressive disorder. Biol Psychiatry 22:1495–1499, 1987

Lesch KP, Laux G, Erb A, et al: Growth hormone (GH) responses to GH-releasing hormone in depression: correlation with GH release following clonidine. Psychiatry Res 25:301–310, 1988

Liebowitz MR, Gorman JM, Fyer AJ, et al: Lactate provocation of panic attacks, II: biochemical and physiological findings. Arch Gen Psychiatry 42:709–719, 1985

Lima DR, Turner P: Propranolol increases reduced beta-receptor function in severely anxious patients. Lancet 2:1505, 1983

Lima L, Carreira I, Morales M, et al: Treatment of patients with panic disorder and an elevated urinary excretion of MHPG using imipramine or clonazepam. Invest Clin 33:39–52, 1992

Malloy PF, Fairbank JA, Keane TM: Validation of a multimethod assessment of posttraumatic stress disorders in Vietnam veterans. J Consult Clin Psychol 51:488–494, 1983

Margraf J, Ehlers A, Roth WT: Mitral valve prolapse and panic disorder: a review of their relationship. Psychosom Med 50:93–113, 1988

Mason JW, Giller EL, Harkness L: Elevation of urinary norepinephrine/cortisol ratio in posttraumatic stress disorder. J Nerv Ment Dis 176:498–502, 1988

Mathew RJ, Ho BT, Kralik P, Taylor D, et al: Catechol-*O*-methyl transferase and catecholamines in anxiety and relaxation. Psychiatry Res 3:85–91, 1980

McCann UD, Penetar DM, Belenky G: Panic attacks in healthy volunteers treated with a catecholamine synthesis inhibitor. Biol Psychiatry 30:413–416, 1991

McFall ME, Murburg MM, Roszell DK, et al: Psychophysiologic and neuroendocrine findings in posttraumatic stress disorder: a review of theory and research. Journal of Anxiety Disorders 3:243–257, 1989

McFall ME, Murburg MM, Ko GN, et al: Autonomic responses to stress in Vietnam combat veterans with posttraumatic stress disorder. Biol Psychiatry 27:1165–1175, 1990

McFall ME, Veith RC, Murburg MM: Basal sympathoadrenal function in posttraumatic distress disorder. Biol Psychiatry 31:1050–1056, 1992

McIntyre IM, Norman TR, Marriott PF, et al: The pineal hormone melatonin in panic disorder. J Affect Disord 12:203–206, 1987

McIntyre IM, Judd FK, Burrows GD, et al: Plasma concentrations of melatonin in panic disorder. Am J Psychiatry 147:462–464, 1990

McLeod DR, Hoehn-Saric R: Perception of physiological changes in normal and pathological anxiety, in Biology of Anxiety Disorders. Edited by Hoehn-Saric R, McLeod DR. Washington, DC, American Psychiatric Press, 1993, pp 223–243

McLeod DR, Hoehn-Saric R, Stefan RL: Somatic symptoms of anxiety: comparison of self-report and physiological measures. Biol Psychiatry 21:301–310, 1986

McLeod DR, Hoehn-Saric R, Zimmerli WD, et al: Treatment effects of alprazolam and imipramine: physiological versus subjective changes in patients with generalized anxiety disorder. Biol Psychiatry 28:849–861, 1990

Middleton HC: Cardiovascular dystonia in recovered panic patients. J Affective Disord 19:229–236, 1990

Munjack DJ, Baltazar PL, DeQuattro B, et al: Generalized anxiety disorder: Some biochemical aspects. Psychiatry Res 32:35–43, 1990

Nesse RM, Cameron OG, Curtis GC, et al: Adrenergic function in patients with panic anxiety. Arch Gen Psychiatry 41:771–776, 1984

Nesse RM, Curtis GC, Thyer BA, et al: Endocrine and cardiovascular responses during phobic anxiety. Psychosom Med 47:320–332, 1985

Norman TR, Kimber NM, Judd FK, et al: Platelet 3H-rauwolscine binding in patients with panic attacks. Psychiatry Res 22:43–48, 1987

Norman TR, Acevedo A, McIntyre IM, et al: A kinetic analysis of platelet monoamine oxidase activity in patients with panic attacks. J Affect Disord 15:127–130, 1988

Nutt DJ: Platelet $alpha_2$-adrenoreceptor binding does not predict brain $alpha_2$-adrenoreceptor function. Br J Pharmacol 24:122–123, 1987

Nutt DJ: Altered central alpha2-adrenoreceptor sensitivity in panic disorder. Arch Gen Psychiatry 46:165–169, 1989

Nutt DJ: The pharmacology of human anxiety. Pharmacol Ther 47:233–266, 1990

Nutt DJ, Fraser SF: Platelet binding studies in panic disorder. J Affect Disord 12:7–11, 1987

Paige SR, Reid GM, Allen MG, et al: Psychophysiological correlates of posttraumatic stress disorder in Vietnam veterans. Biol Psychiatry 27:419–430, 1990

Pallmeyer TP, Blanchard EB, Kolb LC: The psychophysiology of combat-induced post-traumatic stress disorder in Vietnam veterans. Behav Res Ther 24:645–652, 1986

Papp LA, Gorman JM, Liebowitz MR, et al: Epinephrine infusions in patients with social phobia. Am J Psychiatry 145:733–736, 1988

Perry BD, Giller EL, Southwick SM: Altered platelet alpha 2-adrenergic binding sites in posttraumatic stress disorder (letter). Am J Psychiatry 144:1511–1512, 1987

Pitman RK, Orr SP: Twenty-four hour urinary cortisol and catecholamine excretion in combat-related posttraumatic stress disorder. Biol Psychiatry 27:245–247, 1990

Pitman RK, Orr SP, Forgue DF, et al: Psychophysiologic assessment of posttraumatic stress disorder imagery in Vietnam combat veterans. Arch Gen Psychiatry 44:970–975, 1987

Pohl R, Ettedgui E, Bridges M, et al: Plasma MHPG levels in lactate and isoproterenol anxiety states. Biol Psychiatry 22:1127–1136, 1987

Pohl R, Yeragani VK, Balon R, et al: Isoproterenol-induced panic attacks. Biol Psychiatry 24:891–902, 1988a

Pohl R, Yeragani VK, Balon R, et al: The jitteriness syndrome in panic disorder patients treated with antidepressants. J Clin Psychiatry 49:100–104, 1988b

Pyke RE, Greenberg HS: Norepinephrine challenges in panic patients. J Clin Psychopharmacol 6:279–285, 1986

Rahe RH, Karson S, Howard NS, et al: Psychological and physiological assessments on American hostages freed from captivity in Iran. Psychosom Med 52:1–16, 1990

Rapaport MH, Risch SC, Gillin JC, et al: Blunted growth hormone response to peripheral infusion of human growth hormone-releasing factor in patients with panic disorder. Am J Psychiatry 146:92–95, 1989

Rapaport MH, Risch SC, Gillin JC, et al: The effects of physostigmine infusion on patients with panic disorder. Biol Psychiatry 29:658–664, 1991

Rasmussen SA, Goodman WK, Woods SW, et al: Effects of yohimbine in obsessive compulsive disorder. Psychopharmacology (Berl) 93:308–313, 1987

Redmond DE: Studies of the nucleus locus coeruleus in monkeys and hypotheses for neuropsychopharmacology, in Psychopharmacology: The Third Generation of Progress. Edited by Meltzer HY. New York, Raven, 1987, pp 967–975

Robbins TW: Norepinephrine, dopamine, and selective attention: interactions in theory and practice, in Progress in Catecholamine Research, Part B: Central Aspects. Edited by Sandler M, Dahlstrom A, Belmaker RH. New York, Alan R Liss, 1988, pp 169–173

Rogeness GA, Cepeda C, Macedo CA, et al: Differences in heart rate and blood pressure in children with conduct disorder, major depression, and separation anxiety. Psychiatry Res 33:199–206, 1990

Roth WT, Telch MJ, Taylor CB, et al: Autonomic characteristics of agoraphobia with panic attacks. Biol Psychiatry 21:1133–1154, 1986

Roth WT, Telch MJ, Taylor CB, et al: Autonomic changes after treatment of agoraphobia with panic attacks. Psychiatry Res 24:95–107, 1988

Roth WT, Margraf J, Ehlers A, et al: Stress test reactivity in panic disorder. Arch Gen Psychiatry 49:301–310, 1992

Roy-Byrne PP, Uhde TW: Panic disorder and major depression: biological relationships. Psychopharmacol Bull 21:551–554, 1985

Sanderson WC, Rapee RM, Barlow DH: The influence of an illusion of control on panic attacks induced by inhalation of 5.5% carbon dioxide-enriched air. Arch Gen Psychiatry 46:157–162, 1989

Sandler LS, Wilson KG, Asmundson GJG, et al: Cardiovascular reactivity in nonclinical subjects with infrequent panic attacks. Journal of Anxiety Disorders 6:27–39, 1992

Sara SJ, Devauges V, Segal M: Locus coeruleus engagement in memory retrieval and attention, in Progress in Catecholamine Research, Part B: Central Aspects. Edited by Sandler M, Dahlstrom A, Belmaker RH. New York, Alan R Liss, 1988, pp 155–161

Schittecatte M, Charles G, Depauw Y, et al: Growth hormone response to clonidine in panic disorder patients. Psychiatry Res 23:147–151, 1988

Schittecatte M, Ansseau M, Charles G, et al: Growth hormone response to clonidine in male patients with panic disorder untreated by antidepressants. Psychol Med 22:1059–1062, 1992

Sevy S, Papadimitriou GN, Surmont DW, et al: Noradrenergic function in generalized anxiety disorder, major depressive disorder, and healthy subjects. Biol Psychiatry 25:141–152, 1989

Siever LJ, Uhde TW, Silberman EK, et al: Growth hormone response to clonidine as a probe of noradrenergic receptor responsiveness in affective disorder patients and controls. Psychiatry Res 6:171–183, 1982

Siever LJ, Insel TR, Jimerson DC, et al: Growth hormone response to clonidine in obsessive-compulsive patients. Br J Psychiatry 142:184–187, 1983

Southwick SM, Yehuda R, Giller EL, et al: Altered platelet α_2 adrenergic receptor binding sites in borderline personality disorder. Am J Psychiatry 147:1014–1017, 1990

Southwick SM, Krystal JH, Morgan CA, et al: Abnormal noradrenergic function in posttraumatic stress disorder. Arch Gen Psychiatry 50:266–274, 1993

Stanford SC: Central adrenoceptors in response and adaptation to stress, in The Pharmacology of Noradrenaline in the Central Nervous System. Edited by Heal DJ, Marsden CA. New York, Oxford University Press, 1990, pp 379–422

Starkman MN, Cameron OG, Nesse RM, et al: Peripheral catecholamine levels and the symptoms of anxiety: studies in patients with and without pheochromocytoma. Psychosom Med 52:129–142, 1990

Stein JM, Papp LA, Klein DF, et al: Exercise tolerance in panic disorder patients. Biol Psychiatry 32:281–287, 1992a

Stein MB, Tancer ME, Uhde TW: Heart rate and plasma norepinephrine responsivity to orthostatic challenge in anxiety disorders: comparison of patients with panic disorder and social phobia and normal control subjects. Arch Gen Psychiatry 49:311–317, 1992b

Tancer ME, Stein MB, Uhde TW: Effects of thyrotropin-releasing hormone on blood pressure and heart rate in phobic and panic patients: a pilot study. Biol Psychiatry 27:781–783, 1990a

Tancer ME, Stein MB, Uhde TW: Growth hormone (GH) response to clonidine and growth hormone releasing factor (GRF) in normal controls. Psychoneuroendocrinology 15:253–259, 1990b

Tancer ME, Stein MB, Black B, et al: Blunted growth hormone responses to growth hormone-releasing factor and to clonidine in panic disorder. Am J Psychiatry 150:336–337, 1993

Taylor CB, Telch MJ, Havvik D: Ambulatory heart rate changes during panic attacks. J Psychiatr Res 17:261–266, 1983

Theofilopoulos N, Flaskos J, George AJ: Urinary phenylethylamine excretion in phobic and obsessive patients. Human Psychopharmacology 6:43–48, 1991

Trimarco B, Volpe M, Ricciardelli B, et al: Studies of the mechanisms underlying impairment of beta-adrenoceptor-mediated effects in human hypertension. Hypertension 5:584–590, 1983

Tuck ML: The sympathetic nervous system in essential hypertension. Am Heart J 112:877–886, 1986

Tyrer P, Lee I, Alexander J: Awareness of cardiac function in anxious, phobic and hypochondriacal patients. Psychol Med 10:171–174, 1980

Uhde TW, Boulenger JP, Jimerson DC, et al: Caffeine: relationship to human anxiety, plasma MHPG and cortisol. Psychopharmacol Bull 20:426–430, 1984

Uhde TW, Vittone BJ, Siever LJ, et al: Blunted growth hormone response to clonidine in panic disorder patients. Biol Psychiatry 21:1081–1085, 1986

Uhde TW, Stein MB, Vittone BJ, et al: Behavioral and physiological effects of short-term and long-term administration of clonidine in panic disorder. Arch Gen Psychiatry 46:170–177, 1989

Uhde TW, Tancer ME, Black B, et al: Phenomenology and neurobiology of social phobia: comparison with panic disorder. J Clin Psychiatry 52 (suppl 11):31–40, 1991

van der Kolk B, Greenberg M, Boyd H, et al: Inescapable shock, neurotransmitters, and addiction to trauma: toward a psychobiology of post traumatic stress. Biol Psychiatry 20:314–325, 1985

Weissman NJ, Shear MK, Kramer-Fox R, et al: Contrasting patterns of autonomic dysfunction in patients with mitral valve prolapse and panic attacks. Am J Med 82:880–888, 1987

Woods SW, Charney DS, Goodman WK, et al: Carbon dioxide-induced anxiety: behavioral, physiologic, and biochemical effects of carbon dioxide in patients with panic disorders and healthy subjects. Arch Gen Psychiatry 45:43–52, 1988a

Woods SW, Koster K, Krystal JK, et al: Yohimbine alters regional cerebral blood flow in panic disorder. Lancet 2:678, 1988b

Woods SW, Charney DS, Lake J, et al: Carbon dioxide sensitivity in panic anxiety: ventilatory and anxiogenic response to carbon dioxide in healthy subjects and panic anxiety patients before and after alprazolam treatment. Arch Gen Psychiatry 43:900–909, 1986

Woods SW, Charney DS, McPherson CA, et al: Situational panic attacks: behavioral, physiologic, and biochemical characterization. Arch Gen Psychiatry 44:365–375, 1987

Woods SW, Charney DS, Silver JM, et al: Behavioral, biochemical, and cardiovascular responses to the benzodiazepine receptor antagonist flumazenil in panic disorder. Psychiatry Res 36:115–127, 1991

Yehuda R, Southwick S, Giller EL, et al: Urinary catecholamine excretion and severity of PTSD symptoms in Vietnam combat veterans. J Nerv Ment Dis 180:321–325, 1992

Yeragani VK, Balon R, Pohl R, et al: Do higher preinfusion heart rates predict laboratory-induced panic attacks? Biol Psychiatry 22:554–558, 1987a

Yeragani VK, Pohl R, Balon R, et al: Lactate infusions: the role of baseline anxiety and autonomic measures (letter). Psychiatry Res 22:263–264, 1987b

Yeragani VK, Pohl R, Rainey JM, et al: Pre-infusion heart rates and laboratory-induced panic anxiety. Acta Psychiatr Scand 75:51–54, 1987c

Yeragani VK, Balon R, Pohl R: Lactate infusions in panic disorder patients and normal controls: autonomic measures and subjective anxiety. Acta Psychiatr Scand 79:32–40, 1989

Yeragani VK, Balon R, Pohl R, et al: Decreased R-R variance in panic disorder patients. Acta Psychiatr Scand 81:554–559, 1990a

Yeragani VK, Meiri PC, Pohl R, et al: Heart rate and blood pressure changes during postural change and isometric handgrip exercise in patients with panic disorder and normal controls. Acta Psychiatr Scand 81:9–13, 1990b

Yeragani VK, Pohl R, Balon R, et al: Risk factors for cardiovascular illness in panic disorder patients. Neuropsychobiology 23:134–139, 1990c

Yeragani VK, Pohl R, Berger R, et al: Decreased heart rate variability in panic disorder patients: a study of power-spectral analysis of heart rate. Psychiatry Res 46:89–103, 1993

Zahn TP, Nurnberger JI, Berrettini WH, et al: Concordance between anxiety and autonomic nervous system activity in subjects at genetic risk for affective disorder. Psychiatry Res 36:99–110, 1991

17

Noradrenergic Activity in Anorexia Nervosa and Bulimia Nervosa

WALTER H. KAYE, M.D.
THEODORE E. WELTZIN, M.D.

Anorexia nervosa and bulimia nervosa are disorders of unknown etiology that are characterized by alterations of appetitive behavior and distortions of body image (American Psychiatric Association 1994). Eating disorder patients have dysphoric mood (Hudson et al. 1983b) and various forms of hypothalamic-pituitary dysfunction.

The pathophysiology of the eating disorders is poorly understood. The adrenergic neurotransmitter system is one of a number of neurotransmitter systems that may contribute to abnormalities seen in eating disorders. Considerable data in experimental animals and clinical studies in other psychiatric patient groups show that adrenergic neurons contribute to the modulation of appetite, mood, personality variables, neuroendocrine function, and metabolic function. Thus disturbances of norepinephrine (NE) activity could be contributory to the symptoms found in eating disorder patients.

There are several lines of evidence that suggest that noradrenergic dysfunction may be important for understanding the eating disorders. First, patients with eating disorders have an increased occurrence of psychopathological states that have been linked to possible noradrenergic alterations such as depression, anxiety disorders, and alcoholism. Second, antidepressant medications that have been shown to alter noradrenergic systems are effective in the treatment of normal-weight bulimic patients. Finally, measures of noradrenergic activity have been shown to be abnormal in eating disorders.

Subgroups of Patients With Eating Disorders

Considerable literature suggests that certain factors distinguish subgroups of eating disorder patients (Beaumont et al. 1976; Casper et al. 1980; Garfinkel et al. 1980; Garner et al. 1985; Halmi et al. 1981; Strober et al. 1982). These factors include the amount of weight lost, type of pathological eating behavior, and certain psychopathological characteristics (Herzog and Copeland 1985).

The terminology used to differentiate these subgroups has been in flux. The best-known eating disorder is anorexia nervosa, whose most distinguishing characteristic is severe emaciation. Two types of consummatory behavior are seen in anorexia nervosa. Restrictor or fasting anorexic patients—who fit the DSM-IV (American Psychiatric Association 1994) criteria for anorexia nervosa, restrictor subtype—lose weight by pure dieting. Bulimic anorexic patients (who qualify for a DSM-IV diagnosis of both anorexia nervosa, bulimia subtype) also lose weight but have a periodic disinhibition of restraint and engage in bingeing and purging. Compared with restrictor patients, the bulimic subgroup has been characterized as displaying significantly more evidence of premorbid behavioral instability, a higher incidence of premorbid and familial obesity, a greater susceptibility to depression, and a higher incidence of behaviors suggestive of impulse disorder (Beaumont et al. 1976; Casper et al. 1980; Garfinkel et al. 1980; Strober et al. 1982).

The third eating disorder is normal-weight bulimia or bulimia nervosa by DSM-IV criteria. This disorder is at least 10 times more prevalent than anorexia nervosa (Halmi et al. 1981; Pope et al. 1984; Stangler and Printz 1980). These patients periodically binge and purge, usually by vomiting, but never become emaciated. That is, they maintain a body weight above 85% of average body weight (Garner et al. 1985). Normal-weight bulimic patients resemble bulimic anorexic patients in terms of impulsivity and a predisposition to obesity (Garner et al. 1985).

The Relationship of Eating Disorders to Depression and Anxiety

Anorexia Nervosa

Studies of comorbidity of affective disorders in anorexia nervosa have examined the presence of concomitant psychopathology during the acute phase of the illness and following short- and long-term recovery. The widely varying estimates reflect differential methodological and diagnostic procedures across studies. It has been estimated that 21%–91% of anorexic patients have depressive symptoms when they are underweight and malnourished (Eckert et al. 1982; Hendren 1983; Morgan and Russell 1975; Stonehill and Crisp 1977; Theander 1970). Interpretation of these findings must consider that common behavioral accompaniments of starvation (Keys 1950; Strober and Katy 1988) closely resemble the accessory symptoms of primary depressive disorder and thus may inflate depression rating scores in underweight anorexic patients. The presence of depressive symptomatology has also been examined in patients with anorexia nervosa after short- and long-term weight recovery. With duration of follow-up ranging from several months to 4.9 years, an estimated 15%–58% of patients continued to exhibit some degree of depressive disturbance after weight recovery.

The high prevalence of depressive symptoms in anorexic patients mentioned above have generated trials of antidepressant medications. Overall, the few double-blind, placebo-controlled studies have been found to be ineffective or only partially effective. These medications include amitriptyline (Biederman et al. 1985a; Halmi et al. 1986), clomipramine (Crisp et al. 1987; Lacey and Crisp 1980), and lithium (Gross et al. 1981). There have been a few family epidemiology studies of the coaggregation of eating and affective disorders.

In a large and well-controlled family epidemiological study of anorexia nervosa, Strober et al. (1990) found an excess of affective disorders (mainly unipolar) among relatives of only the anorexic probands who were themselves depressed. Biederman et al. (1985b) had similar findings. In contrast, Gershon et al. (1984) found equally high rates of affective disorder among relatives of depressed and nondepressed anorexic probands.

The focus on the possible relationship of depression to anorexia nervosa has tended to obscure investigations of other psychopathology in this illness. In fact, patients with anorexia nervosa have a high prevalence of anxiety disorders including simple and social phobias, panic disorder, and obsessive-compulsive symptoms or disorders (Cantwell et al. 1977; Hsu et al. 1970; Hudson et al. 1983b; Rothenberg 1988; Rowland 1970; Toner et al. 1988). Strober (1980) described the anorexic personality as being markedly obsessional in character makeup, introverted, self-denying, prone to self-abasement with limited spontaneity, overly formalistic, and stereotyped in thinking despite being industrious. Importantly, he noted that these characteristics were ingrained and not related to weight in that they remained present after short-term weight restoration. Casper (1990) reported that women recovered from restricting anorexia nervosa for 8–10 years continued to rate higher on risk avoidance, displayed greater restraint in emotional expression and initiative, and showed greater conformance to authority than age-matched control subjects and had a greater degree of self- and impulse-control than their sisters. Finally, several investigators have noted (Kasvikis et al. 1986; T. Insel, personal communication) that adult women with obsessive-compulsive disorder have an increased incidence of prior anorexia nervosa. Together, these data suggest that anxiety symptoms are prominent in anorexia nervosa and that some anorexic patients may have more in common with obsessive-compulsive disorder patients than depressive disorder patients.

Bulimia Nervosa

A considerable number of studies have found a high incidence of concurrent depressed mood in patients with bulimia nervosa (Gwirtsman et al. 1983; Hatsukami et al. 1984; Herzog 1982; Hudson et al. 1983a, 1983b;

Pope et al. 1983). In addition, family studies of the coaggregation of eating and affective disorders found a high prevalence of affective disorders in the relatives of patients with bulimia nervosa. For example, Kassett et al. (1989) found that first-degree relatives of bulimia nervosa probands had higher rates of major affective disorders than did first-degree relatives of control subjects. Such findings have led investigators to hypothesize that eating disorders are a variant of major affective disorders (Hudson et al. 1983a, 1983b).

The high incidence of depressive symptoms in bulimic patients has prompted trials of antidepressant medication. Placebo-controlled, double-blind trials of desipramine (Blouin et al. 1988; Hughes et al. 1986), isocarboxazid (Kennedy et al. 1986), amitriptyline (Mitchell and Groat 1984), imipramine (Pope et al. 1983), phenelzine (Walsh et al. 1984), and fluoxetine (Freeman et al. 1988) showed that these medications reduce binge eating and purging behavior and/or improve depressive symptoms. Two important points should be made. First, although these medications improve eating behavior, the majority of patients treated with antidepressants continue to exhibit bulimic behavior. Second, antidepressants are associated with improvement of bulimic symptoms in subjects who are not depressed. In summary, data from antidepressant treatment trials suggest some common biological mechanisms between bulimia nervosa and depression, but it cannot be concluded that bulimia nervosa is a subtype of depressive illness.

The focus on the possible relationship of depression to bulimia nervosa has tended to obscure investigations of other psychopathology in this illness. Laessle et al. (1989) found 70% of bulimia nervosa patients had a lifetime diagnosis of anxiety disorders, a figure that was *greater* than the lifetime prevalence (56%) of depression. Hudson et al. (1987) found bulimia nervosa patients had a 43% lifetime prevalence of anxiety disorders in comparison with a 67% prevalence of depression. Studies of bulimia nervosa patients have shown a high incidence of alcohol abuse or dependency, with a range of 16%–49% (Beary et al. 1986; Bulik 1987; Hudson et al. 1983a, 1987; Laessle et al. 1989). For example, Mitchell et al. (1985), in a study of 275 bulimic women, found that 23% acknowledged a history of alcohol abuse and that 18% reported a prior history of treatment for chemical dependency. A high prevalence (33%–83%) of women with bulimia nervosa have at least one close relative with alcoholism (Bulik 1987;

Herzog 1982; Leon et al. 1985; Mitchell et al. 1988; Pyle et al. 1981; Strober et al. 1982). Family epidemiology studies have also found a higher incidence of substance abuse in relatives of patients with bulimia nervosa than in matched control subjects (Hudson et al. 1987; Kassett et al. 1989).

Overview

At best, it remains controversial as to whether eating disorders and major depression share a common diathesis. Critical examination of clinical phenomenology, family history, antidepressant response, biological correlates, course and outcome, and epidemiology yields *limited* support for this hypothesis (Rothenberg 1988; Strober and Katz 1988; Swift et al. 1986). In fact, there is considerable evidence suggesting that eating disorders share some relationship with other disorders. That is, anorexia nervosa may have some relationship to obsessive-compulsive behavior, and bulimia nervosa may share some commonality with anxiety disorders or alcoholism and substance abuse. These questions remain open.

State- Versus Trait-Related Disturbances

Teasing apart cause-and-effect relationships in patients with eating disorders is a major methodological dilemma. It is impractical to study eating disorder patients prospectively. However, studies of eating disorder subjects can be done when they are at a stable and normal weight and dietary intake. At such time, they are more likely to be free of confounding nutritional influences. If neurotransmitter disturbances were found, they are more likely to be trait related. Such studies can be done in anorexia nervosa by studying anorexic patients at various stages in their illness: while underweight and at intervals after weight restoration. Similarly, normal-weight bulimic patients can be studied during pathological eating behavior and after being abstinent from bingeing and purging behavior for some length of time.

Norepinephrine

Noradrenergic function is of much theoretical interest in understanding the pathophysiology of the eating disorders (Leibowitz 1984). It should be noted that several relationships between NE and psychobiological disturbances are possible. First, it is possible that intrinsic defects in NE pathways could contribute to the expression of pathological feeding behavior. When administered into specific areas of the hypothalamus involved in feeding modulation, NE, at near physiological doses, produces eating in satiated animals and potentiates eating in hungry animals (Leibowitz et al. 1985). Such effects are thought to act through an α_2-noradrenergic receptor.

Second, is the possibility that starvation could alter noradrenergic function. Considerable data suggest that nutritional status can affect both peripheral (Esler 1982; Landsberg and Young 1978) and central (Myers and McCaleb 1980; Pirke and Spyra 1982) noradrenergic function. Third, eating disorder patients have disturbances of other systems, such as neuroendocrine and mood, that could potentially contribute to creating, or themselves be secondarily disrupted by, NE dysregulation.

Anorexia Nervosa

It has long been recognized that underweight anorexic patients have hypothermia, bradycardia, and hypotension (Silver 1974; Warren and Vandewiele 1973) and that such findings are probably related to reduced sympathetic nervous system activity. Most studies have found reduced concentrations of 3-methoxy-4-hydroxyphenylglycol (MHPG) or vanillylmandelic acid in 24-hour urine samples in underweight anorexic patients (Abraham et al. 1981; Gerner and Gwirtsman 1981; Gross et al. 1979; Halmi et al. 1978; Johnson et al. 1984; Riederer et al. 1982). Several of these studies have found that noradrenergic metabolites increased after refeeding and weight gain (Abraham et al. 1981; Gross et al. 1979; Halmi et al. 1978; Riederer et al. 1982). Biederman et al. (1984), however, reported urinary MHPG to be decreased before and after weight gain in only those anorexic patients with a major depression.

Measurements of plasma NE have been inconsistent. Earlier studies found underweight anorexic patients had reduced plasma NE (Gross et al. 1979; Luck et al. 1983; Van Loon 1980), with Gross et al. reporting that plasma NE levels normalized after weight gain. More recent studies from two separate groups found that underweight anorexic patients had normal levels of supine plasma NE (Heufelder et al. 1985; Lesem et al. 1989;

Lonati-Galligani and Pirke 1986; Pirke et al. 1985a, 1985b). Moreover, some studies reported that supine plasma NE levels decreased when patients were restudied after a month hospitalization before refeeding and weight gain (Lesem et al. 1989) or after some amount of weight gain and refeeding (Pahl et al. 1985; Pirke et al. 1985b).

The reason for a disassociation between urinary and plasma measures is not certain, but might be explained by the possibility that anorexic patients in different treatment programs might have differences in nutritional status or in other factors. Our group (Kaye et al. 1985a) measured NE and its metabolites in urine, plasma, and cerebrospinal fluid (CSF) in the same anorexic patients when underweight and at intervals during refeeding and weight recovery (although not all measures were done in all patients). We found that supine heart rate and the sum of NE products in urine were significantly reduced in underweight anorexic patients but normalized during refeeding and weight gain. In contrast, supine plasma NE, plasma MHPG, and CSF NE and MHPG were normal in the underweight patients but decreased during recovery. Thus it appears that noradrenergic function, as reflected by heart rate and urinary excretion, is decreased in underweight patients and increases during recovery. For reasons that are not clear, plasma and CSF measures of NE and MHPG show opposite patterns.

Investigators have measured noradrenergic receptor function in anorexia nervosa. Two studies have found that platelet α_2-receptor density was elevated in underweight anorexic patients (Heufelder et al. 1985; Luck et al. 1983), with Heufelder et al. noting that density decreased after weight gain. Luck et al. found that platelet α-receptor affinity was normal, whereas Heufelder et al. found it to be low with no change in the latter study after weight gain. Kaye et al. (1985b) administered clonidine (2 μg/kg iv) to anorexic patients and found that underweight patients had increased α_2-receptor activity and that such activity was reduced after weight gain. Lonati-Galligani and Pirke (1986) found mononuclear leukocyte β_2-density to be reduced, but activity to be normal in underweight anorexic patients. Such changes in receptor function are consistent with reduced sympathetic activity. For example, increased sensitivity of the presynaptic α_2-autoreceptor would enhance inhibition of sympathetic activity (Kaye et al. 1985b).

Although it appears that sympathetic nervous system activity increases during nutritional rehabilitation, it is not certain whether this activity normalizes in the short or long term. Because effects of refeeding and rapid weight gain may persist after short-term (a few months) weight restoration and influence NE measures, we (Kaye et al. 1985a) studied anorexic patients who had been weight recovered and stable for more than 6 months. We found that long-term weight-restored anorexic patients had a reduction of supine and orthostatic plasma NE when compared with control subjects. Pirke and Ploog (1986) also found a reduced orthostatic NE response in similar patients. They reported, however, that these patients had elevated plasma ketones, suggesting that, despite normal weight, some of these patients were dieting or starving at the time of study.

Fewer studies have directly measured indices of central nervous system noradrenergic function in anorexia nervosa. Underweight anorexic patients had normal levels of CSF NE (Kaye et al. 1984a, 1984b) and CSF MHPG (Gerner et al. 1984) with no change in CSF NE levels when restudied a few weeks after target weight was reached. Our group at the National Institute of Mental Health (NIMH) replicated these data in a new group of underweight anorexic patients (W. H. Kaye, H. E. Gwirtsman, M. H. Ebert, unpublished data, June 1984). In addition, we found that CSF MHPG and NE were reduced during refeeding and weight gain. Interestingly, we have also found that long-term weight-recovered anorexic patients have reduced CSF and plasma MHPG and CSF NE (Kaye et al. 1984a). In summary, most investigators postulate that starvation is the most important factor accounting for reduced sympathetic nervous system activity in underweight anorexic patients. However, other psychobiological alterations, such as stress, arousal, increased physical activity, the presence of a depressive or anxiety disorder, macronutrient consumption, cortisol or thyroid function, fluid volume, or electrolyte balance, may also be contributory. The fact that diet and many other factors alter noradrenergic function makes it uncertain whether reduced NE measures in long-term weight-recovered anorexic patients are state or trait related.

Normal-Weight Bulimia Nervosa

Studies from a number of centers have found alterations in noradrenergic activity in normal-weight bulimic patients. Pirke et al. (1985a) studied normal-weight bulimic patients 3–7 days after inpatient admission and found that they had a normal supine plasma NE but a subnormal standing-induced increment in plasma NE. That group (Pirke et al. 1989) studied another cohort of bulimic patients 2–3 weeks after inpatient admission and found they had significantly reduced resting plasma NE levels. Our group (Kaye et al. 1990a) investigated bulimic patients on admission to the hospital and again after 30 days of inpatient hospitalization and observed abstinence from bingeing and vomiting. After 30 days of abstinence, bulimic patients had a reduction of basal plasma NE compared with themselves on admission or compared with healthy control subjects. These data have been replicated in other studies from the NIMH. Patients studied about 3 weeks postadmission have been found to have both a significant reduction and a trend toward a reduction of resting plasma NE (Buckholtz et al. 1988; George et al. 1990).

Several studies have found evidence of altered central noradrenergic activity in bulimic patients. Our group (Kaye et al. 1990b) in studies at the NIMH found that bulimic patients, after 30 days of abstinence, had a reduction of CSF NE levels compared with when they were admitted. In another study, from the University of Pittsburgh and Medical University of South Carolina (Kaye et al. 1990a), bulimic patients (mostly outpatients) had a significant reduction of CSF NE compared with control subjects.

Other studies have found alterations in noradrenergic receptors. Using blood platelets as a model for in vitro assessments of receptor function, Heufelder et al. (1985) demonstrated increased platelet α_2-adrenoceptor number in bulimic patients. Lonati-Galligani and Pirke (1986) found normal cyclic adenosine monophosphate response to isoproterenol stimulation in lymphocytes from recently hospitalized bulimic patients. A study of binge-abstinent hospitalized bulimic patients found increased β-adrenoceptor responsiveness but normal receptor density in blood lymphocytes (Buckholtz et al. 1988). George et al. (1990), using an isoproterenol challenge, found bulimic subjects had a significantly increased chronotropic response (suggesting increased β-adrenergic receptor sensitivity) compared with control subjects. A study assessing the effect of lactate on urges to binge and anxiety symptoms (George et al. 1987) found that, although bulimic patients had increased baseline levels of anxiety, they did not have an increase in urges to binge or panic after a lactate infusion.

The meaning of altered NE activity in bulimia nervosa is not certain. Such alterations could be secondary to factors that affect noradrenergic activity (Potter et al. 1983), such as stress, physical activity, sodium balance, and hormonal status. Many of these factors could be abnormal in bulimic patients.

It is important to consider whether or not disturbed NE function could be related to abnormalities of dietary intake. Considerable data (Esler 1982; Jung et al. 1979; Welle et al. 1980, 1981; Young et al. 1980) in animals and humans suggest that feeding and weight gain increase sympathetic nervous system activity, whereas food restriction and weight loss produces the opposite effect. In fact, plasma NE release was found to be exaggerated when bulimic patients were bingeing in comparison with plasma NE release induced by a large meal in healthy volunteer subjects (Kaye et al. 1990b). Thus it is possible that bingeing transiently stimulates plasma NE and may contribute to findings of normal plasma NE levels near in time to admission. The enormous food consumption during bingeing is of particular contrast with the fact that bulimic patients have been reported to engage in restricted eating when not bingeing (Gwirtsman et al. 1989; Newman et al. 1987; Obarzanek et al. 1987). As noted, there is evidence that a reduction in dietary intake reduces sympathetic nervous system activity. Thus it is possible that reduced NE levels during abstinence are related to restricted caloric intake when not bingeing.

We cannot rule out the possibility that bulimic patients have a trait-related disturbance of noradrenergic function that contributes to pathological regulation of feeding and predisposes them to have reduced metabolism when not bingeing. Such alterations have been postulated to contribute to pathological eating behavior (Leibowitz 1984) and energy metabolism disturbances (Peterson et al. 1988). The possibility that some trait disturbance may predate the onset of pathological eating behavior in bulimic patients is supported by observations that some bulimic patients may adopt binge-vomit cycles to counter a predisposition to develop obesity (Pyle et al. 1981) and that pathological eating begins as a weight-control technique (Chiodo and Latimer 1983).

Future studies that follow abstinent bulimic subjects for a longer period of time will be necessary to determine if noradrenergic disturbances are secondary to cessation of bingeing and eventually normalize, or contribute to the pathogenesis of this disorder. Altered noradrenergic activity during abstinence, whatever the etiology, suggests persistent physiological alterations exist during abstinence. Many bulimic patients resume bingeing and purging despite a variety of treatments (Garner et al. 1987; Pope and Hudson 1986). It is not known whether NE disturbances during abstinence contribute to this high recidivism rate. Moreover, it is not known why antidepressant medications, which in part act on NE systems, reduce bingeing and vomiting behavior. Antidepressants appear to have specific antibulimic properties since they work independently of the presence of depression (Hughes et al. 1986; Walsh et al. 1984). It is worth exploring whether antidepressants exert their antibulimic effects because they correct an intrinsic defect in noradrenergic activity or prevent recidivism by countering alterations in NE activity after bingeing stops.

Conclusion

Disturbances of noradrenergic systems have been shown to occur in acutely ill anorexia nervosa and bulimia nervosa patients. It is likely that dietary abnormalities make a substantial contribution to these abnormalities. Depressive symptoms are common in both illnesses, particularly during the acute stages, and are often reduced with nutritional rehabilitation.

In anorexia nervosa, disturbances of noradrenergic systems appear to persist after long-term weight restoration. A limited amount of data suggests that there is a persistence of affective symptoms in anorexic patients after weight recovery. It is less clear whether there is a family diathesis for depression in anorexia nervosa. Although speculative, it is possible that disturbances of noradrenergic systems could be responsible for the pathogenesis of feeding and behavioral abnormalities in this illness.

Physiological and pharmacological evidence suggests that patients with bulimia nervosa have alterations of NE activity when acutely ill. The fact that antidepressant medications are often useful in reducing bingeing behavior, independent of their effects on mood, suggests that there may be abnormalities in one or more of the pathways that modulate feeding behavior. It is not certain at this time whether these abnormalities, at this stage of illness, are trait or state related since few studies have been done after recovery. Disturbances of NE activity persist in patients studied after a month of abstinence from bingeing and purging. To our knowledge, no studies have been done investigating this neurotransmitter after a longer period of recovery. In addition, it is not clear whether bulimia nervosa has any special link to affective disorders, anxiety disorders, or alcohol and substance abuse.

With reference to eating disorders, our understanding of central nervous system pathophysiology has lagged behind the advances made in understanding many other psychiatric disorders. In part, this may be due to limited knowledge of the mechanisms that modulate appetite and relevant physiological functions. The past decade has seen a rapid increase in preclinical and clinical knowledge about the regulation of appetite. Such knowledge has the potential for offering new insights into understanding the neurobiology of eating disorders.

References

Abraham SF, Beaumont PJV, Cobbin DM: Catecholamine metabolism and body weight in anorexia nervosa. Br J Psychiatry 138:244–247, 1981

American Psychiatric Association: Diagnostic and Statistical Manual of Mental Disorders, 4th Edition. Washington, DC, American Psychiatric Association, 1994

Beary MD, Lacey JH, Merry J: Alcoholism and eating disorders in women of fertile age. Br J Addict 81:685–689, 1986

Beaumont PJV, George GCW, Smart DE: "Dieters" and "vomiters" in anorexia nervosa. Psychol Med 6:617–622, 1976

Biederman J, Herzog DB, Rivinus TM, et al: Urinary MHPG in anorexia nervosa patients with and without a concomitant major depressive disorder. J Psychiatr Res 18:149–160, 1984

Biederman J, Herzog DB, Rivinus TM, et al: Amitriptyline in the treatment of anorexia nervosa: a double-blind study, placebo-controlled study. J Clin Psychopharmacol 5:10–16, 1985a

Biederman J, Rivinus T, Kemper K, et al: Depressive disorders in relatives of anorexia nervosa patients with and without a current episode of nonbipolar major depression. Am J Psychiatry 142:1495–1496, 1985b

Blouin AG, Blouin JF, Perez EL, et al: Treatment of bulimia with fenfluramine and desipramine. J Clin Psychopharmacol 8:261–269, 1988

Buckholtz N, George DT, Davies AO, et al: Lymphocyte beta-adrenergic receptor modification in bulimia. Arch Gen Psychiatry 45:479–482, 1988

Bulik CM: Drug and alcohol abuse by bulimic women and their families. Am J Psychiatry 144:1604–1606, 1987

Cantwell DP, Sturzenburger S, Burroughs J, et al: Anorexia nervosa: an affective disorder? Arch Gen Psychiatry 3:1087–1093, 1977

Casper RC: Personality features of women with good outcome from restricting anorexia nervosa. Psychosom Med 52:156–170, 1990

Casper RC, Eckert ED, Halmi KA, et al: Bulimia: its incidence and clinical importance in patients with anorexia nervosa. Arch Gen Psychiatry 37:1030–1035, 1980

Chiodo J, Latimer PR: Vomiting as a learned weight-control technique in bulimia. J Behav Ther Exp Psychiatry 14:131–135, 1983

Crisp AH, Lacey JH, Crutchfield M: Clomipramine and "drive" in people with anorexia nervosa. Br J Psychiatry 50:355–358, 1987

Eckert ED, Goldberg SC, Halmi KA, et al: Depression in anorexia nervosa. Psychol Med 12:115–122, 1982

Esler M: Assessment of sympathetic nervous function in humans for noradrenaline plasma kinetics. Clin Sci 62:247–254, 1982

Freeman CP, Morris JE, Chesire KE, et al: A double-blind controlled trial of fluoxetine versus placebo for bulimia nervosa (abstract). Third International Conference on Eating Disorders. New York, April 1988

Garfinkel PE, Moldofsky H, Garner DM: The heterogeneity of anorexia nervosa. Arch Gen Psychiatry 37:1036–1040, 1980

Garner DM, Garfinkel PE, O'Shaughnessy M: The validity of the distinction between bulimia with and without anorexia nervosa. Am J Psychiatry 142:581–587, 1985

Garner DM, Fairburn C, Davis R: Cognitive-behavioral treatment of bulimia nervosa: a critical appraisal. Behav Modif 4:398–431, 1987

George DT, Brewerton TD, Jimerson DC: Comparison of lactate-induced anxiety in bulimic patients and healthy controls. Psychiatry Res 21:213–220, 1987

George DT, Kaye WH, Goldstein DS, et al: Altered norepinephrine regulation in bulimia: effects of pharmacological challenge with isoproterenol. Psychiatry Res 33:1–10, 1990

Gerner RH, Gwirtsman HE: Neurochemical abnormalities in anorexia nervosa: similarities to affective disorders. Biol Psychiatry 16:991–995, 1981

Gerner RH, Cohen DJ, Fairbanks L, et al: CSF neurochemistry of women with anorexia nervosa and normal women. Am J Psychiatry 141:1441–1444, 1984

Gershon ES, Schreiber JL, Hamovit JR, et al: Clinical findings in patients with anorexia nervosa and affective illness in their relatives. Am J Psychiatry 141:1419–1422, 1984

Gross HA, Lake CR, Ebert MH, et al: Catecholamine metabolism in primary anorexia nervosa. J Clin Endocrinol Metab 49:805–809, 1979

Gross HA, Ebert MH, Faden VB, et al: A double-blind controlled trial of lithium carbonate in primary anorexia nervosa. J Clin Psychopharmacol 1:376–381, 1981

Gwirtsman HE, Roy-Byrne P, Yager J, et al: Neuroendocrine abnormalities in bulimia. Am J Psychiatry 140:559–563, 1983

Gwirtsman HE, Kaye WH, Obarzanek E, et al: Decreased caloric intake in normal-weight patients with bulimia: comparison with female volunteers. Am J Clin Nutr 49:86–92, 1989

Halmi KA, Dekirmenjian H, Davis JM, et al: Catecholamine metabolism in anorexia nervosa. Arch Gen Psychiatry 35:458–460, 1978

Halmi KA, Falk JR, Schwartz E: Binge-eating and vomiting: a survey of a college population. Psychol Med 11:697–706, 1981

Halmi KA, Eckert ED, LaDu TJ, et al: Anorexia nervosa: treatment efficacy of cyproheptadine and amitriptyline. Arch Gen Psychiatry 43:177–181, 1986

Hatsukami DK, Eckert ED, Mitchell JE, et al: Affective disorder and substance abuse among women with bulimia. Psychol Med 14:701–704, 1984

Hendren RL: Depression in anorexia nervosa. Journal of the American Academy of Child Psychiatry 22:59–62, 1983

Herzog DB: Bulimia in the adolescent. Am J Dis Child 136:985–989, 1982

Herzog DB, Copeland PM: Eating disorders. N Engl J Med 313:295–303, 1985

Heufelder A, Warnhoff M, Pirke KM: Platelet alpha-2 adrenoceptor and adenylate cyclase in patients with anorexia nervosa and bulimia. J Clin Endocrinol Metab 61:1053–1060, 1985

Hsu LKG, Crisp AH, Harding B: Outcome of anorexia nervosa. Lancet 1:61–65, 1970

Hudson JI, Pope HG, Jonas JM, et al: Family history study of anorexia nervosa and bulimia. Br J Psychiatry 142:133–138, 1983a

Hudson JI, Pope HG, Jonas JM, et al: Phenomenologic relationship of eating disorders to major affective disorders. Psychiatry Res 9:345–354, 1983b

Hudson JI, Pope HG, Jonas JM, et al: A controlled family history study of bulimia. Psychol Med 17:883–890, 1987

Hughes PL, Wells LA, Cunningham CJ, et al: Treating bulimia with desipramine. Arch Gen Psychiatry 43:182–186, 1986

Jimerson DC, Lesem MD, Kaye WH, et al: Symptom severity and neurotransmitter studies in bulimia. Psychopharmacology (Berl) 96:S124, 1988

Johnson JL, Leiter LA, Burrow GN, et al: Excretion of urinary catecholamine metabolites in anorexia nervosa: effect of body composition and energy intake. Am J Clin Nutr 40:1001–1006, 1984

Jung RT, Shetty PS, Barrand M, et al: Role of catecholamines in hypotensive response to dieting. BMJ 1:12–13, 1979

Kassett JA, Gershon ES, Maxwell ME, et al: Psychiatric disorders in the first-degree relatives of probands with bulimia nervosa. Am J Psychiatry 11:1468–1471, 1989

Kasvikis YG, Tsakiris F, Marks IM, et al: Past history of anorexia nervosa in women with obsessive compulsive disorder. International Journal of Eating Disorders 5:1069–1075, 1986

Kaye WH, Ebert MH, Raleigh M, et al: Abnormalities in central nervous system monoamine metabolism in anorexia nervosa. Arch Gen Psychiatry 41:350–355, 1984a

Kaye WH, Jimerson DC, Lake CR, et al: Altered norepinephrine metabolism following long-term weight recovery in patients with anorexia nervosa. Psychiatr Res 14:333–342, 1985a

Kaye WH, Gwirtsman HE, Lake CR, et al: Disturbances of norepinephrine metabolism and alpha2 adrenergic receptor activity in anorexia nervosa: relationship to nutritional state. Psychopharmacol Bull 21:419–423, 1985b

Kaye WH, Ballenger JC, Lydiard B, et al: CSF Monoamine levels in normal-weight bulimia: evidence for abnormal noradrenergic activity. Am J Psychiatry 147:225–229, 1990a

Kaye WH, Gwirtsman HE, George DT, et al: Disturbances of noradrenergic systems in normal-weight bulimia: relationship to diet and menses. Biol Psychiatry 27:4–21, 1990b

Kennedy S, Piran N, Garfinkel PE: Isocarboxazede in the treatment of bulimia. Am J Psychiatry 143:1495–1496, 1986

Keys A: The Biology of Human Starvation. Minneapolis, University of Minnesota Press, 1950

Lacey JH, Crisp AH: Hunger, food intake and weight: the impact of clomipramine on a refeeding anorexia nervosa population. Postgrad Med J 56:79–85, 1980

Laessle RG, Wittchen HU, Fichter MM, et al: The significance of subgroups of bulimia and anorexia nervosa: lifetime frequency of psychiatric disorders. International Journal of Eating Disorders 8:569–574, 1989

Landsberg L, Young JB: Fasting, feeding and the regulation of sympathetic activity. N Engl J Med 298:1295–1301, 1978

Leibowitz SF: Noradrenergic function in the medial hypothalamus: potential relation to anorexia nervosa and bulimia, in The Psychobiology of Anorexia Nervosa. Edited by Pirke KM, Ploog D. Berlin, Springer-Verlag, 1984, pp 35–45

Leibowitz SF, Weiss GF, Yee F, et al: Noradrenergic intervention of the paraventricular nucleus: specific role in control of carbohydrate ingestion. Brain Res Bull 14:561–567, 1985

Leon GR, Carroll K, Chernyk B, et al: Binge eating and associated habit patterns within college student and identified bulimic populations. International Journal of Eating Disorders 4:43–57, 1985

Lesem M, George DT, Kaye WH, et al: State-related changes in norepinephrine regulation in anorexia nervosa. Biol Psychiatry 25:509–512, 1989

Lonati-Galligani M, Pirke KM: Beta2-adrenergic receptor regulation in circulating mononuclear leukocytes in anorexia nervosa and bulimia. Psychiatry Res 19:189–198, 1986

Luck P, Mikhailidis DP, Dashwood MR, et al: Platelet hyperaggregability and increased alpha-adrenoceptor density in anorexia nervosa. J Clin Endocrinol Metab 57:911–914, 1983

Mitchell JE, Groat R: A placebo-controlled double-blind trial of amitriptyline in bulimia. J Clin Psychopharmacol 4:186–193, 1984

Mitchell JE, Hatsukami D, Eckert ED, et al: Characteristics of 275 patients with bulimia. Am J Psychiatry 142:482–485, 1985

Mitchell JE, Hatsukami D, Pyle R, et al: Bulimia with and without a family history of drug abuse. Addict Behav 13:245–251, 1988

Morgan H Russell GFM: Value of family background and clinical features as predictors of long-term outcome in anorexia nervosa: four-year follow-up study of 41 patients. Psychol Med 5:355–371, 1975

Myers RD, McCaleb ML: Feeding, satiety signal from intestine triggers brain's noradrenergic mechanism. Science 209:1035–1037, 1980

Newman MM, Halmi KA, Marchi P: Relationship for clinical factors to caloric requirements in subtypes of eating disorders. Biol Psychiatry 22:1253–1263, 1987

Obarzanek E, Lesem M, Jimerson DC: Resting metabolic rate in bulimic patients (NR 204). Paper presented at the annual meeting of the American Psychiatric Association, Chicago, IL, May 9–14, 1987

Pahl J, Pirke KM, Schweiger U, et al: Anorectic behavior, mood, and metabolic and endocrine adaptation to starvation in anorexia nervosa during inpatient treatment. Biol Psychiatry 20:874–887, 1985

Peterson HR, Rothschild M, Weinberg CR, et al: Body fat and the activity of the autonomic nervous system. N Engl J Med 318:1077–1083, 1988

Pirke KM, Ploog D: Psychobiology of anorexia nervosa, in Nutrition and the Brain, Vol 7. Edited by Wurtman RJ, Wurtman JJ. New York, Raven, 1986, pp 167–198

Pirke KM, Spyra B: Catecholamine turnover in the brain and the regulation of luteinizing hormone and corticosterone in starved male rats. Acta Endocrinol (Copenh) 100:168–176, 1982

Pirke KM, Pahl J, Schweiger U, et al: Metabolic and endocrine indices of starvation in bulimia: a comparison with anorexia nervosa. Psychiatry Res 15:33–39, 1985a

Pirke KM, Fichter MM, Pahl J: Noradrenaline, trijodthyronine, growth hormone and prolactin during weight gain in anorexia nervosa. International Journal of Eating Disorders 4:499–509, 1985b

Pirke KM, Eckert M, Ofers B, et al: Plasma norepinephrine response to exercise in bulimia, anorexia nervosa and controls. Biological Psychiatry 25:799–802, 1989

Pope HG, Hudson JI: Antidepressant drug therapy for bulimia: current status. J Clin Psychiatry 47:339–345, 1986

Pope HG, Hudson JI, Jonas JM, et al: Bulimia treated with imipramine: a placebo controlled double-blind study. Am J Psychiatry 140:554–558, 1983

Pope HG, Hudson JI, Yurgelun-Todd D: Anorexia nervosa and bulimia among 300 suburban women shoppers. Am J Psychiatry 141:292–294, 1984

Potter WZ, Muscettola G, Goodwin FK: Source of variance in clinical studies of MHPG, in MHPG: Base Mechanisms and Psychopathology. Edited by Maas JW. New York, Academic Press, 1983, pp 145–165

Pyle RL, Mitchell JE, Eckert ED: Bulimia: a report of 34 cases. J Clin Psychiatry 42:60–64, 1981

Riederer P, Toifl K, Kruzik P: Excretion of biogenic amine metabolites in anorexia nervosa. Clin Chem Acta 123:27–32, 1982

Rothenberg A: Differential diagnosis of anorexia nervosa and depressive illness: a review of 11 studies. Compr Psychiatry 29:427–432, 1988

Rowland C Jr: Anorexia nervosa: a survey of the literature and review of 30 cases. Internal Psychiatry Clinics 7:37–137, 1970

Silver JA: Anorexia nervosa: clinical observations in a successful treatment plan. J Pediatr 84:68–73, 1974

Stangler RS, Printz AM: DSM-III psychiatric diagnosis in a university population. Am J Psychiatry 137:937–940, 1980

Stonehill E, Crisp AH: Psychoneurotic characteristics of patients with anorexia nervosa before and after treatment and at follow-up 4–7 years later. J Psychosom Res 21:187–193, 1977

Strober M: Personality and symptomatological features in young, nonchronic anorexia nervosa patients. J Psychosom Res 24:353–359, 1980

Strober M, Katz JL: Depression in the eating disorders: a review and analysis of descriptive, family, and biological findings, in Diagnostic Issues in Anorexia Nervosa and Bulimia Nervosa. Edited by Garner DM, Garfinkel PE. New York, Brunner/Mazel, 1988, pp 80–111

Strober M, Salkin B, Burroughs J, et al: Validity of the bulimia-restricter criteria in anorexia nervosa. J Nerv Ment Dis 170:345–351, 1982

Strober M, Lampert C, Morrell W, et al: A controlled family study of anorexia nervosa: evidence of familial aggregation and lack of shared transmission with affective disorders. 9:239–253, 1990

Swift WJ, Andrews D, Barklage NE: The relationship between affective disorder and eating disorders: a review of the literature. Am J Psychiatry 143:290–299, 1986

Theander S: Anorexia nervosa: a psychiatric investigation of 84 female patients. Acta Psychiatr Scand Suppl 214:1–194, 1970

Toner BB, Garfinkel PE, Garner DM: Affective and anxiety disorders in the long-term follow-up of anorexia nervosa. Int J Psychiatry Med 18:357–364, 1988

Van Loon GR: Abnormal catecholamine mechanisms in hypothalamic-pituitary disorders. Metabolism 29 (suppl 1):1198–1202, 1980

Walsh BT, Stewart JW, Roose SP, et al: Treatment of bulimia with phenelzine: a double-blind, placebo controlled study. Arch Gen Psychiatry 41:1105–1109, 1984

Warren MP, Vandewiele RL: Clinical and metabolic features of anorexia nervosa. Am J Obstet Gynecol 117:435–449, 1973

Welle S, Lilavivathana U, Campbell RG: Increased plasma norepinephrine concentrations and metabolic rates following glucose ingestion in man. Metabolism 29:806–809, 1980

Welle S, Lilavivathana U, Campbell RG: Thermic effects of feeding in man: increased plasma norepinephrine levels following glucose, but not protein or fat consumption. Metabolism: Clinical and Experimental 30:953–958, 1981

Young JB, Rowe JW, Pallotta JA, et al: Enhanced plasma norepinephrine response to upright posture and oral glucose administration in elderly human subjects. Metabolism: Clinical and Experimental 29:532–539, 1980

18

Adrenergic Changes During Alcohol and Drug Intoxication and Withdrawal

PAUL GLUE, M.D., PH.D.,
DAVID J. NUTT, M.D., PH.D.
MARKKU LINNOILA, M.D., PH.D.

In this chapter, we concentrate on the involvement of adrenergic mechanisms in drug intoxication and withdrawal, looking especially at alcohol, opioids, and psychostimulants. Additionally, we also comment on the limited number of studies investigating other drugs, such as benzodiazepines and phencyclidine. Although the majority of studies to date have been in animals, emphasis is placed on the present state of knowledge in humans. The organization of each section is in the following order: studies on neurotransmitters and metabolites, enzymes and second messengers, receptor binding studies, tests of receptor function, and therapeutic studies using adrenoceptor agonists or antagonists.

Alcohol

Neurotransmitter and Metabolite Concentrations

Perhaps the most extensive studies have been on the effects of alcohol on plasma norepinephrine (NE) and its metabolites 3,4,-dihydroxyphenylglycol (DHPG) or 3-methoxy-4-hydroxyphenylglycol (MHPG). In humans, acute alcohol intake increases urinary, plasma, and cerebrospinal fluid (CSF) MHPG, NE, and epinephrine concentrations (Borg et al. 1981; Brohult et al. 1970; Howes and Reid 1985; Ireland et al. 1984; Perman 1961b), although there are some negative reports (e.g., Heikkonen et al. 1989). The cause of the increased concentrations is probably both increased release and decreased clearance of the catecholamines (Eisenhofer et al. 1983; Perman 1961b). Elevated plasma and CSF concentrations of MHPG and NE persist during chronic alcohol intake (Borg et al. 1981; Sjoquist et al. 1981) and may contribute to the hypertension of chronic alcoholism (Potter et al. 1984). Brain concentrations of MHPG are elevated in chronic alcoholic individuals (Sjoquist et al. 1983), although regional NE concentrations in a small number of alcoholic suicide victims were similar to those of a control group (Moses and Robins 1975). In alcohol withdrawal, there are marked elevations of urinary, plasma and CSF NE, MHPG, and epinephrine concentrations (Borg et al. 1981, 1983; Hawley et al. 1985; C. B. Smith et al. 1989). Evidence has elucidated mechanisms involved in the peripheral increase in NE release. In part, it probably reflects increased central excitation through nonadrenergic mechanisms (see Glue and Nutt 1990), but alterations in central adrenergic function are also involved (see later). The most immediate measures of the altered central NE neuronal activity are increased CSF concentrations of NE and MHPG, although the latter reflects also a peripheral contribution (Kopin et al. 1983). These two variables can serve as an index of central sympathetic overactivity in this condition. Consequences of this include tremor, flushing, and gastrointestinal disturbances. Additionally, it can produce hypokalemia and contribute to hypomagnesemia. The severity of many withdrawal symptoms is positively correlated with concentrations of MHPG in CSF (Hawley et al. 1985).

Of particular interest is the progressive fall in CSF MHPG during abstinence (Borg et al. 1981; Hawley et al. 1985), and a subgroup of abstinent alcoholics have levels below those of appropriately matched control subjects (Borg et al. 1981). Abstinent alcoholics with Korsakoff's psychosis have also been reported by some investigators to have low CSF MHPG concentrations (McEntee et al. 1984). Furthermore, it has been observed that first-degree relatives of alcoholics may show similarly low MHPG concentrations, possibly indicating a biological predisposition to alcoholism (Borg et al. 1984). Perhaps the increased release of central NE by alcohol may contribute to its stimulant or reinforcing properties. Certainly, administration of α-methyl paratyrosine, a catecholamine synthesis inhibitor, abolishes the euphoriant effects of alcohol in the laboratory (Ahlenius et al. 1973). Alternatively, the increase in NE could be an attempt to compensate for the sedative effects of alcohol. This possibility is supported by evidence that α_2-adrenoceptor antagonists, which acutely increase NE release, oppose some of the intoxicating effects of alcohol (Lister et al. 1989).

Findings from animal studies are in agreement with the human data. Acute alcohol administration increases urinary and plasma NE and epinephrine concentrations (Perman 1961a). In the brain, acute and chronic alcohol administration and withdrawal increase NE turnover, with the largest changes occurring in the brain stem (Hunt and Majchrowicz 1974; Pohorecky 1974). The importance of regional differences in the effects of alcohol on brain concentrations of NE and metabolites has been demonstrated. Eisenhofer et al. (1990) showed that brain-stem DHPG concentrations are raised after chronic alcohol administration and during withdrawal. During withdrawal, brain-stem NE turnover is also increased, coinciding with the presence of maximal withdrawal symptoms.

There have been some recent interesting studies on the possible role of epinephrine in alcohol intoxication, tolerance, and withdrawal. It has been postulated that in the brain, the role of epinephrine is generally inhibitory, possibly via α_2-adrenoceptors (see Mefford 1988). Rats given an enzyme inhibitor preventing epinephrine synthesis become resistant to the intoxicating effects of alcohol (Mefford et al. 1990). After acute alcohol administration, brain-stem and hypothalamic epinephrine concentrations are reduced, whereas only hypothalamic epinephrine levels are reduced by chronic alcohol and during withdrawal, with the magnitude of this reduction significantly correlating with blood alcohol concentrations (Linnoila et al. 1987) (Table 18–1).

Enzymes and Second Messengers

α_2-, β_1-, and β_2-adrenoceptors are linked to adenylate cyclase. The activity of this enzyme may be assessed by quantifying basal levels of its metabolic product, cyclic adenosine monophosphate (cAMP) or by the magnitude of cAMP responses after stimulation by a range of agonists. Acute intoxicating doses of alcohol in humans significantly reduce CSF levels of cAMP within 24 hours (Orenberg et al. 1976), although lymphocyte cAMP levels are not altered within 4 hours of smaller doses of alcohol (Heikkonen et al. 1989). Chronic alcoholics have reduced plasma or lymphocyte levels of cAMP during abstinence (Diamond et al. 1987; Lykouras et al. 1988). Stimulation studies by agonists in animal brains after chronic alcohol administration reveal reduced α- and β-stimulated adenylate cyclase activity (French et al. 1975; Rommelspacher and Strauss 1987; Saito et al. 1987; Thurman et al. 1980).

Early in alcohol withdrawal, increased cAMP responses to NE are seen in mouse cortex (French et al. 1975), with normalization of responses to β-agonists 24–48 hours after withdrawal (Saito et al. 1987). In humans, CSF cAMP levels are within the normal range after 1 month of abstinence (Orenberg et al. 1976). However, stimulated adenylate cyclase responses have been suggested to be still abnormal in long-term abstinent alcoholics (Tabakoff et al. 1988). Although cAMP levels appear to be state dependent, the abnormal adenylate cyclase activity in long-term abstinent alcoholics appears to be a more enduring marker. Whether this is a cause or consequence of alcoholism is unclear and needs to be studied further (for review, see Hoffman and Tabakoff 1990). Not all adrenoceptor-linked second messenger systems are altered by alcohol. For instance, responses of the α_1-adrenoceptor-linked phosphatidyl 4,5-bisphosphate (PIP_2) to NE in mouse cortex are not altered by chronic alcohol, whereas cholinergic-stimulated PIP_2 responses are decreased (Hoffman et al. 1986).

Abnormalities relating to the biogenic monoamine degrading enzyme monoamine oxidase (MAO) have also been noted in the platelets of alcoholics. Low levels of platelet MAO activity have been reported in depen-

Table 18–1. Alcohol effects on animals and humans

	Animals				Humans			
	Acute	**Chronic**	**Withdrawal**	**Abstinence**	**Acute**	**Chronic**	**Withdrawal**	**Abstinence**
Neurotransmitter levels								
Urine NE					↑		↑	
Urine epinephrine	↑				↑			
Plasma NE	↑				↑	↑	↑	
Plasma epinephrine	↑				NC/↑		↑	
CSF NE					↑	↑	↑	
Brain NE	↑[a]	↑[a]	↑[a]			NC		
Brain eprinephrine	↓[a]	↓[a]						
Metabolism levels								
Plasma MHPG					↑	↑	↑	NC
CSF MHPG					↑	↑	↑	↓
Brain MHPG		↑[a]	↑[a]			↑		
Second messengers/enzymes								
Adenylate cyclase/cAMP								
Periphery					NC/↓	↓	↓	NC/↓
Brain		↓	↑	NC				
Platelet MAO						↓	↓	NC/↓
Brain MAO						NC/↓		NC

continued

Table 18–1. Alcohol effects on animals and humans *(continued)*

	Animals				Humans			
	Acute	**Chronic**	**Withdrawal**	**Abstinence**	**Acute**	**Chronic**	**Withdrawal**	**Abstinence**
Receptor binding								
Platelet α							↓/NC	
Lymphocyte β					NC			NC
Brain β	NC	NC/↓[a]	NC/↓[a]	↑			↓[a]	NC[a]
Brain α		NC/↓[a]						
Challenge tests								
β-Agonists					NC			NC
α_2-Agonists			↓					
Growth hormone							↓	↓
Hypotension							↓	NC
Hypothermia			NC				↓	NC
Sedation							↓	NC
MHPG fall							↓	NC

Note. ↑ = increased activity/levels; ↓ = reduced activity/levels; NE = norepinephrine; NC = no change from control values; CSF = cerebrospinal fluid; MHPG = 3-methoxy-4-hydroxyphenylglycol; cAMP = cyclic adenosine monophosphate; MAO = monoamine oxidase.

[a]Regional effect.

dent, withdrawing, and recently abstinent alcoholics (Alexopoulos et al. 1981; Major and Murphy 1978; Sullivan et al. 1990; von Knorring et al. 1985; Wiberg et al. 1977). Sullivan et al. noted that Type II alcoholics had lower levels of activity than control subjects and Type I alcoholics, although enzyme activity in the latter group was also lower than in control subjects. In long-term abstinent alcoholics, platelet MAO activity is similar to that of a control group, but MAO inhibition by alcohol is significantly higher in the platelets of alcoholics (Tabakoff et al. 1988). Therefore, platelet MAO activity also appears to be a state marker, although the latter finding in alcoholics of persistently greater inhibition of MAO by alcohol may represent an inherited vulnerability factor for alcoholism. There is disagreement as to whether brain MAO activity is altered in alcoholism. Reduced (Gottfries et al. 1975) or normal (Grote et al. 1974) MAO activity has been reported in alcoholic suicide victims, with normal MAO activity noted in the brains of abstinent alcoholics (Tabakoff et al. 1985). Additional factors such as duration of abstinence prior to death, depression, or suicidal behavior in the subjects studied may complicate interpretation of these data (Table 18–1).

Receptor Binding Studies

As described elsewhere (Nelesen and Ziegler, Chapter 7, this volume; Smith and Hollingsworth, Chapter 4, this volume), peripheral adrenoceptors exist on platelets (α_2), on lymphocytes (β_2), and on blood vessels and the heart (α_1, α_2, β). These can be studied in several ways. Receptor number and affinity can be estimated by binding of radioligands, such as ^{3}H-rauwolscine (α_2-antagonist), ^{3}H-UK-14304 (high-affinity α_2-agonist), ^{128}I-cyanopindolol, and ^{3}H-dihydroalprenolol (β-antagonists). C. B. Smith et al. (1989) showed reduced platelet α_2-adrenoceptor number in alcoholics during early withdrawal, which would support findings from receptor challenge studies showing reduced responsiveness to clonidine (see later). Another study of platelet α_2-binding in alcoholics during withdrawal found binding parameters not to be significantly different from those of a control group (Nutt et al. 1987). β-Receptors can be assessed by ligand binding to lymphocytes. Acute or chronic alcohol in humans does not alter lymphocyte β-adrenoceptor binding (Heikkonen et al. 1989; Padovani et al. 1987). Significant negative correlations are seen in healthy volunteers between plasma NE concentrations and β-receptor number, probably indicating β-down-regulation as a consequence of higher plasma NE concentrations (Padovani et al. 1987). It might be predicted that such an association would be more marked in alcoholics, who would have persistently elevated NE concentrations; however, this is not seen in recently abstinent alcoholics (Padovani et al. 1987). The loss of the negative association between plasma NE levels and β-adrenoceptor numbers may indicate receptor subsensitivity secondary to persistently elevated levels of NE. Alternatively, alcohol may disrupt adaptive processes controlling receptor regulation (see later in this chapter).

A direct way of studying the effects of alcohol on central adrenoceptors is by measurement of receptor binding in postmortem brain tissue. Although an early study investigating central β-adrenoceptors using antagonist binding in human brains found no differences between alcoholics and control subjects (Tabakoff et al. 1985), a more recent study comparing intoxicated and abstinent alcoholics showed that the former lacked high-affinity agonist binding in the cortex (Valverius et al. 1989a). This finding is supported by additional work from this group, who reported similar observations in mouse brain cortex and hippocampus during alcohol withdrawal (Valverius et al. 1987, 1989b). The regional alcohol-induced loss of the high-affinity state of the β-adrenoceptor, which is linked to the G_S-protein and to adenylate cyclase, may support the state-dependent abnormalities in adenylate cyclase responses (see earlier). It is also of interest that these changes appear to normalize with abstinence. The only caveat in these studies is the use of iodocyanopindolol as the antagonist radioligand. It has a high affinity for serotonin receptors as well.

There is extensive animal literature concerning the effects of alcohol administration and withdrawal on antagonist binding to adrenoceptors in various brain regions. Methodological differences between studies (e.g., animal species, receptor binding assays, route and dose of alcohol administration used) complicate comparison of results. In particular, studies using alcohol inhalation techniques to induce dependence have had limited success in producing cortical receptor changes. Chronic alcohol and alcohol withdrawal reduce β-adrenoceptor binding in mouse cerebral cortex; the changes are in β_2- but not β_1-receptors (Banerjee et al. 1978a; Rabin et al. 1980). Shortly after withdrawal, increased β-adrenoceptor numbers may be seen (Banerjee et al. 1978a; Kuriyama et al. 1981). α_2-Adrenoceptor density in hypothalamus is reduced by chronic alcohol administration (Strickland and Wooles 1989). However, there are several studies that do not report any changes in α- or β-adrenoceptor binding in a number of brain regions (Hunt and Dalton 1981; Kuriyama et al. 1981; Rommelspacher and Strauss 1987; Turkka et al. 1990) (Table 18–1).

Receptor Challenge Paradigms

The effects of alcohol on adrenoceptor function can also be studied by challenge tests. Peripheral β-receptor function can be assessed by infusions of agonists, such as isoprenaline (β only) or NE (α and β). Studies in healthy volunteers show reduced α-adrenoceptor-mediated cardiovascular responses after single doses of alcohol (Eisenhofer et al. 1984). In alcoholics, George (T. D. George, unpublished observations) found no alterations in isoprenaline sensitivity when infusions were performed after 3–5 weeks of abstinence.

In theory, it is possible to challenge each of the adrenoceptor subtypes in brain; this has been done in animals. However, in humans, this approach has limitations. β-Agonists do not cross the blood-brain barrier easily, and their peripheral activities on the cardiovascular system limit the doses that can be administered. β-Antagonists could be used, since there are studies that show measurable responses (e.g., growth hormone release; see Checkley 1980). However, there are no reports of using this strategy in alcoholics. α_1-Agonists have problems similar to those of β-agonists, having poor penetration through the blood-brain barrier and causing peripheral hypertension. Neither α_1-agonists nor -antagonists appear to have been used in challenge paradigms. Thus, the only central adrenoceptor that has been explored using agonist challenge paradigms has been the α_2-adrenoceptor, particularly by using clonidine. This drug was initially marketed as a centrally acting antihypertensive and has since been used to treat alcohol and opioid withdrawal (see later in this chapter). In withdrawal states, it appears to act by reducing NE outflow due to overactivity of the locus coeruleus. This effect is centrally mediated and involves the agonist-induced hyperpolarization of neurons by activation of the inhibitory presynaptic α_2-adrenoceptor (see Cedarbaum and Aghajanian 1977).

Interest in α_2-adrenoceptor function in alcoholics comes from several lines of evidence. First, there is an extensive literature on the complex effects of alcohol on NE turnover (see earlier; for review, see Nutt and Glue 1986). Chronic ethanol has been shown to produce a pronounced depletion of brain epinephrine concentrations. Since epinephrine is a potent agonist at the presynaptic inhibitory α_2-adrenoceptor, its depletion could have a profound effect on neuronal activity (Mefford 1988). The second indication of alcohol/α_2-adrenoceptor interaction comes from clinical observations. There is a clear association between alcohol ingestion and hypertension (see Potter et al. 1984). Sympathetic overactivity is a cardinal feature of alcohol withdrawal, and NE is the major transmitter of this branch of the autonomic nervous system. Finally, alcoholic dementias are well recognized, and there is preliminary evidence of noradrenergic neuronal loss in some of these disorders (McEntee et al. 1984) and that α-adrenergic agonists may be of use in treating them (see later).

Clonidine and other agonists such as lofexidine and UK-14304 have a range of central actions, all of which appear to be mediated through α_2-adrenoceptors (for reviews, see Checkley 1980; Glue and Nutt 1988). Each of these actions can be assessed in alcoholics and, in comparison with control responses, give an index of receptor sensitivity. The various responses are summarized in Table 18–1.

During alcohol withdrawal, there are reduced responses in a number of paradigms. Growth hormone release is completely blunted in almost 80% of patients (Nutt et al. 1988). The hypotensive, hypothermic, and sedative effects of clonidine are also markedly attenuated in these patients (Glue and Nutt 1987; Glue et al. 1989a, 1989b; Nutt et al. 1988). There is also some indication of reduced responsiveness in terms of inhibition of plasma MHPG. Taken together, these data suggest a global subsensitivity of α_2-receptors in alcoholics in withdrawal compared with control subjects. The apparently anomalous findings that anxious and dysphoric mood and the severity of tremor are alleviated by clonidine can be explained in terms of baseline artifact; that is, alcoholics have high initial ratings of depression, anxiety, and tremor, whereas control subjects' scores are usually zero for these measures. The fact that clonidine effectively treated withdrawal symptoms probably led to a secondary amelioration of mood. After 5 weeks of abstinence, only the growth hormone response to clonidine was persistently blunted, all other responses having returned to normal (Glue et al. 1989a, 1989b). A study in early (1–3 weeks) and late (5–6 weeks) abstinence has not confirmed the blunted growth hormone in this group of alcoholics (Muller et al. 1989), and further studies are required to clarify this area. There is some preliminary evidence that heavy social drinking may alter α_2-adrenoceptor sensitivity, as reduced growth hormone responses to clonidine have been reported in such subjects while drinking and after 1 month's abstinence (Matussek et al. 1984).

What does the blunted growth hormone response to clonidine tell us about adrenoceptor function in withdrawal? One possibility is that central α_2-adrenoceptors are subsensitive. This offers an explanation for the withdrawal syndrome, since defective presynaptic inhibitory α_2-adrenoceptors would lead to excessive uncontrolled firing of neurons of the locus coeruleus. However, there is another possibility, which is that the noradrenergic system is being excessively driven for some other reason (e.g., increased corticotropin-releasing factor or increased *N*-methyl-D-aspartate [NMDA] overactivity; for review, see Glue and Nutt 1990). This could competitively limit the activity of clonidine in severe withdrawal (B. J. Robinson et al. 1989). In other situations in which there is evidence for NE overactivity, such as essential hypertension, elevated baseline measures (e.g., MHPG and blood pressure) are more sensitive to the effects of clonidine (Goldstein et al. 1985; Reid 1983). This does not extend to postsynaptic sensitivity, as growth hormone responses are decreased in this condition. The persistence of a blunted growth hormone response during abstinence may be an indicator of hypothalamic damage and may also provide insight into depressive or anxiety disorders associated with abstinence.

Testing of adrenoceptor function in animals, apart from second messenger responses, is limited. Hoffman and Tabakoff (1977) showed normal hypothermic responses to clonidine in withdrawing mice, suggesting intact α_2-adrenoceptor function. However, the considerable dose of clonidine used (0.5 mg) may have led to stimulation of postsynaptic α_2-adrenoceptors. Postsynaptic α_2-adrenoceptor sensitivity may be reduced in dependent and withdrawing rats, using as an index the magnitude of the response of protein kinase inhibition to high doses of clonidine (Szmigielski et al. 1989) (Table 18–1).

Therapeutic Studies

Three aspects of alcoholism may be ameliorated by adrenergic medications: alcohol withdrawal symptoms, maintenance of abstinence, and the alcohol amnestic syndrome. Agonists of both α- and β-adrenoceptors have been shown to be of benefit in alcohol withdrawal (for review, see Castaneda and Cushman 1989). The α_2-agonists clonidine and lofexidine are more effective than placebo in reducing hyperautonomic symptoms in withdrawal (Bjorkvist 1975; Cushman et al. 1985; Wilkins et al. 1983), and appear to be as effective as benzodiazepines (Baumgartner and Rowen 1987). β-Adrenoceptor blockers have also been shown to be useful in reducing withdrawal symptoms (Kraus et al. 1989; Sellers et al. 1977). Both types of medications appear to be working via a noradrenergic mechanism (Nutt et al. 1988; Sellers et al. 1976). However, in patients with severe withdrawal symptoms, both clonidine and β-blockers may be ineffective and even dangerous because they lack anticonvulsant potency (Jacobs et al. 1983; Liskow and Reed 1986; B. J. Robinson et al. 1989).

The Toronto group has attempted to reduce alcohol consumption in heavy social drinkers using serotonin reuptake inhibitors (for review, see Naranjo and Sellers 1989). A modest reduction in alcohol intake has been noted. However, studies in rats using the noradrenergic tricyclic desipramine have shown similar results to those reported for the serotonergic antidepressants (Z. Amit, personal communication, June 1990). As selective serotonergic and noradrenergic antidepressants do not show specificity in their alteration of CSF monoamine metabolites during chronic administration (Potter et al. 1985), it is possible that there may be an element of adrenergic control in reducing the urge to consume alcohol.

A third area where adrenergic drugs may be beneficial is in the treatment of the alcoholic amnestic syndrome. McEntee and colleagues reported lowered CSF MHPG levels in Korsakoff's psychosis and that amnestic scores in such patients improved after treatment with clonidine (McEntee and Mair 1978, 1980; McEntee et al. 1984). A more recent study also reported significant clinical and other improvements in patients with alcoholic amnestic syndrome after modafinil, an α_1-agonist (Saletu et al. 1990). It is possible that the effects of clonidine reported above may be due to α_1- rather than α_2-agonist effects. Ongoing work with the selective serotonin reuptake inhibitor fluvoxamine also appears promising in this disorder (Martin et al. 1989). However, the magnitude of improvement reported in all of these studies is modest, and further work is needed before such treatment can be recommended.

Opiates

Neurotransmitters and Metabolites

Acute administration of morphine to rats has regional effects on central NE turnover. By using accumulation of MHPG as an index of NE turnover, increases in MHPG are seen in whole brain, hypothalamus, cerebellum, and brain stem, but not in cortex or striatum (Attila 1989; Gomes et al. 1985; Lopachin and Reigle 1978; Roffman et al. 1977, 1979; Sugrue 1974). Although NE turnover is increased, brain-stem NE and adrenal epinephrine and NE concentrations are reduced after acute morphine (Maynert and Klingman 1962), due to an inhibitory effect of morphine on locus coeruleus cell firing (Aghajanian 1978; Korf et al. 1974). Some variability between studies may be due to different experimental conditions; for example, the dose of morphine used, the presence of anesthetics, and whether animals are prehandled may affect tissue responses to morphine (Gunne et al. 1969; for review, see Ahtee and Attila 1987; Attila 1989). Tolerance to these effects develops after chronic morphine (Maynert and Klingman 1962; Reigle and Huff 1980; Roffman et al. 1977, 1979). Plasma NE and epinephrine concentrations are unaltered by chronic morphine, although NE responses to sympathetic nerve stimulation are reduced in morphine-tolerant rats (Leung et al. 1989).

Opiate withdrawal may be produced by stopping drug administration, or precipitated by opiate antagonists. Using the former technique, MHPG levels in the hypothalamus and cerebellum are reduced after short-term morphine (Roffman et al. 1977), although after longer administration NE turnover increases, especially in the cortex (Ahtee et al. 1989; Attila and Ahtee 1983, 1984). Urinary epinephrine and NE excretion is markedly increased within 72 hours of withdrawal, and subsequent NE excretion is lower than that of control animals (Sloan and Eisenman 1968). Naloxone-precipitated withdrawal produced substantial increases in plasma epinephrine, NE, and MHPG (Delle et al. 1990; Swann et al. 1983); brain MHPG (Crawley et al. 1979; Esposito et al. 1987; Swann et al. 1983; for review, see Roth et al. 1982); and NE turnover (Laverty and Roth 1980). This appears to be due to increased firing of NE neurons in the locus coeruleus (Aghajanian 1978).

After acute doses of opiates in humans, plasma and urinary NE, epinephrine, and catecholamine metabolites increase, indicating increased NE turnover, whereas after chronic morphine or heroin, levels tend to return to normal (Eisenman et al. 1969; Manner et al. 1987; Matussek and Hoehe 1989; Meyer et al. 1978; Schildkraut et al. 1977; Weil-Malherbe et al. 1965). In these studies, inconsistent changes are seen after gradual drug withdrawal, although urinary epinephrine may be elevated. In a more detailed study, DeLeon-Jones et al. (1983) compared urinary catecholamine metabolites in subjects abruptly or gradually withdrawn from methadone. During chronic methadone, urinary metanephrine levels increase and fall significantly during withdrawal. Naltrexone-precipitated withdrawal in methadone-addicted persons increases plasma MHPG concentrations, which are significantly correlated with ratings of abstinence symptoms (Charney et al. 1982b, 1983a). Similar results are seen in morphine-dependent monkeys after antagonist-precipitated withdrawal (Roth et al. 1982) (Table 18–2).

Enzymes and Second Messengers

cAMP responses to acute morphine administration are complex, being dose- and region-dependent. In some areas (e.g., cortex, cerebellum), cAMP levels initially rise, then fall below baseline levels; in other areas (e.g., medulla, hypothalamus), levels decline after acute morphine (Clouet et al. 1975). There is a tendency for adenylate cyclase activity to increase after chronic morphine (Clouet et al. 1975; Duman et al. 1988; Kuriyama et al. 1978; Merali et al. 1975; Nathanson and Redmond 1981; Nestler et al. 1989), and cAMP responses to NE are increased (Llorens et al. 1978) or reduced (Kuriyama et al. 1978; Naito and Kuriyama 1973). In withdrawal, striatal cAMP levels remain elevated, whereas adenylate cyclase activity is reduced (Merali et al. 1975; Nathanson and Redmond 1981). It seems that opiates inhibit cAMP production acutely, with compensatory increases occurring during tolerance, and rebound overactivity when opiates are stopped. In support of this is the finding that withdrawal symptoms are worsened by intracerebroventricular injection of cAMP and improved by drugs that enhance its breakdown (Collier and Francis 1975).

In heroin withdrawal in humans, the dose-response curve of adrenaline-induced inhibition of platelet adenylate cyclase is shifted to the left

(Garcia-Sevilla et al. 1987). This is consistent with the finding that platelet adrenoceptor numbers are increased (see later). Unfortunately, these studies were conducted with independent control groups, and it would be of interest to repeat this work longitudinally, during methadone maintenance and withdrawal (Table 18–2).

Receptor Binding Studies

Acute doses of morphine do not alter α- or β-adrenoceptor binding in rat brain (Hamburg and Tallman 1981). α_1-Adrenoceptors are unaltered by chronic morphine administration (Hamburg and Tallman 1981; Rokosz-Koslowska et al. 1988). There are conflicting findings on the effect of chronic morphine on brain α_2-adrenoceptors, with increased, unaltered, or decreased receptor density having been found (Hamburg and Tallman 1981; C. B. Smith et al. 1989; T. L. Smith et al. 1983; Ulibarri et al. 1987; Vicentini et al. 1983), probably due to differences in methodology between studies. During morphine withdrawal, α_2-adrenoceptor density has been reported to be both increased (Ulibarri et al. 1987) and decreased (Hamburg and Tallman 1981; C. B. Smith et al. 1989).

β-Adrenoceptor binding in rat cortex is increased (Llorens et al. 1978; Moises and Smith 1987, 1989; Swann et al. 1983) or unaltered by chronic morphine administration (Kuriyama et al. 1981). There is a reduction in cortical and brain-stem β-adrenoceptors during morphine withdrawal (Hamburg and Tallman 1981; Moises and Smith 1987, 1989), although there are discordant reports (Kuriyama et al. 1981). The changes in β-adrenoceptor number are associated with changes in sensitivity to stimulation with β-agonists.

In human subjects, the only receptor binding studies have been of platelet α_2-adrenoceptors during withdrawal. There is an increase in the density of platelet α_2-adrenoceptor agonist binding sites, with the increase following a similar time course to abstinence symptoms and being significantly correlated with withdrawal symptom severity. In addition, measures of functional activity (adrenaline-induced aggregation responses) are increased (Garcia-Sevilla et al. 1985, 1987). These findings contrast with those in the brain, and they could be construed as reflecting an attempt to reduce NE release by increasing presynaptic α_2-adrenoceptor sensitivity (Table 18–2).

Table 18–2. Opioid effects on animals and humans

	Animals				Humans			
	Acute	**Chronic**	**Withdrawal**	**Abstinence**	**Acute**	**Chronic**	**Withdrawal**	**Abstinence**
Neurotransmitter levels								
Urine NE					↑	NC		
Urine epinephrine					↑	NC		
Plasma NE		NC	↑					
Plasma epinephrine		NC	↑					
Adrenal NE	↓	NC						
Adrenal epinephrine	↓	NC						
Brain NE	↓[a]							
Metabolism levels								
Urine NE metabolites					↑	NC/↑	↓	
Plasma MHPG			↑				↑	
CSF MHPG			↑					
Brain MHPG	↑[a]	NC[a]	↑/NC/↓[a]					
Second messengers/enzymes								
Adenylate cyclase/cAMP								
Periphery							↑	
Brain	↑/↓[a]	↑/↓[a]	↑[a]					

Receptor binding				
Platelet α				↑
Brain β	NC	↑[a]	↓[a]	
Brain α_2	NC	↓/NC/↑[a]	↓/↑[a]	
Brain α_1	NC	NC		
Challenge tests				
β-Agonists		↑	↓	
α-Agonists			↓	
Growth hormone				NC
Hypothermia		NC		

Note. ↑ = increased activity/levels; ↓ = reduced activity/levels; NE = norepinephrine; NC = no change from control values; MHPG = 3-methoxy-4-hydroxyphenylglycol; CSF = cerebrospinal fluid; cAMP = cyclic adenosine monophosphate.

[a]Regional effect.

Receptor Challenge Paradigms

Brambilla et al. (1984) carried out clonidine challenge testing in nonwithdrawing heroin-addicted persons and found that a range of hormonal responses are identical to those of a control group, supporting other evidence that there is no cross-tolerance between opiate and α_2-adrenoceptors (Freedman and Aghajanian 1985). Sensitivity of hippocampal slices to clonidine indicates functional subsensitivity of α_2-adrenoceptors during morphine withdrawal (C. B. Smith et al. 1989). Altered agonist sensitivity of β-adrenoceptors has been reported, with supersensitivity during dependence and subsensitivity during withdrawal. Moises and Smith (1987, 1989) investigated cortical neuronal sensitivity to iontophoretically applied NE or isoprenaline during morphine dependence and withdrawal and showed increased sensitivity to both agonists during dependence and reduced sensitivity during withdrawal. This could explain the increase in NE during withdrawal and is similar to changes seen during alcohol withdrawal. Nathanson and Redmond (1981) showed differences in cAMP responses to isoprenaline during morphine dependence and withdrawal.

Delle et al. (1990) studied baroreceptor function in rats undergoing precipitated morphine withdrawal, examining the effects of phenylephrine on baroreceptor-induced inhibition of sympathetic nerve activity. They found that during withdrawal, renal sympathetic nerve activity was reduced, but adrenal and lumbar sympathetic nerve activity was increased (Table 18–2).

Therapeutic Studies

Clonidine has become well established in the management of opiate withdrawal symptoms since the early reports of its efficacy in humans (Gold et al. 1978) and in animals (Fielding et al. 1978; Roth et al. 1982). It is effective not only in nonprecipitated (i.e., spontaneous) withdrawal (Charney et al. 1981), but can also be used to cover naloxone-precipitated withdrawal in accelerated detoxification regimes (Charney et al. 1982b). Improvement is seen mainly in autonomic withdrawal symptoms, with relatively little effect on restlessness, insomnia, or anxiety. Other α_2-agonists such as guanfacine have been used effectively to treat acute opiate withdrawal (Schubert et al. 1984), although there are much less data on the use of these agents.

Psychostimulants

Despite cocaine and amphetamine having similar behavioral, psychological, and cardiovascular effects (Bejerot 1970; see Johanson and Fischman 1989), their pharmacology is different. Amphetamine-like drugs cause the release of catecholamines and block their reuptake, whereas cocaine and similarly acting drugs (e.g., methylphenidate) act mainly through reuptake inhibition (Arnold et al. 1977; Axelrod et al. 1962; Burgess and Tessel 1980; Chiueh and Kopin 1978; Hadfield et al. 1980; Holmes and Rutledge 1976), although cocaine has additional local anesthetic properties. Most work on the neurochemistry of the psychostimulants has investigated their effects on central dopamine systems (for review, see Gold et al. 1985). In contrast to alcohol and opioids, the adrenergic effects of psychostimulant administration and withdrawal have been barely researched, with very little information from human studies.

Neurotransmitters and Metabolites

Neurotransmitter and metabolite changes after cocaine administration are complex. In rats, regional NE concentrations increase immediately after administration, with reduced regional or whole brain concentrations by 20 minutes afterward (Pradhan et al. 1978; S. E. Robinson and Hambrecht 1988; Scheel-Kruger 1972). Electrophysiological studies have revealed the reason for the rapid changes in NE levels, as both amphetamine and cocaine dose-dependently reduce cell firing in the locus coeruleus (Graham and Aghajanian 1971; Pitts and Marwah 1986, 1987). Preadministration of α_2-adrenoceptor antagonists blocks this effect of cocaine, indicating that the reduction in firing rate is due to an effect of increased synaptic NE concentrations on the inhibitory α_2-adrenoceptor. Rate of disappearance of NE metabolites from the brain tissue is increased after acute cocaine. However, NE turnover is increased by acute cocaine, as indicated by metabolite concentrations or metabolite/NE ratios (S. E. Robinson and Hambrecht 1988; Roffman et al. 1978; Scheel-Kruger 1972). Chronic cocaine has also been reported to reduce NE concentrations in several brain structures (Roy et al. 1978; Taylor and Ho 1977), although these results may reflect the time at which samples were obtained (> 20 minutes after the last cocaine administration). NE concentrations are increased in rat hypothalamus 24 hours after the last dose of chronic cocaine, but not in other brain areas (Yeh 1987).

The effects of acute and chronic amphetamine on brain NE are similar to those of cocaine, with regional reduction in NE concentrations (Calderini et al. 1975; Ellison et al. 1978; Gunne and Lewander 1968; Morgan and Gibb 1980; Scheel-Kruger 1972). CSF NE concentrations in monkeys are increased in a dose-dependent manner by amphetamine (Ziegler et al. 1979). During amphetamine withdrawal and early abstinence, NE and normetanephrine concentrations are reduced in several regions of rat brain (Tonge 1974). After long-term amphetamine abstinence, regional NE concentrations in rats are similar to those of control subjects (Ellison et al. 1978). Brain MHPG concentrations are increased after acute amphetamine, unaltered by chronic amphetamine, and reduced in withdrawal (Calderini et al. 1975; Roffman et al. 1978).

In the periphery, acute intravenous cocaine increases plasma epinephrine and NE concentrations in monkeys (Nahas et al. 1988), probably due to increased release from the adrenals (Chiueh and Kopin 1978). An increase in adrenal catecholamine synthesis may contribute to the persistent increases in rat adrenal NE and epinephrine concentrations after chronic cocaine (Dixon et al. 1989; Gunne and Jonsson 1964). However, cardiac and splenic NE concentrations are reduced for as long as 1 week after cocaine withdrawal (St. John-Allan et al. 1989). Whether this is due to transmitter depletion or reduced turnover is not clear.

In contrast to the animal studies, there is a paucity of data on the effects of psychostimulants on plasma NE, epinephrine, or MHPG in humans. In detailed early work, Liddell and Weil-Malherbe (1953) showed that amphetamines altered plasma epinephrine concentrations in a complex manner, with levels rising immediately, then falling below baseline by 20 minutes, and then returning to baseline by 60 minutes. Acute doses of amphetamine increase plasma NE concentrations in a dose-dependent manner (Mikkelsen et al. 1981; Rosenblatt et al. 1979). Intravenous cocaine does not alter plasma NE in humans in doses that produce the expected psychological and cardiovascular effects (Sherer 1988). Indeed, in this latter study, there is a tendency for plasma MHPG to decrease 1 hour after cocaine administration. Such a fall in MHPG after acute cocaine has been reported previously in primates (Post et al. 1976) and may be caused by reduced access of synaptic NE to neuronal MAO. Chronic but not acute fenfluramine administration reduced plasma NE by up to 70%, with a return toward pretreatment levels 1 week after stopping the treatment (de la Vega et al. 1977).

Peripheral indices of NE activity are low during psychostimulant withdrawal and abstinence. Urinary MHPG levels are reduced early in abstinence from amphetamine, this fall being correlated with increases in depression ratings (Watson et al. 1972). In cocaine withdrawal and early abstinence, plasma MHPG concentrations are low and continue to fall over the first few weeks of abstinence (Demer et al. 1989; Tennant 1985). In this regard, cocaine abstinence would appear to produce opposite actions to the stopping of tricyclic antidepressants, in that the latter situation is associated with increases in plasma MHPG (Charney et al. 1982a), plasma melatonin release (Cowen et al. 1985), and anxiety but not depression (Charney et al. 1982a). Why withdrawal from drugs with similar uptake blocking effects could produce such different neurochemical actions is an area that urgently awaits further study (Table 18–3).

Enzymes and Second Messengers

Tyrosine hydroxylase activity is reduced regionally after acute and chronic amphetamine, with reduced activity in the caudate persisting into abstinence (Ellison et al. 1978; Morgan and Gibb 1980). However, these changes are likely to reflect changes in dopamine rather than NE systems. Chronic amphetamine reduces MAO activity in guinea pig brain, although activity over the first 2–3 weeks of abstinence is greater than normal (Utena et al. 1959). Although short-term amphetamine administration reduces cAMP responses to NE in rat brain (Mobley et al. 1979), increased cAMP responses to NE occur in cocaine and amphetamine withdrawal (Banerjee et al. 1979). The latter finding supports receptor binding data that indicate β-adrenoceptor supersensitivity during withdrawal (Table 18–3).

Receptor Binding Studies

Noradrenergic systems have not been shown to be of major importance in mediating the rewarding properties of psychostimulants (Davis et al. 1975; De Wit and Wise 1977; Risner and Jones 1976; Roberts et al.

Table 18–3. Psychostimulant effects on animal and humans

	Animals				**Humans**			
	Acute	**Chronic**	**Withdrawal**	**Abstinence**	**Acute**	**Chronic**	**Withdrawal**	**Abstinence**
Neurotransmitter levels								
Plasma NE	↑				NC/↑[a]	↓		
Plasma epinephrine	↑				↑			
Adrenal NE	NC	↑						
Adrenal epinephrine	NC	↑						
Cardiac NE		↓		↓				
CSF NE	↑							
Brain NE	↑/↓[b]	↓[c]	↑/NC/↓[c]	↓				
Metabolism levels								
Urine MHPG							↓	↓
Plasma MHPG	↓				↓		↓	↓
Brain MHPG	↑[c]	NC	↓					
Second messengers/enzymes								
Adenylate cyclase/cAMP								
Brain		↓	↑					
Brain MAO		↓	↓	↑				
Tyrosine hydroxylase	↓	↓	↓	↓				

Receptor binding				
Placental β				↓
Cardiac β		NC		
Brain β	↑	↑/NC	↑	
Brain α_2		↑		
Brain α_1	NC	NC		
Challenge tests				
α- , β-Agonists		↓	↑	
α_2-Agonists		↓		

Note. ↑ = increased activity/levels; ↓ = reduced activity/levels; NE = norepinephrine; NC = no change from control values; CSF = cerebrospinal fluid; MHPG = 3-methoxy-4-hydroyphenylglycol; cAMP = cyclic adenosine monophosphate; MAO = monoamine oxidase.

[a]Cocaine and amphetamine differ.

[b]Regional and time-related effect.

[c]Regional effect.

1977). This is supported by studies showing that reinforcing potency is correlated with binding potency of compounds to the dopamine transporter site, not to adrenoceptors (Ritz and Kuhar 1989; Ritz et al. 1987). However, cardiovascular complications of cocaine use seem to be particularly associated with its adrenergic effects (Nahas et al. 1988; see Billman 1990). For instance, cardiovascular responses to cocaine are enhanced by pretreatment with yohimbine (Wilkerson 1989). Preadministration of cocaine has been shown to potentiate the hypertensive or tachycardic effects caused by administered NE or by sympathetic nerve stimulation (Jain et al. 1990; Pals et al. 1968). Tolerance develops rapidly to the former effect, which can be prevented by pretreatment with α_2-adrenoceptor antagonists.

Increased β-adrenoceptor binding density is seen in rat brain 12 hours after a single dose of cocaine, but not amphetamine (Banerjee et al. 1978b, 1979). Similar and persistent increases in β-adrenoceptor number are seen after chronic cocaine and during withdrawal from chronic cocaine or amphetamine (Banerjee et al. 1978b, 1979). It is possible that the increase in receptor number is secondary to the reduction in NE, but as mentioned earlier it is paradoxical that these receptor changes are the opposite to those found after chronic administration of tricyclic antidepressants. There are some studies that have not confirmed these findings, possibly because drugs were not given for a sufficient duration or because of methodological differences (Reisine et al. 1980; Sellinger-Barnette et al. 1980). α_2-Adrenoceptor numbers are increased, and α_1-adrenoceptor numbers unaltered by short-term amphetamine administration (Reisine et al. 1980). Cardiac β-adrenoceptor numbers are unaltered by chronic cocaine administration (Avakian et al. 1987).

The only receptor binding data in humans in this time frame are in a study reporting reduced placental β-adrenoceptor density in mothers who used cocaine during pregnancy (Wang and Schnoll 1987). Unfortunately, this study gives no details of duration and frequency of cocaine use or when the last dose was taken (Table 18–3).

Receptor Challenge Paradigms

The only challenge study somewhat related to adrenergic functions to have been carried out in those addicted to psychostimulants used the dexamethasone suppression test (DST) in recently abstinent cocaine-abusing persons (Gawin and Kleber 1985). Abnormal (nonsuppressor) results were noted in 8 of 19 subjects, who did not differ clinically from those with normal DST responses. Unfortunately, although there is a noradrenergic component to the DST, other factors such as stress can complicate interpretation of any results, and as such the DST is not an adequate test of central adrenoceptor function.

Adrenoceptor subsensitivity after chronic cocaine has been demonstrated in animals. Cardiac responses to epinephrine are reduced after chronic cocaine (Avakian et al. 1990), although heart β-adrenoceptor density is unaltered (see earlier). A reduction in the inhibitory effect of clonidine on electrically stimulated NE efflux in cat cerebral arteries has been reported (Balfagon and Marin 1989). Although epinephrine is a nonselective agonist, the reduced response to clonidine in the latter study indicates α_2-adrenoceptor subsensitivity. Early in cocaine withdrawal, cardiac responses to epinephrine are exaggerated, indicating enhanced receptor sensitivity (Avakian et al. 1990) (Table 18–3).

Therapeutic Studies

To date, there is no treatment of choice for psychostimulant dependency or withdrawal. The tricyclic antidepressant desipramine has been suggested to reduce cocaine craving, although 1–2 weeks of treatment is needed for this effect to appear (Fischman et al. 1978; Gawin et al. 1989a; for review, see Johanson and Fischman 1989). Other nonadrenergic drugs, such as the dopamine antagonist flupenthixol (Gawin et al. 1989b) or the mixed opiate agonist/antagonist buprenorphine (Kosten et al. 1989), have also been reported to reduce cocaine intake. From the peripheral adrenoceptor challenge studies described earlier, α_2-agonists such as clonidine should worsen cocaine withdrawal symptoms such as depression and anergia, and α_2-antagonists may be of value in symptomatic treatment. In treatment of cocaine dependence or withdrawal, it is important that treatments do not augment the toxic effects of cocaine. Two studies have examined the effects of adrenergic agonists and antagonists on acute cocaine toxicity; drugs that reduced NE activity (α_2-agonists, β- , and α_1-antagonists) lower mortality rates, whereas α_2-antagonists do not affect cocaine lethality (Derlet and Albertson 1990;

Jackson et al. 1990). A possible mechanism for this may be due to the former compounds preventing or reducing the cardiotoxic effects of cocaine.

Other Drugs

Benzodiazepines

Benzodiazepines have relatively little effect on adrenoceptor function, except in withdrawal. Acute doses of benzodiazepines tend to reduce plasma NE, epinephrine, and MHPG concentrations (Charney et al. 1983b, 1986; Hossman et al. 1980; Nutt and Molyneux 1987; Risby et al. 1989; Stratton and Halter 1985; Ziegler et al. 1984), although it is unlikely that there are changes after chronic administration (Charney and Heninger 1985; Sheehan et al. 1984; for review, see Hsiao and Potter 1990). Similar findings have been reported in animals, with reduced NE turnover after acute benzodiazepines and probably unaltered turnover after chronic administration (Rastogi et al. 1978a, 1978b; Taylor and Laverty 1969). cAMP activity and β-adrenoceptor density are unaltered after chronic benzodiazepines (Gallagher et al. 1985; Sethy and Hodges 1982; Stanford et al. 1986), and platelet α_2- and β-adrenoceptor density are unaltered as well (Nutt et al. 1985).

Increased noradrenergic activity is seen during benzodiazepine withdrawal in humans, with increased urinary (Peturrson et al. 1983) and plasma MHPG concentrations (Nutt and Molyneux 1987); in addition, precipitated withdrawal has been shown to elevate plasma NE concentrations (Duka et al. 1986). Similar findings have also been reported in animals on precipitated withdrawal (Grant et al. 1985). However, α_2-adrenoceptor sensitivity is not reduced during withdrawal, as demonstrated by the exaggerated fall in MHPG after clonidine (Nutt and Molyneux 1987). Other α_2-adrenoceptor-mediated responses are essentially unaltered by acute or chronic diazepam, or during withdrawal (Nutt 1986). There have been reports that clonidine is useful in treating benzodiazepine withdrawal in humans (Keshavan and Crammer 1985) and also in reducing withdrawal hyperactivity in animals (Kuchandy and Kulkarni 1986). However, a more recent clinical study has failed to demonstrate a usefulness for clonidine in benzodiazepine withdrawal (Joyce et al. 1990).

Phencyclidine

Little work has been carried out on the adrenergic effects of phencyclidine. Acutely, phencyclidine reduces whole brain NE concentrations (Tonge and Leonard 1969) and blocks synaptosomal NE uptake (Garey and Heath 1976; R. C. Smith et al. 1977). Chronic phencyclidine increases brain-stem NE (Lew 1989).

Discussion

As apparent from this chapter, there is extensive evidence for an involvement of adrenoceptors in the central nervous system actions of drugs of abuse and, perhaps more importantly, in the reactions of the central nervous system to such drugs. However, it should be pointed out that compelling as these findings are, other neurotransmitters are most certainly also involved in these processes. For instance, alcohol appears to have actions on both inhibitory and excitatory neurotransmitters (for review, see Glue and Nutt 1990). Research has described an action of alcohol to potentiate both gamma-aminobutyric acid-A ($GABA_A$) receptor function (Suzdak et al. 1986) and to inhibit NMDA receptor activity (Lovinger et al. 1989). Acutely alcohol acts as a functional agonist of the former and antagonist of the latter. The end result of these effects is probably increased inhibition and reduced excitation in many parts of the central nervous system after acute administration of alcohol. Alcohol withdrawal seems to represent the opposite situation with reduced chloride flux through the $GABA_A$ receptor complex (Morrow et al. 1988) and perhaps increased agonist activity at the excitatory amino acid receptors, particularly in parts of the hippocampus (Morgan et al. 1992). These changes could produce increased central nervous system excitation during alcohol withdrawal.

The changes in biogenic monoamine functions during alcohol intoxication and withdrawal are likely due to a combination of direct effects of alcohol on the function of monoamine receptors and indirect effects of alcohol via the excitatory and inhibitory amino acid receptor systems. The indirect effects are probably predominant and provide a strong rationale for therapeutic intervention in alcohol withdrawal with $GABA_A$ receptor functional agonists such as benzodiazepines and perhaps in the future with excitatory amino acid receptor functional antagonists. As mentioned previously, the therapeutic role for drugs that inhibit NE outflow (e.g., clonidine) is limited by their inability to prevent alcohol withdrawal seizures. However, there may be a future role for drugs that augment central NE (e.g., desipramine) in maintenance treatment of abstinent alcoholics.

In contrast to alcohol, the interactions of opioids with NE processes would appear to be more direct, via receptors located on catecholamine neurons (North and Williams 1985), where these receptors may coregulate ion channel functions with α_2-adrenoceptors. Because opioid withdrawal does not involve a high risk of seizures, clonidine and probably other more specific α_2-adrenoceptor agonists can be safely used to suppress autonomous symptoms of opioid withdrawal without additional anticonvulsant medication (Gold et al. 1978). Withdrawal from tricyclic antidepressants would also appear to be a primary disorder of NE function, probably due to subsensitive inhibitory tone at α_2-adrenoceptors.

The apparent differences in monoamine functions during withdrawal from cocaine and tricyclic antidepressants outlined await further research to elucidate the basic mechanisms involved.

During the past three decades, much has been learned concerning adrenergic changes during alcohol and drug intoxication and withdrawal. A large number of descriptive and correlational studies have contributed to this knowledge. The main methodologies used have been quantification of monoamines and their metabolites in various body fluids and tissue samples, in vitro receptor binding, and careful clinical observations. In the future, direct measurement of receptor function and monoamine turnover in vivo with various imaging techniques and quantification of intracellular signaling functions will further elucidate the role of adrenergic changes associated with alcohol and other substance abuse.

References

Aghajanian GK: Tolerance of locus coeruleus neurones to morphine and suppression of withdrawal response to clonidine. Nature 276:186–188, 1978

Ahlenius S, Carlsson A, Engel J, et al: Antagonism by alpha-methyl paratyrosine of the ethanol-induced stimulation and euphoria in man. Clin Pharmacol Ther 14:586–591, 1973

Ahtee L, Attila LMJ: Cerebral monoamine neurotransmitters in opioid withdrawal and dependence. Medical Biology 65:113–119, 1987

Ahtee L, Attila LMJ, Carlson KR, et al: Changes in brain monoamine metabolism during withdrawal from chronic oral self-administration of morphine and in response to a morphine challenge in the withdrawn state. J Pharmacol Exp Ther 249:303–310, 1989

Alexopoulos GS, Lieberman KW, Frances R, et al: Platelet MAO during the alcohol withdrawal syndrome. Am J Psychiatry 138:1254–1255, 1981

Arnold EB, Molinoff PB, Rutledge CO: The release of endogenous norepinephrine and dopamine from cerebral cortex by amphetamine. J Pharmacol Exp Ther 202:544–557, 1977

Attila LMJ: Effects of acute morphine administration on noradrenaline turnover and metabolism in various brain parts of control handled rats. Pharmacol Toxicol 64:202–209, 1989

Attila LMJ, Ahtee L: Cerebral dopamine and noradrenaline turnover and effects of morphine test dose in rats withdrawn from 20 days' morphine treatment. Medical Biology 61:249–257, 1983

Attila LMJ, Ahtee L: Acceleration of cerebral noradrenaline turnover after morphine withdrawal and its retardation by acute morphine administration in rats. Archives of Pharmacology 327:193–200, 1984

Avakian EV, Dunlap CV, Zoltoski RK, et al: Effects of chronic cocaine administration on cardiac beta-receptor concentration in rats. Proc West Pharmacol Soc 30:299–301, 1987

Avakian EV, Leroy M, St John-Allan K, et al: Effect of chronic cocaine administration and cocaine withdrawal on coronary flow rate and heart rate responses to epinephrine and cocaine in isolated perfused rat hearts. Life Sci 46:1569–1574, 1990

Axelrod J, Hertting G, Potter L: Effect of drugs on the uptake and release of 3-H-norepinephrine in the rat heart. Nature 194:297, 1962

Balfagon G, Marin J: Modulation of noradrenaline release from cat cerebral arteries by presynaptic alpha-2 adrenoceptors: effect of chronic treatment with desipramine and cocaine. Gen Pharmacol 20:289–294, 1989

Banerjee SP, Sharma VK, Khanna JM: Alterations in beta-adrenergic receptor binding during ethanol withdrawal. Nature 276:407–409, 1978a

Banerjee SP, Sharma VP, Kung LS, et al: Amphetamine induces beta-adrenergic receptor supersensitivity. Nature 271:380–381, 1978b

Banerjee SP, Sharma VK, Kung-Cheung LS, et al: Cocaine and d-amphetamine induce changes in central beta-adrenoceptor sensitivity: effects of acute and chronic drug treatment. Brain Res 175:119–130, 1979

Baumgartner GR, Rowen RC: Clonidine versus chlordiazepoxide in the management of acute alcohol withdrawal syndrome. Arch Intern Med 147:1223–1226, 1987

Bejerot N: A comparison of the effects of cocaine and synthetic central stimulants. Br J Addict 65:35–37, 1970

Billman GE: Mechanisms responsible for the cardiotoxic effects of cocaine. FASEB J 4:2469–2475, 1990

Bjorkvist SE: Clonidine in alcohol withdrawal. Acta Psychiatr Scand 52:256–263, 1975

Borg S, Kvande H, Sedvall G: Central norepinephrine metabolism during alcohol intoxication in addicts and healthy volunteers. Science 213:1135–1137, 1981

Borg S, Kvande H, Mossberg D, et al: Central nervous noradrenaline metabolism in alcoholics during long-term abstinence. Alcohol 18:321–323, 1983

Borg S, Kvande H, Mossberg D, et al: Central norepinephrine metabolism during alcohol intoxication in man: relationship to blood alcohol concentration, in Catecholamines, Vol B. Edited by Usdin E. New York, Alan R Liss, 1984, pp 153–158

Brambilla F, Lampertico M, Panerai AE, et al: Effect of clonidine on the secretion of anterior pituitary hormones in heroin addicts and normal volunteers. Psychiatry Res 13:295–304, 1984

Brohult J, Levi L, Reichard H, et al: Urinary excretion of adrenal hormones in man. Acta Medica Scandinavica 188:5–13, 1970

Burgess SK, Tessel ER: Epinephrine is released from hypothalamus by depolarizing agents and amphetamine. Brain Res 194:259–262, 1980

Calderini G, Morselli PL, Garattini S: Effects of amphetamine and fenfluramine on brain noradrenaline and MOPEG-SO4. Eur J Pharmacol 34:345–350, 1975

Castaneda R, Cushman P: Alcohol withdrawal: a review of clinical management. J Clin Psychiatry 50:278–284, 1989

Cedarbaum JM, Aghajanian GK: Catecholamine receptors on locus coeruleus neurons: pharmacological characterization. Eur J Pharmacol 44:375–385, 1977

Charney DS, Heninger GR: Noradrenergic function and the mechanism of action of antianxiety treatment, 1: the effect of long-term alprazolam treatment. Arch Gen Psychiatry 42:458–467, 1985

Charney DS, Sternberg DE, Kleber HD, et al: The clinical use of clonidine in abrupt withdrawal from methadone. Arch Gen Psychiatry 38:1273–1277, 1981

Charney DS, Heninger GR, Sternberg DE, et al: Abrupt discontinuation of tricyclic antidepressant drugs: evidence for noradrenergic hyperactivity. Br J Psychiatry 141:377–386, 1982a

Charney DS, Riodan CE, Kleber HD, et al: Clonidine and naltrexone: safe effective and rapid treatment of abrupt withdrawal from methadone therapy. Arch Gen Psychiatry 39:1327–1332, 1982b

Charney DS, Redmond DE, Galloway MP, et al: Naltrexone precipitated opiate withdrawal in methadone addicted human subjects: evidence for noradrenergic hyperactivity. Life Sci 35:1263–1272, 1983a

Charney DS, Heninger GR, Redmond DE: Yohimbine-induced anxiety and increased noradrenergic function in humans: effects of diazepam and clonidine. Life Sci 33:19–29, 1983b

Charney DS, Breier A, Jatlow P, et al: Behavioral, biochemical and blood pressure responses to alprazolam in healthy subjects: interactions with yohimbine. Psychopharmacology (Berl) 88:133–140, 1986

Checkley SA: Neuroendocrine tests of monoamine function in man: a review of basic theory and its application to the study of depressive illness. Psychol Med 10:35–53, 1980

Chiueh CC, Kopin IJ: Centrally mediated release by cocaine of endogenous epinephrine and norepinephrine from the sympathoadrenal medullary system of unanesthetized rats. J Pharmacol Exp Ther 205:148–154, 1978

Clouet DH, Gold GJ, Iwatsubo K: Effects of narcotic analgesic drugs on the cyclic adenosine 3′5′-monophosphate-adenylate cyclase system in rat brain. Br J Pharmacol 54:541–548, 1975

Collier HOJ, Francis DL: Morphine abstinence is associated with increasing brain cyclic AMP. Nature 255:159–162, 1975

Cowen PJ, Green AR, Grahame-Smith DG, et al: Plasma melatonin during desmethylimipramine treatment: evidence for changes in noradrenergic transmission. Br J Clin Pharmacol 19:799–805, 1985

Crawley JN, Laverty R, Roth RH: Clonidine reversal of increased norepinephrine metabolite levels during morphine withdrawal. Eur J Pharmacol 57:247–250, 1979

Cushman P, Forbes R, Lerner R, et al: Alcohol withdrawal syndromes: clinical management with lofexidine. Alcoholism: Clinical and Experimental Research 9:103–108, 1985

Davis WM, Smith SG, Khalsa JH: Noradrenergic role in the self-administration of morphine or amphetamine. Pharmacol Biochem Behav 3:477–484, 1975

de La Vega CE, Slater S, Ziegler MG, et al: Reduction in plasma norepinephrine during fenfluramine treatment. Clin Pharmacol Ther 21:216–221, 1977

De Wit H, Wise RA: Blockade of cocaine reinforcement in rats with the dopamine receptor blocker pimozide, but not with the noradrenergic blockers phentolamine or phenoxybenzamine. Can J Psychol 31:195–203, 1977

DeLeon-Jones FA, Davis JM, Inwang EE, et al: Excretion of catecholamine metabolites during methadone maintenance and withdrawal. Arch Gen Psychiatry 40:841–847, 1983

Delle M, Ricksten S-E, Haggendal J, et al: Regional changes in sympathetic nervous system activity and baroreceptor reflex function and arterial plasma levels of catecholamines, renin and vasopressin during naloxone-precipitated morphine withdrawal in rats. J Pharmacol Exp Ther 253:646–654, 1990

Demer JL, Volkow ND, Ulrich I, et al: Eye movements in cocaine abusers. Psychiatry Res 29:123–136, 1989

Derlet RW, Albertson TE: Acute cocaine toxicity: antagonism by agents interacting with adrenoceptors. Pharmacol Biochem Behav 36:225–231, 1990

Diamond I, Wrubel B, Estrin W, et al: Basal and adenosine receptor-stimulated levels of cAMP are reduced in lymphocytes from alcoholic patients. Proc Natl Acad Sci U S A 84:1413–1416, 1987

Dixon WR, Lau B, Chang APL, et al: Effect of oral self-administration of cocaine on adrenal catecholamine levels and cardiovascular parameters in the conscious rat. Proc West Pharmacol Soc 32:231–234, 1989

Duka T, Ackenheil M, Noderer J, et al: Changes in noradrenaline plasma levels and behavioral responses induced by benzodiazepine agonists with the benzodiazepine antagonist Ro 15-1788. Psychopharmacology (Berl) 90:351–357, 1986

Duman RS, Tallman JF, Nestler EJ: Acute and chronic opiate-regulation of adenylate cyclase in brain: specific effects in locus coeruleus. J Pharmacol Exp Ther 246:1033–1039, 1988

Eisenhofer G, Lambie DG, Johnson RH, et al: Effects of ethanol on plasma catecholamines and norepinephrine clearance. Clin Pharmacol Ther 34:143–147, 1983

Eisenhofer G, Lambie DG, Johnson RH: Effects of ethanol ingestion on alpha-adrenoceptor-mediated circulatory responses in man. Br J Clin Pharmacol 18:581–586, 1984

Eisenhofer G, Szaba G, Hoffman PL: Opposite changes in turnover of noradrenaline and dopamine in the CNS of ethanol-dependent mice. Neuropharmacology 29:37–45, 1990

Eisenman AJ, Sloan JW, Martin WR, et al: Catecholamine and 17-hydroxycorticosteroid excretion during a cycle of morphine dependence in man. J Psychiatr Res 7:19–28, 1969

Ellison G, Eison MS, Huberman HS, et al: Long-term changes in dopaminergic innervation of caudate nucleus after continuous amphetamine administration. Science 201:276–278, 1978

Esposito E, Kruszewska A, Ossowska G, et al: Noradrenergic and behavioral effects of naloxone injected into the locus coeruleus of morphine-dependent rats and their control by clonidine. Psychopharmacology (Berl) 93:393–396, 1987

Fielding S, Wilker J, Hynes M, et al: A comparison of clonidine and morphine for antinociceptive and antiwithdrawal actions. J Pharmacol Exp Ther 207:899–905, 1978

Fischman MW, Foltin RW, Nestadt G, et al: Effects of desipramine maintenance on cocaine self-administration by humans. J Pharmacol Exp Ther 253:760–770, 1978

Freedman JE, Aghajanian GK: Opiate and alpha-2-adrenoceptor responses of rat amygdaloid neurons: co-localization and interactions during withdrawal. J Neurosci 5:3016–3024, 1985

French SW, Palmer DS, Narod ME, et al: Noradrenergic sensitivity of the cerebral cortex after chronic ethanol ingestion and withdrawal. J Pharmacol Exp Ther 194:319–326, 1975

Gallagher DW, Malcolm AB, Anderson SA, et al: Continuous release of diazepam: electrophysiological, biochemical and behavioral consequences. Brain Res 342:26–36, 1985

Garcia-Sevilla JA, Ugedo L, Ulibarri I, et al: Platelet alpha-2 adrenoceptors in heroin addicts during withdrawal and after treatment with clonidine. Eur J Pharmacol 114:365–374, 1985

Garcia-Sevilla JA, Ulibarri I, Ugedo L, et al: Alpha-2 adrenoceptor-mediated inhibition of platelet adenylate cyclase activity in heroin addicts in abstinence. Psychopharmacology (Berl) 92:320–323, 1987

Garey RE, Heath RG: Effects of phencyclidine on the uptake of ^{3}H-catecholamines by rat striatal and hypothalamic synaptosomes. Life Sci 18:1105–1110, 1976

Gawin FH, Kleber HD: Neuroendocrine findings in chronic cocaine abusers: a preliminary report. Br J Psychiatry 147:569–573, 1985

Gawin FH, Kleber HD, Byck R, et al: Desipramine facilitation of initial cocaine abstinence. Arch Gen Psychiatry 46:117–121, 1989a

Gawin FH, Allen D, Humblestone B: Outpatient treatment of "crack" cocaine smoking with flupenthixol decanoate. Arch Gen Psychiatry 46:322–325, 1989b

Glue P, Nutt D: Clonidine in alcohol withdrawal: a pilot study of differential symptom responses following I.V. clonidine. Alcohol 22:161–166, 1987

Glue P, Nutt D: Clonidine challenge testing of alpha-2-adrenoceptor function in man: the effects of mental illness and psychotropic medication. Journal of Psychopharmacology 2:119–137, 1988

Glue P, Nutt B: Overexcitement and disinhibition: dynamic neurotransmitter interactions in alcohol withdrawal. Br J Psychiatry 157:491–499, 1990

Glue P, Sellman JD, Joyce PR, et al: The hypothermic response to clonidine is absent in alcohol withdrawal but returns in abstinence. Biol Psychiatry 124:102–104, 1989a

Glue P, Sellman JD, Nicholls MG, et al: Studies of alpha-2 adrenoceptor function in abstinent alcoholics. Br J Addict 84:97–102, 1989b

Gold MS, Redmond DE, Kleber HD: Clonidine in opiate withdrawal. Lancet 1:929–930, 1978

Gold MS, Washton AM, Dackis CA: Cocaine abuse: neurochemistry, phenomenology, and treatment. NIDA Res Monogr 61:130–150, 1985

Goldstein DS, Levinson PD, Zimlichman R, et al: Clonidine suppression testing in essential hypertension. Ann Intern Med 102:42–49, 1985

Gomes C, Svensson TH, Trolin G: Effects of morphine on central catecholamine turnover, blood pressure and heart rate in the rat. Archives of Pharmacology 294:141–147, 1985

Gottfries CG, Oreland L, Wiberg A, et al: Lowered monoamine oxidase activity in brains from alcoholic suicides. J Neurochem 25:667–673, 1975

Graham AW, Aghajanian GK: Effects of amphetamine on single cell activity in a catecholamine nucleus, the locus coeruleus. Nature 234:100–102, 1971

Grant SJ, Galloway MP, Major R, et al: Precipitated diazepam withdrawal elevates noradrenergic metabolism in primate brain. Eur J Pharmacol 107:127–132, 1985

Grote SS, Moses SG, Robins E, et al: A study of selected catecholamine metabolizing enzymes: a comparison of depressive suicides and alcoholic suicides with controls. J Neurochem 23:791–802, 1974

Gunne L-M, Jonsson J: Effects of cocaine administration on brain, adrenal and urinary adrenaline and noradrenaline in rats. Psychopharmacology (Berl) 6:125–129, 1964

Gunne L-M, Lewander T: Brain catecholamines during chronic amphetamine intoxication. Res Publ Assoc Res Nerv Ment Dis 46:106–116, 1968

Gunne L-M, Jonsson J, Fuxe K: Effects of morphine intoxication on brain catecholamine neurons. Eur J Pharmacol 5:338–342, 1969

Hadfield MG, Mott DEW, Ismay JA: Cocaine: effect of in vivo administration on synaptosomal uptake of norepinephrine. Biochem Pharmacol 29:1861–1863, 1980

Hamburg M, Tallman JF: Chronic morphine administration increases the apparent number of alpha-2-adrenergic receptors in rat brain. Nature 291:493–495, 1981

Hawley RJ, Major LF, Schulman EA, et al: Cerebrospinal 3-methoxy-4-hydroxyphenylglycol and norepinephrine levels in alcohol withdrawal. Arch Gen Psychiatry 42:1056–1062, 1985

Heikkonen E, Mäki T, Kontula K, et al: Effect of acute ethanol intake and hangover on the levels of plasma and urinary catecholamines and lymphocytic beta-adrenergic receptors. Alcoholism: Clinical and Experimental Research 13:20–24, 1989

Hoffman PL, Tabakoff B: Alterations in dopamine receptor sensitivity by chronic ethanol treatment. Nature 268:551–553, 1977

Hoffman PL, Tabakoff B: Ethanol and guanine nucleotide binding proteins: a selective interaction. FASEB J 4:2612–2622, 1990

Hoffman PL, Moses F, Luthin GR, et al: Acute and chronic effects of ethanol on receptor-mediated phosphatidylinositol 4,5-biphosphate breakdown in mouse brain. Mol Pharmacol 30:13–18, 1986

Holmes JC, Rutledge CO: Effects of the d- and l-isomers of amphetamine on uptake, release and catabolism of norepinephrine, dopamine and 5-hydroxytryptamine in several regions of rat brain. Biochem Pharmacol 25:447–451, 1976

Hossman V, Maling TJB, Hamilton C, et al: Sedative and cardiovascular effects of clonidine and nitrazepam. Clin Pharmacol Ther 28:167–176, 1980

Howes LG, Reid JL: Changes in plasma noradrenaline, 3,4-dihydroxyphenylethylene glycol and blood pressure following acute alcohol administration. Br J Pharmacol 20:267–268, 1985

Hsiao JK, Potter WZ: Mechanisms of action of antipanic drugs, in Clinical Aspects of Panic Disorder. Edited by Ballenger JC. New York, Wiley-Liss, 1990, pp 297–317

Hunt WA, Dalton TK: Neurotransmitter-receptor binding in various brain regions in ethanol-dependent rats. Pharmacol Biochem Behav 14:733–739, 1981

Hunt WA, Majchrowicz E: Alterations in the rate of turnover of brain norepinephrine and dopamine in alcohol-dependent rats. J Neurochem 23:549–552, 1974

Ireland MA, Vandongen R, Davidson L, et al: Acute effects of moderate alcohol consumption on blood pressure and plasma catecholamines. Clin Sci 66:643–648, 1984

Jackson HC, Ball DM, Nutt DJ: Noradrenergic mechanisms do not appear to be involved in cocaine-induced seizures and lethality. Life Sci 47:353–359, 1990

Jacobs MS, Zilm DH, Macleod SM, et al: Propranolol-associated confused states during alcohol withdrawal. J Clin Psychopharmacol 32:185–187, 1983

Jain RK, Jain MK, Bachenheimer LC, et al: Factors determining whether cocaine will potentiate the cardiac effects of neurally released norepinephrine. J Pharmacol Exp Ther 252:147–153, 1990

Johanson C-E, Fischman M: The pharmacology of cocaine related to its abuse. Pharmacol Rev 41:3–52, 1989

Joyce EM, Moodley P, Keshavan MS, et al: Failure of clonidine treatment in benzodiazepine withdrawal. Journal of Psychopharmacology 4:42–45, 1990

Keshavan MS, Crammer JL: Clonidine in benzodiazepine withdrawal. Lancet 1:1325–1326, 1985

Kopin IJ, Gordon EK, Jimerson DC, et al: Relation between plasma and cerebrospinal fluid levels of 3-methoxy-4-hydroxyphenylglycol. Science 219:73–75, 1983

Korf J, Bunney BS, Aghajanian GK: Noradrenergic neurons: morphine inhibition of spontaneous activity. Eur J Pharmacol 25:165–169, 1974

Kosten TR, Kleber HD, Morgan C: Role of opioid antagonists in treating intravenous cocaine abuse. Life Sci 44:887–892, 1989

Kraus ML, Gottlieb LD, Horwitz RJ, et al: Randomized clinical trial of atenolol in alcohol withdrawal. N Engl J Med 313:905–909, 1989

Kuchandy J, Kulkarni SK: Reversal by alpha-2-agonists of diazepam withdrawal hyperactivity in rats. Psychopharmacology (Berl) 90:198–202, 1986

Kuriyama K, Nakagawa K, Naito K, et al: Morphine-induced changes in cyclic AMP metabolism and protein kinase activity in the brain. Jpn J Pharmacol 28:73–84, 1978

Kuriyama K, Muramatsu M, Aiso M, et al: Alteration in beta-adrenergic receptor binding in brain, lung and heart during morphine and alcohol dependence and withdrawal. Neuropharmacology 20:659–666, 1981

Laverty R, Roth RH: Clonidine reverses the increased norepinephrine turnover during morphine withdrawal in rats. Brain Res 182:482–485, 1980

Leung CMK, Dai S, Ogle CW: Arterial catecholamine levels in morphine-treated rats subjected to sympathetic nerve stimulation. Br J Pharmacol 96:888–894, 1989

Lew GM: Changes in norepinephrine concentration following chronic administration of phencyclidine (PCP) to genetically hypertensive and normotensive rats. Gen Pharmacol 20:571–573, 1989

Liddell DW, Weil-Malherbe H: The effects of methedrine and of lysergic acid diethylamide on mental processes and on the blood adrenaline level. J Neurol Neurosurg Psychiatry 16:7–13, 1953

Linnoila M, Mefford I, Nutt D, et al: Alcohol withdrawal and noradrenergic function. Ann Intern Med 107:875–889, 1987

Liskow B, Reed J: Atenolol for alcohol withdrawal. N Engl J Med 314:783–785, 1986

Lister RG, Durcan MJ, Nutt DJ, et al: Attenuation of ethanol intoxication by alpha-2-adrenoceptor antagonists. Life Sci 44:111–119, 1989

Llorens C, Martres MP, Baudry M, et al: Hypersensitivity to noradrenaline in cortex after chronic morphine: relevance to tolerance and dependence. Nature 274:603–605, 1978

Lopachin RM, Reigle TG: The effects of several narcotic analgesics on brain levels of 3-methoxy-4-hydroxyphenylethylene glycol sulphate in the rat. J Pharmacol Exp Ther 207:151–158, 1978

Lovinger DM, White G, Weight FF: Ethanol inhibits NMDA-activated ion current in hippocampal neurons. Science 243:1721–1724, 1989

Lykouras L, Markianos M, Moussas G: Plasma cyclic AMP in non-abstinent chronic alcoholics: relation to clinical parameters. Drug Alcohol Depend 21:7–9, 1988

Major LF, Murphy DL: Platelet and plasma amine oxidase activity in alcoholic individuals. Br J Psychiatry 132:548–554, 1978

Manner T, Kanto J, Scheinin H, et al: Meptazinol and pentazocine: plasma catecholamines and other effects in healthy volunteers. Br J Clin Pharmacol 24:689–697, 1987

Martin PR, Adinoff B, Eckardt MJ, et al: Effective pharmacotherapy of alcoholic amnestic disorder with fluvoxamine. Arch Gen Psychiatry 46:617–621, 1989

Matussek N, Hoehe M: Investigations with the specific mu-opiate agonist fentanyl in depressive patients: growth hormone, prolactin, cortisol, noradrenaline and euphoric responses. Neuropsychobiology 21:1–8, 1989

Matussek N, Ackenheil M, Herz M: The dependence of the clonidine growth hormone test on alcohol drinking habits and the menstrual cycle. Psychoneuroendocrinology 9:173–177, 1984

Maynert EW, Klingman GI: Tolerance to morphine, 1: effects on catecholamines in the brain and adrenal glands. J Pharmacol Exp Ther 135:285–295, 1962

McEntee WJ, Mair RG: Memory impairment in Korsakoff's psychosis: a correlation with brain noradrenergic activity. Science 202:905–907, 1978

McEntee WJ, Mair RG: Memory enhancement in Korsakoff's psychosis by clonidine: further evidence for a noradrenergic deficit. Ann Neurol 7:466–470, 1980

McEntee WJ, Mair RG, Langlais PJ: Neurochemical pathology in Korsakoff's psychosis: implications for other cognitive disorders. Neurology 34:648–652, 1984

Mefford I: Epinephrine in the mammalian brain. Prog Neuropsychopharmacol Biol Psychiatry 12:365–388, 1988

Mefford IN, Lister RG, Ota M, et al: Antagonism of ethanol intoxication in rats by inhibitors of phenylethanolamine n-methyltransferase. Alcoholism: Clinical and Experimental Research 14:53–57, 1990

Merali Z, Singhal RL, Herdina PD, et al: Changes in brain cyclic AMP metabolism and acetylcholine and dopamine during narcotic dependence and withdrawal. Life Sci 16:1889–1894, 1975

Meyer RE, Schildkraut JJ, Mirin SM, et al: Opiates, catecholamines, behavior, and mood. Psychopharmacology (Berl) 56:327–333, 1978

Mikkelsen E, Lake CR, Brown GL, et al: The hyperactive child syndrome: peripheral sympathetic nervous system function and the effect of *d*-amphetamine. Psychiatry Res 4:157–169, 1981

Mobley PL, Sanders-Bush E, Smith HE, et al: Modification of the noradrenergic cyclic AMP generating system in the rat limbic forebrain by amphetamine: role of its hydroxylated metabolites. Archives of Pharmacology 306:267–273, 1979

Moises HC, Smith CB: Changes in cortical beta-adrenergic receptor density and neuronal sensitivity to norepinephrine accompany morphine dependence and withdrawal. Brain Res 400:110–126, 1987

Moises HC, Smith CB: Electrophysiological responsiveness to isoproterenol in rat hippocampal slices correlates with changes in beta-adrenergic receptor density induced by chronic morphine treatment. Brain Res 485:67–78, 1989

Morgan ME, Gibb JW: Short-term and long-term effects of methamphetamine on biogenic amine metabolism in extra-striatal dopaminergic nuclei. Neuropharmacology 19:989–995, 1980

Morgan P, Nadi S, Karandian J, et al: Mapping rat brain structures activated during ethanol withdrawal: role of glutamate and NMDA receptors. Eur J Pharmacol 225:217–223, 1992

Morrow AL, Suzdak PD, Karanian JW, et al: Chronic ethanol administration alters gamma-amino butyric acid, pentobarbital and ethanol-medicated ^{36}CI-uptake in cerebral cortical synaptoneurosomes. J Pharmacol Exp Ther 246:158–164, 1988

Moses SG, Robins E: Regional distribution of norepinephrine and dopamine in the brains of depressive suicides and alcoholic suicides. Psychopharmacology Communications 1:327–337, 1975

Muller N, Hoehe M, Klein HE, et al: Endocrinological studies in alcoholics during withdrawal and after abstinence. Psychoendocrinology 14:113–123, 1989

Nahas GG, Trouve R, Manger WM, et al: Cocaine cardiac toxicity and endogenous catecholamines, in Progress in Catecholamine Research, Part B: Central Aspects. Edited by Sandler M, Dahlstrom A, Belmaker RH. New York, Alan R Liss, 1988, pp 457–462

Naito K, Kuriyama K: Effect of morphine administration on adenyl cyclase and 3′5′-cyclic nucleotide phosphodiesterase. Jpn J Pharmacol 23:274–276, 1973

Naranjo CA, Sellers EM: Serotonin uptake inhibitors attenuate ethanol intake in problem drinkers, in Recent Developments in Alcoholism, Vol 7. Edited by Galanter M. New York, Plenum, 1989, pp 255–266

Nathanson JA, Redmond DE: Morphine withdrawal causes subsensitivity of adrenergic receptor response. Life Sci 28:1353–1360, 1981

Nestler EJ, Terwilliger R, Beitner D: Regulation by chronic clonidine of adenylate cyclase and cyclic AMP-dependent protein kinase in the rat locus coeruleus. Life Sci 45:1073–1080, 1989

North RA, Williams JT: On the potassium conductance increased by opioids in rat locus coeruleus neurons. J Physiol (Lond) 364:365–380, 1985

Nutt D: Interaction between diazepam and noradrenergic function in man. Human Psychopharmacology 1:35–40, 1986

Nutt D, Glue P: Monoamines and alcohol. Br J Addict 81:327–338, 1986

Nutt D, Molyneux S: Benzodiazepines, plasma MHPG and alpha-2-adrenoceptor function in man. Int Clin Psychopharmacol 2:151–157, 1987

Nutt D, Fraser S, Gosden B, et al: Benzodiazepines and human platelet receptor binding. Br J Clin Pharmacol 19:554–555, 1985

Nutt D, Glue P, Stewart A: Platelet monoamine receptor binding in alcoholics during withdrawal. Br J Addict 82:1253–1255, 1987

Nutt D, Glue P, Molyneux S, et al: Alpha-2 adrenoceptor function in alcohol withdrawal: a pilot study of the effects of IV clonidine in alcoholics and normals. Alcoholism: Clinical and Experimental Research 12:14–18, 1988

Orenberg EK, Zarcone VP, Renson JF, et al: The effects of ethanol ingestion on cyclic AMP, homovanillic acid and 5-hydroxyindoleacetic acid in human cerebrospinal fluid. Life Sci 19:1669–1672, 1976

Padovani A, Govoni S, Battaini F, et al: Alcohol impairs age-dependent adaptation of human lymphocyte beta-adrenergic receptors. Eur J Clin Invest 17:511–514, 1987

Pals DT, Fulton RW, Masucci FD: Angiotensin, cocaine and desipramine: comparison of effects on blood pressure responses to norepinephrine, tyramine and phenylephrine in the pithed rat. J Pharmacol Exp Ther 162:85–91, 1968

Perman ES: Effect of ethanol and hydration on the urinary excretion of adrenaline and noradrenaline and on the blood sugar of rats. Acta Physiol Scand 51:68–74, 1961a

Perman ES: Observations on the effect of ethanol on the urinary excretion of histamine, 5-hydroxyindole acetic acid, catecholamines and 17-hydroxycorticosteroids in man. Acta Physiol Scand 51:62–67, 1961b

Peturrson H, Bond PA, Smith B, et al: Monoamine metabolism during chronic benzodiazepine treatment and withdrawal. Biol Psychiatry 18:207–213, 1983

Pitts DK, Marwah J: Effects of cocaine on the electrical activity of single noradrenergic neurons from locus coeruleus. Life Sci 38:1229–1234, 1986

Pitts DK, Marwah J: Electrophysiological actions of cocaine on noradrenergic neurons in rat locus ceruleus. J Pharmacol Exp Ther 240:345–351, 1987

Pohorecky LA: Effects of ethanol on central and peripheral noradrenergic neurons. J Pharmacol Exp Ther 189:380–391, 1974

Post RM, Kopanda RT, Black KE: Progressive effects of cocaine on behavior and central amine metabolism in rhesus monkeys: relationship to kindling and psychosis. Biol Psychiatry 11:403–419, 1976

Potter JF, Bannan LT, Beevers DG: Alcohol and hypertension. Br J Addict 79:365–372, 1984

Potter WZ, Scheinin M, Golden RN, et al: Selective antidepressants and cerebrospinal fluid: lack of specificity on norepinephrine and serotonin metabolites. Arch Gen Psychiatry 42:1171–1177, 1985

Pradhan S, Roy SN, Pradjan SN: Correlation of behavioural and neurochemical effects of acute administration of cocaine in rats. Life Sci 22:1737–1744, 1978

Rabin RA, Wolfe BB, Dibner MD, et al: Effects of ethanol administration and withdrawal on neurotransmitter receptor systems in C57 mice. J Pharmacol Exp Ther 213:491–496, 1980

Rastogi RB, Lapierre YD, Singhal RL: Effect of a new benzodiazepine bromazepam on locomotor performance and brain monoamine metabolism. J Neural Transm 42:251–262, 1978a

Rastogi RB, Lapierre YD, Singhal RL: Synaptosomal uptake of norepinephrine and 5-hydroxytryptamine and synthesis of catecholamines during benzodiazepine treatment. Can J Physiol Pharmacol 56:777–784, 1978b

Reid JL: Central and peripheral autonomic control mechanisms, in Autonomic Failure. Edited by Bannister R. Oxford, England, Oxford University Press, 1983, pp 17–35

Reigle TG, Huff JW: Single dose tolerance to the effects of morphine on brain 3-methoxy-4-hydroxyphenylethylene glycol sulfate. Biochem Pharmacol 29:2249–2251, 1980

Reisine TD, U'Prichard D, Wiech NL, et al: Effects of combined administration of amphetamine and iprindole on brain adrenergic receptors. Brain Res 188:587–592, 1980

Risby ED, Hsiao JK, Gold GJ, et al: Intravenous alprazolam challenge in normal subjects: biochemical, cardiovascular and behavioral effects. Psychopharmacology (Berl) 99:508–514, 1989

Risner ME, Jones BE: Role of noradrenergic and dopaminergic processes in amphetamine self-administration. Pharmacol Biochem Behav 5:477–482, 1976

Ritz MC, Kuhar MJ: Relationship between self-administration of amphetamine and monoamine receptors in brain: comparison with cocaine. J Pharmacol Exp Ther 248:1010–1017, 1989

Ritz MC, Lamb RJ, Goldberg SR, et al: Cocaine receptors on dopamine transporters are related to self-administration of cocaine. Science 237:1219–1223, 1987

Roberts DCS, Corcoran ME, Fibiger HC: On the role of ascending catecholaminergic systems in intravenous self-administration of cocaine. Pharmacol Biochem Behav 6:615–620, 1977

Robinson BJ, Robinson GM, Maling TJB, et al: Is clonidine useful in the treatment of alcohol withdrawal? Alcoholism: Clinical and Experimental Research 13:95–98, 1989

Robinson SE, Hambrecht KL: The effect of cocaine on hippocampal cholinergic and noradrenergic metabolism. Brain Res 457:383–385, 1988

Roffman M, Cassens G, Schildkraut JJ: The effect of acute and chronic administration of morphine on norepinephrine turnover in rat brain regions. Biochem Pharmacol 26:2355–2358, 1977

Roffman M, Cassens G, Schildkraut JJ: Effects of the d- and l-isomers of amphetamine on the levels of 3-methoxy-4-hydroxyphenylglycol sulfate in whole rat brain and rat brain regions. Biochem Pharmacol 27:1774–1777, 1978

Roffman M, Reigle T, Orsulak P, et al: Further studies on the effects of morphine on the levels of MHPG-SO_4 in rat brain. Neuropharmacology 18:483–487, 1979

Rokosz-Koslowska A, Antkiewicz-Michaluk L, Vetulani J: High doses of morphine do not change characteristics of the cerebral catecholamine receptors in the rat. J Pharm Pharmacol 40:387–392, 1988

Rommelspacher H, Strauss S: Down-regulation of cerebral beta-adrenoceptors induced by long term administration of antidepressants is prevented by ethanol. Alcohol 1 (suppl):697–701, 1987

Rosenblatt JE, Lake CR, Van Kammen DP, et al: Interactions of amphetamine, pimozide and lithium on plasma norepinephrine and dopamine-beta-hydroxylase in schizophrenic patients. Psychiatry Res 1:45–52, 1979

Roth RH, Elsworth JD, Redmond DE: Clonidine suppression of noradrenergic hyperactivity during morphine withdrawal by clonidine: biochemical studies in rodents and primates. J Clin Psychiatry 43:42–46, 1982

Roy SN, Bhattacharyya AK, Pradhan S, et al: Behavioural and neurochemical effects of repeated administration of cocaine in rats. Neuropharmacology 17:559–564, 1978

Saito T, Lee JM, Hoffman PL, et al: Effects of chronic ethanol treatment on the beta-adrenergic receptor-coupled adenylate cyclase system of mouse cerebral cortex. J Neurochem 48:1817–1822, 1987

Saletu B, Saletu M, Grünberger J, et al: On the treatment of the alcoholic organic brain syndrome with an alpha-adrenergic agonist modafinil: double-blind, placebo-controlled clinical, psychometric and neurophysiological studies. Prog Neuropsychopharmacol Biol Psychiatry 14:195–214, 1990

Scheel-Kruger J: Behavioural and biochemical comparison of amphetamine derivatives, cocaine, benztropine, and tricyclic antidepressant drugs. Eur J Pharmacol 18:63–73, 1972

Schildkraut JJ, Meyer RE, Orsulak PJ, et al: Catecholamine metabolism during heroin use. Am J Psychiatry 134:534–537, 1977

Schubert H, Fleischhacker WW, Meise U, et al: Preliminary results of guanfacine treatment of acute opiate withdrawal. Am J Psychiatry 141:1271–1273, 1984

Sellers EM, Degani NC, Zilm DH, et al: Propranolol-decreased noradrenaline excretion and alcohol withdrawal. Lancet 1:94–95, 1976

Sellers EM, Zilm DH, Degani NC, et al: Comparative effects of propranolol and chlordiazepoxide in alcohol withdrawal. J Stud Alcohol 38:2096–2108, 1977

Sellinger-Barnette MM, Mendels J, Frazer A: The effect of psychoactive drugs on beta-adrenergic receptor binding sites in rat brain. Neuropharmacology 19:447–454, 1980

Sethy VH, Hodges DH: Role of beta-adrenergic receptors in the antidepressant activity of alprazolam. Res Commun Chem Pathol Pharmacol 36:329–332, 1982

Sheehan DV, Coleman JH, Greenblatt DJ, et al: Some biochemical correlates of panic attacks with agoraphobia, and their response to a new treatment. J Clin Psychopharmacol 4:66–75, 1984

Sherer MA: Intravenous cocaine: psychiatric effects, biological mechanisms. Biol Psychiatry 24:863–864, 1988

Sjoquist B, Borg S, Kvande H: Catecholamine-derived compounds in urine and after long-standing intoxication. Substance and Alcohol Actions/Misuse 2:63–72, 1981

Sjoquist B, Perdahl E, Winblad B: The effects of alcoholism on salsolinol and biogenic amines in human brain. Drug Alcohol Depend 12:15–23, 1983

Sloan JW, Eisenman AJ: Long persisting changes in catecholamine metabolism following addiction to and withdrawal from morphine. Res Publ Assoc Res Nerv Ment Dis 46:96–105, 1968

Smith CB, Moises HC, Spengler RN, et al: Changes in α_2-adrenoceptor number and function in brains of morphine-dependent rats. Eur J Pharmacol 161:111–119, 1989

Smith RC, Meltzer HY, Arora RC, et al: Effects of phencyclidine on ^{3}H-catecholamine and 3H-serotonin uptake in synaptosomal preparation from rat brain. Biochem Pharmacol 26:1435–1443, 1977

Smith TL, Vickers AE, Brendel K, et al: Influence of chronic ethanol treatment on alpha-1 adrenergic and vasopressin receptor-stimulated phosphatidylinositol synthesis in isolated rat hepatocytes. Biochem Pharmacol 32:3059–3063, 1983

St. John-Allan K, Vulliet PR, Avakian EV: Effects of chronic cocaine administration on rat cardiac and splenic norepinephrine levels. Proc West Pharmacol Soc 32:61–63, 1989

Stanford SC, Little HJ, Nutt DJ, et al: Effects of chronic treatment with benzodiazepine receptor ligands on cortical adrenoceptors. Eur J Pharmacol 129:181–184, 1986

Stratton JR, Halter JB: Effect of a benzodiazepine (alprazolam) on plasma epinephrine and norepinephrine levels during exercise stress. Am J Cardiol 56:136–139, 1985

Strickland JA, Wooles WR: Blood ethanol level and physiologic measurements during ethanol-induced hypertension. Alcohol 6:109–114, 1989

Sugrue MF: The effects of acutely administered analgesics on the turnover of noradrenaline and dopamine in various regions of the rat brain. Br J Pharmacol 52:159–165, 1974

Sullivan JL, Baezinger JC, Wagner DL, et al: Platelet MAO in subtypes of alcoholism. Biol Psychiatry 27:911–922, 1990

Suzdak PD, Glowa JR, Crawley JN, et al: A selective imidazobenzodiazepine antagonist of ethanol in the rat. Science 243:1243–1247, 1986

Swann AC, Elsworth JD, Charney DS, et al: Brain catecholamine metabolites and behavior in morphine withdrawal. Eur J Pharmacol 86:167–175, 1983

Szmigielski A, Szmigielski H, Wejman I: The effect of prolonged ethanol administration on central alpha-2 adrenoceptors sensitivity. J Pharm Pharmacol 41:263–272, 1989

Tabakoff B, Hoffman PL, Valverius P, et al: Characteristics of receptors and enzymes in brains of human alcoholics. Alcohol 2:419–423, 1985

Tabakoff B, Hoffman PL, Lee JM, et al: Differences in platelet enzyme activity between alcoholics and nonalcoholics. N Engl J Med 318:134–139, 1988

Taylor D, Ho BT: Neurochemical effects of cocaine following acute and repeated injection. J Neurosci Res 3:95–101, 1977

Taylor KM, Laverty R: The effects of chlordiazepoxide, diazepam, and nitrazepam on catecholamine metabolism in regions of the rat brain. Eur J Pharmacol 8:296–301, 1969

Tennant FS: Effect of cocaine dependence on plasma phenylalanine and tyrosine levels and on urinary MHPG excretion. Am J Psychiatry 142:1200–1201, 1985

Thurman RG, Harden TK, Winn K: Cyclic amp and beta-adrenergic receptors during the development of physical dependence on ethanol in the rat. Adv Exp Med Biol 126:145–155, 1980

Tonge SR: Noradrenaline and 5-hydroxytryptamine metabolism in six areas of rat brain during post-amphetamine depression. Psychopharmacology (Berl) 38:181–186, 1974

Tonge SR, Leonard BE: The effects of some hallucinogenic drugs upon the metabolism of 5-hydroxytryptamine in the brain. Life Sci 8:805–812, 1969

Turkka J, Gurguis G, Karanian J, et al: Effects of chronic exposure to ethanol alone and in combination with desipramine on beta-adrenoceptors of rat brain. Eur J Pharmacol 177:171–179, 1990

Ulibarri I, Garcia-Sevilla JA, Ugedo L: Modulation of brain alpha-2 adrenoceptor and mu-opiod receptor densities during morphine dependence and spontaneous withdrawal in rats. Archives of Pharmacology 336:530–537, 1987

Utena H, Ezoe T, Kato N, et al: Effects of chronic administration of methamphetamine in enzymic patterns in brain tissue. J Neurochem 4:161–169, 1959

Valverius P, Hoffman PL, Tabakoff B: Effect of ethanol on mouse cerebral cortical beta-adrenergic receptors. Mol Pharmacol 32:217–222, 1987

Valverius P, Borg S, Valverius MR, et al: Beta-adrenergic receptor binding in brain of alcoholics. Exp Neurol 105:280–286, 1989a

Valverius P, Hoffman PL, Tabakoff B: Hippocampal and cerebellar beta-adrenergic receptors and adenylate cyclase are differentially altered by chronic ethanol ingestion. J Neurochem 52:492–497, 1989b

Vicentini LM, Miller RJ, Robertson MJ: Chronic opiate treatment does not modify alpha-2-adrenergic receptors in rat cerebral cortex, kidney and in the neurotumor cell line NCB20. Eur J Pharmacol 95:265–270, 1983

von Knorring A-L, Bohman M, Von Knorring L, et al: Platelet MAO activity as a biological marker in subgroups of alcoholism. Acta Psychiatr Scand 72:51–58, 1985

Wang CH, Schnoll SH: Prenatal cocaine use associated with down regulation of receptors in human placenta. Neurotoxicol Teratol 9:301–304, 1987

Watson R, Hartmann E, Schildkraut JJ: Amphetamine withdrawal: affective state, sleep patterns, and MHPG excretion. Am J Psychiatry 129:263–269, 1972

Weil-Malherbe H, Smith ERB, Fraser HF: Plasma catecholamine levels and urinary excretion of catecholamines and metabolites in two human subjects during a cycle of morphine addiction and withdrawal. Biochem Pharmacol 14:1621–1633, 1965

Wiberg A, Gottfries CG, Oreland L: Low platelet monoamine oxidase activity in human alcoholics. Medical Biology 55:181–186, 1977

Wilkerson RD: Cardiovascular effects of cocaine: enhancement by yohimbine and atropine. J Pharmacol Exp Ther 248:57–61, 1989

Wilkins AJ, Jenkins WJ, Steiner JA: Efficacy of clonidine in treatment of alcohol withdrawal state. Psychopharmacology (Berl) 81:78–80, 1983

Yeh SY: Effects of repeated cocaine administration on brain monoamines in rats. Federation Proceedings 46:404, 1987

Ziegler MG, Lake CR, Ebert MH: Norepinephrine elevations in cerebrospinal fluid after *d*- and *l*-amphetamine. Eur J Pharmacol 57:127–133, 1979

Ziegler MG, Ludwig L, Klotz U: Stress protective effects during steady state conditions of bromazepam. Pharmacopsychiatry 17:194–198, 1984

Index

Page numbers printed in **boldface** type refer to tables or figures.